STUDENT PROBLEM SET WITH SOLUTIONS

Steven G. Conahan
State University of New York at Buffalo

ELECTRIC CIRCUIT ANALYSIS

THIRD EDITION

David E. Johnson
John L. Hilburn
Johnny R. Johnson
Peter D. Scott

JOHN WILEY & SONS, INC.
New York • Chichester • Weinheim • Brisbane • Singapore • Toronto

This book was printed and bound by Bradford & Bigelow.

The paper in this book was manufactured by a mill whose forest management programs include sustained yield harvesting of its timberlands. Sustained yield harvesting principles ensure that the number of trees cut each year does not exceed the amount of new growth.

This book is printed on acid-free paper.

Printed in the United States of America

10 9 8 7 6 5 4

ISBN 0-471-36724-9

Contents

Chapter 1

Introduction

1.1 Definitions and Units

1.1 Express the following values in milliamperes: (a) 5 μA, (b) 37 pA, (c) 2 kA, (d) 0.5 MA.

1.2 In a certain compact disc player, a digital to analog converter chip updates the output voltage 4×10^4 times per second. (a) What is the time in seconds between two consecutive updates? (b) What is this time in microseconds?

1.2 Charge and Current

1.3 Let $f(t)$ in the graph shown below be the charge $q(t)$ in coulombs that has entered the positive terminal of an element as a function of time. Find (a) the total charge that has entered the terminal between 1 and 3s, (b) the total charge that has entered the terminal between 4 and 7s, and (c) the current entering the terminal at 1s, 4s and 6s.

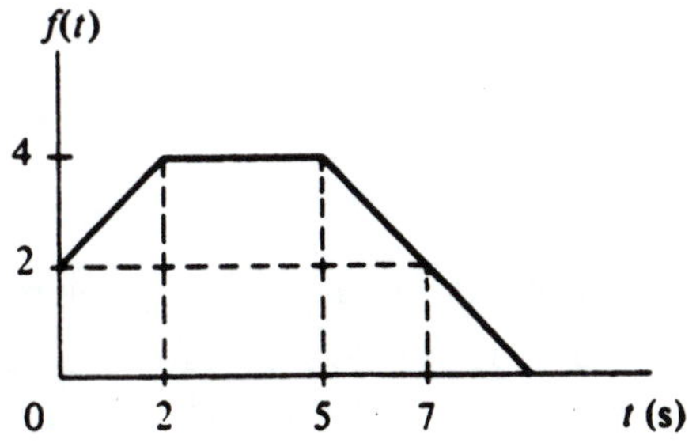

PROBLEM 1.3

1.4 Let $f(t)$ in Prob. 1.3 be the current $i(t)$ in milliamperes entering an element terminal as a function of time. Find (a) the charge that enters the terminal between 0 and 7s and (b) the rate at which the charge is entering at $t = 5$s.

1.5 The current entering a terminal is given by

$$i(t) = \begin{cases} 1 - e^{-3t} \text{ A}, & t > 0 \\ 0, & \text{otherwise.} \end{cases}$$

Find (a) the total charge entering the terminal between $t = 0$ and $t = 5$s, and (b) the total charge entering the terminal between $t = 0$ and $t = T$ s (for some time $T > 0$).

1.6 Let $f(t)$ in the graph shown below be the current $i(t)$ in milliamperes entering an element terminal as a function of time. Find (a) the charge that enters the terminal between 0 and 10s and (b) the rate at which the charge is entering at $t = 2$s and $t = 4$s.

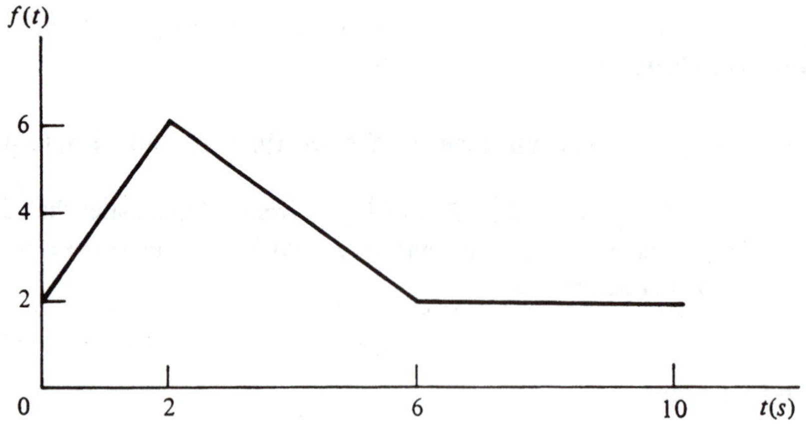

PROBLEM 1.6

1.7 Let $f(t)$ in Prob. 1.6 be the charge $q(t)$ in coulombs that has entered the positive terminal of an element as a function of time. Find (a) the total charge that has entered the terminal between 1 and 2s, (b) the total charge that has entered between 6 and 10s, and (c) the current entering the terminal at 1s, 4s, and 8s.

1.3 Voltage, Energy, and Power

1.8 The current entering the positive terminal of a device is given by $i(t) = 10\sin(101\pi\ t)$ mA and the voltage across the terminals is $v = 20.2\pi$ V for $0 \le t \le 1$ s. Find the energy delivered to the device during the time period $-\infty < t \le 1$ s if $w(0) = 2$ mJ.

1.9 Find the current into the positive terminal if $v = 4$ V and the element is
(a) absorbing power of $p = 28$ mW
(b) absorbing power of $p = -4$ W
(c) delivering to the external circuit power $p = 16$ mW
(d) delivering to the external circuit power $p = -12$ μW

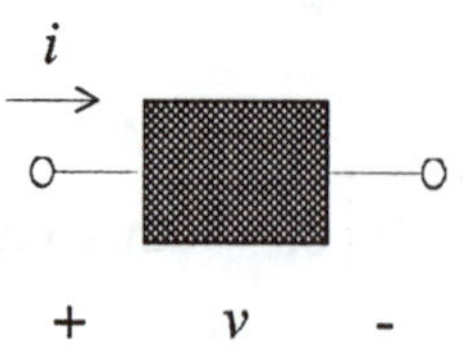

PROBLEM 1.9

1.10 The current entering the positive terminal of a certain two-terminal element is given by $i(t) = 2e^{-0.5t}$ A for $t > 1$s. The energy delivered to the element at $t = 1$s is 100 mJ. Find an equation for the energy delivered to the element $w(t)$ for any $t > 1$s if the voltage across the element during this time period is (a) $v = 10$V, (b) $v = 3i$ V, and (c) $v = di/dt$ V.

1.11 Find the power delivered to an element at $t = \pi$ seconds if the charge entering the positive terminal is $q = 8\sin(t/4)$ C and the voltage is $v = 6\cos(t/4)$ V.

1.12 The power delivered to an element is $p = 6e^{-4t}$ W and the charge entering the positive terminal is $q = -20e^{-2t}$ mC. Find (a) the voltage across the element and (b) the energy delivered to the element between 0 and 250 ms.

1.13 The voltage across an element is 6 V and the charge q entering the positive terminal is as shown. Find (a) the total charge and (b) the total energy delivered to the element between 1 and 10 ms.

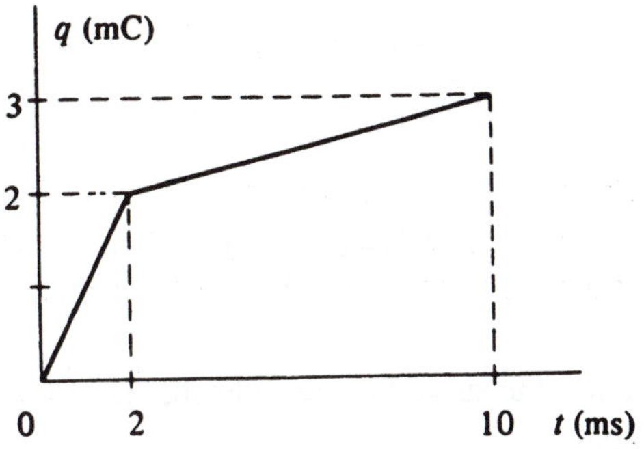

PROBLEM 1.13

1.14 The graph below shows the current i in A versus t in seconds into the positive terminal of an element whose voltage is given by $v = 10di/dt$ V. Find p at 1s and at 4s.

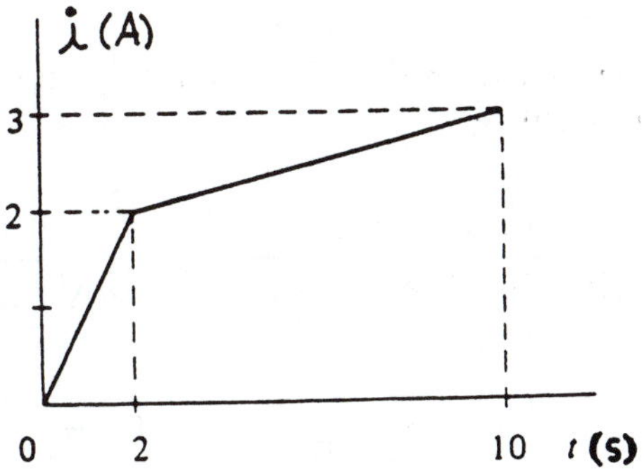

PROBLEM 1.14

1.4 Passive and Active Elements

1.15 In a certain car, the alternator supplies a charging current into the positive terminal of the battery (as shown below) of 30 mA for each kilometer per hour (kph) of speed. That is,

$$i = (30 \text{ mA}) \times (\text{speed in kph}).$$

If the car is traveling at 100 kph,
(a) How much charging current is being supplied to the battery?
(b) What is the energy supplied to the 12 V battery after driving 2 hours at this speed?
(c) What is the total charge delivered to the battery after driving 2 hours at this speed?

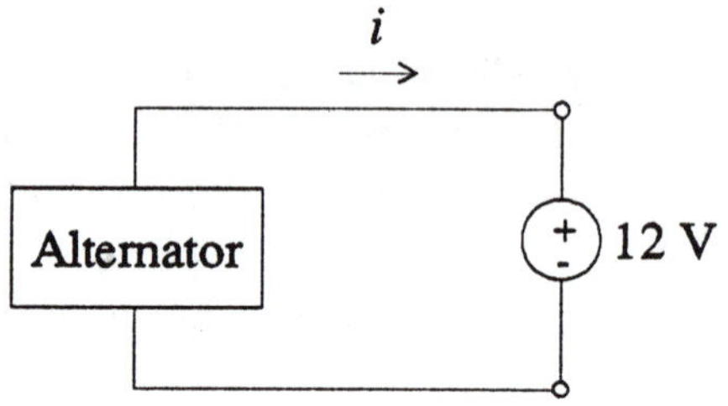

PROBLEM 1.15

1.16 How fast would someone need to drive the car in Prob. 1.15 in order to supply the same charge to the battery in only 10 minutes?

1.17 An electric circuit is shown in the diagram below. (a-g) Find the power absorbed by each element (Hint: remember to use the passive sign convention). (h) What is the sum of all the powers absorbed by each element in the circuit? Does this obey conservation of energy/power?

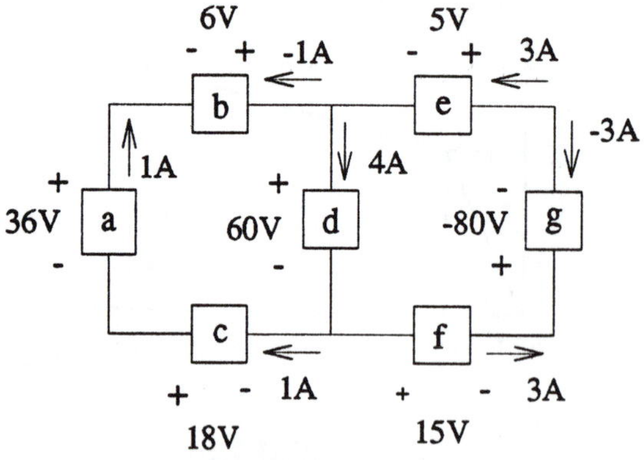

PROBLEM 1.17

1.18 The charge entering the positive terminal of a device is given by $q = \frac{1}{12}t^2 - 20t + 5$ mC. If the voltage across the terminals varies linearly from 5 to 10 V as t varies from 0 to 10s, find (a) the power absorbed in this time period as a function of t and (b) the total energy delivered during this time period.

1.19 If the current *leaving* the positive terminal of an element is $i = \frac{1}{2}t^2 - e^{-6t}$ A and the potential difference between the positive and negative terminals is 10 V, find the energy delivered to the element between $0 < t \le 5$ minutes.

1.20 Let the current entering the positive terminal of an element be given by

$$i(t) = \begin{cases} 0 \text{ A}, & \text{for } t < 0 \\ 3\cos(4t) - 3 \text{ A}, & \text{for } t \ge 0. \end{cases}$$

If the voltage is $v = 2di/dt$ V, find the energy delivered to the element for all time.

1.21 In Prob. 1.20 find the (a) the total charge delivered to the element at $t = \pi/4$ s and (b) the power absorbed at $t = \pi/8$ s.

1.22 The current entering the positive terminal of a device is given by

$$i(t) = \begin{cases} 0 \text{ A}, & \text{for } t < 0 \\ 6\sin(2t), & \text{for } t \ge 0. \end{cases}$$

(a) If the voltage is $v = 4di/dt$ V, show that the energy delivered to the device is nonnegative for all time (i.e. the device is passive).

(b) Repeat part (a) if $v = 2\int_{-\infty}^{t} i dt$ V.

CHAPTER 1 SOLUTIONS

1.1 (a) $1\,mA = 1\times10^{-3}A$, $1\,\mu A = 1\times10^{-6}A$

$\Rightarrow 1000\,\mu A = 1\,mA$. Thus,

$5\,\mu A \times \left(\frac{1\,mA}{1\times10^{3}\,\mu A}\right) = 0.005\,mA$

(b) $37\,pA \times \left(\frac{1\,mA}{1\times10^{9}\,pA}\right) = 3.7\times10^{-8}\,mA$

(c) $2\,kA \times \left(\frac{1\,mA}{1\times10^{-6}\,kA}\right) = 2\times10^{6}\,mA$

(d) $0.5\,MA \times \left(\frac{1\,mA}{1\times10^{-9}\,MA}\right) = 5\times10^{8}\,mA$

1.2 (a) $1/(4\times10^{4}\text{ updates/s}) = 2.5\times10^{-5}$ s/update

(b) $25\,\mu s$

1.3 From the graph:

(a) $q_T = q(3) - q(1) = 4 - 3 = 1\,C$

(b) $q_T = q(7) - q(4) = 2 - 4 = -2\,C$

(c) $i = dq/dt$ = slope of q curve

$i(1) = 1A$, $i(4) = 0A$, $i(6) = -1A$

1.4 (a) $q_T = \int_0^7 i(t)\,dt$ = area under $i(t)$

$= 2\times7 + 2\times5 = 24\,mC$

(b) rate $= dq/dt = i(t)$, $i(5) = 4\,mA$

1.5 (a) $q_T = \int_0^5 i\,dt = \int_0^5 (1 - e^{-3t}A)\,dt$

$= \frac{14}{3} + \frac{e^{-15}}{3}\ C$

(b) $q_T = \int_0^T i\,dt = \left(T + \frac{e^{-3T}}{3} - \frac{1}{3}\right) C$

1.6 (a) $q_T = \int_0^{10} i\,dt = 2\times10 + \frac{1}{2}(6)(4)$

$= 32\,mC$

(b) rate $= dq/dt = i(t) \Rightarrow i(2) = 6\,mA$

$i(4) = 4\,mA$

1.7 (a) From the graph,

$q_T = q(2) - q(1) = 6 - 4 = 2C$

(b) $q_T = q(10) - q(6) = 2 - 2 = 0\,C$

(c) $i = \frac{dq}{dt}$ = slope of q curve

$i(1) = \frac{6-2}{2} = \underline{2A}$

$i(4) = \frac{2-6}{6-2} = \underline{-1A}$

$i(8) = \underline{0A}$

1.8 $W(t) - W(t_0) = \int_{t_0}^{t} v i\,d\tau$

$W(1s) = W(0s) + \int_0^1 v i\,dt$

$= 2mJ + \int_0^1 (20.2\pi\,V)[10\sin 101\pi t\ mA]\,dt$

$= 2mJ + 202\pi \int_0^1 \sin(101\pi t)\,dt$

$= 4mJ$

1.9 (a) $p = vi \Rightarrow i = 7\,mA$

(b) $p = vi \Rightarrow i = -1\,A$

(c) $p = -vi \Rightarrow i = -1\,mA$

(d) $p = -vi \Rightarrow i = 3\,\mu A$

1.10 (a) $W(t) = W(1) + \int_1^t (10V)(2e^{-0.5\tau})\,d\tau$

$= 100mJ + 10\,(-4e^{-0.5\tau})\Big|_1^t\ J$

$= 24.36 - 40e^{-0.5t}\ J$

(b) $W(t) = 0.1 + \int_1^t (6e^{-0.5\tau})(2e^{-0.5\tau})\,d\tau$

$= 4.51 - 12e^{-t}\ J$

(c) $W(t) = 0.1 + \int_1^t (-e^{-0.5\tau})(2e^{-0.5\tau})\,d\tau$

$= -0.64 + 2e^{-t}\ J$

1.11 $i(t) = dq/dt = 2\cos(t/4)\ A$

$p(\pi) = vi = (6\cos\pi/4)(2\cos\pi/4)$

$= 12/2 = \underline{6W}$

1.12 (a) $i = dq/dt = 40e^{-2t}\ mA$

$v = P/i = \frac{6e^{-4t}\,W}{40e^{-2t}\,mA} = 150e^{-2t}V$

(b) $w = \int_0^{250ms} p\,dt = \int_0^{.25} 6e^{-4t}\,dt$

$= 1.5(1 - e^{-1})\ J$

1.13 (a) $q_T = q(10ms) - q(1ms) = 3 - 1 = \underline{2mC}$

(b) $i = \frac{dq}{dt} = \begin{cases} \frac{2-0}{2-0} = 1A, & 0 \le t < 2ms \\ \frac{3-2}{10-2} = \frac{1}{8}A, & 2ms < t \le 10ms \end{cases}$

1.13 cont.

$$W_T = \int_{10^{-3}}^{10\times10^{-3}} vi\,dt$$
$$= \int_{10^{-3}}^{2\times10^{-3}} (6)(1)dt + \int_{2\times10^{-3}}^{10\times10^{-3}} 6\left(\tfrac{1}{8}\right)dt$$
$$= (6+6)10^{-3} = \underline{12\text{ mJ}}$$

1.14
$$v = \begin{cases} 10\text{ V} & 0 \le t < 2\text{s} \\ 10/8\text{ V} & 2\text{s} < t < 10\text{s} \end{cases}$$
$$p(1) = v(1)i(1) = \underline{10\text{ W}}$$
$$p(4) = v(4)i(4) = \tfrac{10}{8}\left(2+\tfrac{2}{8}\right) = \underline{2.813\text{ W}}$$

1.15 (a) $i = (30\text{ mA})\times 100 = \underline{3\text{A}}$

(b) $W = \int_0^{2(3600)} vi\,dt = \int_0^{7200} (12)(3)dt$
$= \underline{259.2\text{ kJ}}$

(c) $q = \int_0^{7200} i(t)dt = \int_0^{7200} 3dt = \underline{21.6\text{ kC}}$

1.16 $q = 21.6\text{kC} = \int_0^{10(60)} i\,dt = 600 i$

$\Rightarrow i = 36\text{A} \Rightarrow \text{Speed} = \frac{i}{30\text{mA/kph}} = \underline{1200\text{kph}}$

1.17 (a) $p = (36\text{V})(-1\text{A}) = -36\text{W}$

(b) $p = (6\text{V})(-1\text{A}) = -6\text{W}$

(c) $p = (18\text{V})(-1\text{A}) = -18\text{W}$

(d) $p = (60\text{V})(4\text{A}) = 240\text{W}$

(e) $p = (5\text{V})(3\text{A}) = 15\text{W}$

(f) $p = (15\text{V})(3\text{A}) = 45\text{W}$

(g) $p = (-80\text{V})(3\text{A}) = -240\text{W}$

(h) $\Sigma p = 0$, Yes - agrees with conserv. of power.

1.18 (a) $i = \frac{dq}{dt} = \left(\frac{1}{6}t - 20\right)\text{mA}$

$v = \left(\frac{1}{2}t + 5\right)\text{V}$

$\cdot\ p(t) = vi = \left(\frac{1}{12}t^2 - \frac{55}{6}t - 100\right)\text{mW}$

1.18 cont. (b) $w = \int_0^{10} p(t)\,dt$

$= \left(\frac{1}{36}t^3 - \frac{55}{12}t^2 - 100t\right)\Big|_0^{10}$

$= -1430.5\text{ mJ} \approx \underline{-1.43\text{ J}}$

1.19 $w = \int_0^{5(60)} p(t)dt = \int_0^{300} vi\,dt$

$= \int_0^{300} 10\text{V}\left(e^{-6t} - \frac{1}{2}t^2\text{A}\right)dt$

$= -\frac{10}{6}\left[e^{-1800} + 2.7\times10^7\right]\text{J}$

$\approx -45\text{ MJ}$

1.20 Case 1, $t<0$: $i(t) = 0 \Rightarrow W(t) = 0$

Case 2, $t \ge 0$: $p(t) = v(t)\,i(t)$

$\Rightarrow p(t) = -72\sin 4t\cos 4t + 72\sin 4t$

$W(t) = \int_{-\infty}^{t} p(t)dt = \int_0^t p(t)dt$

$= -\frac{72}{8}\sin^2 4\tau\Big|_0^t + \frac{72}{4}\left[-\cos 4\tau\right]_0^t$

$= -9\sin^2 4t - 18\cos 4t + 18\text{ J}$

1.21 (a) $q_T = \int_{-\infty}^{\pi/4} i(t)dt = \int_0^{\pi/4} i(t)dt$

$= \left(\frac{3\sin 4t}{4} - 3t\right)\Big|_0^{\pi/4} = -\frac{3\pi}{4}\text{ C}$

(b) $p\left(\frac{\pi}{8}\right) = -72\left[\sin 4\left(\frac{\pi}{8}\right)\cos 4\left(\frac{\pi}{8}\right) - \sin 4(\pi/8)\right]$

$= 72\text{W}$

1.22 (a) For $t<0$, $i = 0 \Rightarrow W(t) = 0$ ✓

For $t \ge 0$, $w = \int_0^t (48\cos 2t)(6\sin 2t)dt$

$= 72\sin^2 2t \ge 0$ ✓

(b) $v = 2\int_0^t 6\sin 2t = -6\cos 2t + 6\text{ V}$

$w = \int_0^t (-6\cos 2t + 6)(6\sin 2t)dt$

$= -9\sin^2 2t - 18\cos 2t + 18$

$= 9(\cos^2 2t - 1 - 2\cos 2t + 2)$

$= 9(\cos 2t - 1)^2 \ge 0$ ✓

Chapter 2

Resistive Circuits

2.1 Kirchhoff's Laws

2.1 Find i_1, i_2, and i_3 in the following circuit using only KCL.

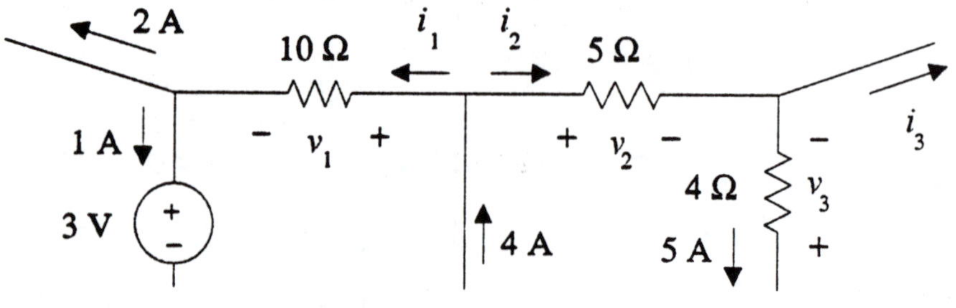

PROBLEM 2.1

2.2 Find i_3, i_4, i_6, and i_7 in the following circuit if $i_1 = -1$ A, $i_2 = 5$ A, and $i_5 = 7$ A.

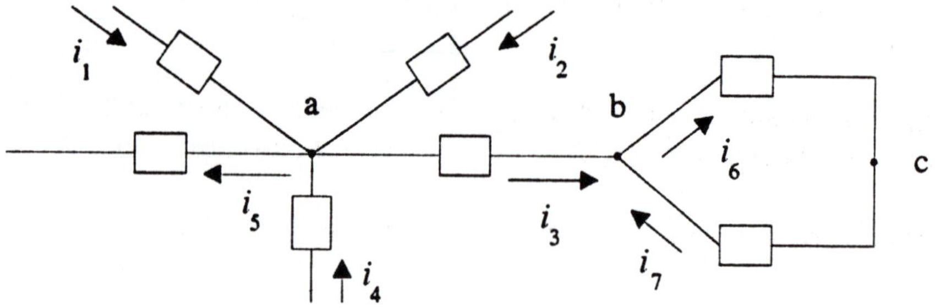

PROBLEM 2.2

2.3 Find v_1, v_2, and v_3 in the following circuit.

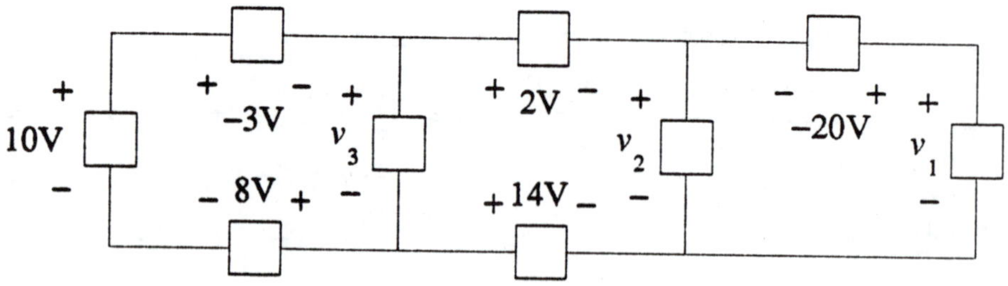

PROBLEM 2.3

2.4 (a) If $v_1 = v_2$, find v_1.
(b) If $v_1 = 2v_2$, find v_1.
(c) If $v_2 = 2v_1$, find v_1.
(d) If $v_2 = 10v_1$, find v_1.

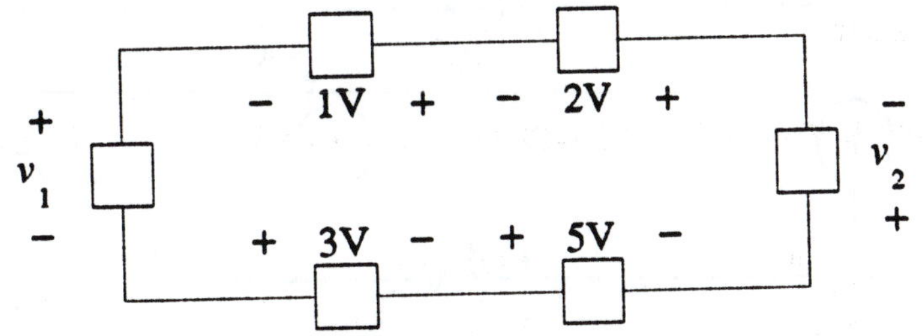

PROBLEM 2.4

2.2 Ohm's Law

2.5 A 2 kΩ resistor is connected to a battery and 6 mA flow. What current will flow if the battery is connected to a 600 Ω resistor?

2.6 A resistor connected to a 12 V source carries a current of 60 mA. Find the resistance and the power dissipated by the resistor. If a 400 Ω resistor is inserted in series in the circuit, find the voltage across it.

2.7 A toaster is essentially a resistor that becomes hot when it carries a current. If a toaster is dissipating 400 W at a current of 2 A, find its voltage and its resistance.

2.8 If a toaster with a resistance of 10 Ω is operated at 100 V for 8 seconds, find the energy it uses.

2.9 Find i_1, i_2, and v.

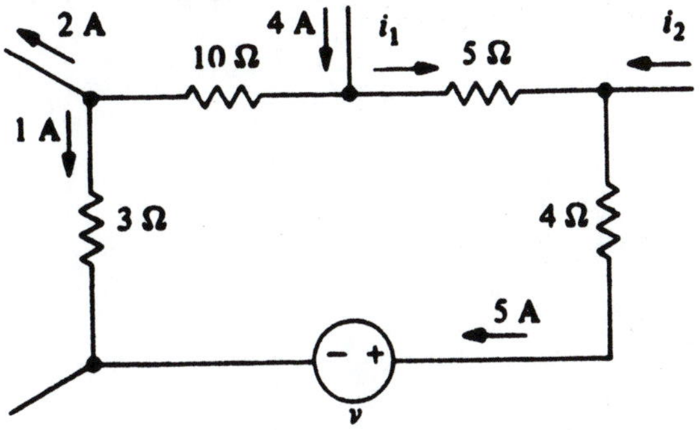

PROBLEM 2.9

2.10 Find v.

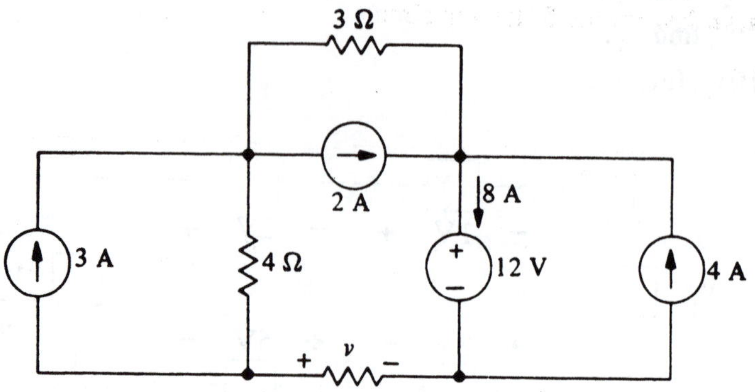

PROBLEM 2.10

2.11 Find i and v_{ab}.

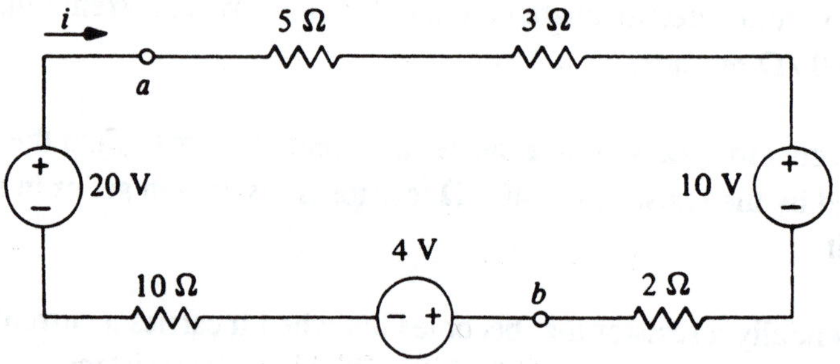

PROBLEM 2.11

2.12 Find v_1, v_2, and v_3 in Prob. 2.1.

2.13 If all the elements in Prob. 2.2 are resistors and the magnitude of the voltage drops across the resistors R_1 and R_2 which currents i_1 and i_2 flow (respectively) is 3 V, find the value of R_1 and R_2.

2.3 Equivalent Subcircuits

2.14 Find the terminal law for the following circuit.

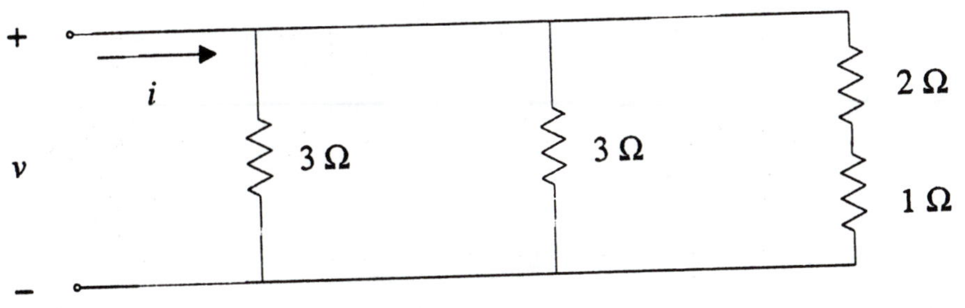

PROBLEM 2.14

2.15 Find the terminal law for the following circuit.

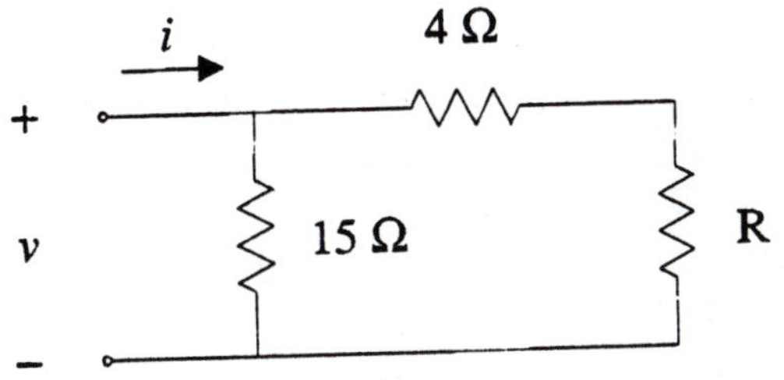

PROBLEM 2.15

2.16 Find the value of R in Prob. 2.15 that gives the circuit a terminal law of $v = 6i$.

2.17 Find a circuit containing just a single element that is equivalent to the circuit in Prob. 2.14.

2.4 Series Equivalents and Voltage Division

2.18 (a) Reduce the circuit shown to two equivalent elements, (b) find i, and (c) find the instantaneous power absorbed by each of the three sources and each of the three resistors in the original circuit.

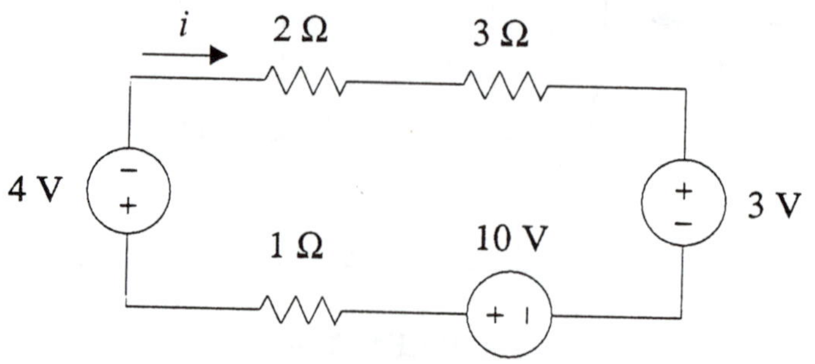

PROBLEM 2.18

2.19 Find R_2 if $v_2 = \cos 9t - \frac{1}{4}e^{-7t}$ V, $R_1 = 5\Omega$, $R_3 = 13\Omega$, and $R_4 = 2\Omega$.

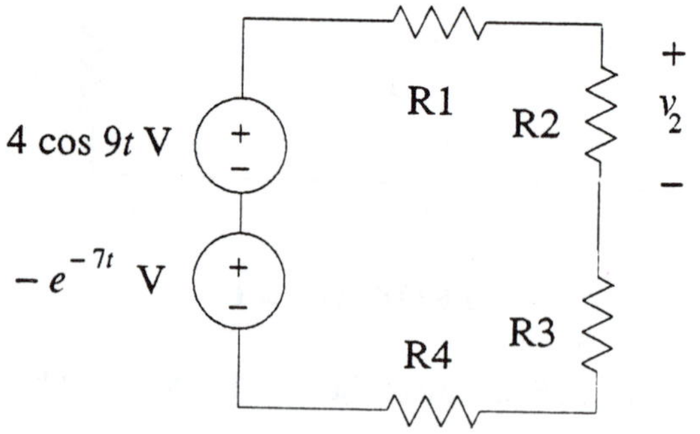

PROBLEM 2.19

2.20 Find v_s so that $i_s = 1$ A.

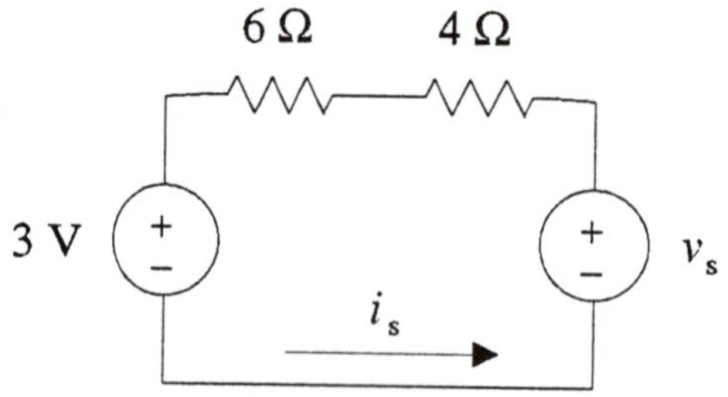

PROBLEM 2.20

2.21 Find v using voltage division.

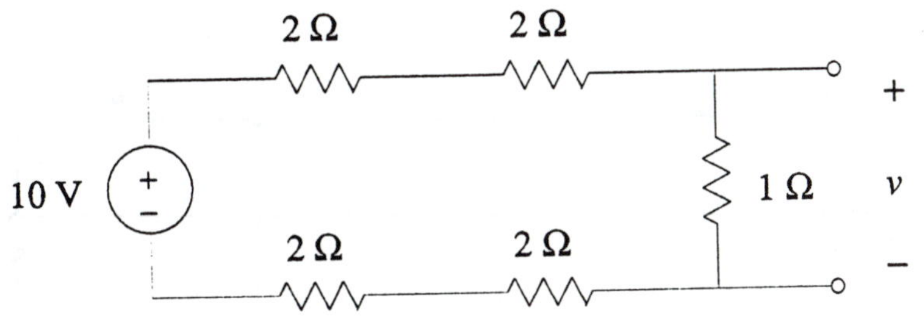

PROBLEM 2.21

2.22 (a) Find i_s when $v_s = 1$ V.
(b) Find i_s when $v_s = 2$ V.
(c) Find v_1 when $v_s = 1$ V.
(c) Find v_1 when $v_s = 2$ V.

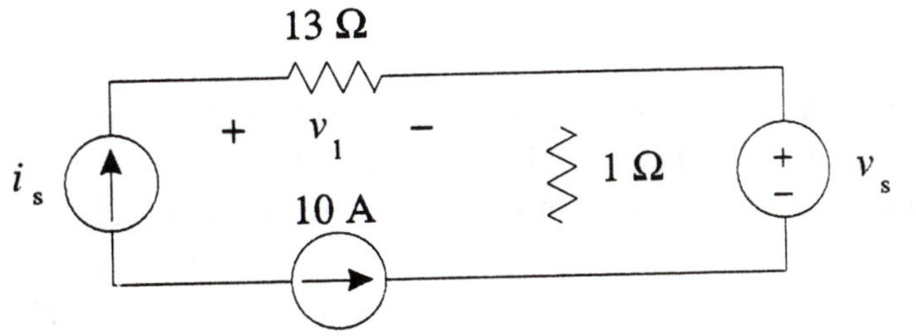

PROBLEM 2.22

2.23 A 10 V source in series with several resistors carries a current of 50 mA. If a 300 Ω resistor is inserted in series in the circuit, find the resulting current.

2.24 A 50 V source delivers 500 mW to two resistors, R_1 and R_2, connected in series. If the voltage across R_1 is 10 V, find R_1 and R_2.

2.25 A 50 V source and two resistors, R_1 and R_2, are connected in series. If $R_2 = 3R_1$, find the voltages across the two resistors.

2.5 Parallel Equivalents and Current Division

2.26 Find i and the power delivered to the 3 Ω resistor in the circuit shown.

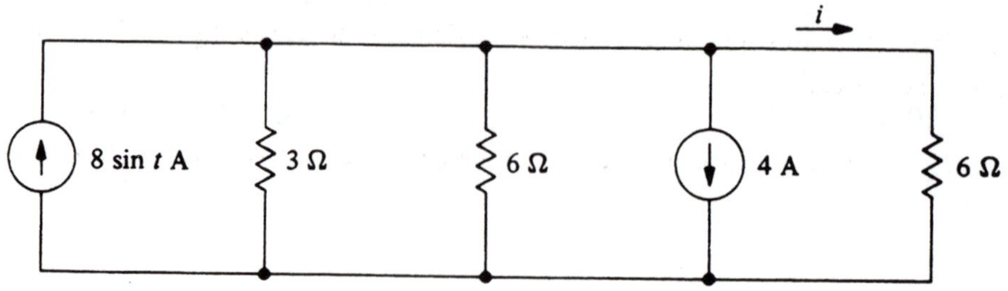

PROBLEM 2.26

2.27 A 5 A current source, a 20 Ω resistor, and a 30 Ω resistor are connected in parallel. Find the voltage, current, and minimum power rating of each resistor.

2.28 A 40 Ω resistor, a 120 Ω resistor, and a resistor R are connected in parallel to form an equivalent resistance of 15 Ω. Find R and the current it carries if an 8 A current source is connected to the combination.

2.29 A current divider consists of 10 resistors in parallel. Nine of them have an equal resistance of 30 kΩ and the tenth is a 10 kΩ resistor. Find the equivalent resistance of the divider, and if the total current entering the divider is 20 mA, find the current through the 10 kΩ resistor.

2.30 A 40 Ω resistor, a 60 Ω resistor, and a resistor R are connected in parallel to form an equivalent resistance of 8 Ω. Find R and the current it carries if a 6 A current source is connected in parallel to the combination.

2.31 Find i_1, i_2, i_3, i_4, and i_5 in the following circuit.

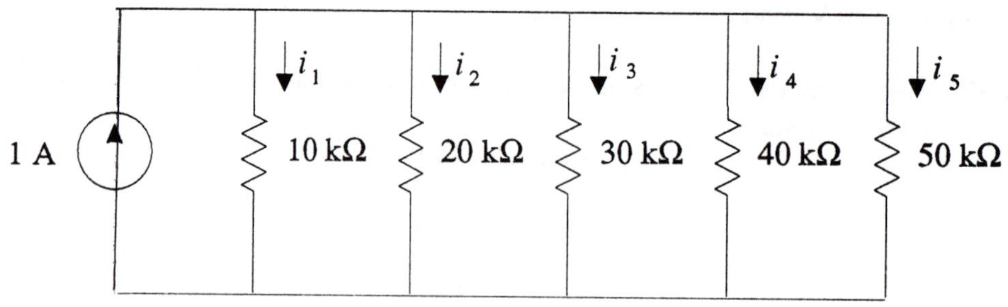

PROBLEM 2.31

2.6 Thevenin and Norton Equivalents

2.32 Find the Norton equivalent of a Thevenin equivalent circuit consisting of a 10 V source in series with a 5 Ω resistor.

2.33 Find the Thevenin equivalent of a Norton equivalent circuit consisting of a 3 A source in parallel with a 17 Ω resistor.

2.34 Find the Thevenin and Norton equivalents of the circuit to the left of terminals *a* and *b*.

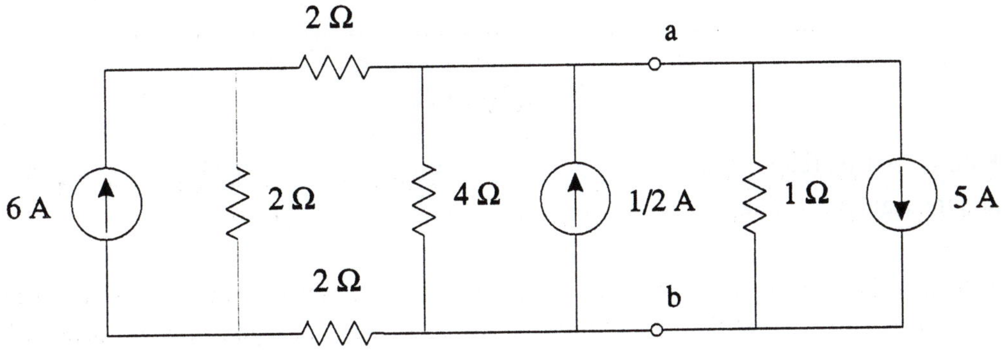

PROBLEM 2.34

2.35 Find the Thevenin and Norton equivalent circuits for the circuit shown.

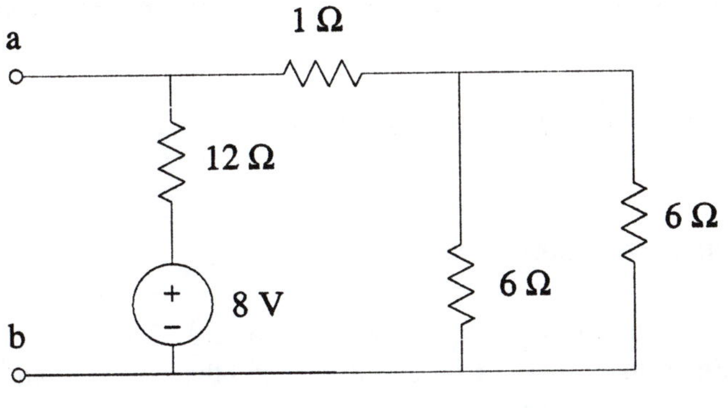

PROBLEM 2.35

2.36 Find the Thevenin and Norton equivalent circuits for the circuit shown.

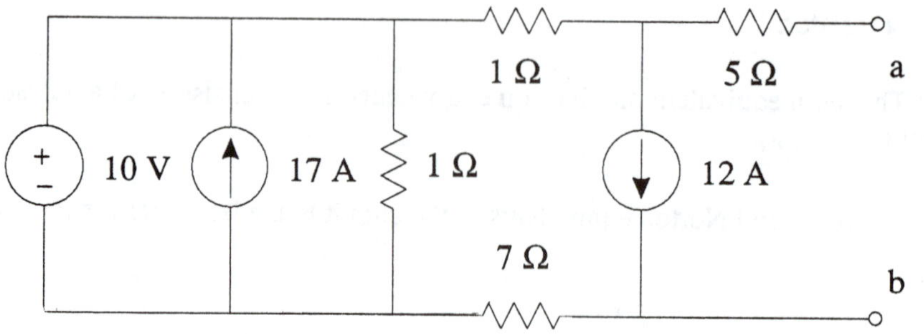

PROBLEM 2.36

2.37 Find i_1 by first finding the Norton equivalent circuits to the left of the line *ab* and the right of the line *cd*.

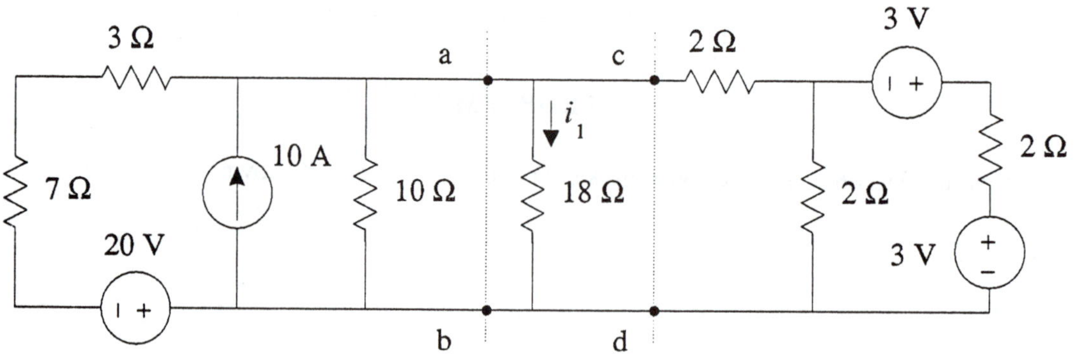

PROBLEM 2.37

2.7 Practical Sources and Resistors

2.38 The open-circuit voltage of a battery is measured to be 9.12 V. When a 15 Ω resistor is placed across its terminals, the voltage drops to 8.86 V. Find the Thevenin and Norton equivalents of the battery and the maximum current that it can supply.

2.39 Find an equation for instantaneous power delivered to a load resistor R_L connected to the terminals of the battery in Prob. 2.38. Using calculus, find the maximum power which can be delivered to R_L and the value of R_L required for this case.

2.40 A source which can be modeled as a 1.5 V independent voltage source in series with a 20 Ω resistor is used to provide a 10 mA current to a 1 Ω resistor. Find a way to do this using a shunt resistor across the terminals of the source. Find a second way to do this using a resistor in series with the load resistor. Which of these two methods is more efficient and why?

2.41 A source provides 1 A to a 1 Ω resistor placed across its terminals. When a 10 Ω resistor is instead placed across the terminals, 500 mA flow through it. Find the Thevenin and Norton equivalent circuit models for this source. Sketch the load current, i_L, versus R_L for $0 \le R_L \le 100\ \Omega$.

2.42 Determine the color codes for resistors having the following resistance ranges: (a) 2.97 - 3.63 MΩ, (b) 1.14 - 1.26 kΩ, and (c) 50.4 - 61.6 Ω.

2.43 Find the resistance range of resistors having color bands of (a) brown, black, red, silver; (b) red, violet, yellow, silver; and (c) blue, gray, silver, gold.

2.8 Ammeters, Voltmeters, and Ohmmeters

2.44 An Arsonval meter has a full-scale current of 5 mA and a resistance of 9.8 Ω. If a 0.2 Ω parallel resistor is used in the following figure, what is i_{FS}? What is the voltage across the meter?

2.45 A 36,000 Ω/V voltmeter has a full-scale voltage of 90 V. What current flows in the meter when measuring 45 V?

2.46 Two 3.3 kΩ resistors are connected in series across a 50 V source. What voltage will the voltmeter of Prob. 2.45 measure across one of the 3.3 kΩ resistors?

2.47 The Arsonval meter of Prob. 2.44 is used for the ohmmeter of the following figure. What value of series resistance is required if $E = 5$ V? What value of unknown resistance will cause a one-third full-scale deflection?

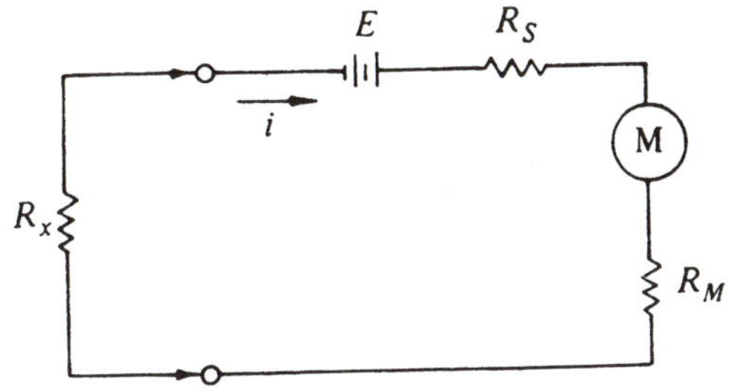

PROBLEM 2.47

2.9 Design of Resistive Circuits

2.48 Design a voltage divider to provide 2, 5, 15, and 35 V, all with a common negative terminal, from
a 40 V source. The source is to deliver 80 mW of power.

2.49 A voltage divider is to be constructed with a 60 V source and a number of 10 kΩ resistors. Find the minimum number of resistors required if the output voltage is (a) 40 V and (b) 30 V.

2.50 A voltage divider is to be constructed with a 50 V source and a number of 10 kΩ resistors. If the output voltage is to be 20 V, find the minimum number of resistors that can be used. Draw the circuit.

2.51 Find all the possible values of equivalent resistance that can be obtained by someone having three 6Ω resistors.

2.52 Design a current divider out of 10 S conductances which when supplied with a 1 A input current yields currents of 1/3 A, 1/4 A, 1/5 A, 1/6 A, and 1/20 A.

2.53 Design a 60 Ω load resistance to dissipate 5 W dc power using a minimum number of 100 Ω resistors each rated at 1/4 W. Draw the circuit.

2.54 Design a practical electrical source out of ideal sources and/or resistors which delivers at least 10 A to a 100 Ω load but is voltage-limited to 1500 V.

2.55 Using 2 Ω resistors, all 1/4W at 5% tolerance, design a 4 Ω (5% tolerance) load that can dissipate at least 1 W.

2.1 KCL: $\Sigma i_{entering} = 0$

$\Rightarrow i_1 - 2A - 1A = 0 \Rightarrow \underline{i_1 = 3A}$

$4A - i_1 - i_2 = 0 \Rightarrow \underline{i_2 = 4A - 3A = 1A}$

$i_2 - i_3 - 5A = 0 \Rightarrow \underline{i_3 = 1A - 5A = -4A}$

2.2 KCL at node c: $i_6 = i_7$

KCL at node b: $i_3 + i_7 = i_6 \Rightarrow \underline{i_3 = 0}$

KCL at node a: $i_4 = i_5 + i_3 - i_1 - i_2$

$\underline{i_4 = 7A + 0 + 1A - 5A = 3A}$

i_6 and i_7 can be any value such that $i_6 = i_7$

2.3 KVL around leftmost loop:

$v_3 + 8v - 10v + (-3v) = 0 \Rightarrow \underline{v_3 = 5V}$

KVL around center loop:

$v_2 - 14V - 5V + 2V = 0 \Rightarrow \underline{v_2 = 17V}$

KVL around rightmost loop:

$v_1 - v_2 - (-20V) = 0 \Rightarrow \underline{v_1 = -3V}$

2.4

(a) KVL: $v_1 + 3V + 5V + v_1 + 2V + 1V = 0$

$\Rightarrow \underline{v_1 = -11/2\ V}$

(b) KVL: $v_1 + 3V + 5V + \frac{1}{2}v_1 + 2V + 1V = 0$

$\Rightarrow \underline{v_1 = -22/3\ V}$

(c) KVL:

$v_1 + 8V + 10v_1 + 3V = 0 \Rightarrow \underline{v_1 = -1V}$

2.5 $v = iR = (2 \times 10^3 \Omega)(6 \times 10^{-3} A)$

$= 12\ V$;

$i_{600\Omega} = \frac{v}{R} = \frac{12V}{600\Omega} = \underline{20mA}$

2.6 $R = \frac{v}{i} = \frac{12V}{60mA} = \underline{200\Omega}$

$P = vi = (12V)(60mA) = \underline{720mW}$

$i = \frac{v}{R_1 + R_2} = \frac{12V}{200 + 400\Omega} = 20mA$;

$v_{400\Omega} = iR = (20mA)(400\Omega)$

$= \underline{8V}$

2.7

$v = \frac{P}{i} = \frac{400W}{2A} = \underline{200V}$;

$R = \frac{v}{i} = \frac{200V}{2A} = \underline{100\Omega}$

2.8 $P = \frac{v^2}{R} = \frac{(100V)^2}{10\Omega} = 1000W$;

$\Delta W = \int_0^8 p\,dt = \int_0^8 1000\,dt = \underline{8.0\ kJ}$

2.9

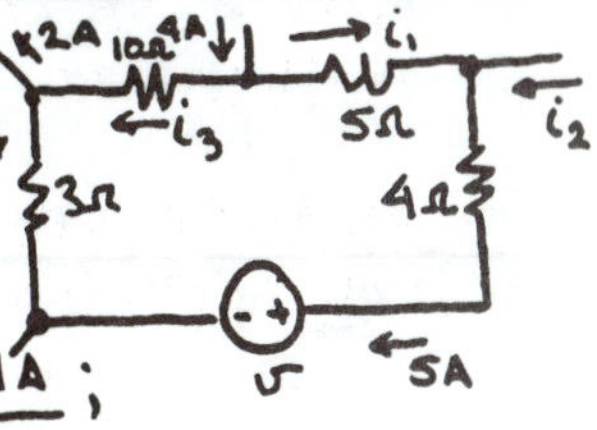

By KCL,

$i_3 = 2A + 1A = 3A$;

$i_1 = 4A - i_3$

$= 4A - 3A = \underline{1A}$;

$i_2 = 5A - i_1 = 5A - 1A = \underline{4A}$;

By KVL, $v = -(5)(4) - (1)(5) + (3)(10)$

$+ (1)(3) = \underline{8V}$

2.10 By KCL; $i_{3\Omega} = 8A - 2A - 4A = 2A$ to the right

$i_{4\Omega} = 2A - 3A + 2A = 1A$ up

By KVL; $v = (4)(1) + (3)(2) + 12$

$= \underline{22V}$

2.11

By KVL;

$(5 + 3 + 2 + 10)i + 10V + 4V - 20V = 0$

$i = \frac{6V}{20\Omega} = \underline{0.3A}$

$v_{ab} = (5 + 3 + 2)i + 10V = \underline{13V}$

2.12 $v_1 = (10\Omega)(i_1) = (10\Omega)(3A) = \underline{30V}$

$v_2 = (5\Omega)(i_2) = (5\Omega)(1A) = \underline{5V}$

$v_3 = (4\Omega)(-5A) = \underline{-20V}$

2.13 $R_1 = \frac{3V}{1A} = 3\Omega$, $R_2 = \frac{3V}{5A} = 0.6\Omega$

2.14 $R_{eq} = 3 \| 3 \| [2+1] = \frac{1}{\frac{1}{3} + \frac{1}{3} + \frac{1}{3}}$

$\Rightarrow R_{eq} = 1\Omega$

$v = R_{eq} i = (1\Omega)\, i$

Terminal Law: $\underline{v = i}$

2.15 $R_{eq} = 15 \| (4+R) = \frac{15(4+R)}{15+4+R}$

$\Rightarrow R_{eq} = \frac{60+15R}{19+R}$

$v = R_{eq} i = \left[\frac{60+15R}{19+R}\right] i$

2.16 Pick $R=6$:

$R_{eq} = 60 + 15(6)/(19+6) = 6\Omega$

2.17 $R_{eq} = 1\Omega \Rightarrow$

2.18 Add voltage sources in series:

$V_{TOT} = -4v + 10v - 3v = +3v$

Add resistors in series: $1+2+3 = 6\Omega$

(a)

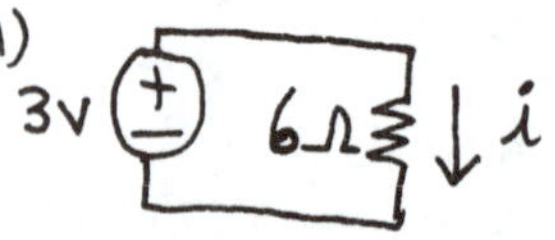

(b) $i = \frac{V}{R} = \frac{3V}{6\Omega}$

$\Rightarrow i = \frac{1}{2}A$

(c) $P = iv$

4V source: $P = (\frac{1}{2}A)(4V) = 2W$

3V source: $P = (\frac{1}{2}A)(3V) = 1.5W$

10V source: $P = (-\frac{1}{2}A)(10V) = -5W$

1Ω resistor: $P = i^2R = \frac{1}{4}(1) = \frac{1}{4}W$

2Ω resistor: $P = i^2R = \frac{1}{4}(2) = \frac{1}{2}W$

3Ω resistor: $P = i^2R = \frac{1}{4}(3) = \frac{3}{4}W$

2.19 Voltage sources in series add.

Using voltage division:

$v_2 = (4\cos 9t - e^{-7t})\left[\frac{R_2}{R_1+R_2+R_3+R_4}\right]$

$= (4\cos 9t - e^{-7t})(R_2/[20+R_2])$

$= \cos 9t - \frac{1}{4}e^{-7t}$

$\Rightarrow \frac{R_2}{20+R_2} = \frac{1}{4} \Rightarrow R_2 = 6.667\Omega$

2.20 KVL: $v_s - 3v - i_s(6\Omega) - i_s(4) = 0$

$i_s = 1 \Rightarrow v_s - 3 - 6 - 4 = 0$

$\Rightarrow v_s = 13V$

2.21

$v = 10v\left[\frac{1}{2+2+1+2+2}\right]$

$\Rightarrow v = \frac{10}{9}V = 1.\bar{1}V$

2.22 (a) $i_s = -10A$ to obey KCL

(b) $i_s = -10A$ to obey KCL

(c) Since $i_s = -10A$ no matter what,

$v_1 = i_s(13\Omega) = -130V$

(d) $v_1 = -130V$ (same as c)

2.23 R_1 = original resistance

$= \frac{10V}{50mA} = 200\Omega$

$i = \frac{v}{R_1 + 300\Omega} = \frac{10V}{500\Omega} = 20mA$

2.24 $R_1 + R_2 = \frac{v^2}{P} = \frac{(50V)^2}{500mW} = 5k\Omega$

By voltage division,

$10V = \frac{R_1}{R_1+R_2}(50V)$

$\Rightarrow R_1 = \frac{10V}{50V}(R_1+R_2) = \frac{10(5k)}{50} = 1k\Omega$

$\Rightarrow R_2 = 5k\Omega - 1k\Omega = 4k\Omega$

2.25 v_i = voltage across R_i, $i = 1, 2$

By voltage division:

$v_1 = \frac{R_1}{R_1+R_2}(50) = \frac{50R_1}{R_1+3R_1} = \frac{50}{4} = 12.5V$

$v_2 = 50 - v_1 = 37.5V$

2.26

Equivalent current source has current $8\sin t - 4A$, up.

By current division,

$i = \frac{(1/6)}{(1/3)+(1/6)+(1/6)}(8\sin t - 4)$

$= 2\sin t - 1A$

$i_{3\Omega} = \frac{1/3}{1/3+1/6+1/6}(8\sin t - 4) = 4\sin t - 2$

$P_{3\Omega} = (i_{3\Omega})^2(3) = 12(2\sin t - 1)^2 W$

2.27 By current division

$$i_{20\Omega} = \frac{1/20}{1/20 + 1/30}(5A) = \underline{3A}$$

$$i_{30\Omega} = \frac{1/30}{1/20 + 1/30}(5A) = \underline{2A}$$

$$v_{20\Omega} = v_{30\Omega} = (20)(3) = \underline{60V}$$

$$P_{20\Omega} = (60V)(3A) = \underline{180W}$$

$$P_{30\Omega} = (60V)(2A) = \underline{120W}$$

2.28 $G_p = \frac{1}{15} = \frac{1}{40} + \frac{1}{120} + \frac{1}{R}$; $R = \underline{30\Omega}$

By current division,

$$i_R = \frac{1/R}{1/15}(8) = \underline{4A}$$

2.29 R_1 = equivalent resistance of 9 equal R's

$= 30/9 = 10/3 \, k\Omega$

R_p = divider resistance

$$= \frac{10/3(10)}{10/3 + 10} = \underline{2.5 \, k\Omega}$$

By current division,

$$i_{10} = \frac{10/3}{10/3 + 10}(20mA) = \underline{5mA},$$ current in the tenth resistor.

2.30 $G_P = \frac{1}{8} = \frac{1}{40} + \frac{1}{60} + \frac{1}{R}$; $R = \underline{12\Omega}$

By current division,

$$i_R = \frac{1/R}{1/8}(6) = \underline{4A}$$

2.31

$$i_1 = (1A)\frac{1/10k}{1/10k + 1/20k + 1/30k + 1/40k + 1/50k}$$

$$= \frac{1}{1 + 1/2 + 1/3 + 1/4 + 1/5} = \underline{\frac{60}{137}A}$$

$$i_2 = (1A)\frac{1/2}{1 + 1/2 + 1/3 + 1/4 + 1/5} = \underline{\frac{30}{137}A}$$

$$i_3 = (1A)\frac{1/3}{1 + 1/2 + 1/3 + 1/4 + 1/5} = \underline{\frac{20}{137}A}$$

$$i_4 = (1A)\frac{1/4}{1 + 1/2 + 1/3 + 1/4 + 1/5} = \underline{\frac{15}{137}A}$$

$$i_5 = (1A)\frac{1/5}{1 + 1/2 + 1/3 + 1/4 + 1/5} = \underline{\frac{12}{137}A}$$

2.32 $R_N = R_T = 5\Omega$

$$i_N = v_T / R_T = \frac{10V}{5\Omega} = 2A$$

2.33 $R_T = R_N = 17\Omega$

$$v_T = R_N i_N = (17)(3) = 51V$$

2.34

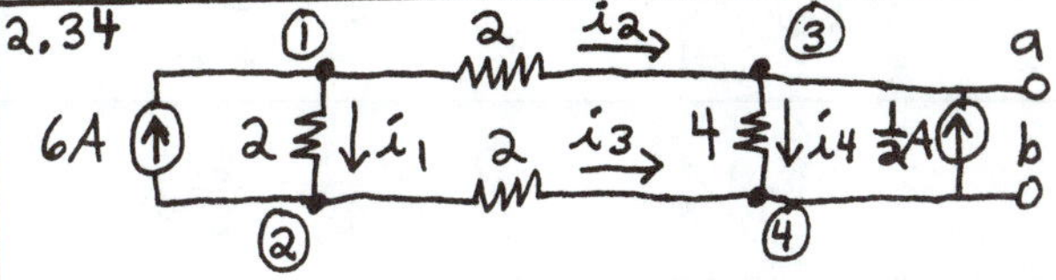

KCL at ①: $6 = i_1 + i_2$

KCL at ②: $i_1 = 6 + i_3$

KCL at ③: $i_2 + 1/2 = i_4$

KCL at ④: $i_3 + i_4 = \frac{1}{2}$

$i_1 = 17/4$ A

$i_2 = 7/4$ A

$i_3 = -7/4$ A

$i_4 = 9/4$ A

$$V_{oc} = 4(i_4) = 4(9/4) = \underline{9V = V_T}$$

To find R_T, set current sources to zero:

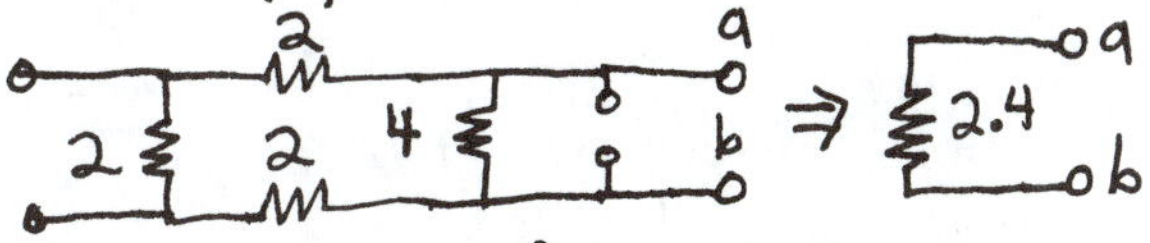

$\underline{R_T = 2.4\Omega} = R_N$

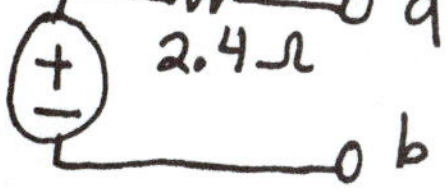

$\underline{I_N = 3.75A}$

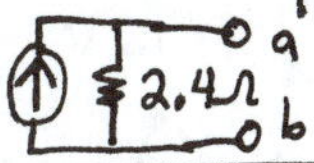

2.35 To find R_T, Set sources to zero:

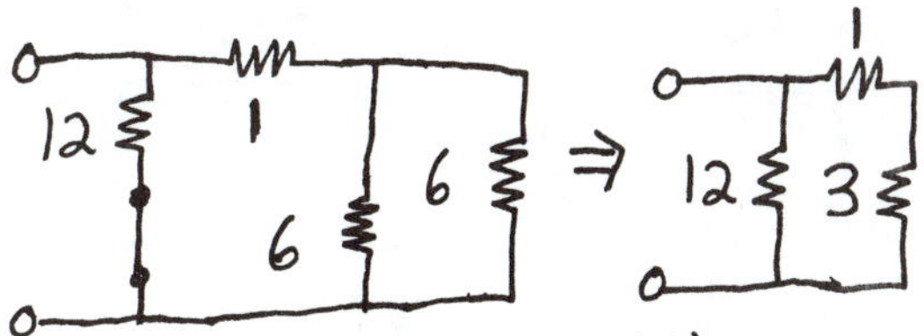

$$R_T = R_N = 12 \| 4 = \frac{12(4)}{12+4} = \underline{3\Omega}$$

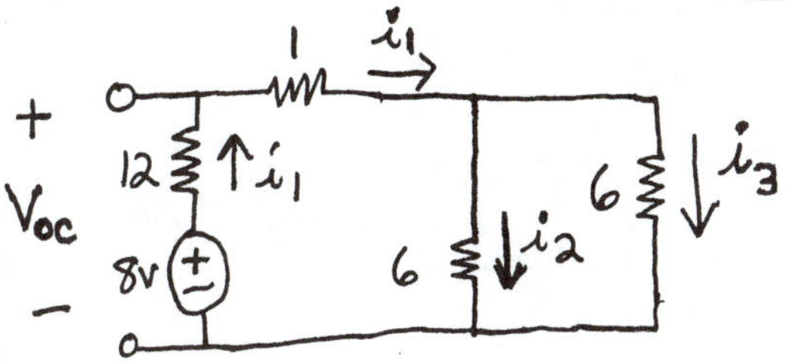

2.35 (cont'd)

Simplifying:

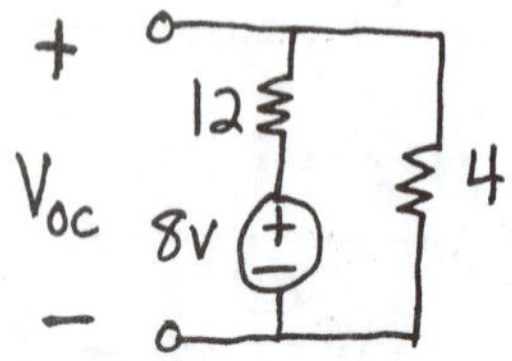

Voltage division

$$V_{oc} = 8v\left[\frac{4}{12+4}\right] = 2V = V_T$$

$$I_N = V_T/R_N = 2v/3\Omega = \frac{2}{3}A$$

2.36 Zero the sources to find R_T:

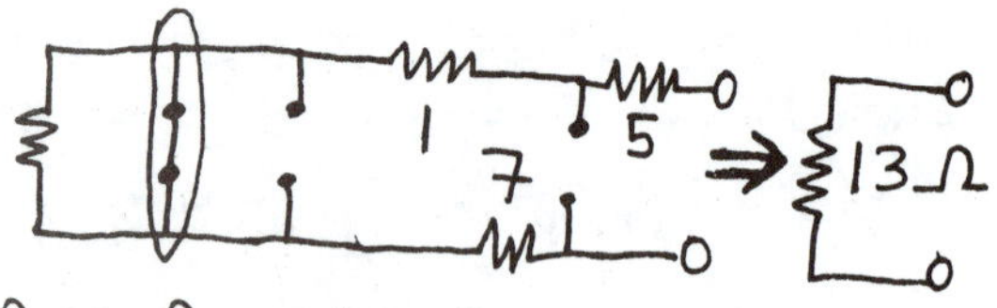

$R_T = R_N = 13\Omega$

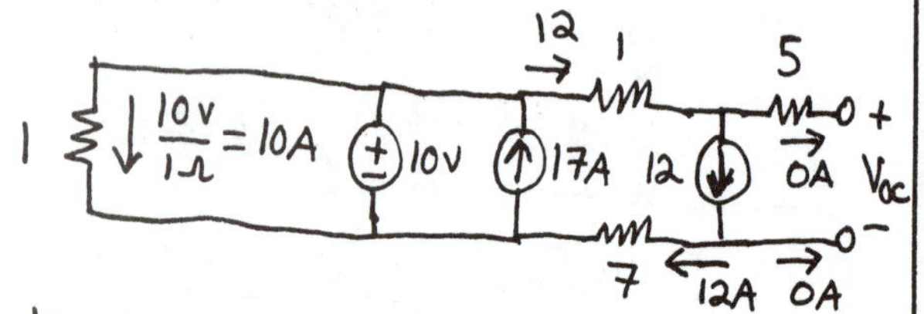

KVL: $V_{oc} + 12(7) - 10V + 12(1) = 0$

$\Rightarrow V_{oc} = V_T = -86V$

$\Rightarrow I_N = V_T/R_N = -6.6154A$

2.37 To the left

Set sources to zero to find R_N:

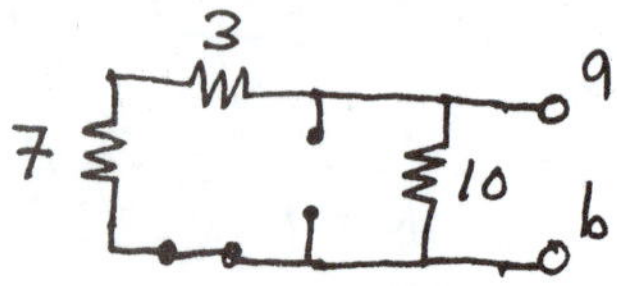

$R_N = 10\|10 = 5\Omega$

Find I_{sc}:

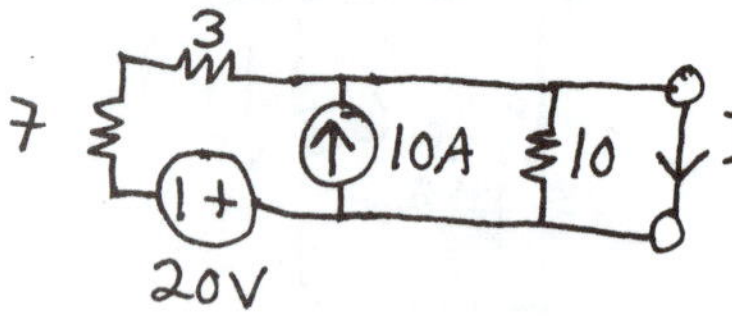

$I_{sc} = 10A$

2.37 (cont'd)

To the right Set sources to zero:

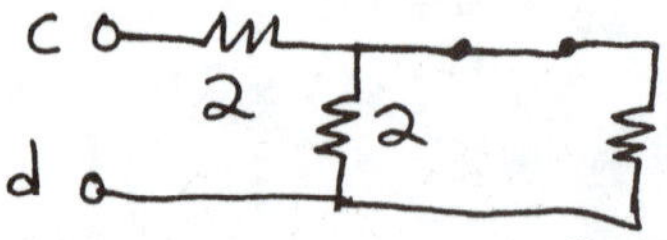

$\Rightarrow R_N = 3\Omega$

Find I_{sc}:

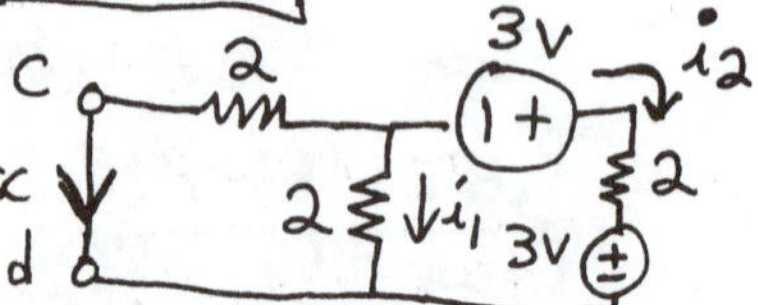

KCL: $I_{sc} + i_1 + i_2 = 0$

KVL: $2I_{sc} - 2i_1 = 0 \Rightarrow i_1 = I_{sc}$

$\Rightarrow I_{sc} = -\frac{1}{2}i_2$

KVL: $-3v + 2i_2 + 3v - 2I_{sc} = 0$

$\Rightarrow i_2 = I_{sc}$ and $I_{sc} = -\frac{1}{2}i_2$

$\Rightarrow I_{sc} = 0$

Redraw the original circuit:

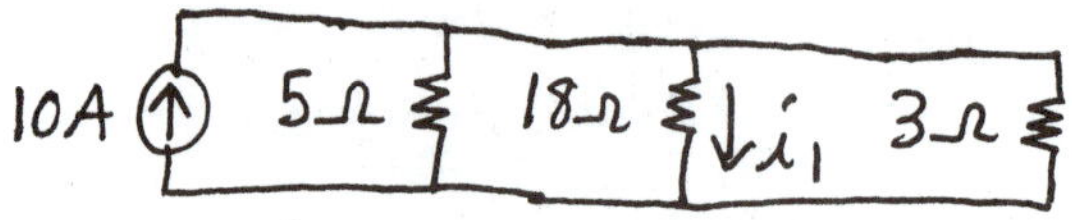

Current division:

$$i_1 = (10A)\frac{1/18}{1/5 + 1/18 + 1/3} = (10A)\frac{15}{159} = \frac{50}{53}A$$

2.38 $V_{oc} = V_T = 9.12V$

$i = 8.86V/15\Omega = 0.590667A$

KVL: $8.86V - 9.12V + (0.590667)R_T = 0$

$\Rightarrow R_T = 0.4402\Omega = R_N$

$I_N = V_T/R_T = 20.72A$ = Maximum current

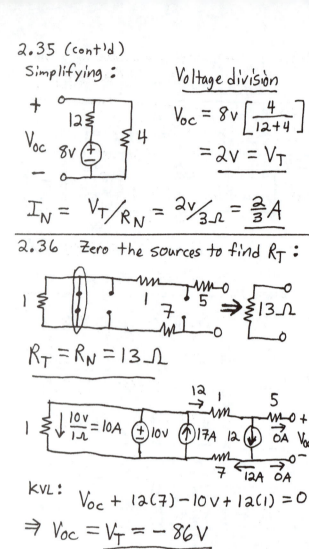

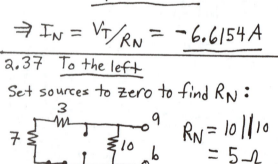

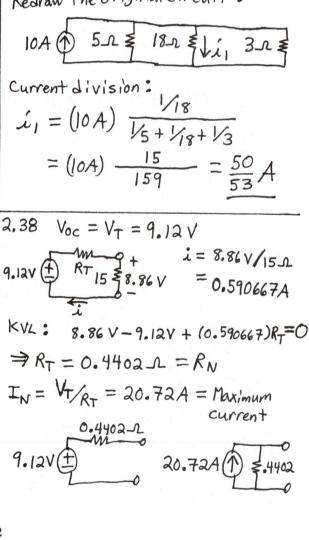

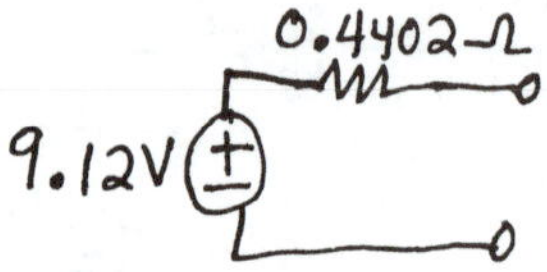

2.39

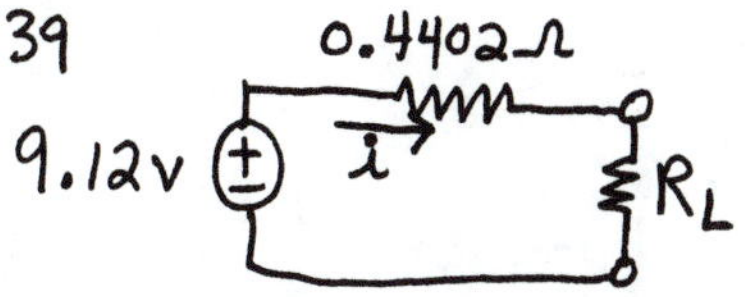

$$P_L = i^2 R_L = \left[\frac{9.12V}{0.4402 + R_L}\right]^2 R_L$$

$$P_L = \frac{83.1744\,R_L}{R_L^2 + 0.8804 R_L + 0.1938}$$

To maximize P_L, differentiate with respect to R_L:

$$\frac{dP_L}{dR_L} = \frac{83.1744}{R_L^2 + 0.8804 R_L + 0.1938} + 83.1744 R_L \left(-[2R_L + .8804][R_L^2 + .8804 R_L + 0.1938]^{-2}\right)$$

$$= 1 - \frac{2R_L^2 + 0.8804 R_L}{R_L^2 + 0.8804 R_L + 0.1938} = 0$$

$\Rightarrow R_L^2 + 0.88 R_L + 0.1938 = 2R_L^2 + 0.88 R_L$

$\Rightarrow R_L^2 = 0.1938 \Rightarrow \underline{R_L = 0.4402\,\Omega}$

$\therefore\ P_{L,max} = \underline{47.24\ W}$

2.40

(a)

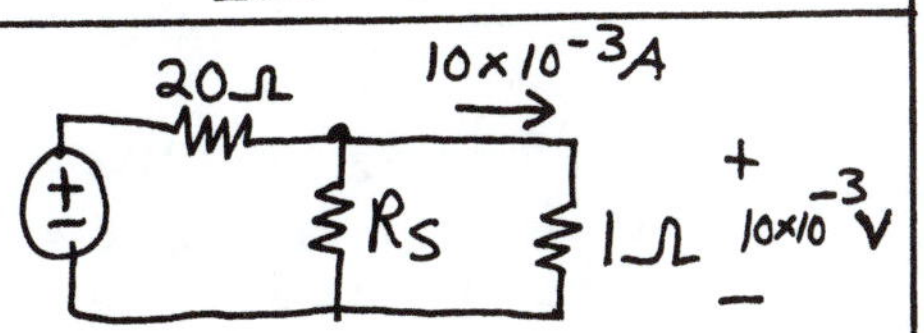

Voltage division:

$$10\times10^{-3}\ V = 1.5\ V\left[\frac{(1\|R_S)}{(1\|R_S) + 20}\right]$$

$$= 1.5\ R_S / [21 R_S + 20]$$

$\Rightarrow \underline{R_S = 0.1550\,\Omega}$ "Shunt"

$$P_{LOAD} = i_{eq}^2 R_{eq} = \left[10\times10^{-3} A + \frac{10\times10^{-3}\,V}{0.1550}\right]^2 (1\|R_S)$$

$= 0.745\ mW$

(b)

$1.5V = 10mA\,(21 + R_2)$

$\Rightarrow \underline{R_2 = 129\,\Omega}$ "Series"

$$P_{LOAD} = i^2 R_{eq} = (10\times10^{-3})^2(130\,\Omega)$$

$= 13\ mW$

$\therefore$ First method (shunt resistor) is more efficient.

2.41

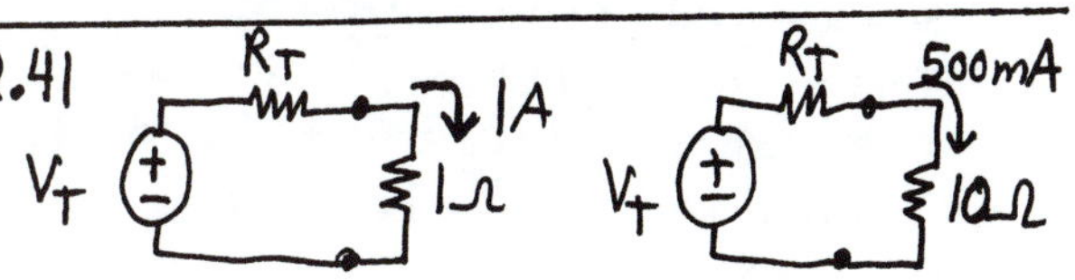

(a) KVL left ckt: $V_T = (R_T + 1)(1A)$

KVL right ckt: $V_T = \frac{1}{2}(R_T + 10)$

$\Rightarrow \underline{R_T = R_N = 8\,\Omega}$, $\underline{V_T = 9V}$,

$\underline{I_N = V_T/R_T = \frac{9}{8} A}$

(b) $i_L = \dfrac{9V}{8\Omega + R_L}$

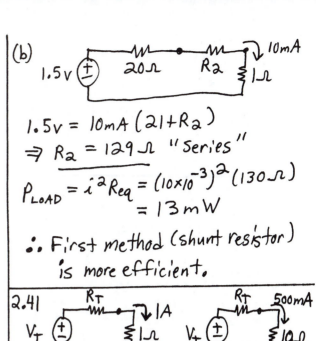

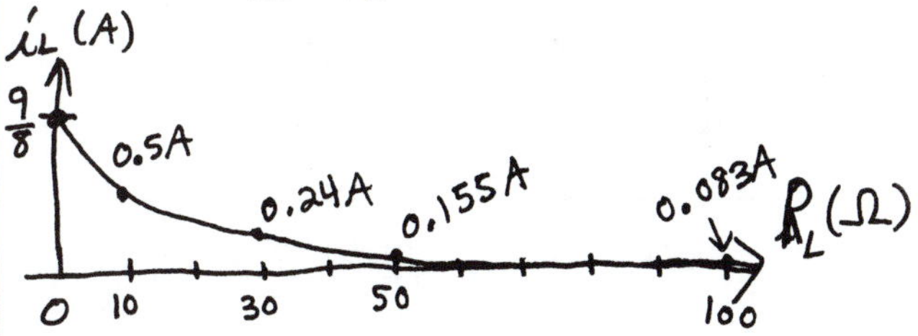

2.42 (a) Tolerance $= \frac{3.63-2.97}{2} = 0.33\,M\Omega$

Nominal resistance $= 2.97 + 0.33 = 3.3\,M\Omega$

% tolerance $= \frac{0.33}{3.3} \times 100 = 10\%$

$(10a+b)10^c = 3.3\,M\Omega$; $c = 5$

$a = b = 3$; Color code is

<u>orange-orange-green-silver</u>

(b) Tolerance $= \frac{1.26-1.14}{2} = 0.06\,K\Omega$

Nominal resistance $= 1.14 + 0.06 = 1.2\,K\Omega$

% tolerance $= \frac{.06}{1.2} \times 100 = 5\%$

$(10a+b)10^c = 1.2\,K\Omega$; $c = 2$, $a = 1, b = 2$ color code is

<u>Brown-Red-Red-Gold</u>.

(c) Tolerance $= \frac{1}{2}(61.6-50.4) = 5.6\,\Omega$

Nominal Resistance $= 50.4 + 5.6 = 56\,\Omega$

% tolerance $= \frac{5.6}{56} \times 100 = 10\%$

$(10a+b)10^c = 56\,\Omega$; $c = 0$, $a = 5$, $b = 6$: color code is

<u>Green-Blue-Black-silver</u>

2.43 (a) $R = (1\times10+0)\times10^2 \pm 10\%$

$= 1000 \pm 100$

resistance range $=$ <u>900 − 1100 Ω</u>

(b) $R = (2\times10+7)\times10^4 \pm 10\%$

$= 270\,k\Omega \pm 27\,k\Omega$

resistance range $=$ <u>243 − 297 KΩ</u>

(c) $R = (6\times10+8)\times10^{-2} \pm 5\%$

$= 680 \pm 34\,m\Omega$

resistance range $=$ <u>646 − 714 mΩ</u>

2.44

$$i_{FS} = \frac{(R_m+R_p)I_{FS}}{R_p} = \frac{(9.8+0.2)5}{0.2} = \underline{250\,mA}$$

2.45

$$i = \frac{v}{v_{FS}(36{,}000\,\Omega/V)} = \frac{45}{(90)(36{,}000)} = 13.9\,\mu A$$

2.46

$$R_{eq} = \frac{(3.3\,K\Omega)(90)(36000)}{3.3\,K\Omega + 3.24\,M\Omega} = 3{,}297\,\Omega$$

$$i = \frac{50V}{3.3K\Omega + 3297\Omega} = 7.58\,mA$$

$$v = R_{eq}\,i = (3297)(7.58mA) = \underline{24.99\,V}$$

2.47

$I_{FS} = 5mA$, $R_M = 9.8\,\Omega$

$(5mA)(9.8 + R_s) = 5V$; $R_s = \underline{990.2\,\Omega}$

$$R_x = \frac{5V}{(5mA)/3} - (9.8 + 990.2) = \underline{2\,K\Omega}$$

2.48

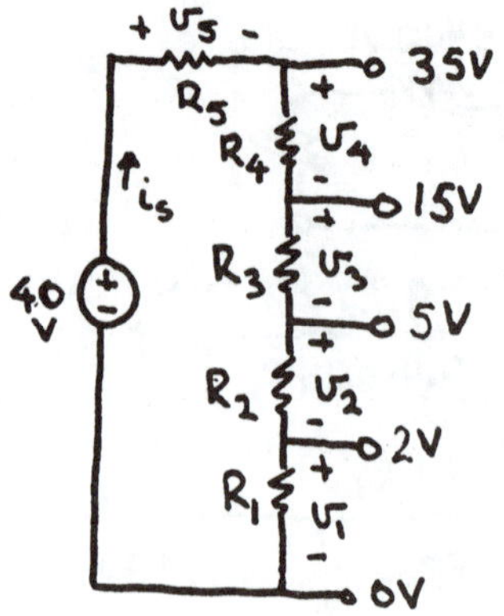

$$i_s = \frac{P_s}{v_s} = \frac{80mW}{40V} = 2mA$$

$v_1 = V = R_1 i_s$

$R_1 = \frac{2V}{2mA} = \underline{1\,K\Omega}$

$v_2 = 5 - 2V = R_2 i_s$

$R_2 = \frac{3V}{2mA} = \underline{1.5\,K\Omega}$

$v_3 = 15 - 5V = R_3 i_s$

$R_3 = \frac{10V}{2mA} = \underline{5\,K\Omega}$

$v_4 = 35 - 15V = R_4 i_s$; $R_4 = \frac{20V}{2mA} = \underline{10\,K\Omega}$

$v_S = 40 - 35V = R_5 i_s$; $R_5 = \frac{5V}{2mA} = \underline{2.5\,K\Omega}$

2.49

R_1 and R_2 are the voltage divider resistances with output taken across R_2. By voltage division, $\frac{v_{out}}{60} = \frac{R_2}{R_1+R_2}$.

(a) $v_{out} = 40V$, $\therefore 40(R_1+R_2) = 60R_2$ or $4R_1 = 2R_2$. Thus $R_2 = 2R_1$ and <u>three</u> $10K\Omega$ resistors are required using two in series for R_2.

2.50

R_1 and R_2 are the voltage divider resistances with the output taken across R_2. By voltage division, $\frac{R_2}{R_1+R_2} = \frac{v_{out}}{50}$

$v_{out} = 20V$; $\therefore 50(R_2) = 20(R_1+R_2)$

Thus $R_2 = \frac{3}{2}R_1$, Let $R_2 = 10K\Omega$

Then $R_1 = 15K\Omega = 10K\Omega + \frac{10K\Omega}{2}$

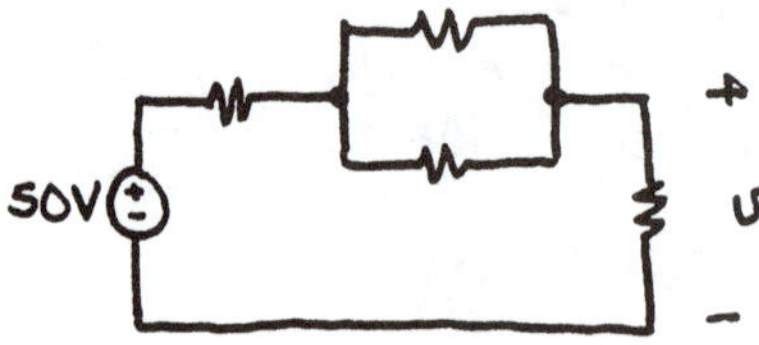

<u>4 resistors</u> required

2.51 (1) $\underline{6\,\Omega}$, one resistor
(2) $\underline{12\,\Omega}$, (6+6) two resistors in series
(3) $\underline{18\,\Omega}$, (6+6+6), all three in series.
(4) $\underline{3\,\Omega}$, (6/2), two resistors in ||.
(5) $\underline{2\,\Omega}$, (6/3), all three in parallel.
(6) $\underline{9\,\Omega}$, $(6+6/2)$, one in series, with the parallel connection of the other two.
(7) $\underline{4\,\Omega}$, $\left(\frac{(6+6)(6)}{6+6+6}\right)$, one resistor in parallel with the series connection of the other two.

2.52

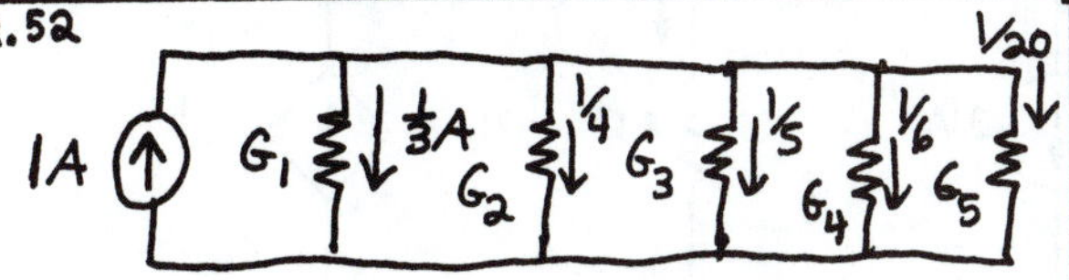

If $G_1+G_2+G_3+G_4+G_5 = 60$

$G_1 = \frac{1}{3}(60) = 20$, $G_2 = \frac{1}{4}(60) = 15$

$G_3 = \frac{1}{5}(60) = 12$, $G_4 = \frac{1}{6}(60) = 10$

$G_5 = \frac{1}{20}(60) = 3$

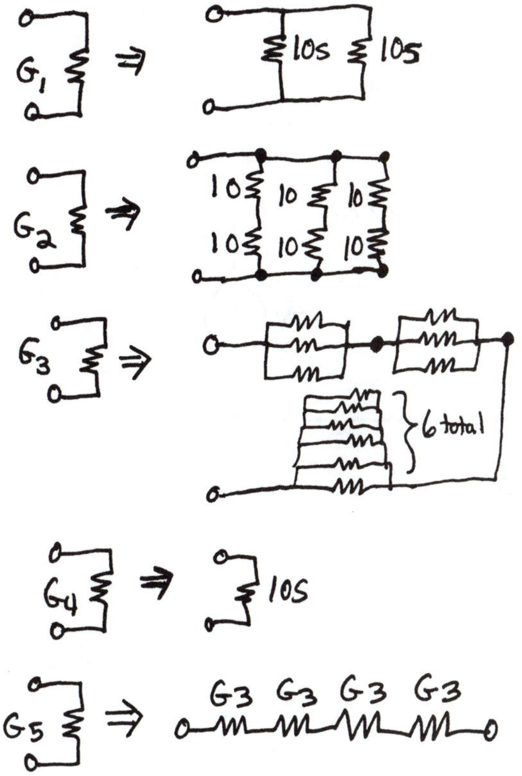

2.53 To dissipate 5 W, at least 5 W × 4 resistors / W = $\underline{20 \text{ resistors}}$ must be used.

Here is a 40-resistor solution:

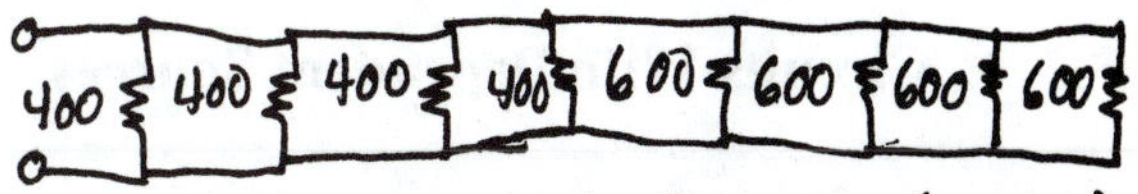

The 400 Ω resistance is 4 × 100 Ω (series).
The 600 Ω resistance is 6 × 100 Ω (series).
This dissipates 10 W (maximum)

2.54 $V_{oc} = V_T = 1500\,V$

$$\frac{1500\,V}{R_T + 100\,\Omega} \geq 10\,A \Rightarrow R_T \leq 50\,\Omega$$

One possibility:

2.55 To dissipate 1 W, at least 4 resistors need to be used.

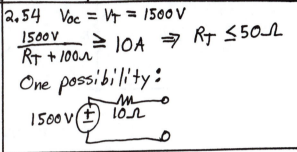

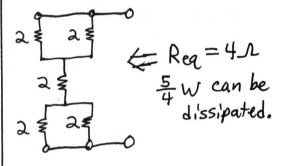

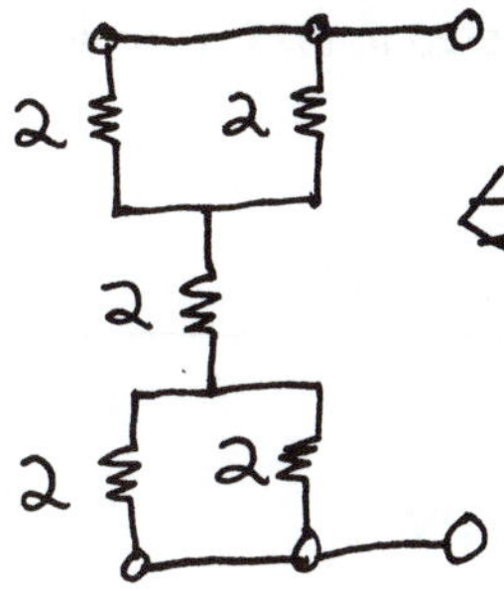

⇐ $R_{eq} = 4\,\Omega$

$\frac{5}{4}$ W can be dissipated.

Chapter 3

Dependent Sources and Op Amps

3.2 Circuits With Dependent Sources

3.1 Find v if $R = 3\ \Omega$.

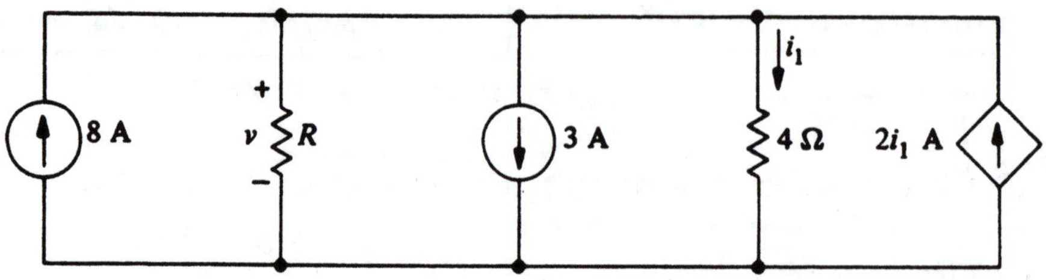

PROBLEM 3.1

3.2 Find v in Prob. 3.1 if $R = 5\ \Omega$.

3.3 Find the power delivered to the 6 Ω resistor.

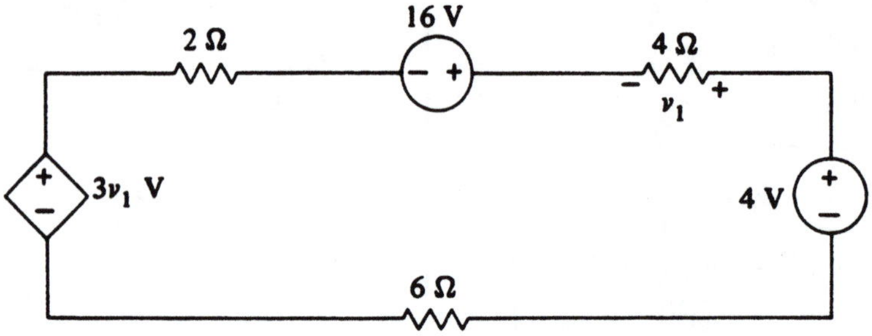

PROBLEM 3.3

3.4 Find i_1 if $R_1 = 1.5\ \Omega$, $R_2 = 2\ \Omega$, and $i_g = 3$ A.

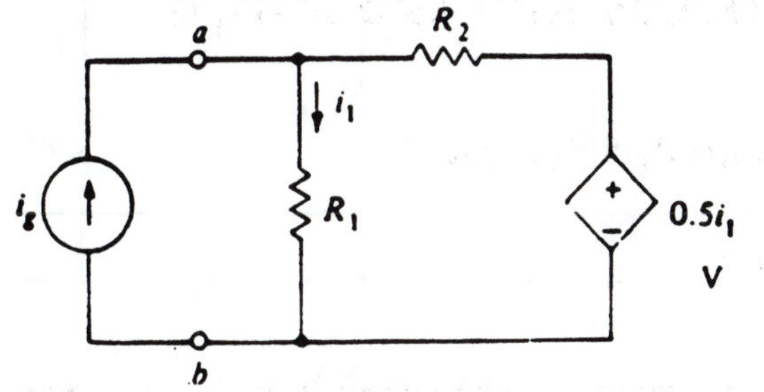

PROBLEM 3.4

3.5 In Prob. 3.4, find the resistance seen by the source looking into terminals *a-b*.

3.6 In Prob. 3.4, let $R_1 = R_2 = 0.1\ \Omega$. Find the resistance seen by the source.

3.7 Find i if $R = 15\ \Omega$.

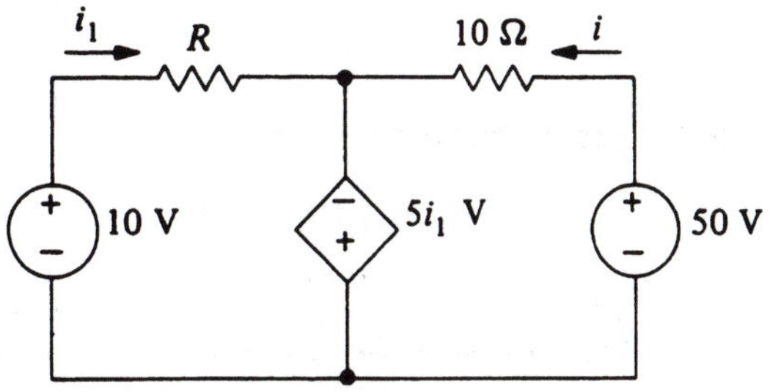

PROBLEM 3.7

3.8 Find R in Prob. 3.7 so that $i = 10$ A.

3.9 Find i and v_1.

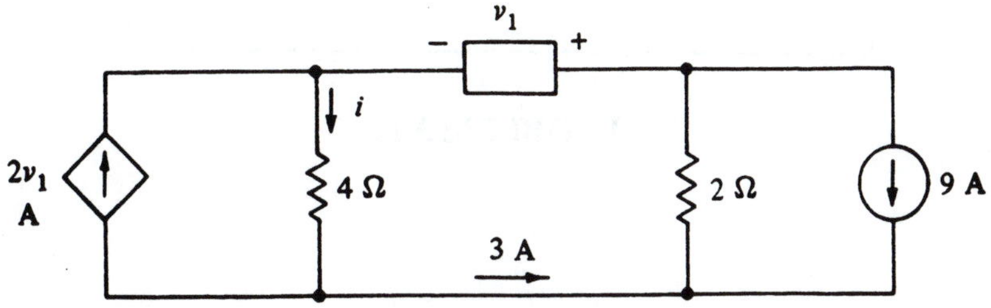

PROBLEM 3.9

3.10 Find i.

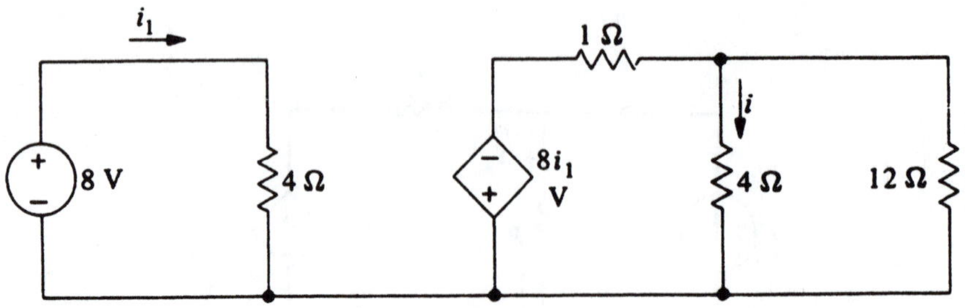

PROBLEM 3.10

3.11 Find (a) the equivalent resistance, in terms of R_1, R_2, and α, seen by the source v_g, and (b) the current if $R_1 = 4\ \text{k}\Omega$, $R_2 = 2\ \text{k}\Omega$, and $v_g = 20$ V.

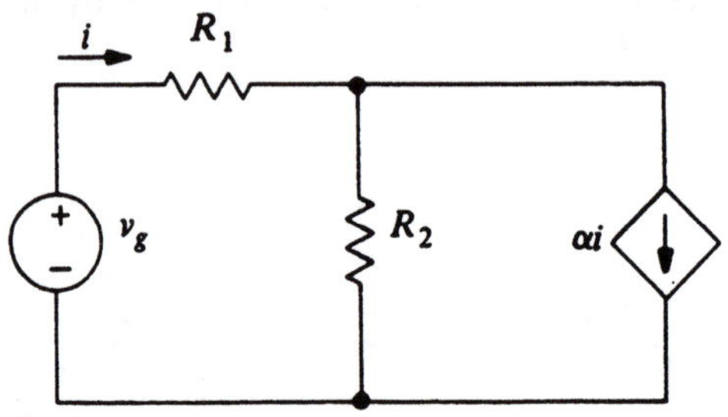

PROBLEM 3.11

3.12 Find v.

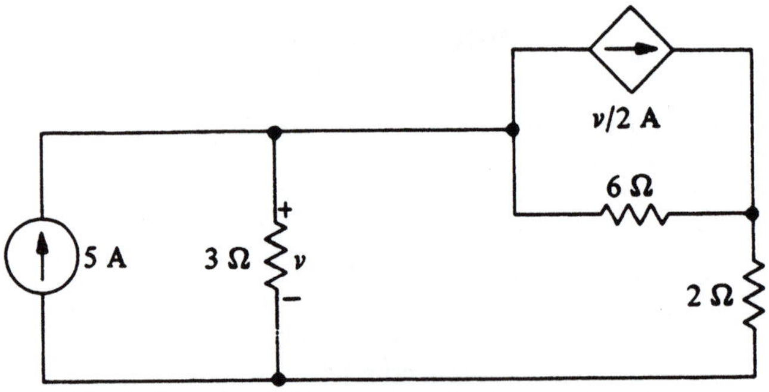

PROBLEM 3.12

3.13 Find v_1 and the power delivered to the 8 Ω resistor.

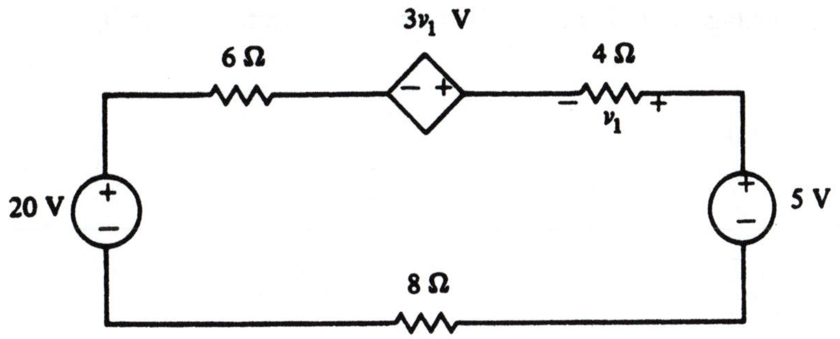

PROBLEM 3.13

3.14 Find v.

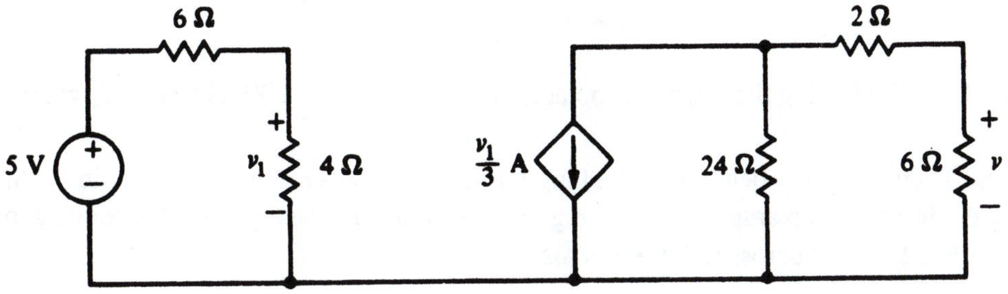

PROBLEM 3.14

3.15 Find the Thevenin and Norton equivalent circuits looking into the right of terminals a-b in Prob. 3.4. Draw the new equivalent circuit.

3.16 Find the Thevenin and Norton equivalent circuits seen by the 10 V source in Prob. 3.7.

3.17 Find the Thevenin and Norton equivalent circuits seen by the 5 A source in Prob. 3.12.

3.3 Operational Amplifiers

3.18 Using the ideal voltage amplifier model with open-loop gain $A = 50000$, find v_a and v_b.

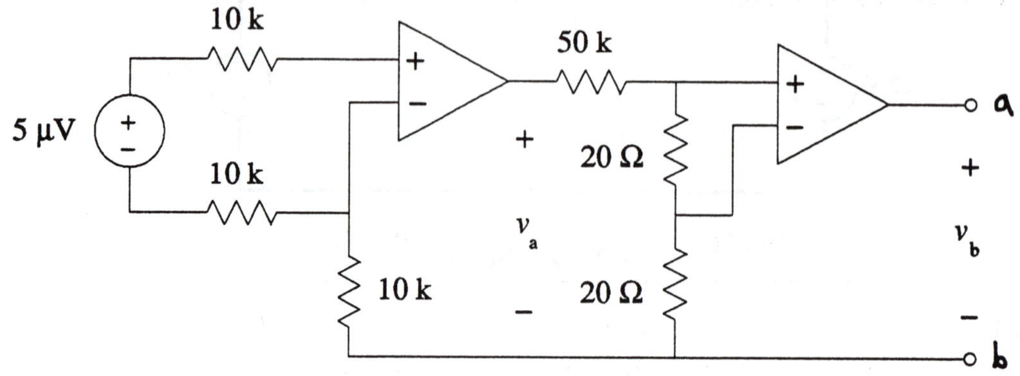

PROBLEM 3.18

3.19 Repeat Prob. 3.18 using the improved op amp model with $R_i = 100\ \text{k}\Omega$ and $R_o = 10\ \Omega$.

3.20 Find the Thevenin equivalent circuit looking into the terminals *a-b* of the circuit in Prob. 3.19 (using the improved op amp model). Using this result, determine the current through a load resistor $R_L = 100\ \Omega$ placed across these terminals.

3.4 Role of Negative Feedback

3.21 Determine the voltage transfer ratio v_2 / v_1 of the circuit shown using (a) the ideal voltage amplifier model with $A = 1000$ and (b) the improved op amp model with $A = 1000$, $R_i = 1\ \text{M}\Omega$, and $R_o = 30\ \Omega$.

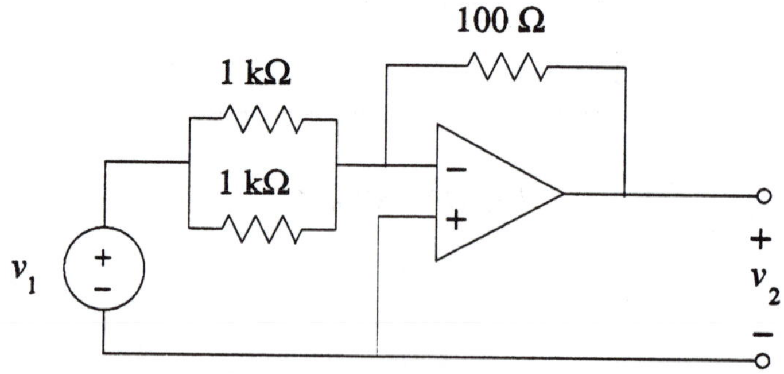

PROBLEM 3.21

3.22 Repeat Prob. 3.21 in general for an open-loop gain A.
What is the voltage transfer ratio in the limit as $A \to \infty$?

3.23 Find an equation relating v_o, v_1, and v_2 using the improved op amp model with $R_i = 1\ \text{M}\Omega$, $R_o = 30\ \Omega$, and $A = 100{,}000$.

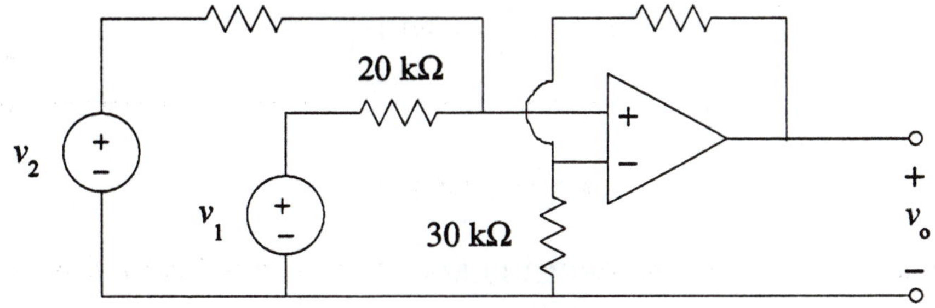

PROBLEM 3.23

3.24 How high must the open-loop gain be in the voltage follower circuit so that a change of $\pm 10\%$ in A leads to a change of only $\pm 0.1\%$ in v_2? Use the improved op amp model with $R_i = 1\ \text{M}\Omega$ and $R_o = 30\ \Omega$.

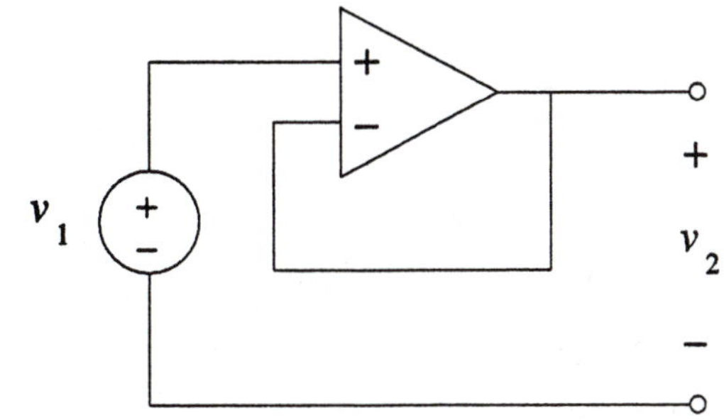

PROBLEM 3.24

3.5 Op Amp Building Block Circuits

3.25 (a) Find the voltage transfer equation for the following circuit using the ideal voltage amplifier model. (b) Explain how the gain of this circuit can be increased by 10 times by replacing just one resistor.

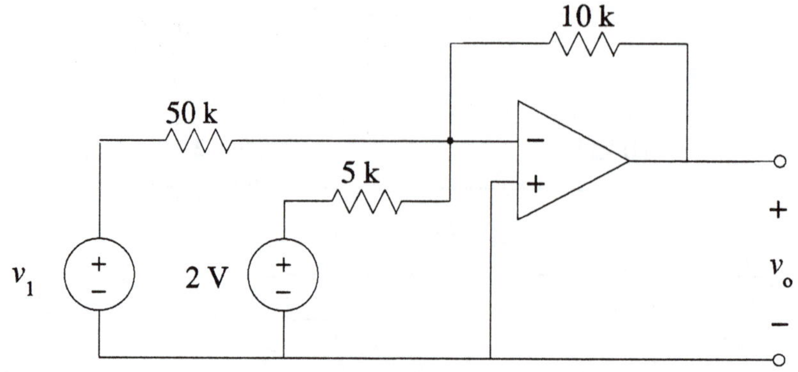

PROBLEM 3.25

3.26 Find the voltage transfer equation for the following circuit.

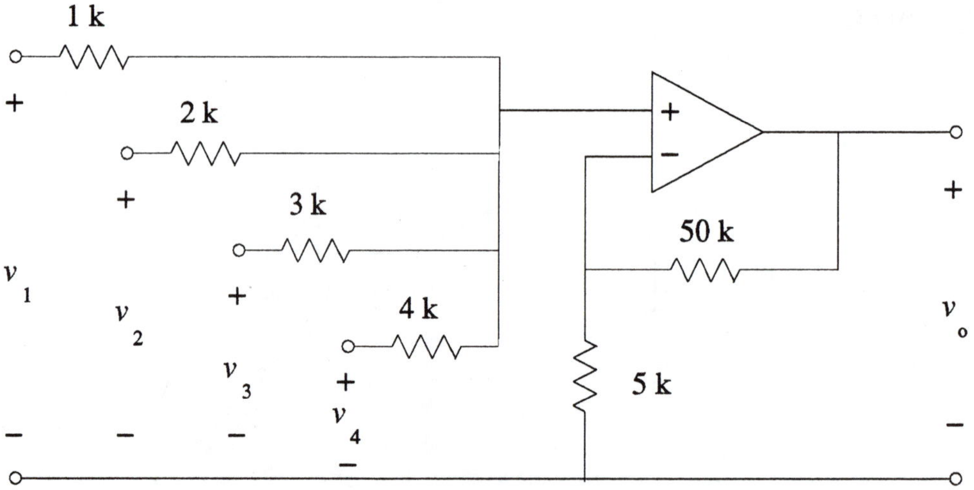

PROBLEM 3.26

3.27 Find the voltage transfer equation for the following circuit.

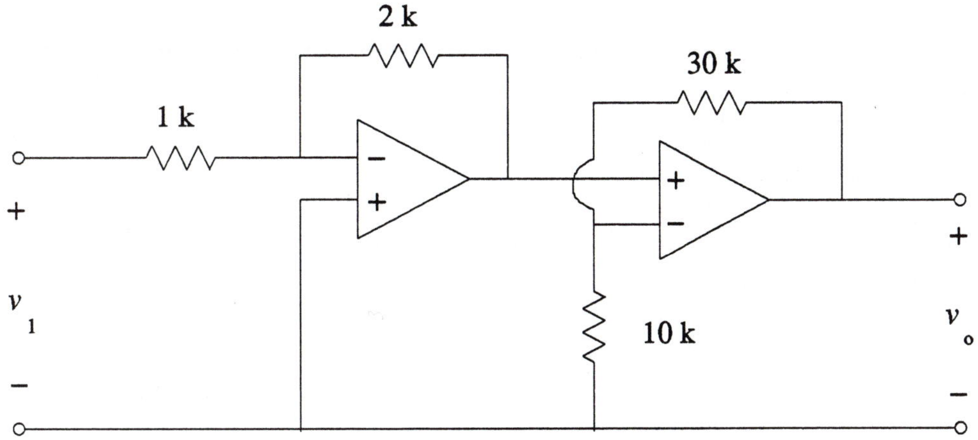

PROBLEM 3.27

3.28 Find the voltage transfer equation for the following circuit, sometimes called a general summer.

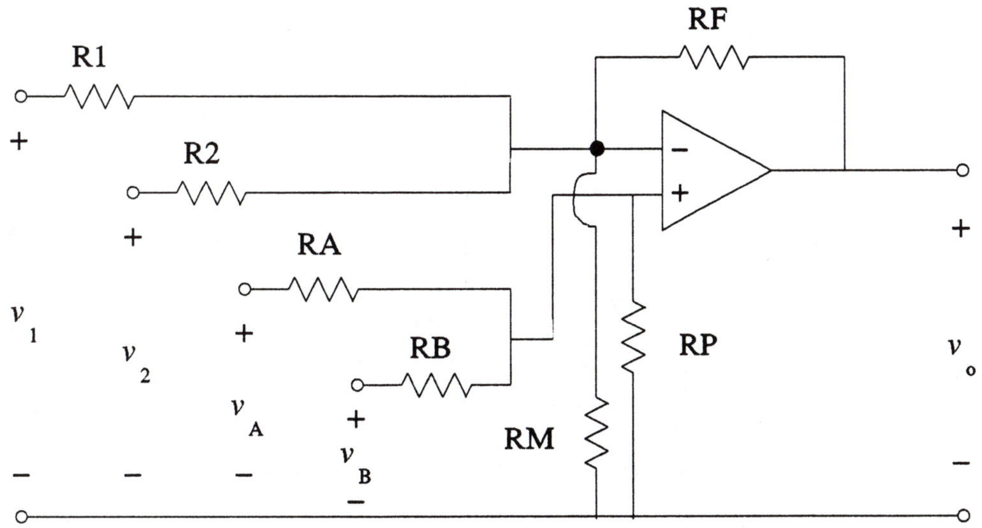

PROBLEM 3.28

3.6 Interconnecting Op Amp Building Blocks

3.29 (a) Find the voltage transfer equation. (b) Find the voltage transfer equation when the follower is replaced with a short circuit between terminals *a* and *b*.

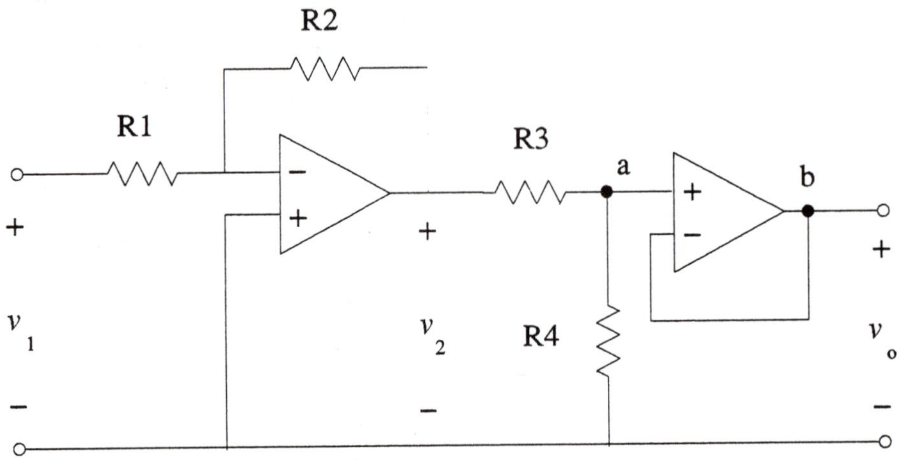

PROBLEM 3.29

3.30 (a) Find the voltage transfer equation for the following circuit.
(b) Pick R_1, R_2, R_3, R_4, R_A, and R_F so that $v_o = 0$ V.

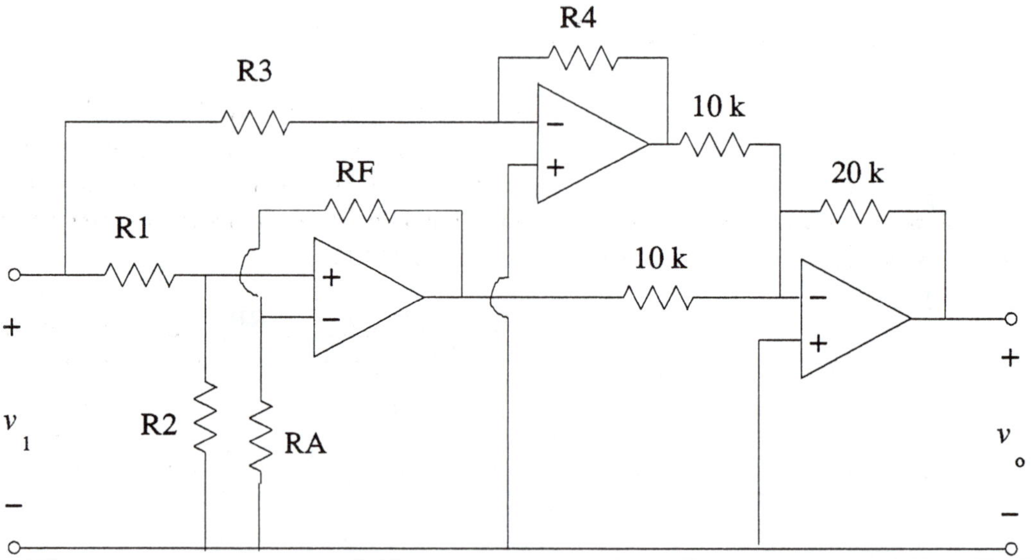

PROBLEM 3.30

3.31 Find the voltage transfer equation for the circuit shown.

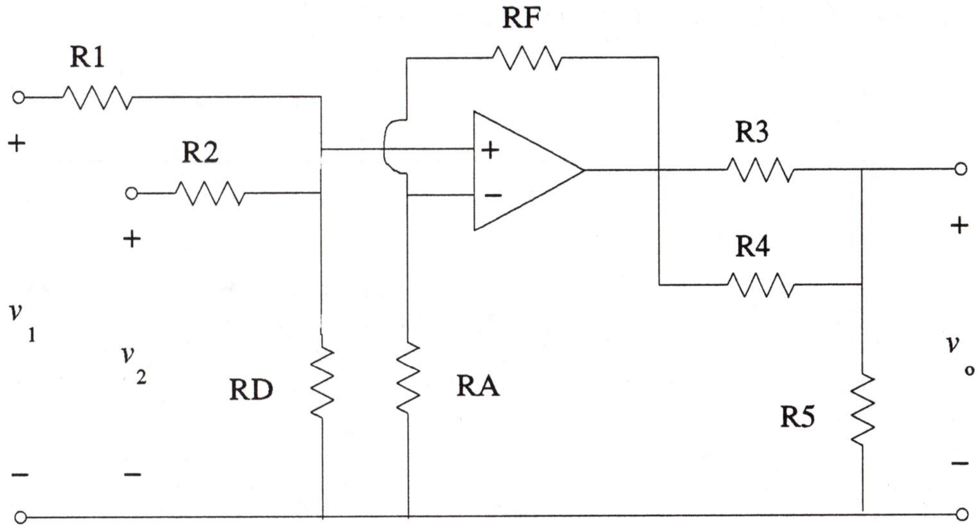

PROBLEM 3.31

3.32 Design a circuit with the voltage transfer equation $v_o = -3v_1 + 10v_2 - \frac{10}{3}v_3$.

3.33 Design a circuit with the voltage transfer equation $v_o = \frac{1}{5}(v_1 - 2v_2)$.

3.8 Design of Simple Op Amp Circuits

3.34 A potentiometer is a variable resistor often used in the volume control circuit of an audio amplifier. Potentiometers (or "pots") may be adjusted from 0 Ω to some maximum rated value by turning a knob or sliding a control, for example. (a) Design a volume controller which can adjust an input voltage v_1 between 0 V and $-100v_1$ using a 20 kΩ pot. (b) Design a volume controller which outputs a voltage between v_1 and $+100v_1$ using a 20 kΩ pot.

3.35 Design an analog computer using op amps that will compute the negated average of 5 voltages v_1, v_2, v_3, v_4, and v_5 using a single op amp in the inverting summer configuration.

3.36 Design a circuit that has the voltage transfer equation $v_o = 5v_1 - \frac{2}{3}v_2 + 5\left(v_3 - \frac{6}{5}v_4\right)$.

CHAPTER 3 SOLUTIONS

3.1 By KCL, $8 - v/R - 3 - i_1 + 2i_1 = 0$,

$i_1 = \frac{v}{4}$, $v = \frac{5}{1/R - 1/4} = \underline{60V}$.

3.2 $v = \frac{5}{1/5 - 1/4} = \underline{-100V}$.

3.3 i leaves the + terminal of the 16-V source ∴ $i = -\frac{v_1}{4}$,

By KVL,

$16 + v_1 - 4 + 6\frac{v_1}{4} + 3v_1 + 2\frac{v_1}{4} = 0$

$v_1 = \frac{12}{-\frac{3}{2} - \frac{1}{2} - 4} = -2V$, $i = -\frac{2}{4} = -\frac{1}{2}$

$P_{6\Omega} = 6i^2 = \underline{1.5W}$

3.4 By KCL, $i_g - i_1$ is the current to the right in R_2; By KVL around the right loop,

$R_2(i_g - i_1) + 0.5i_1 - R_1 i_1 = 0$

$i_1 = \frac{(2)(3)}{2 + 1.5 - 0.5} = \underline{2A}$

3.5 $v_{a-b} = i_1 R_1 = (2)(1.5) = 3V$

$R_{a-b} = \frac{v_{a-b}}{i_g} = \frac{3V}{3A} = \underline{1\Omega}$

3.6 $i_1 = \frac{(\frac{1}{10})3}{\frac{1}{10} + \frac{1}{10} - 0.5} = -1A$, $v_{a-b} = (-1)(\frac{1}{10}) = -\frac{1}{10}V$

$R_{a-b} = \frac{-\frac{1}{10}V}{3A} = \underline{-\frac{1}{30}\Omega}$.

3.7 By KVL around the left loop,

$10V - Ri_1 + 5i_1 = 0$, $i_1 = 1A$;

By KVL around the right loop,

$50 - 10i + 5i_1 = 0$, $i = \frac{55}{10} = \underline{5.5A}$

3.8 using KVL equation of Prob. 3.7

$i_1 = \frac{50 - 10(10)}{-5} = 10A$;

$R = \frac{10 + 5(10)}{10} = \underline{6\Omega}$.

3.9 By KCL the current going up through the 2-Ω resistor is 12A.

By KCL, $i = 3 + 2v_1$,

By KVL around the center loop

$v_1 + (12)2 + 4(3 + 2v_1) = 0$

$v_1 = \frac{-36}{1+8} = \underline{-4V}$

$i = 3 - 4(2) = \underline{-5A}$

3.10 By KVL around the left loop

$8 = i_1 4$, $i_1 = 2A$.

By combining the 4Ω & 12Ω to make a 3Ω resistor. The voltage across the 4Ω resistor is, by voltage division.

$v_{4\Omega} = \frac{3}{1+3}(-8i_1) = -12V$

$i = \frac{v_{4\Omega}}{4\Omega} = \underline{-3A}$

3.11 By KCL the current down through

(a) R_2 is $i - \alpha i$,

By KVL around the right loop.

$v_g - R_1 i - R_2(i - \alpha i) = 0$

$v_g = i[R_1 + R_2(1 - \alpha)]$

R_g = resistance seen by v_g

$= \frac{v_g}{i} = \underline{R_1 + R_2(1 - \alpha)\ \Omega}$

(b) using a current divider,

$i = \frac{R_2\ (i\alpha)}{R_2 + R_1 + R_g}$, the solving for α

∴ $\alpha = 2$, $i = \frac{v_g}{R_g} = \frac{20\ V}{4 + 2(1-2)k\Omega} = \underline{10mA}$

3.12 Current downward in 3-Ω resistor is $v/3$. By KCL the current down in 2-Ω resistor is $5 - v/3$. Current to the right in 6Ω resistor is, by KCL,

$5 - v/3 - v/2 = 5 - \frac{5v}{6}$, By KVL around the loop containing only resistors.

$v - 6(5 - 5v/6) - 2(5 - v/3) = 0$

∴ $v = \underline{6V}$.

3.13 i leaves + terminal of 20-V source. ∴ $i = -v_1/4$, By KVL

$20 - 6(-\frac{v_1}{4}) + 3v_1 + v_1 - 5 - 8(-\frac{v_1}{4}) = 0$

∴ $v_1 = \underline{-2V}$, $i = -\frac{(-2)}{4} = \frac{1}{2}A$

$P_{8\Omega} = 8i^2 = \underline{2W}$.

3.14 By voltage division,

$v_1 = \frac{4}{6+4}(15) = 2V$. Let i be the current entering + terminal of v. Then by current division

$i = \frac{24}{24 + 6 + 2}(-\frac{v_1}{3}) = -\frac{1}{2}A$

∴ $v = 6(-\frac{1}{2}) = -\underline{3V}$.

3.15 In Prob. 3.5, it was found that the source sees a resistance of $1\,\Omega = R_T = R_N$.

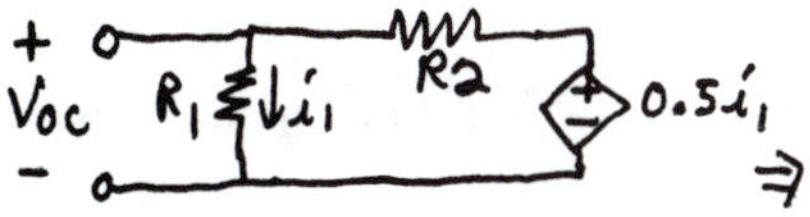

No sources $\Rightarrow i_1 = 0$
$\Rightarrow V_{oc} = 0$

$I_{sc} = 0$ also.

$\Rightarrow$ Thevenin and Norton equivalents are simply a $1\,\Omega$ resistor:

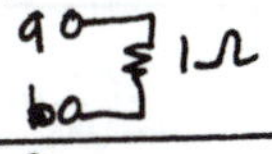

3.16

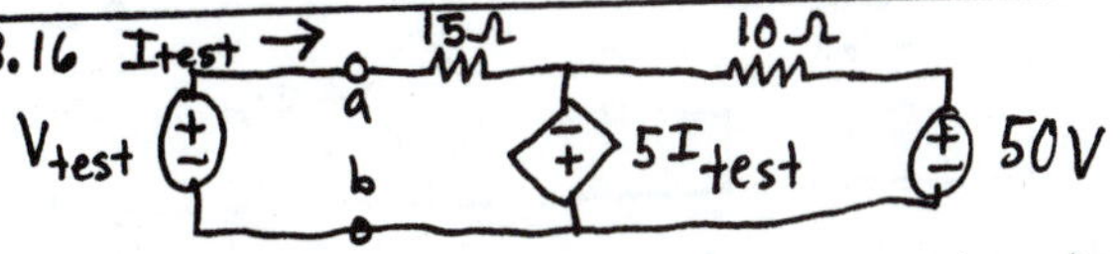

KVL left loop: $V_{test} + 5I_{test} - 15\,\Omega(I_{test}) = 0$

$\Rightarrow V_{Test} = 10 I_{Test} \Rightarrow R_T = R_N = 10\,\Omega$

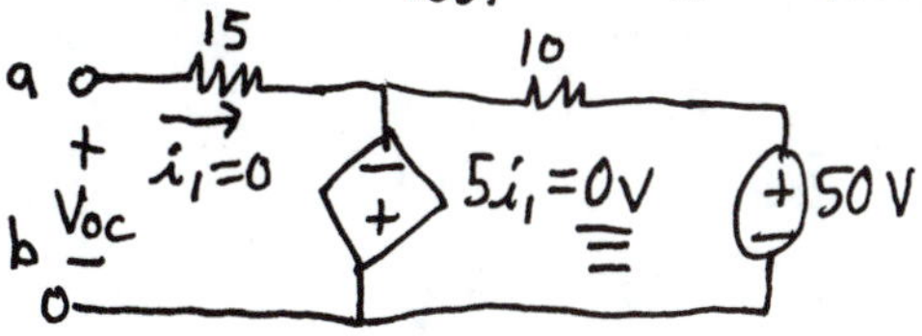

$V_{oc} = 0\text{ V} \Rightarrow$

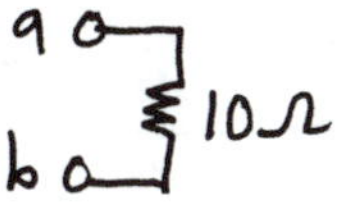

Thevenin and Norton equivalent

3.17

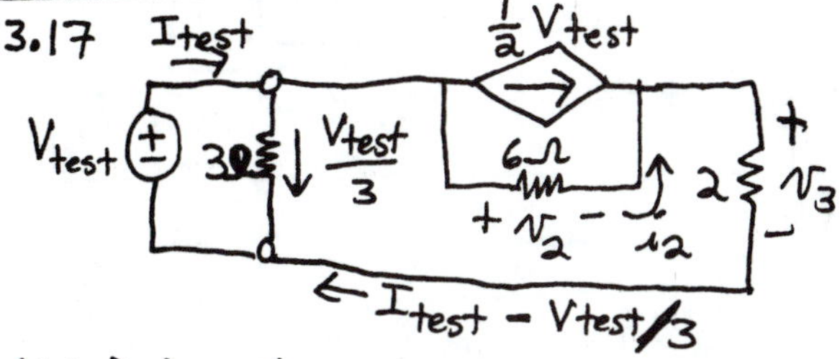

$KCL \Rightarrow i_2 = \left(I_{test} - \frac{V_{test}}{3}\right) - \frac{1}{2}V_{test}$

$= I_{test} - 5/6\, V_{test}$

$\Rightarrow v_2 = 6I_{test} - 5V_{test} \Rightarrow v_3 = V_{test} - 6I_{test} + 5V_{test}$

$\frac{v_3}{2\,\Omega} = I_{test} - \frac{V_{test}}{3} \Rightarrow$

$\frac{V_{test}}{2} - 3I_{test} + \frac{5}{2}V_{test} = I_{test} - \frac{V_{test}}{3}$

$\Rightarrow \frac{V_{test}}{I_{test}} = \frac{12}{10} = 1.2\,\Omega = R_T = R_N$

No sources $\Rightarrow V_{oc} = 0\text{ V} \Rightarrow$

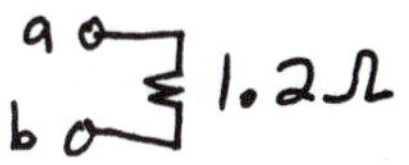

3.18

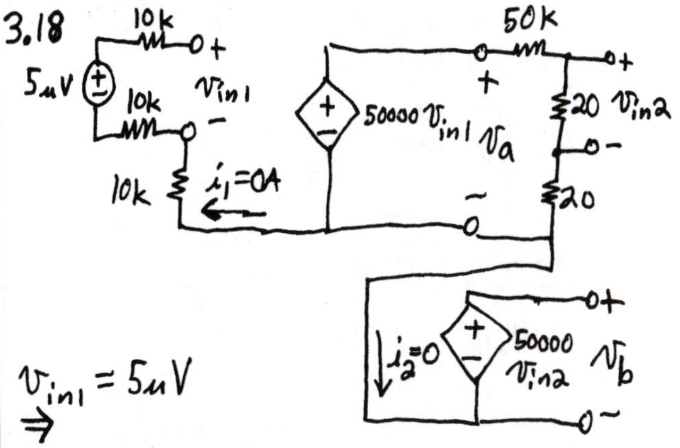

$v_{in1} = 5\,\mu V$

$\Rightarrow$

$v_a = 50000(5\,\mu V) = \underline{0.25\text{ V}}$

$v_{in2} = v_a\left(\frac{20}{50k + 20 + 20}\right) = (3.9968 \times 10^{-4})\, v_a$

$= 9.9920 \times 10^{-5}\text{ V}$

$\Rightarrow v_b = 50000(9.9920 \times 10^{-5}) = \underline{4.996\text{ V}}$

3.19

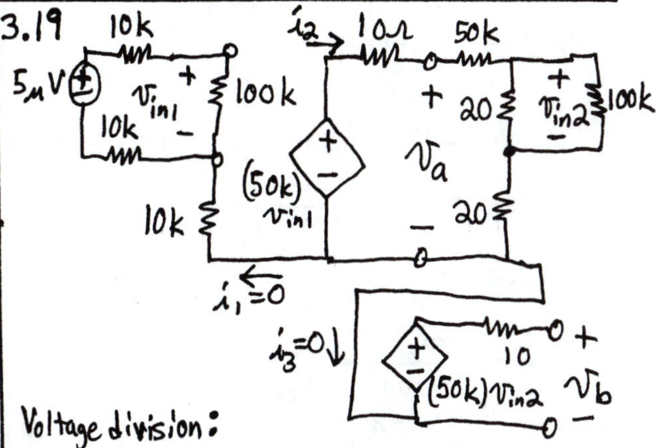

Voltage division:

$v_{in1} = 5\,\mu V\left[\frac{100k}{120k}\right] = 4.1667\,\mu V$

$50000\, v_{in1} = 0.208\overline{3}$

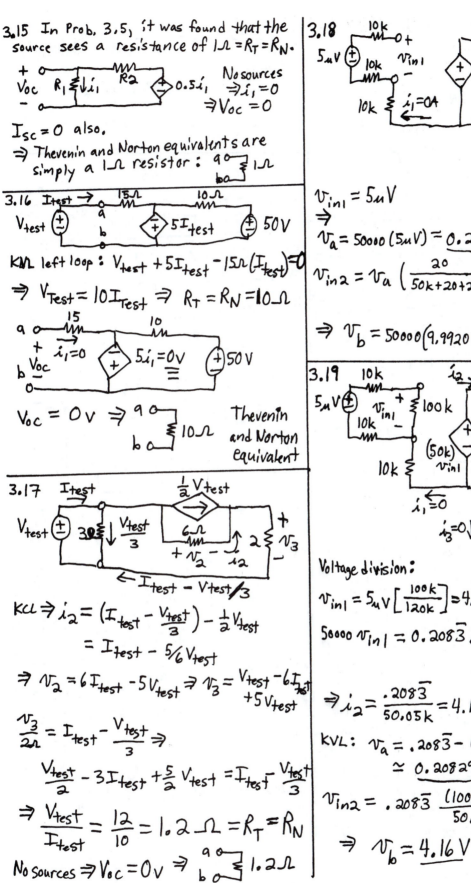

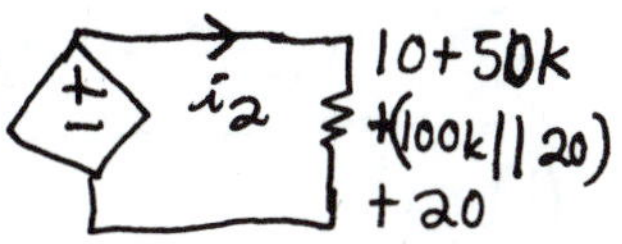

$= 50.05\,k$

$\Rightarrow i_2 = \frac{.208\overline{3}}{50.05k} = 4.1625 \times 10^{-6}\text{ A}$

KVL: $v_a = .208\overline{3} - (10\,\Omega)(4.1625\,\mu A)$

$\simeq \underline{0.20829\text{ V}}$

$v_{in2} = .208\overline{3}\ \frac{(100k\|20)}{50.05k} = 8.323 \times 10^{-5}$

$\Rightarrow v_b = \underline{4.16\text{ V}}$

3.20 Since $i_3 = 0A$ in Prob. 3.19, the circuit looks as follows:

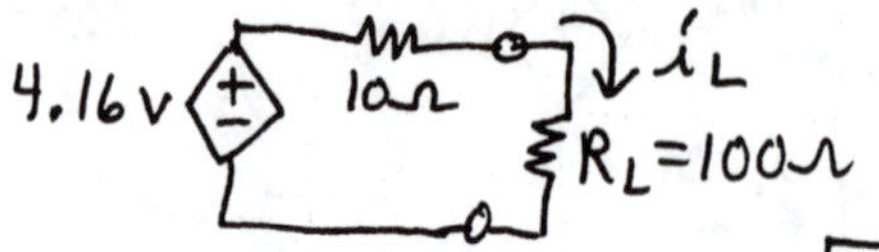

Thevenin equivalent: 4.16V

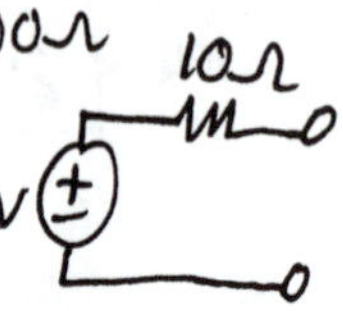

$$i_L = \frac{4.16V}{110Ω} = 0.0378A = \underline{37.8mA}$$

3.21(a)

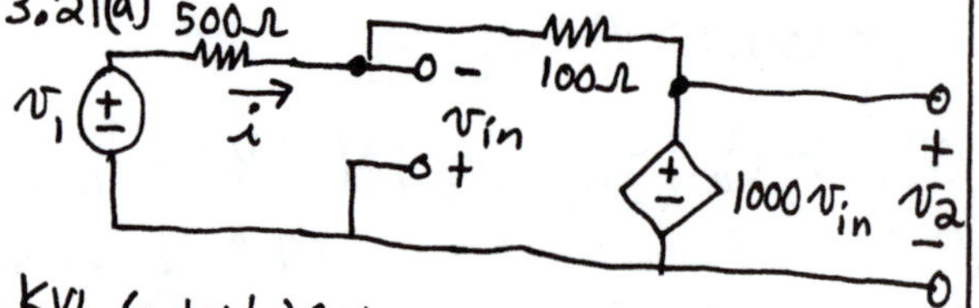

KVL (outside): $1000v_{in} - v_1 + i(500+100) = 0$

KVL (left): $-v_1 + i(500) - v_{in} = 0$

$$\Rightarrow i = \frac{v_{in}+v_1}{500}$$

KVL (right): $1000v_{in} + v_{in} + i(100) = 0$

$$\Rightarrow v_{in} = -i\left[\frac{100}{1001}\right] = -.0999001i$$

$$v_2 = 1000v_{in} \Rightarrow v_{in} = \frac{v_2}{1000}$$

$$\Rightarrow i = \frac{v_2}{500000} + \frac{v_1}{500} \quad \text{Substituting:}$$

$$\Rightarrow v_2 - v_1 + \left[\frac{v_2}{500000} + \frac{v_1}{500}\right](600) = 0$$

$$\Rightarrow \underline{v_2 = -0.19976\,v_1}$$

(b)

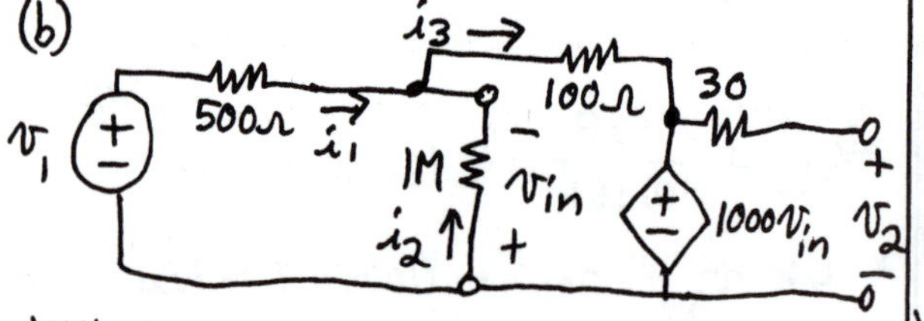

KVL (left): $v_1 + i_2(1M) - i_1(500) = 0$

KVL (center): $1000v_1 + i_2(1M) + i_3(100) = 0$

$$i_2 = \frac{v_{in}}{1M}, \Rightarrow v_2 = v_1 - 500i_1 - 100i_3$$

KCL: $i_1 + i_2 = i_3 \Rightarrow i_1 = i_3 - v_2/(1\times10^9)$

Solving,

$$\Rightarrow i_3 = -0.01001\,v_2$$

$$\Rightarrow \underline{v_2 = -0.19976\,v_1}$$

3.22

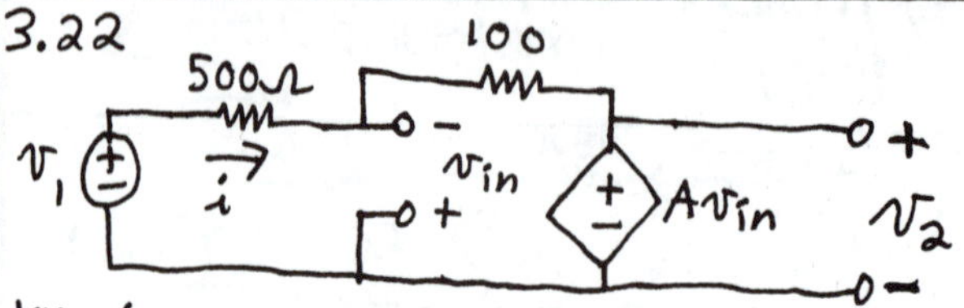

KVL (outside): $Av_{in} - v_1 + i(500+100) = 0$

KVL (left): $-v_1 + i(500) - v_{in} = 0$

KVL (right): $Av_{in} + v_{in} + i(100) = 0$

$$\Rightarrow i = \frac{v_{in}+v_1}{500}, \quad v_{in} = -i\left(\frac{100}{A+1}\right)$$

$$v_2 = Av_{in} \Rightarrow v_{in} = v_2/A$$

$$\Rightarrow v_2\left[1 + 1.2/A\right] = v_1\left[-0.2\right]$$

$$\Rightarrow \boxed{v_2 = -\frac{0.2}{(1+1.2/A)}v_1}$$

As $A \to \infty$, $\boxed{v_2 = -0.2\,v_1}$

3.23

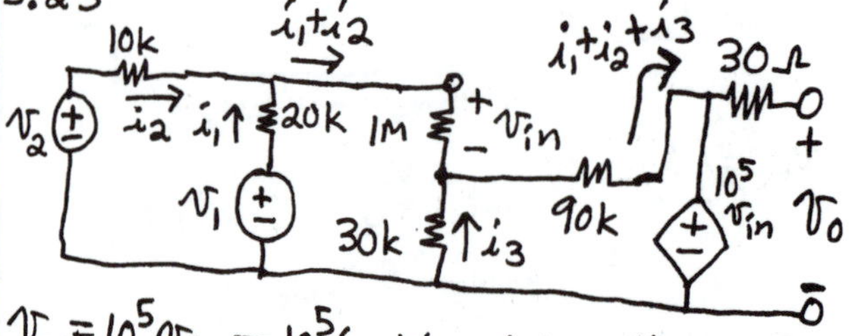

$$v_0 = 10^5 v_{in} = 10^5(1M)(i_1+i_2) = 10^{11}(i_1+i_2)$$

KVL: $v_1 - v_2 + 10000i_2 - 20000i_1 = 0$

KVL: $v_1 + i_3(30000) = 20000i_1 + (i_1+i_2)10^6$

KVL: $30000i_3 + 90000(i_1+i_2+i_3) + 10^5 v_{in} = 0$

$$\Rightarrow i_3 = -\frac{3}{4}(i_1+i_2) - 10^5 v_{in}$$

$$\Rightarrow i_1 = \frac{v_1 - v_2 + 10000i_2}{20000}$$

$$\Rightarrow i_2 = \left[v_0 + (5\times10^6)(v_2 - v_1)\right]/(1.5\times10^{11})$$

3.23 (cont'd)

Solving $\Rightarrow v_0[-30k - 1\times10^{-5}] = \frac{2}{3}(v_1 - v_2) - v_1$

$$\Rightarrow v_0 \simeq \frac{-\frac{1}{3}v_1 - \frac{2}{3}v_2}{-30000}$$

$$\Rightarrow \boxed{v_0 = 1.1\overline{1}\times10^{-5}\,v_1 + 2.2\overline{2}\times10^{-5}\,v_2}$$

3.24

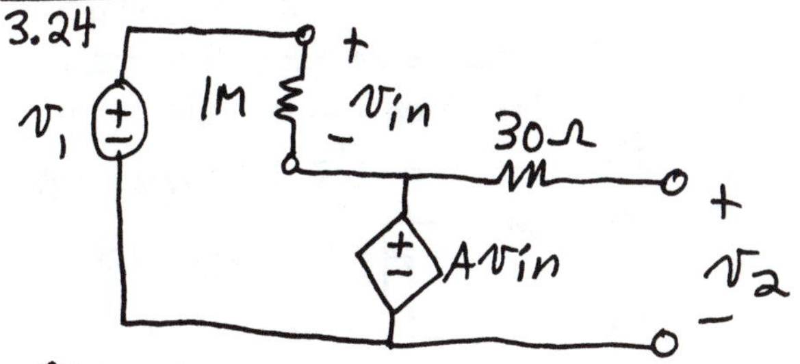

$v_2 = A v_{in}$

KVL: $v_1 - Av_{in} - v_{in} = 0$

$\Rightarrow v_{in} = v_1/[A+1]$

$\Rightarrow \boxed{v_2 = \left(\frac{A}{A+1}\right)v_1} \Rightarrow A \geq 1110$ to meet spec.

3.25 (a)

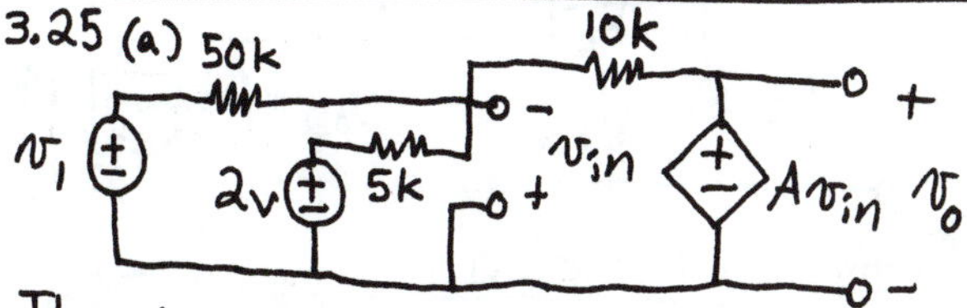

Thevenin equiv to the left of v_{in} terminals:

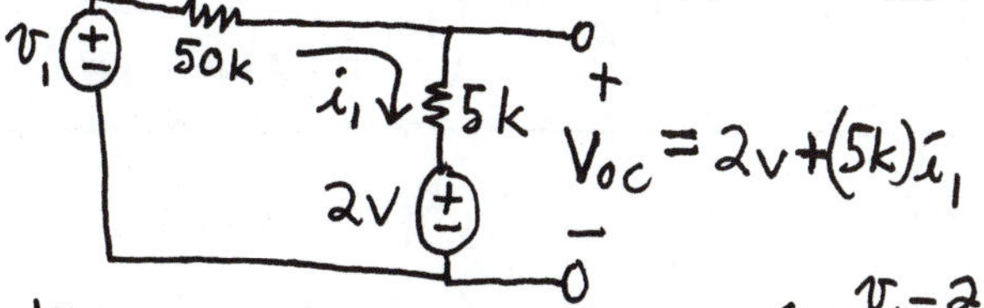

KVL: $v_1 - 2 - i_1(55000) = 0 \Rightarrow i_1 = \frac{v_1 - 2}{55000}$

$\Rightarrow V_{oc} = V_T = 2v + 0.0\overline{909}(v_1 - 2)$

kill independent srcs $\Rightarrow R_T = 50k \| 5k = 4545.\overline{45}\,\Omega$

New circuit:

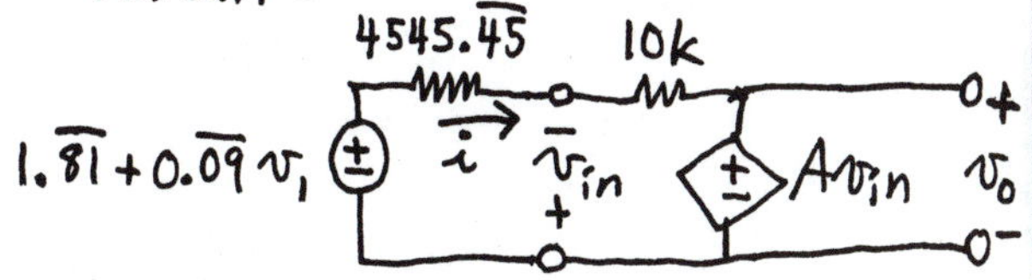

KVL: $i = \dfrac{1.\overline{81} + 0.\overline{09}v_1 - v_0}{14545.\overline{45}\,\Omega}$

KVL: $v_{in} + 10000i + Av_{in} = 0$

Solving: $\boxed{v_0 = -4v - 0.2v_1}$

(b) By replacing the 10k feedback resistor in the inverting summer configuration with a 100k resistor, the gain of the circuit will be increased 10 times

$[\Rightarrow v_0 = -40v - 2.0v_1]$

3.26 Noninverting summer:

$$v_0 = \left(1 + \frac{50k}{5k}\right)\left[\frac{R_T}{1k}v_1 + \frac{R_T}{2k}v_2 + \frac{R_T}{3k}v_3 + \frac{R_T}{4k}v_4\right]$$

$R_T = R_1 \| R_2 \| R_3 \| R_4 = 1k \| 2k \| 3k \| 4k$

$$= \frac{1}{\left(\frac{1}{1k} + \frac{1}{2k} + \frac{1}{3k} + \frac{1}{4k}\right)} = 480\,\Omega$$

$\therefore \boxed{v_0 = 5.28v_1 + 2.64v_2 + 1.76v_3 + 1.32v_4}$

3.27 Left op amp circuit is an inverting amp with output $v_{01} = -2v_1$

Right op amp circuit is a noninverting amp with output $v_0 = \left(1 + \frac{30}{10}\right)v_{01}$

$= [\,4\,][-2v_1]$

$\Rightarrow \boxed{v_0 = -8v_1}$

3.28 To solve this problem more easily, superposition should be used (see Sec. 4.2)

① Set all inputs to zero except v_1:

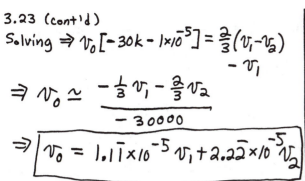

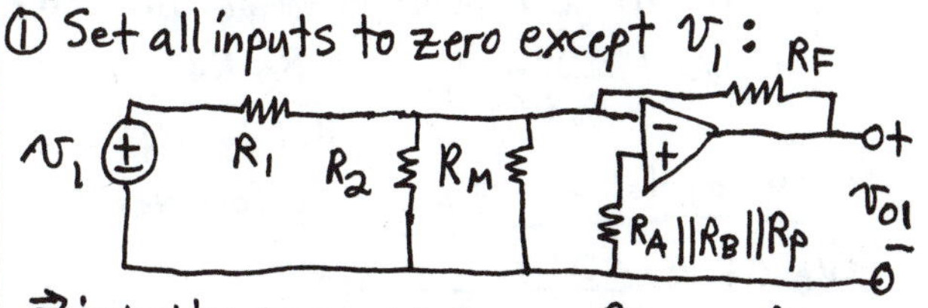

$\Rightarrow$ inverting summer: $v_{01} = -\frac{R_F}{R_1}v_1 - \frac{R_F}{R_2}(0) - \frac{R_F}{R_M}(0)$

$\Rightarrow v_{01} = -\frac{R_F}{R_1}v_1$

3.28 (cont'd)

② Set all inputs to zero except v_2:

$v_{o2} = -\frac{R_F}{R_2} v_2$

③ Set all to zero except v_A:

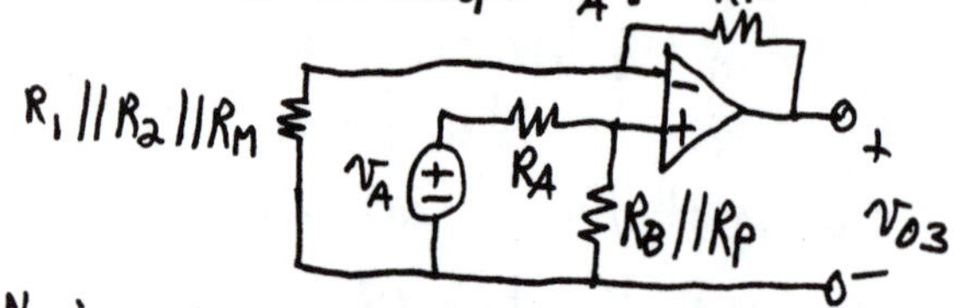

Noninverting amp: Let $R_T \equiv R_1 || R_2 || R_M$

$$v_{o3} = \left(1 + \frac{R_F}{R_T}\right)\left[v_A \frac{R_B || R_P}{R_A + (R_B || R_P)}\right]$$

$$\Rightarrow v_{o3} = \left(1 + \frac{R_F}{R_T}\right)\left[\frac{R_B R_P}{R_A(R_B + R_P) + R_B R_P}\right] v_A$$

④ Set all sources to zero except v_B:

Like above: $v_{o4} = \left(1 + \frac{R_F}{R_T}\right)\left[\frac{R_A R_P}{R_B(R_A + R_P) + R_A R_P}\right]$

$$\therefore\ v_o = v_{o1} + v_{o2} + v_{o3} + v_{o4}$$

$$v_o = -\frac{R_F}{R_1} v_1 - \frac{R_F}{R_2} v_2 + \left(1 + \frac{R_F}{R_T}\right)\left[\frac{R_B R_P\, v_A + R_A R_P\, v_B}{R_A R_B + R_A R_P + R_B R_P}\right]$$

3.29 (a) Output of left op amp circuit (inverting amp) is $v_2 = -\frac{R_2}{R_1} v_1$

Input to voltage follower is voltage divided:

$$v_a = v_2\left(\frac{R_4}{R_3 + R_4}\right) = -\frac{R_2 R_4}{R_1(R_3 + R_4)} v_1$$

Output of follower = input to follower:

$$\boxed{v_o = -\frac{R_2 R_4}{R_1(R_3 + R_4)} v_1}$$

Note: no dependence on R_L!

(b) If short ckt between a and b,

$$v_o = v_2 \frac{(R_4 || R_L)}{R_3 + (R_4 || R_L)}$$

$$= -\frac{R_2}{R_1} \cdot \frac{(R_4 || R_L)}{R_3 + (R_4 || R_L)}$$

$$\therefore\ \boxed{v_o = -\frac{R_2 R_4 R_L}{R_1[R_3(R_4 + R_L) + R_4 R_L]}}$$

3.30 (a) Rightmost op amp circuit:

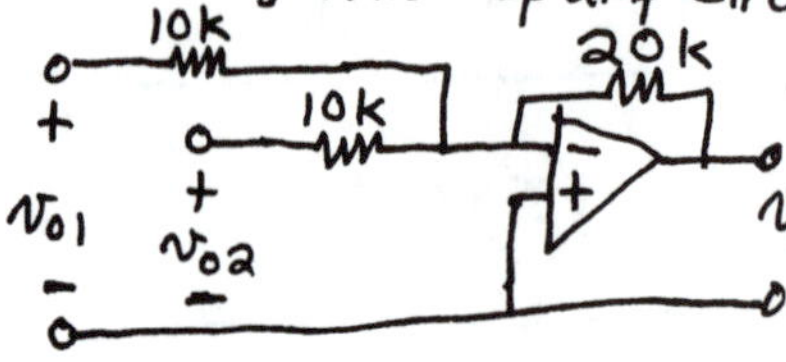

Inv. summer

$$v_o = -\frac{20k}{10k} v_{o1} - \frac{20k}{10k} v_{o2}$$

$$\Rightarrow v_o = -2(v_{o1} + v_{o2})$$

Top inverting amp:

$$v_{o1} = -\frac{R_4}{R_3} v_1$$

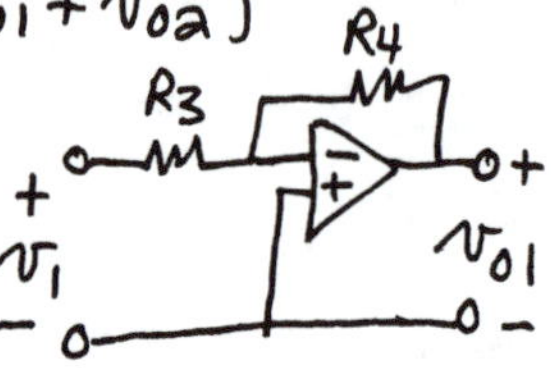

Left noninverting amp:

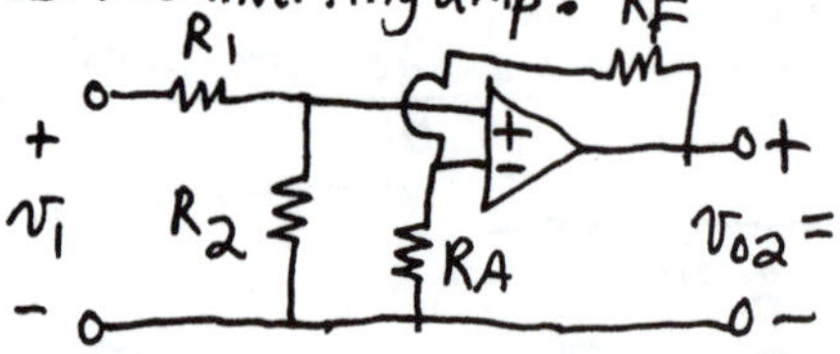

$$v_{o2} = \left(1 + \frac{R_F}{R_A}\right)\left[\frac{R_2}{R_1 + R_2} v_1\right]$$

Putting it all together:

$$\boxed{v_o = 2\left[\frac{R_4}{R_3} - \left(1 + \frac{R_F}{R_A}\right)\left(\frac{R_2}{R_1 + R_2}\right)\right] v_1}$$

(b) Pick R_1, R_2, R_A, R_F, R_3, and R_4 so that

$$\frac{R_4}{R_3} = \left(1 + \frac{R_F}{R_A}\right)\left(\frac{R_2}{R_1 + R_2}\right)$$

to make $v_o = 0$ V.

Example: Set all equal to 10 kΩ.

3.31

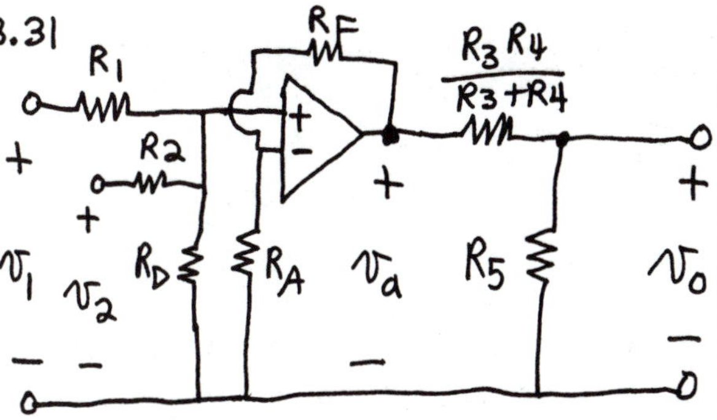

Voltage divider at right:

$$v_0 = v_a \frac{R_5}{R_5 + \left[\frac{R_3R_4}{R_3+R_4}\right]}$$

$$\Rightarrow v_0 = v_a \left[\frac{R_5(R_3+R_4)}{R_5(R_3+R_4)+R_3R_4}\right]$$

Noninverting summer:

$$v_a = \left(1+\frac{R_F}{R_A}\right)\left[\frac{R_T}{R_1}v_1 + \frac{R_T}{R_2}v_2 + \frac{R_T}{R_D}(0)\right]$$

$$R_T = R_1 \| R_2 \| R_D = \frac{R_1R_2R_D}{R_2R_D + R_1R_D + R_1R_2}$$

Substituting:

$$v_0 = \left[\frac{R_5(R_3+R_4)}{R_3R_4+R_3R_5+R_4R_5}\right]\left[1+\frac{R_F}{R_A}\right]\left[\frac{R_T}{R_1}v_1 + \frac{R_T}{R_2}v_2\right]$$

3.32 (There are many solutions)

$$v_0 = -3v_1 + 10v_2 - \frac{10}{3}v_3$$

Using two inverting amplifiers:

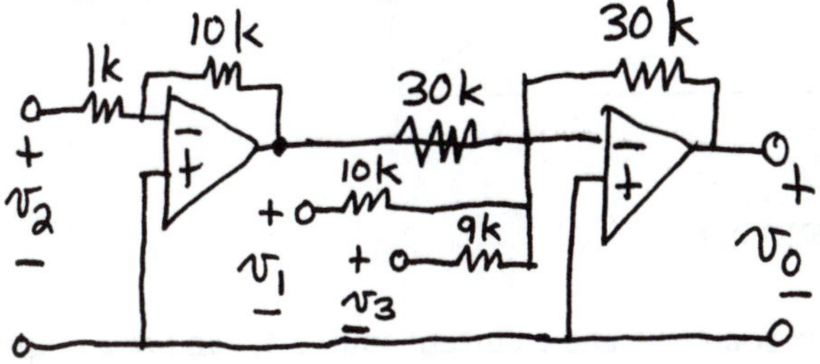

Inverting amp gain = −10

Inverting summer Gains: −1, −3, $-\frac{10}{3}$

3.33 (There are many solutions)

$$v_0 = \frac{1}{5}v_1 - \frac{2}{5}v_2$$

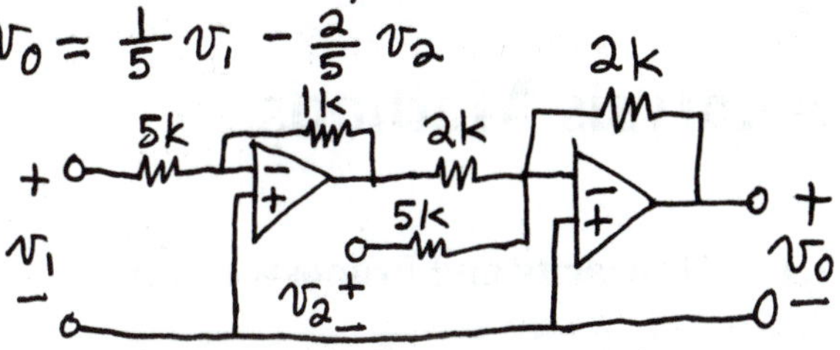

Inv. amp gain = $-\frac{1}{5}$

Inv. sum, gains −1, $-2/5$

3.34

(a)

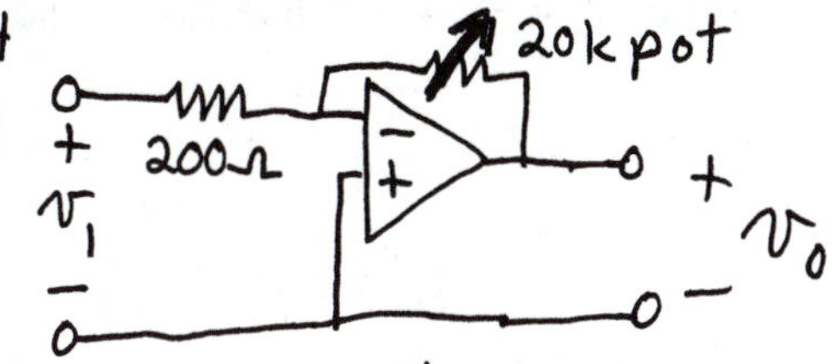

(b)

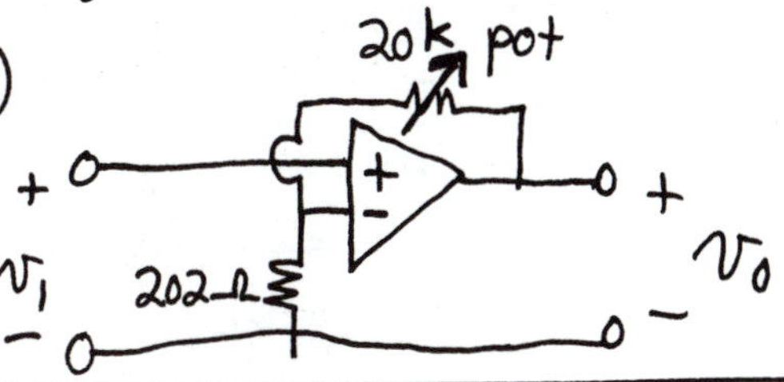

3.35 (Many solns.)

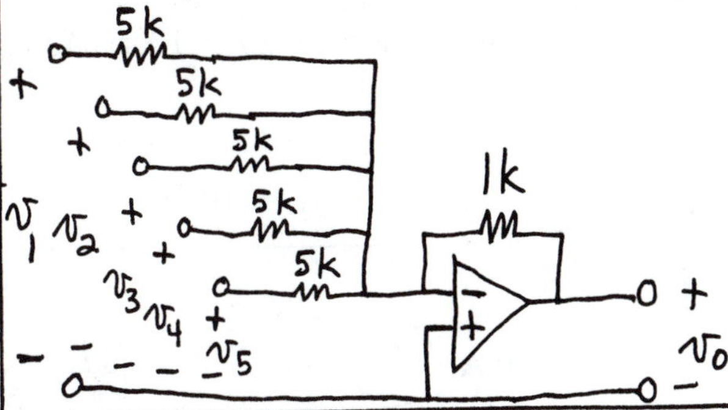

3.36 (Many solns.)

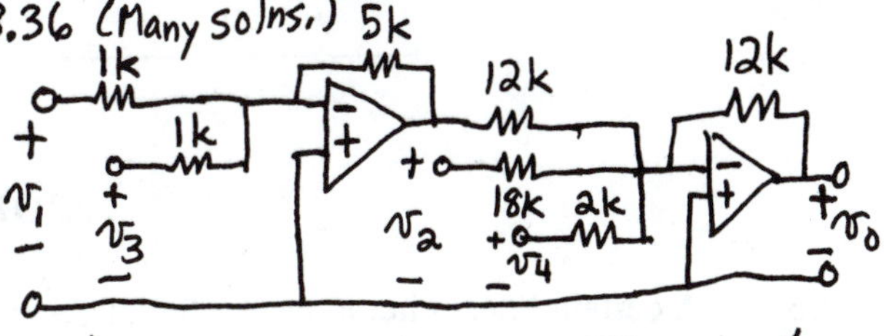

Inv. summer gains −5, −5

Inv. summer gains −1, $-\frac{2}{3}$, −6

Chapter 4

Analysis Methods

4.1 Linearity and Proportionality

4.1 (a) Find v_s if $v_o = 1$ V.

(b) If $v_s = 100$ V, use linearity to find v_o .

(c) If $v_s = -2$ V, use linearity to find v_o .

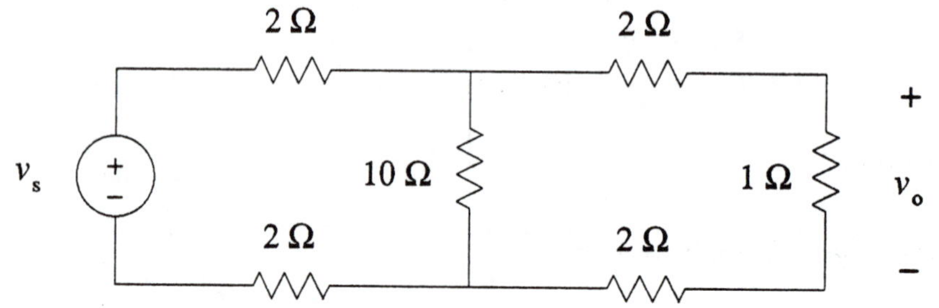

PROBLEM 4.1

4.2 Find v_1 by first assuming $i_1 = 1$ A and using proportionality.

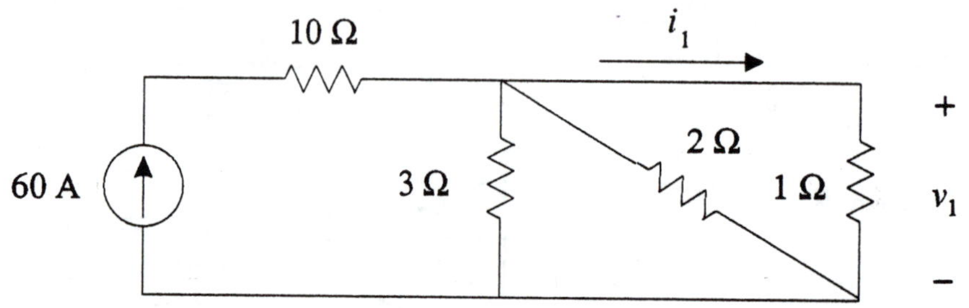

PROBLEM 4.2

4.3 A certain linear circuit has two inputs, a voltage source v_1 and a current source i_1 , and one output i_o . When $v_1 = 50$ V and $i_1 = 30$ A, the output is measured to be $i_o = -5$ mA. When $v_1 = -2$ V and $i_1 = 3$ A, the output is measured to be $i_o = +5$ mA. (a) Find an equation relating the inputs and outputs using linearity. (b) Design an op amp circuit to realize this input-output relationship.

4.4 (a) Find i_o.

(b) Find i_o if the values of the current and voltage sources are 9 A and 12 V, respectively.

(c) Find i_o if the values of the current and voltage sources are −1.5 A and −2 V, respectively.

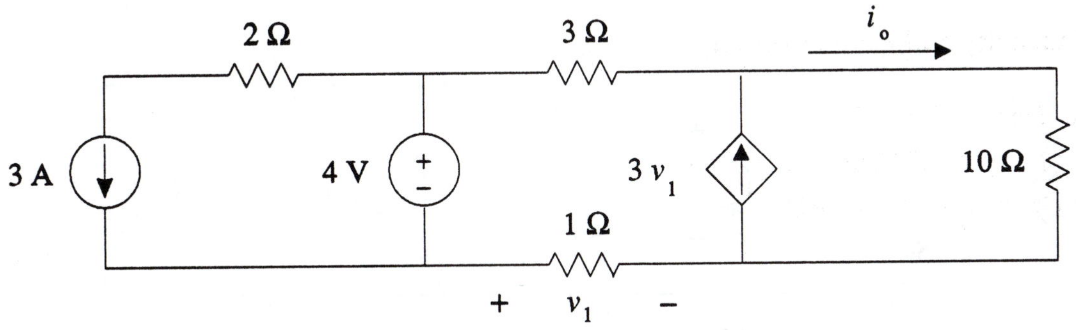

PROBLEM 4.4

4.2 Superposition

4.5 Solve Prob. 4.4(a) using superposition.

4.6 Find i_o using superposition.

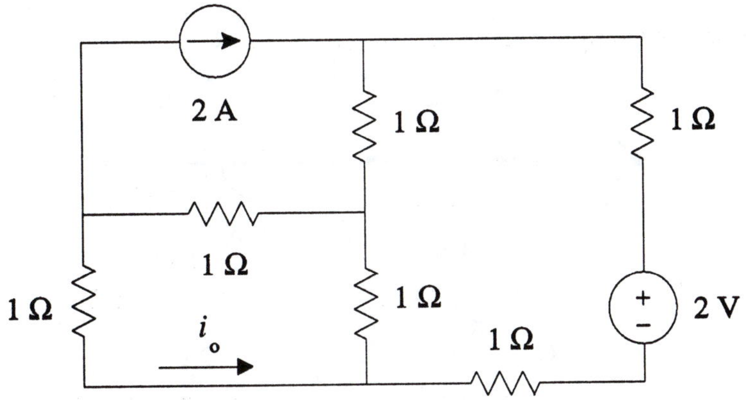

PROBLEM 4.6

4.7 Find v_o using superposition.

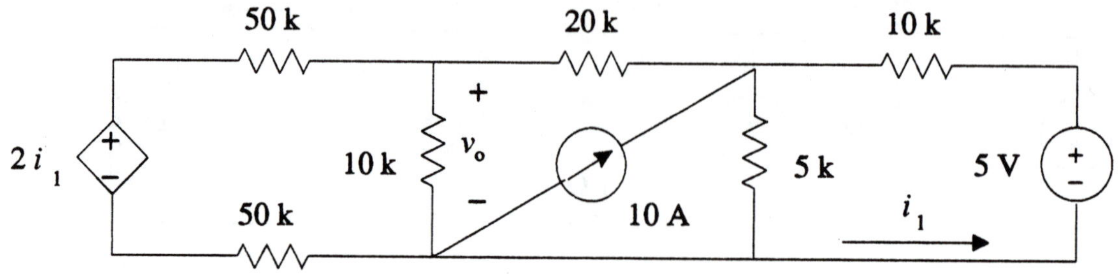

PROBLEM 4.7

4.8 Find v_o using superposition.

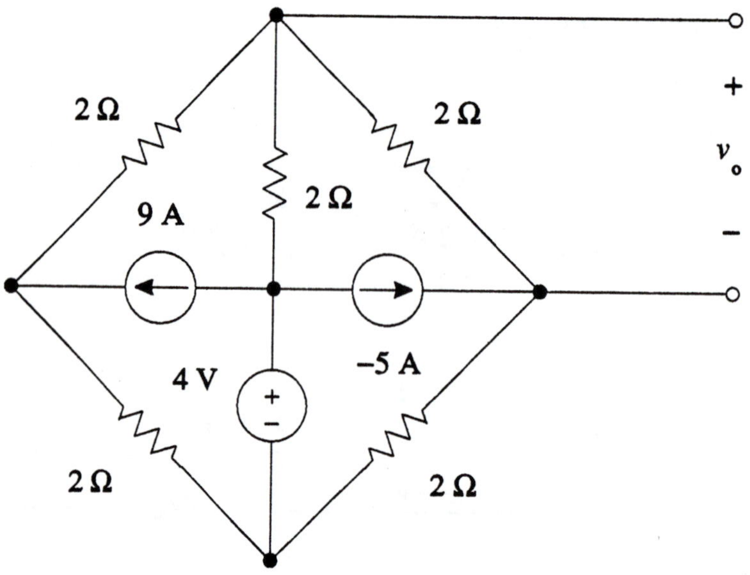

PROBLEM 4.8

4.3 Nodal Analysis

4.9 Find v using nodal analysis.

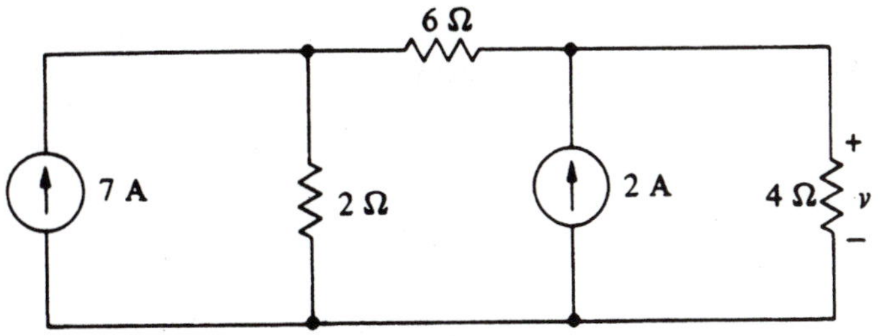

PROBLEM 4.9

4.10 Find i using nodal analysis.

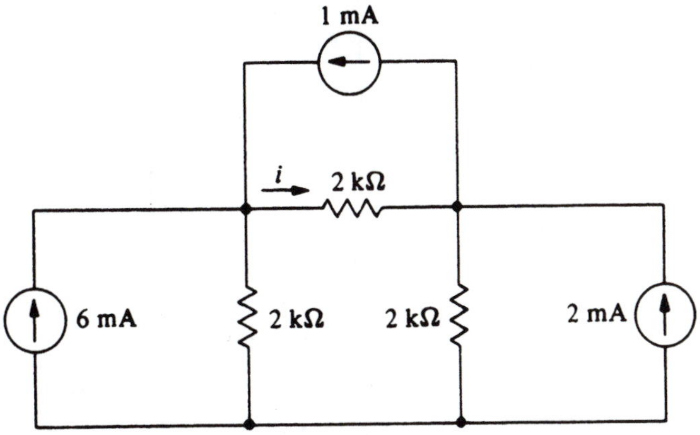

PROBLEM 4.10

4.11 Find the power delivered to the 8 Ω resistor using nodal analysis.

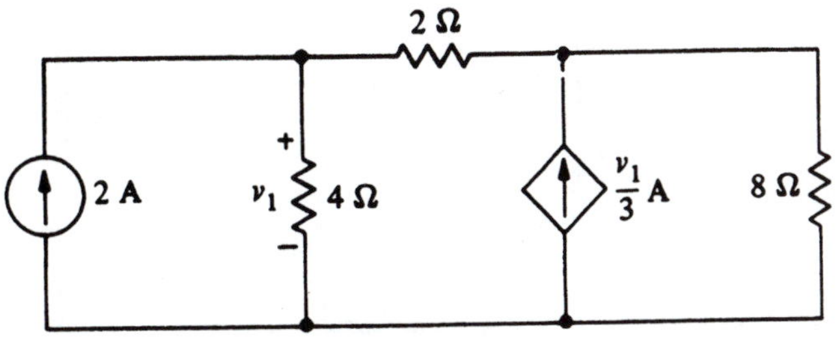

PROBLEM 4.11

4.12 Using nodal analysis, find v_1 and v_2 if $R_1 = 2\ \Omega$, $R_2 = 1\ \Omega$, $R_3 = 4\ \Omega$, $i_{g1} = 14$ A, and $i_{g2} = 7$ A.

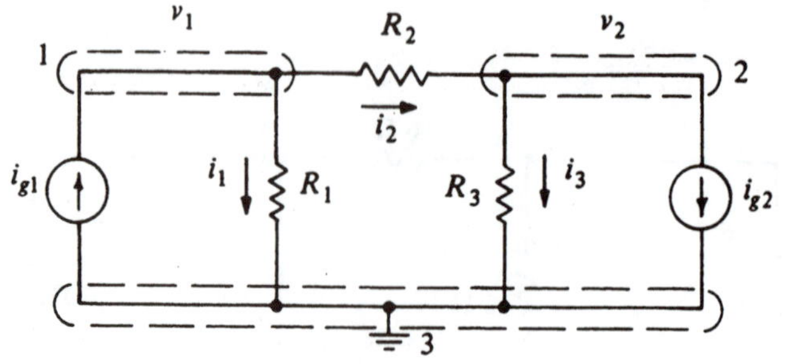

(a)

PROBLEM 4.12

4.13 Using nodal analysis, find v_1, v_2, and i.

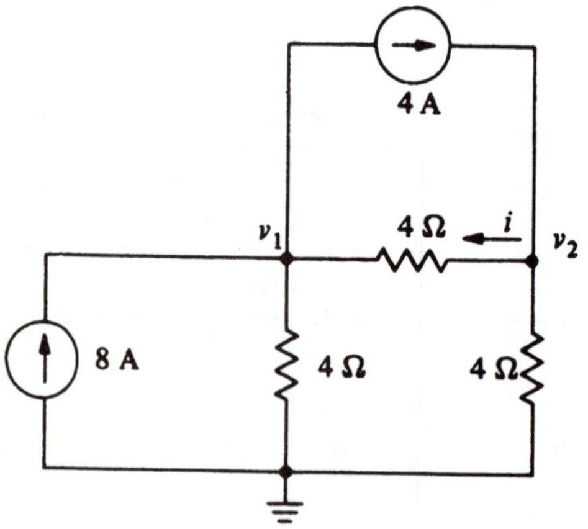

PROBLEM 4.13

4.14 Using nodal analysis, find v_1, v_2, and v_3.

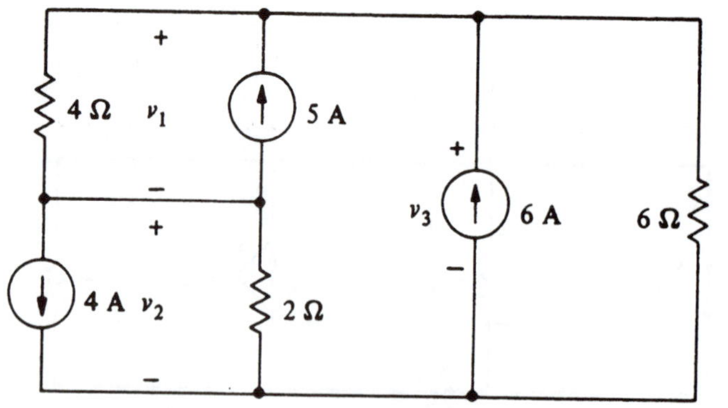

PROBLEM 4.14

4.15 Find v and i using nodal analysis.

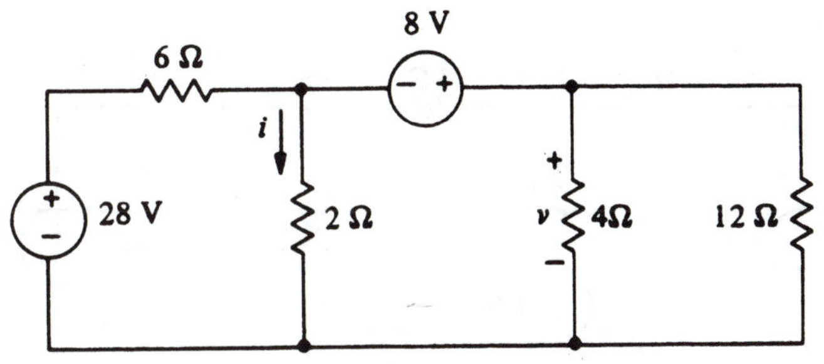

PROBLEM 4.15

4.16 Find i_1 using nodal analysis.

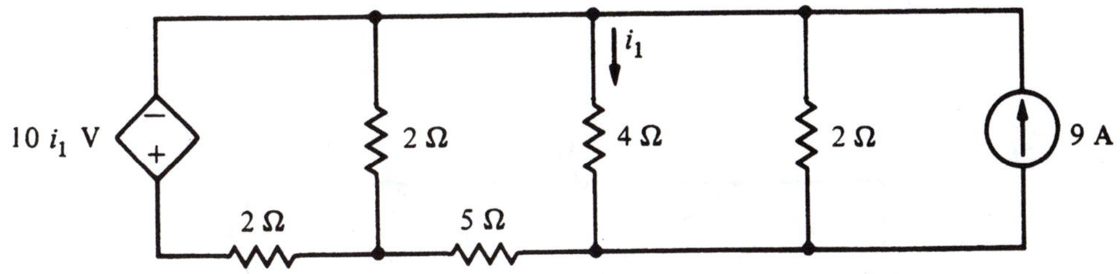

PROBLEM 4.16

4.4 Circuits Containing Voltage Sources

4.17 Find v and i using nodal analysis.

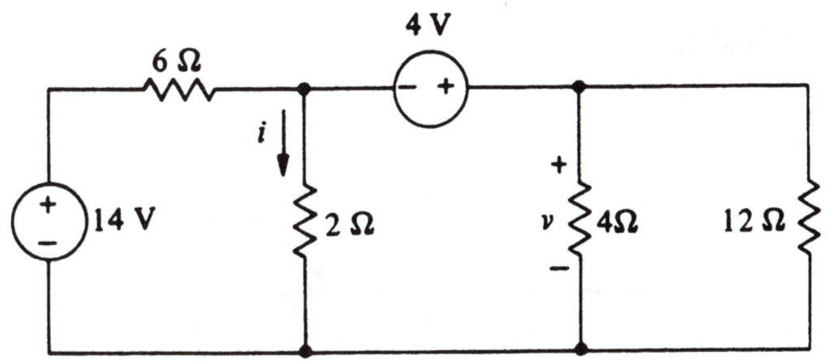

PROBLEM 4.17

4.18 Find v using nodal analysis.

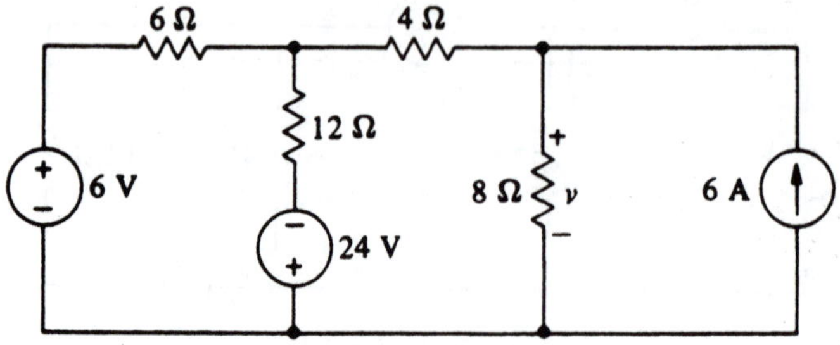

PROBLEM 4.18

4.19 Find v using nodal analysis.

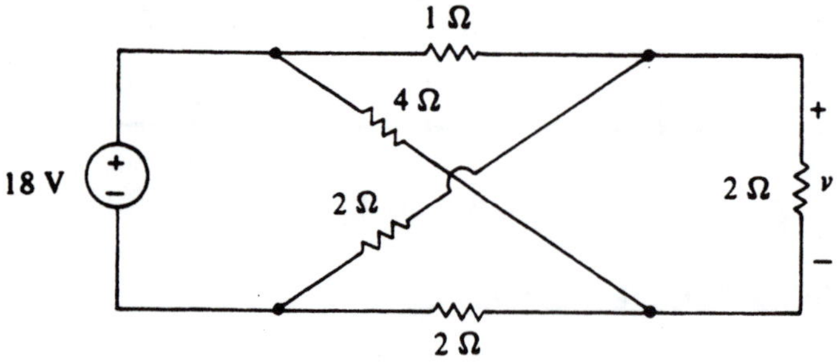

PROBLEM 4.19

4.20 Using nodal analysis, find i_1.

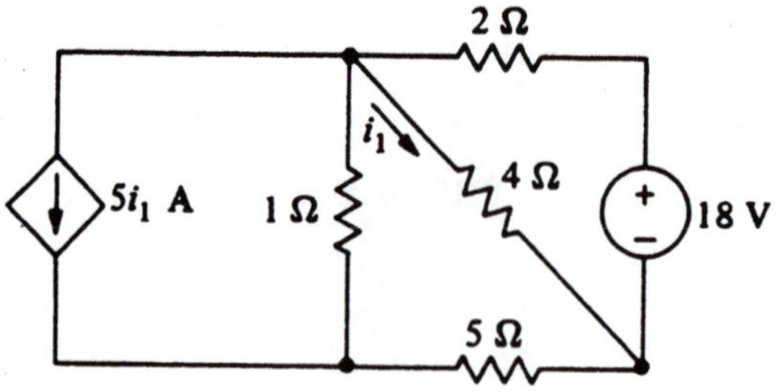

PROBLEM 4.20

4.21 Using nodal analysis, find i if element x is a 4 Ω resistor.

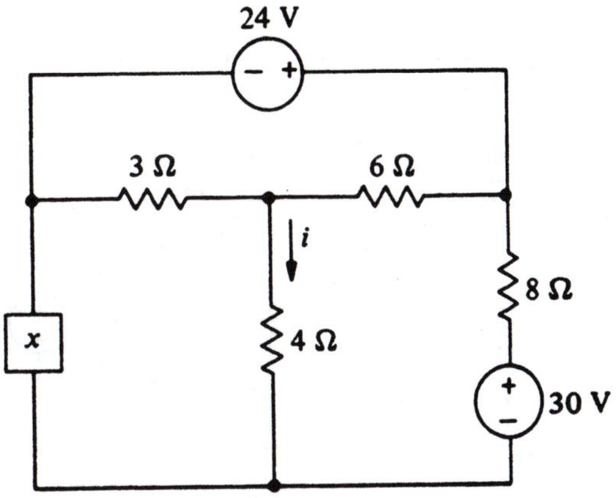

PROBLEM 4.21

4.22 Find i in Prob. 4.21 if element x is a 7 A independent current source directed upward.

4.23 Find i in Prob. 4.21 if element x is a dependent voltage source of $5i$ V with the positive terminal at the top.

4.24 Find v_2 using nodal analysis if $G_1 = 2$ S, $G_2 = 0.5$ S, $G_3 = 0.25$ S, $G_4 = 1$ S, $G_5 = 1$ S, $\beta = 5$, and $v_g = 4$ V.

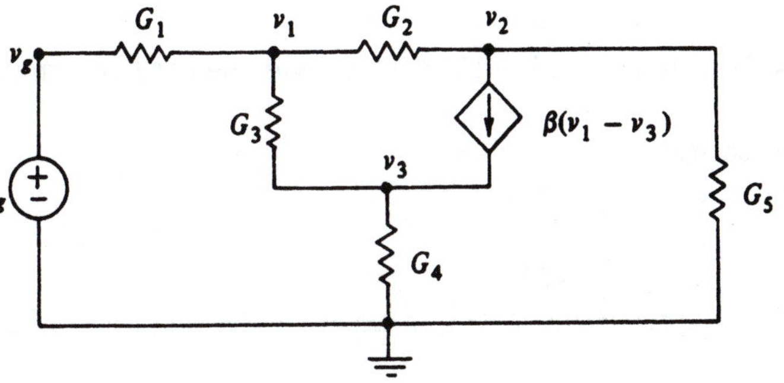

PROBLEM 4.24

4.25 Using nodal analysis, find v_1 in Prob. 4.24.

4.5 Mesh Analysis

4.26 Using mesh analysis, find i_1 and i_2 if $R_1 = 3\ \Omega$, $R_2 = 12\ \Omega$, $R_3 = 6\ \Omega$, $v_{g1} = 51\text{ V}$, and $v_{g2} = 6\text{ V}$.

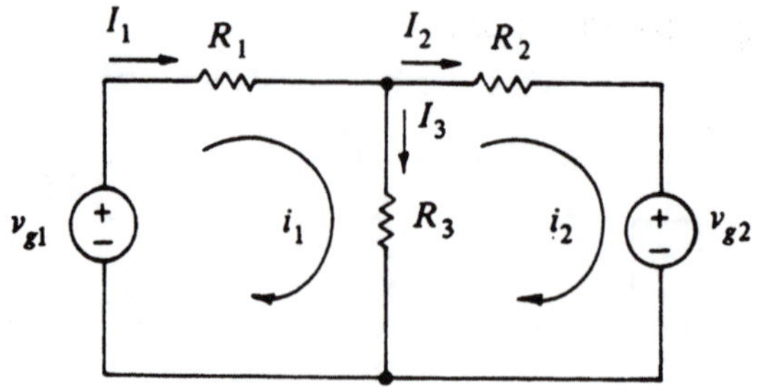

PROBLEM 4.26

4.27 Using mesh analysis, find i_1 and i_2 if element x is a $6\ \Omega$ resistor.

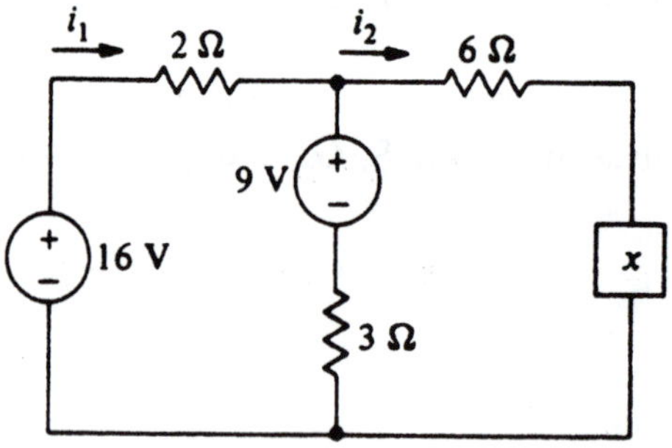

PROBLEM 4.27

4.28 Repeat Prob. 4.27 if element x is a 12 V source with the positive terminal at the bottom.

4.29 Find i using mesh analysis.

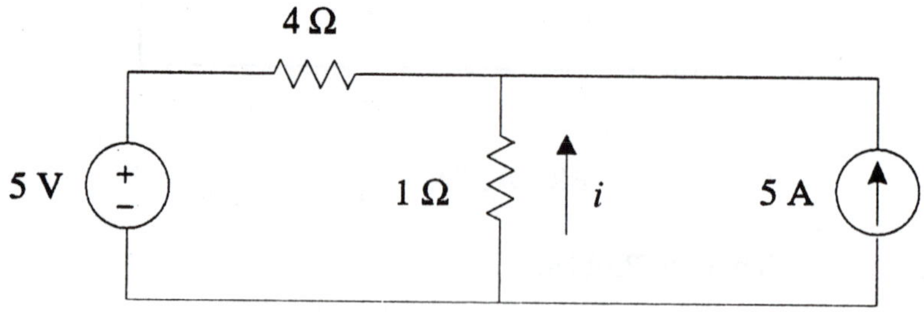

PROBLEM 4.29

4.30 Solve Prob. 4.2 using mesh analysis.

4.31 Solve Prob. 4.6 using mesh analysis.

4.32 Solve Prob. 4.8 using mesh analysis.

4.6 Circuits Containing Current Sources

4.33 Using mesh analysis, find v_1.

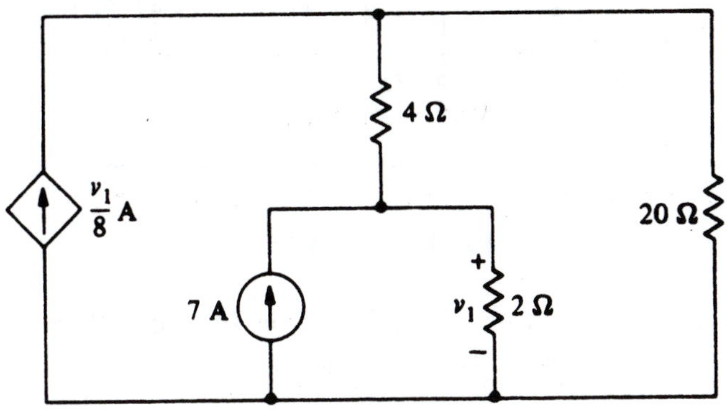

PROBLEM 4.33

4.34 Find the power absorbed in the 3 kΩ resistor using mesh analysis.

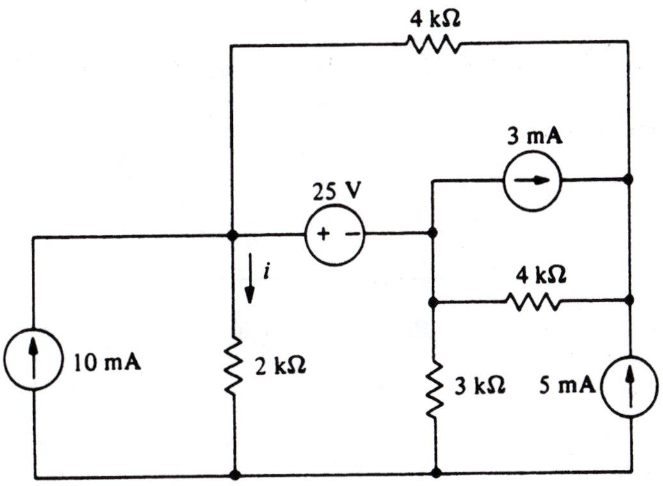

PROBLEM 4.34

4.35 Find the current in the 3 Ω resistor using loop analysis.

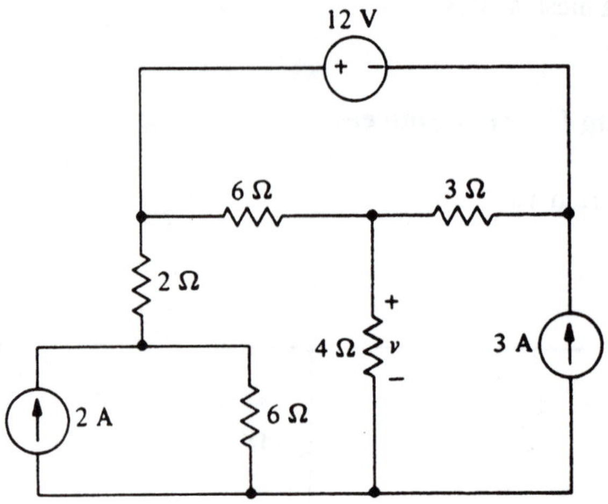

PROBLEM 4.35

4.36 If $i_{g1} = 2$ A, $i_{g2} = 8$ A, $v_{g3} = 24$ V, $R_1 = 1\ \Omega$, $R_2 = 3\ \Omega$, and $R_3 = 4\ \Omega$, find the power dissipated in R_3.

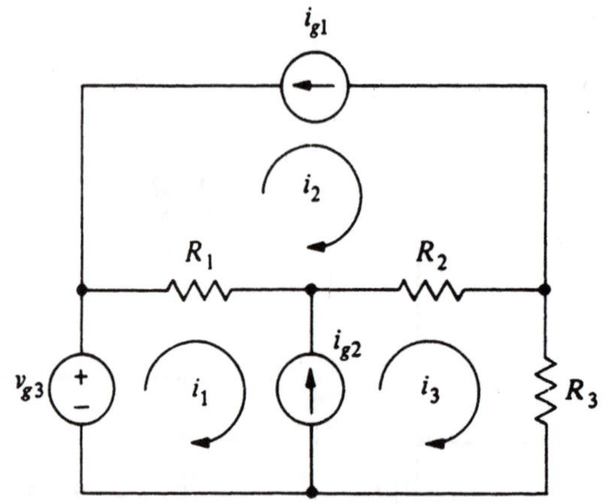

PROBLEM 4.36

4.37 Find i using mesh analysis.

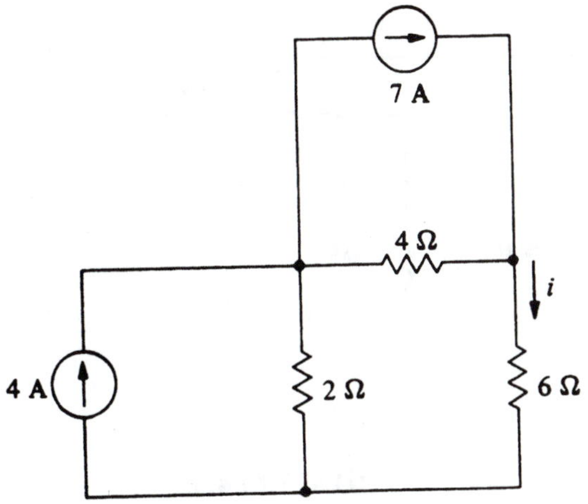

PROBLEM 4.37

4.7 Duality

4.38 Find the dual of the circuit in Prob. 4.2.

4.39 Find the dual of the circuit in Prob. 4.4.

4.40 Find the dual of the circuit in Prob. 4.6.

4.41 Find the dual of the circuit in Prob. 4.7.

4.42 Find the dual of the circuit in Prob. 4.8.

4.8 Virtual Short Principle for Op Amps

4.43 Find i.

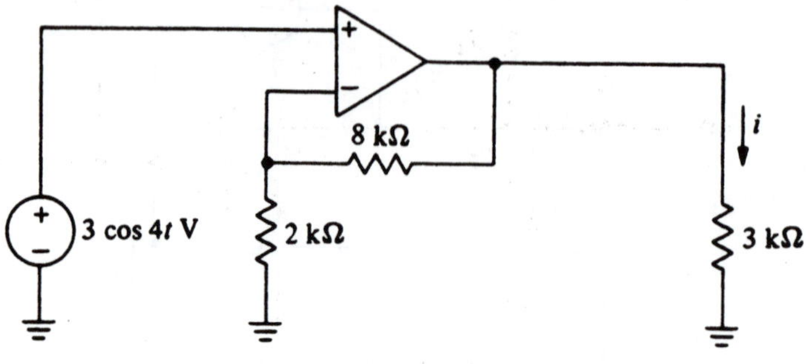

PROBLEM 4.43

4.44 Find v if $v_g = 6\cos(2t)$ V.

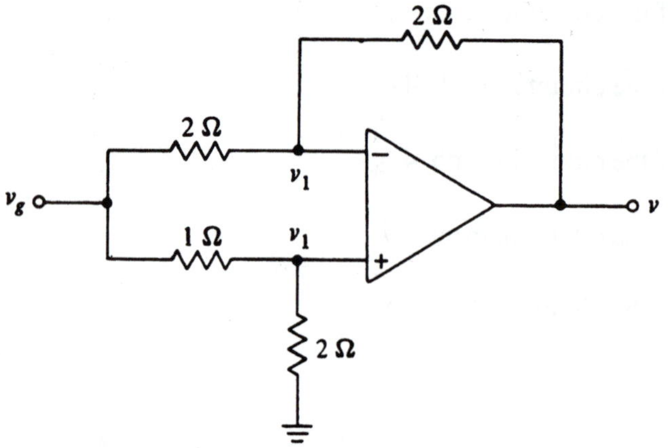

PROBLEM 4.44

4.45 Find v_2 for $G_1 = 0.25\,\text{S}$, $G_2 = 0.5\,\text{S}$, $G_3 = 1\,\text{S}$, $G_4 = 2\,\text{S}$, $G_5 = 0.5\,\text{S}$, and $v_1 = 8\cos(1000t)\,\text{V}$.

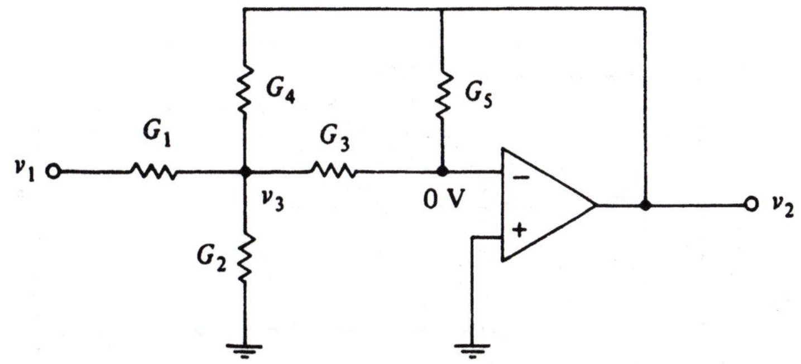

PROBLEM 4.45

4.46 Find v_3 if $v_g = 8\,\text{V}$.

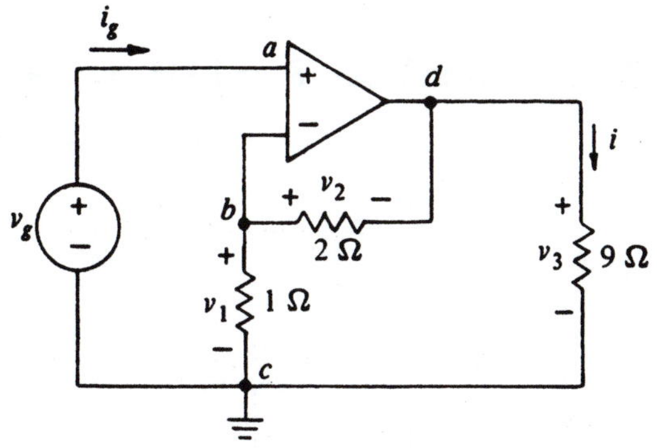

PROBLEM 4.46

4.47 Find the current in the 5 kΩ resistor if $v_g = 5$ V.

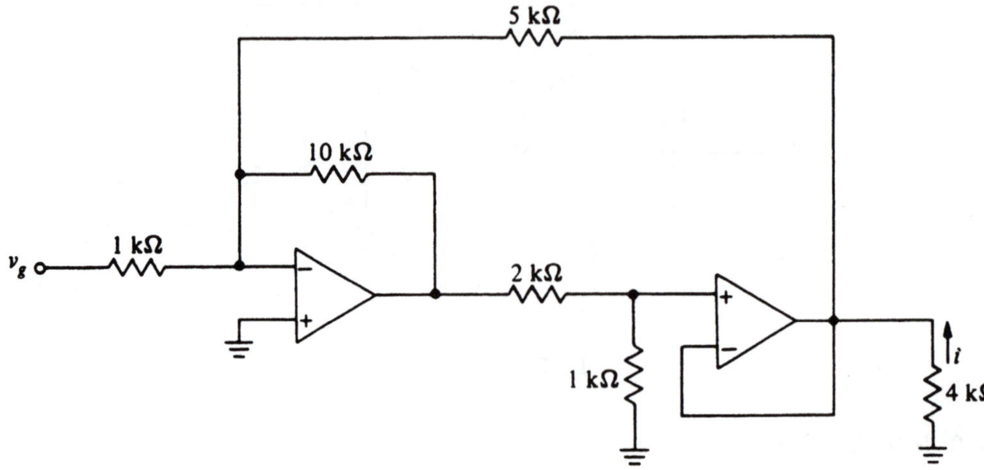

PROBLEM 4.47

4.48 Find v_o if $R_1 = 1.2$ kΩ, $R_2 = 4.7$ kΩ, and $R_3 = 1$ kΩ, $v_1 = 4$ V, and $v_2 = 2\cos(377t)$ V.

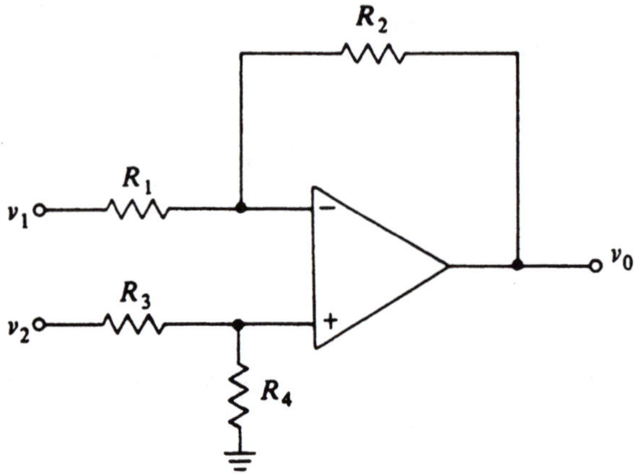

PROBLEM 4.48

4.49 Find i.

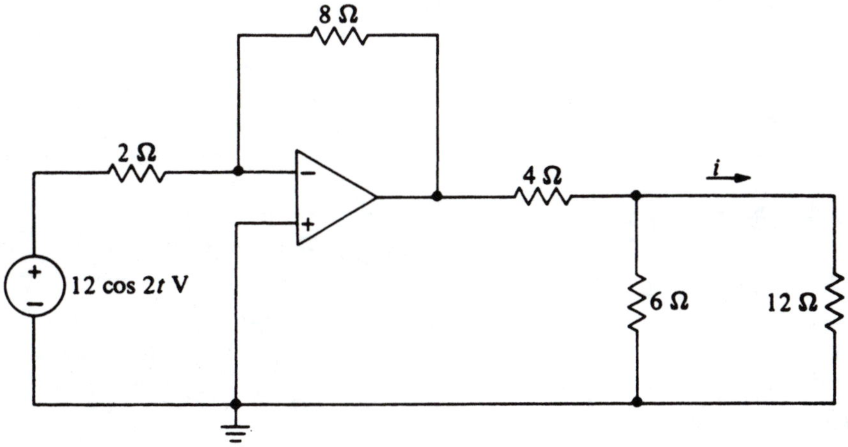

PROBLEM 4.49

4.9 Computer-Aided Circuit Analysis Using SPICE

4.50 Solve Prob. 4.2 using SPICE.

4.51 Solve Prob. 4.8 using SPICE.

4.52 Solve Prob. 4.20 using SPICE.

4.53 Solve Prob. 4.24 using SPICE.

4.54 Solve Prob. 4.44 using SPICE (use the improved op amp model from Chapter 3 with $A = 10^5$, $R_i = 1\ \text{M}\Omega$, and $R_o = 30\ \Omega$).

4.55 Solve Prob. 4.47 using SPICE (use the improved op amp model from Chapter 3 with $A = 10^5$, $R_i = 1\ \text{M}\Omega$, and $R_o = 30\ \Omega$).

CHAPTER 4 SOLUTIONS

4.1 (a)

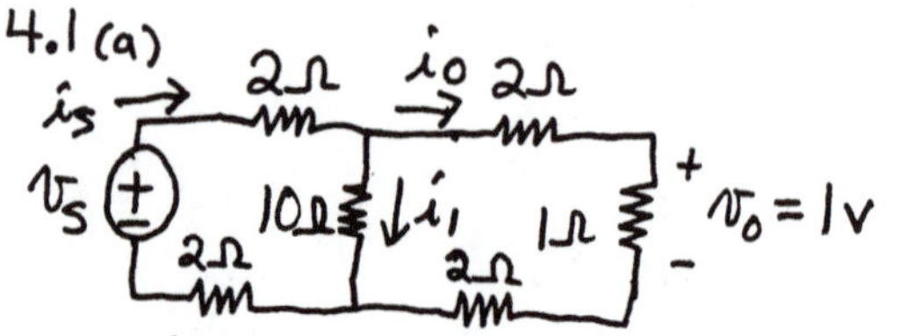

$i_0 = \frac{v_0}{1\Omega} = 1A$, $i_s = i_1 + i_0$ (KCL)

KVL (right loop): $i_0(2+1+2) = i_1(10)$

$\Rightarrow i_1 = 0.5A \Rightarrow i_s = 1.5A$

KVL (left loop): $v_s = 2i_s + 10i_1 + 2i_s$

$\therefore \boxed{v_s = 11V}$

(b) If $v_s = 100V = \frac{100}{11}(11V)$

$\Rightarrow v_0 = \frac{100}{11}(1V) = \underline{9.\overline{09}\ V}$

(c) If $v_s = -2V = -\frac{2}{11}(11V)$

$\Rightarrow v_0 = -\frac{2}{11}(1V) = \underline{0.\overline{18}\ V}$

4.2 Assume $i_1 = 1A$:

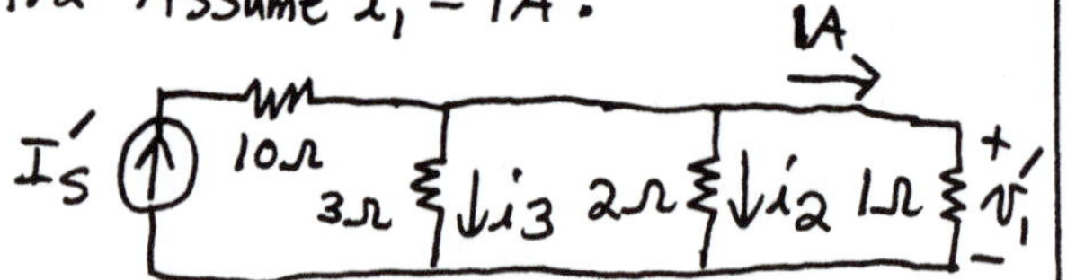

$v_1' = (1A)(1\Omega) = 1V \Rightarrow i_2 = \frac{1V}{2\Omega} = \frac{1}{2}A$

$i_3 = \frac{1V}{3\Omega} = \frac{1}{3}A$

KCL: $I_s' = i_1 + i_2 + i_3 = 1.8\overline{3}A$

However, $I_s = 60A = 32.\overline{72}(I_s')$

$\Rightarrow i_1 = (32.\overline{72})(1A) = 32.\overline{72}A$

$\Rightarrow \boxed{v_1 = i_1(1\Omega) = 32.\overline{72}V}$

4.3 Linear equation: $i_0 = Av_1 + Bi_1$

(a)

$$\Rightarrow \begin{cases} A(50) + B(30) = -5\times10^{-3} \\ A(-2) + B(3) = 5\times10^{-3} \end{cases}$$

$$\Delta = \begin{vmatrix} 50 & 30 \\ -2 & 3 \end{vmatrix} = 150 + 60 = 210$$

$$A = \frac{\begin{vmatrix} -5\times10^{-3} & 30 \\ 5\times10^{-3} & 3 \end{vmatrix}}{\Delta} = -7.857\times10^{-4}$$

$$B = \frac{\begin{vmatrix} 50 & -5\times10^{-3} \\ -2 & 5\times10^{-3} \end{vmatrix}}{\Delta} = 1.143\times10^{-3}$$

$$\therefore \boxed{i_0 = (1.143\times10^{-3})i_1 - (7.857\times10^{-4})v_1}$$

(b) (Many solutions)

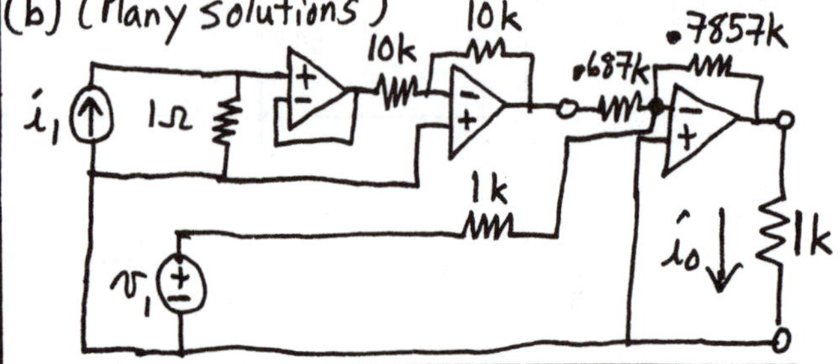

4.4 (a)

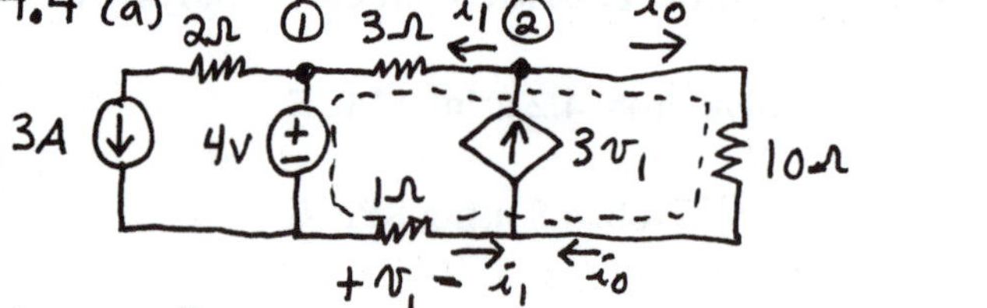

KCL at ②: $i_1 = 3v_1 - i_0$

KVL around dashed loop: $4V = 3\Omega(i_0 - 3v_1) + 10i_0 + 1\Omega(i_0 - 3v_1)$

$\Rightarrow 4 = 14i_0 - 12v_1$

$v_1 = (3v_1 - i_0)1\Omega \Rightarrow i_0 = 2v_1$

$\Rightarrow \boxed{v_1 = \frac{1}{4}V} \Rightarrow \boxed{i_0 = \frac{1}{2}A}$

(b) Sources are both multiplied by 3

$\Rightarrow$ outputs are multiplied by 3

$\Rightarrow \boxed{i_0 = \frac{3}{2}A,\ v_1 = \frac{3}{4}V}$

(c) Sources are both multiplied by -0.5

$\Rightarrow \boxed{i_0 = -\frac{1}{4}A,\ v_1 = -\frac{1}{8}V}$

4.5 ① Kill current source (open circuit):

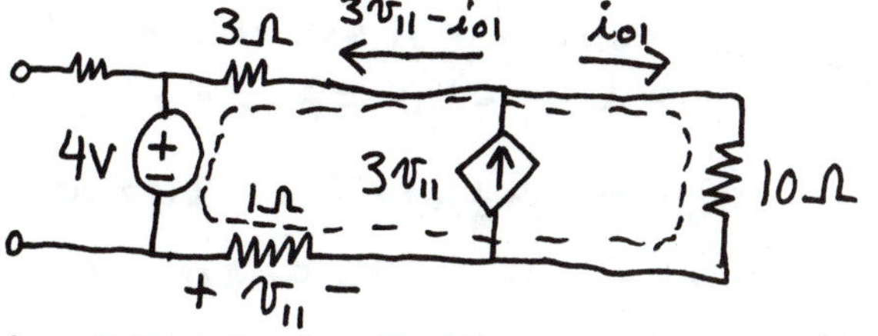

Ohms Law: $v_{11} = 1\Omega\,(3v_{11} - i_{01}) \Rightarrow i_{01} = 2v_{11}$

KVL around dashed loop:

$4v = 3(i_{01} - 3v_{11}) + 10 i_{01} - v_{11}$

$\Rightarrow v_{11} = \frac{1}{4}V \Rightarrow i_{01} = \frac{1}{2}A$

② Kill voltage source (short circuit)

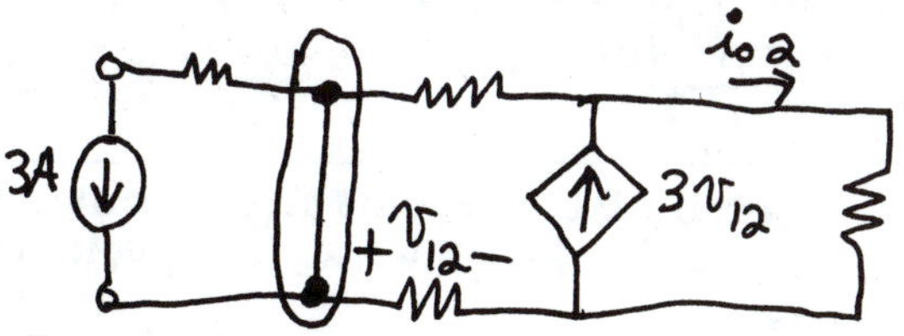

Due to short circuit, no current reaches right half of circuit $\Rightarrow v_{12} = 0,\ i_{02} = 0$

$\therefore \boxed{i_0 = i_{01} + i_{02} = \frac{1}{2}A,\ v_1 = v_{11} + v_{12} = \frac{1}{4}V}$

4.6 ① Short ckt the voltage source:

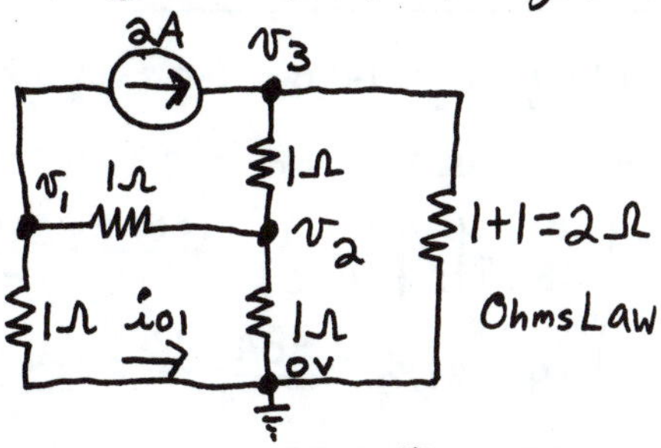

Ohms Law: $i_{01} = \frac{v_1}{1\Omega}$

KCL: $2A = \frac{v_3 - v_2}{1\Omega} + \frac{v_3}{2\Omega} \Rightarrow \underline{2 = \frac{3}{2}v_3 - v_2}$

KCL: $2A + \frac{v_1}{1\Omega} + \frac{v_1 - v_2}{1\Omega} = 0$

$\Rightarrow \underline{2v_1 - v_2 = -2}$

KCL: $v_3 - v_2 = v_2 - v_1 + v_2$

$\Rightarrow \underline{v_3 = 3v_2 - v_1}$

Solving: $v_1 = -\frac{10}{11} \Rightarrow \boxed{i_{01} = -\frac{10}{11}A}$

② Open ckt the 2A source:

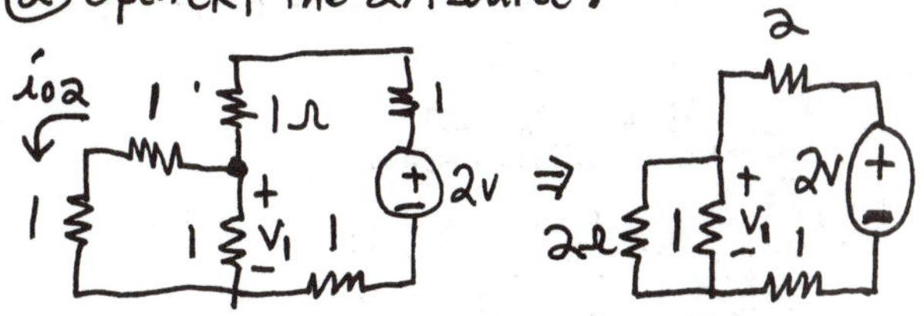

Voltage division: $v_1 = 2v\left(\frac{2/3}{11/3}\right) = \frac{4}{11}$

$\Rightarrow \boxed{i_{02} = \frac{v_1}{2\Omega} = \frac{2}{11}A}$

$\therefore \boxed{i_0 = i_{01} + i_{02} = -\frac{8}{11}A}$

4.7 ① Kill current sources

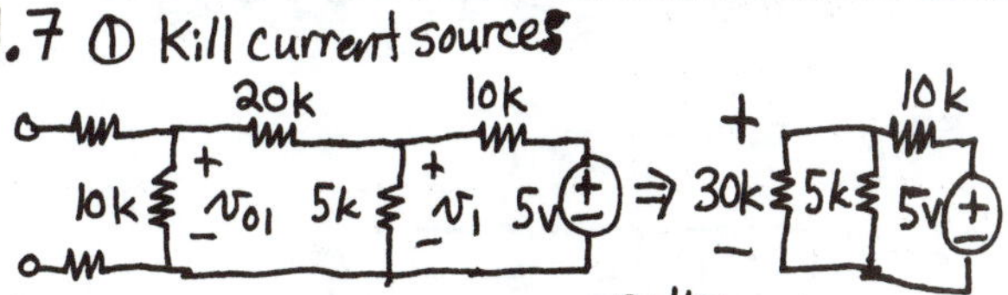

Voltage division: $v_1 = 5v\left(\frac{30\|5}{10 + (30\|5)}\right)$

$\Rightarrow v_1 = 5v\left(\frac{4.2857k}{14.2857k}\right) = 1.5\ V$

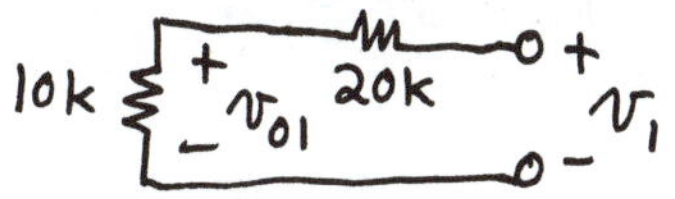

Voltage division $v_{01} = v_1\left(\frac{10k}{30k}\right)$

$\Rightarrow v_{01} = 0.5V$

② Kill voltage source:

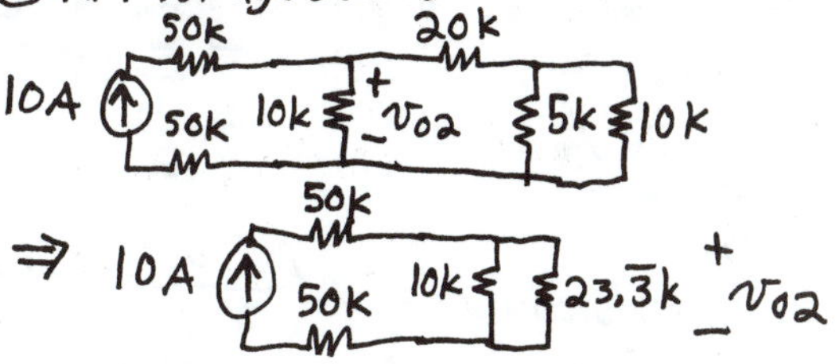

$v_{02} = (10A)(10k \| 23.\bar{3}k)$

$= 10A(7k) = 70000V$

$\therefore v_0 = v_{01} + v_{02} = \boxed{70000.5\ V}$

4.8 ① Kill current sources:

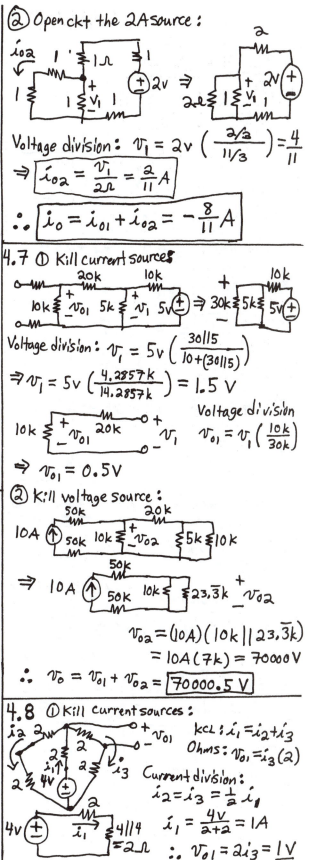

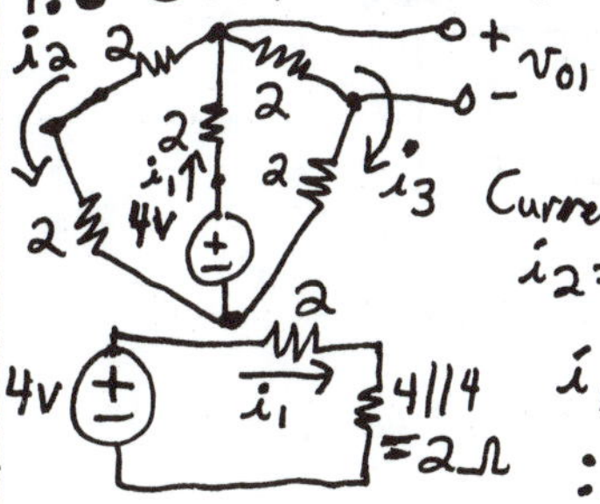

KCL: $i_1 = i_2 + i_3$

Ohms: $v_{01} = i_3(2)$

Current division:

$i_2 = i_3 = \frac{1}{2}i_1$

$i_1 = \frac{4V}{2+2} = 1A$

$\therefore v_{01} = 2i_3 = \underline{1V}$

4.8 (cont'd) ② Kill 4v and −5A sources:

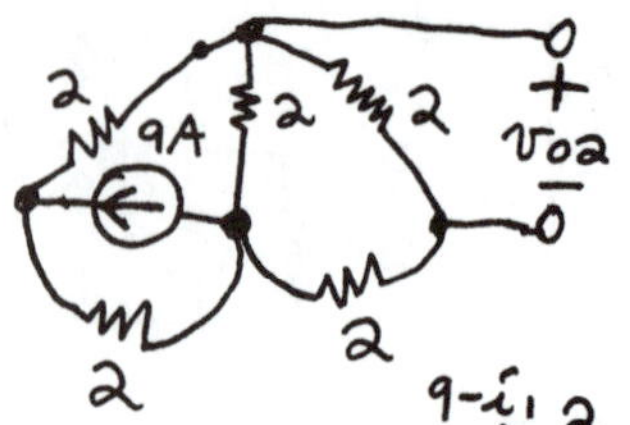

⇒ 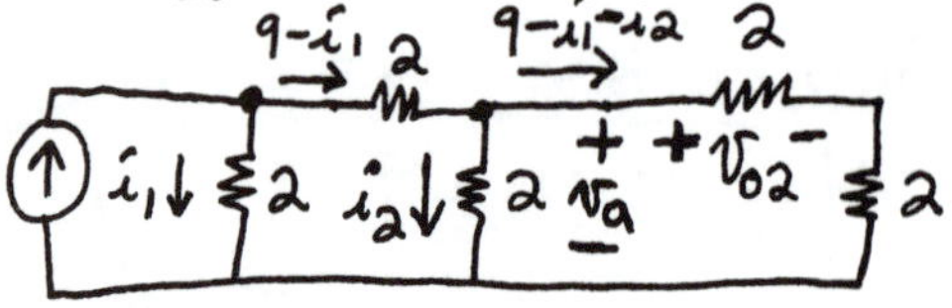

$v_{o2} = (2\Omega)[9-i_1-i_2]$

Current div: $i_2 = 2(9-i_1-i_2)$
(or Ohms Law)

$\Rightarrow i_2 = 6-\frac{2}{3}i_1$

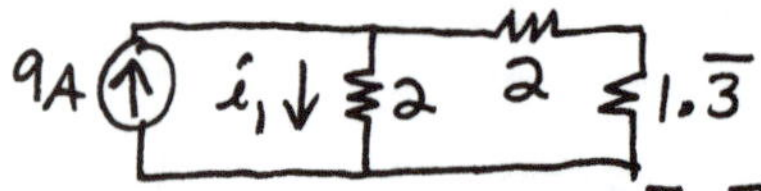

Current div: $i_1 = \left[\frac{3.\bar{3}}{2+3.\bar{3}}\right]9A = 5.625A$

$\Rightarrow v_{o2} = 2\Omega\,(9-5.625-6+\frac{2}{3}(5.625))$
$= \underline{2.25\text{ V}}$

③ Kill 4v and 9A sources:

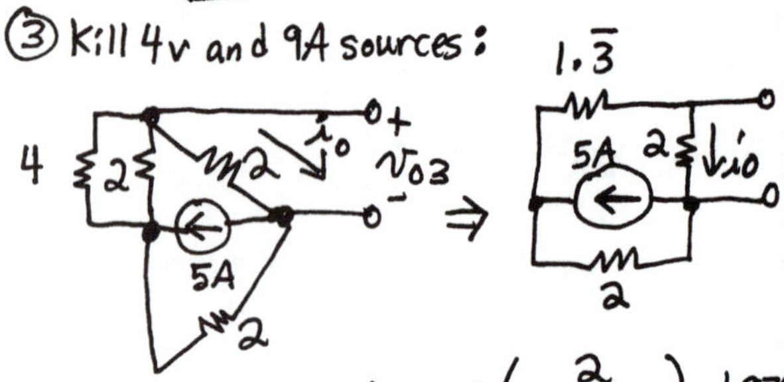

Current division: $i_o = 5A\left(\frac{2}{2+3.\bar{3}}\right) = 1.875$

$\Rightarrow v_{o3} = (2\Omega)(1.875A) = \underline{3.75\text{ V}}$

$\therefore v_o = v_{o1} + v_{o2} + v_{o3}$
$= 1.0\text{ V} + 2.25\text{ V} + 3.75\text{ V}$

$v_o = \underline{\underline{7.0\text{ V}}}$

4.9 With v_1 & v as node voltages, at terminals of 6-Ω resistor, KCL,

$(\frac{1}{2}+\frac{1}{6})v_1 - \frac{1}{6}v = 7$ or $4v_1 - v = 42$

$(-\frac{1}{6})v_1 + (\frac{1}{6}+\frac{1}{4})v = 2$ or $-2v_1 + 5v = 24$

Adding, $9v = 90$ and $v = \underline{10\text{ V}}$.

4.10 Let the node voltages be v_1 and v_2 at the terminals of the 2-kΩ resistor. The node equations are:

$(\frac{1}{2}+\frac{1}{2})v_1 - \frac{1}{2}v_2 = 6+1$ or $2v_1 - v_2 = 14$

$-\frac{1}{2}v_1 + (\frac{1}{2}+\frac{1}{2})v_2 = 2-1$ or $-v_1 + 2v_2 = 2$

Adding, $3v_2 = 18$ and $v_2 = 6\text{V}$,

$v_1 = \frac{14+v_2}{2} = 10\text{V}$, $i = \frac{v_1 - v_2}{2\text{k}\Omega} = \underline{2\text{ mA}}$.

4.11 Let v_2 be the voltage across the 2-Ω resistor. The node equations are:

$(\frac{1}{4}+\frac{1}{2})v_1 - \frac{1}{2}v_2 = 2$ or $3v_1 - 2v_2 = 8$

$(-\frac{1}{2}-\frac{1}{3})v_1 + (\frac{1}{2}+\frac{1}{8})v_2 = 0$ or $-4v_1 + 3v_2 = 0$

$v_2 = \begin{vmatrix} 3 & 8 \\ -4 & 0 \end{vmatrix} \Big/ \begin{vmatrix} 3 & -2 \\ -4 & 3 \end{vmatrix} = \frac{(4)(8)}{9-8} = 32\text{V}$

$P_{8\Omega} = \frac{v_2^2}{8} = \underline{128\text{ W}}$.

4.12 $(\frac{1}{2}+1)v_1 - v_2 = 14$ or $3v_1 - 2v_2 = 28$

$-v_1 + (1+\frac{1}{4})v_2 = -7$ or $-4v_1 + 5v_2 = -28$

$v_1 = \begin{vmatrix} 28 & -2 \\ -28 & 5 \end{vmatrix} \Big/ \begin{vmatrix} 3 & -2 \\ -4 & 5 \end{vmatrix} = \frac{140-56}{15-8} = \underline{12\text{V}}$

$v_2 = \begin{vmatrix} 3 & 28 \\ -4 & -28 \end{vmatrix} \Big/ 7 = \frac{-84+112}{7} = \underline{4\text{V}}$.

4.13 $(\frac{1}{4}+\frac{1}{4})v_1 - \frac{1}{4}v_2 = 8-4$ or $2v_1 - v_2 = 16$

$(-\frac{1}{4})v_1 + (\frac{1}{4}+\frac{1}{4})v_2 = 4$ or $-v_1 + 2v_2 = 16$

adding, $3v_1 = 48$ and $v_1 = \underline{16\text{V}}$

$v_2 = \frac{16+v_1}{2} = \underline{16\text{V}}$ $i = \frac{v_2 - v_1}{4\Omega} = \underline{0\text{ A}}$

4.14 $\left(\frac{1}{4}+\frac{1}{6}\right)v_3 - \frac{1}{4}v_2 = 6+5$ or $-3v_2+5v_3=132$

$\left(\frac{1}{2}+\frac{1}{4}\right)v_2 - \frac{1}{4}v_3 = -5-4$ or $3v_2 - v_3 = -36$

adding, $4v_3 = 96$ and $v_3 = \underline{24V}$,

$v_2 = \frac{-36+v_3}{3} = \underline{-4V}$,

$v_1 = v_3 - v_2 = \underline{28V}$.

4.15 The node voltages are 28, $v-8$, and v. KCL for the supernode containing the 8-V source yeilds

$\frac{v-8-28}{6} + \frac{v-8}{2} + \frac{v}{4} + \frac{v}{12} = 0$ or

$12v = 120 \therefore v = \underline{10V}$ and

$i = (v-8)/2 = \underline{1A}$.

4.16 with v_1 and v_2 as node voltages, at terminals of 5-Ω resistor. The node equations are:

$v_1\left(\frac{1}{4}+\frac{1}{2}+\frac{1}{5}\right) - v_2\frac{1}{5} = -9$

$v_1\left(-\frac{1}{5}\right) + v_2\left(\frac{1}{5}+\frac{1}{2}+\frac{1}{2}\right) - \frac{10i_1}{2} = 0$

since $i_1 = \frac{-v_1}{4}$, the node equations become,

$19v_1 - 4v_2 = -180$, $21v_1 + 24v_2 = 0$

Adding six times the first equation

$135v_1 = -1{,}080 \therefore v_1 = -8V$

$i_1 = -\frac{(-8V)}{4} = \underline{2A}$.

4.17 The node voltages are 14, $v-4$, and v. KCL for the supernode containing the 4V source yeilds

$\frac{v-4-14}{6} + \frac{v-4}{2} + \frac{v}{4} + \frac{v}{12} = 0$ or

$12v = 60 \therefore v = \underline{5V}$ and

$i = (v-4)/2 = \underline{0.5A}$

4.18 with v and v_1 as node voltages, at the terminals of the 4-Ω. The node equations are:

$v\left(\frac{1}{8}+\frac{1}{4}\right) - v_1\frac{1}{4} = 6$ or $3v - 2v_1 = 48$

$-\frac{v}{4} + v_1\left(\frac{1}{12}+\frac{1}{6}+\frac{1}{4}\right) = \frac{6}{6} - \frac{24}{12}$

$-3v + 6v_1 = -12$, Adding equation

$4v_1 = 36V$, $v = \frac{48+2v_1}{3} = \underline{22V}$

4.19 with v_1 and v_2 as node voltages at terminals of 2-Ω resistor. The node equations are

$v_1\left(\frac{1}{2}+\frac{1}{2}+1\right) - \frac{v_2}{2} = \frac{18}{1}$

$-\frac{v_1}{2} + v_2\left(\frac{1}{4}+\frac{1}{2}+\frac{1}{2}\right) = \frac{18}{4}$ or

$4v_1 - v_2 = 36$, $-2v_1 + 5v_2 = 18$

Adding twice the second equation

$9v_2 = 72$, or $v_2 = 8V$, $v_1 = \frac{36+v_2}{4} = 11V$

$v = v_1 - v_2 = \underline{3V}$.

4.20 with v_1 and v_2 as node voltages at terminals of 1-Ω resistor and $i_1 = v_1/4$. The node equations are

$v_1\left(1+\frac{1}{4}+\frac{1}{2}+\frac{5}{4}\right) - v_2 = \frac{18}{2}$ or $12v_1 - 4v_2 = 36$

$-v_1\left(1+\frac{5}{4}\right) + v_2\left(\frac{1}{5}+1\right) = 0$ or $-45v_1 + 24v_2 = 0$

Adding 6 times the first equation

$27v_1 = 216$ or $v_1 = 8V \therefore i_1 = 8/4 = \underline{2A}$.

4.21 The node voltages are v_1, v_2 and $v_2 - 24$ with $i = v_1/4$. The node equations are:

$v_1\left(\frac{1}{3}+\frac{1}{4}+\frac{1}{6}\right) - \frac{v_2}{6} - \frac{(v_2-24)}{3} = 0$

$\frac{v_2-30}{8} + \frac{v_2}{6} - \frac{v_1}{6} + \frac{v_2-24}{x} + \frac{v_2-24}{3} - \frac{v_1}{3} = 0$

or $9v_1 - 6v_2 = 96$

$-12v_1 + 21v_2 = 426$

$v_1 = \begin{vmatrix} 96 & -6 \\ 426 & 21 \end{vmatrix} \Big/ \begin{vmatrix} 9 & -6 \\ -12 & 21 \end{vmatrix} = \frac{4572}{117} = 39.08V$

$\therefore i = \frac{39.08}{4} = \underline{9.77A}$

4.22 Node equations are using node voltages of Prob. 4.13.

$9v_1 - 6v_2 = 96$,

$\frac{v_2-30}{8} + \frac{v_2-v_1}{6} + \frac{v_2-24-v_1}{3} = 7$ or

$-12v_1 + 15v_2 = 289$

$v_1 = \begin{vmatrix} 96 & -6 \\ 289 & 15 \end{vmatrix} \Big/ \begin{vmatrix} 9 & -6 \\ -12 & 15 \end{vmatrix} = \frac{3174}{63} = 50.38V$

$\therefore i = \frac{50.38V}{4\Omega} = \underline{12.60A}$

4.23 The node voltages are 30, $5i$, $5i+24$ an v_1 of Prob. 4.13. $i = v_1/4$. The node equation is:

$$\frac{v_1-5i}{3}+\frac{v_1}{4}+\frac{v_1-(5i+24)}{6}=0$$

$3v_1 = 96$, $v_1 = 32\text{V}$

$i = \frac{32}{4} = \underline{8\text{A}}$.

4.24. The node equation are

$v_1(2+0.5+0.25)-v_2(0.5)-v_3(0.25)=4(2)$

or $11v_1 - 2v_2 - v_3 = 32$

$-v_1(0.5)+v_2(0.5+1)+5(v_1-v_3)=0$

or $9v_1 + 3v_2 - 10v_3 = 0$

$-v_1(0.25)-5(v_1-v_3)+v_3(0.25+1)=0$

or $-21v_1 + 25v_3 = 0$

$$v_2 = \frac{\begin{vmatrix} 11 & 32 & -1 \\ 9 & 0 & -10 \\ -21 & 0 & 25 \end{vmatrix}}{\begin{vmatrix} 11 & -2 & -1 \\ 9 & 3 & -10 \\ -21 & 0 & 25 \end{vmatrix}} = \underline{-0.606}\ \text{V}$$

4.25

$$v_1 = \frac{\begin{vmatrix} 32 & -2 & -1 \\ 0 & 3 & -10 \\ 0 & 0 & 25 \end{vmatrix}}{\begin{vmatrix} 11 & -2 & -1 \\ 9 & 3 & -10 \\ -21 & 0 & 25 \end{vmatrix}} = \underline{3.03\text{V}}$$

4.26 Apply KVL to each mesh loop,

$i_1(3+6) - 6i_2 = 51$ or $9i_1 - 6i_2 = 51$

$-i_1 6 + i_2(6+12) = -6$ or $-6i_1 + 18i_2 = -6$

Adding 3 times the first equation

$21i_1 = 147$ or $i_1 = \underline{7\text{A}}$.

$i_2 = \frac{-6+6i_1}{18} = \underline{2\text{A}}$.

4.27 The mesh equations are

$i_1(2+3) - 3i_2 = 16-9$ or $5i_1 - 3i_2 = 7$

$-3i_1 + i_2(6+6+3) = 9$ or $-3i_1 + 15i_2 = 9$

Adding 5 times the first

$22i_1 = 44$ or $i_1 = \underline{2\text{A}}$.

$i_2 = \frac{9+3i_1}{15} = \underline{1\text{A}}$.

4.28 The second mesh equation is

$-3i_1 + i_2(6+3) = 9+12$ or $-3i_1 + 9i_2 = 21$

Adding 3 times the first of Prob 4.26

$12i_1 = 42$ or $i_1 = \underline{3.5\text{A}}$.

$i_2 = \frac{21+3i_1}{9} = \underline{3.5\text{A}}$.

4.29

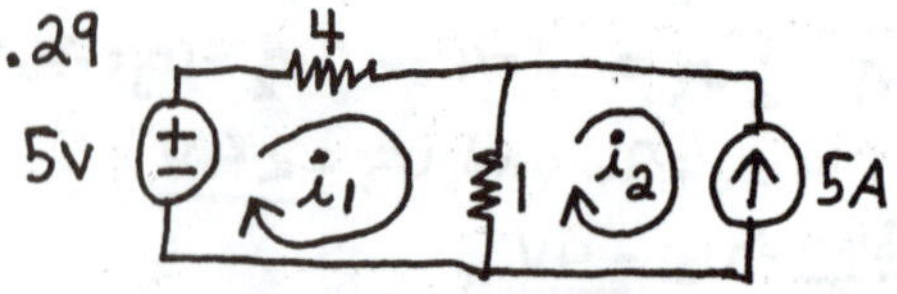

Current src: $i_2 = -5\text{A}$

Left mesh: $i_1(4+1) - i_2(1) = 5\text{ V}$

$\Rightarrow 5i_1 + 5 = 5 \Rightarrow i_1 = 0$

$\therefore \boxed{i = i_2 - i_1 = -5\text{A}}$

4.30

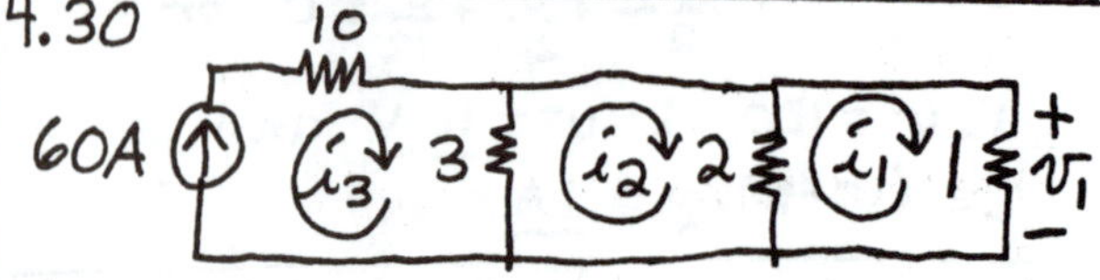

Current src: $i_3 = 60$

Center mesh: $2(i_2 - i_1) + 3(i_2 - 60) = 0$

Right mesh: $1i_1 + 2(i_1 - i_2) = 0$

Solving $\Rightarrow i_2 = \frac{3}{2}i_1 \Rightarrow i_1 = 32.\overline{72}\text{A}$

$\therefore \boxed{v_1 = i_1(1\Omega) = 32.\overline{72}\text{V}}$

4.31

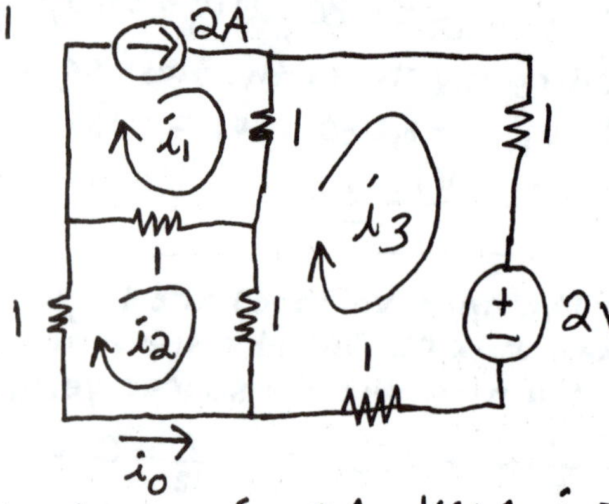

Current src: $i_1 = 2\text{A}$, KCL: $i_o = -i_2$

Mesh 2: $i_2(1+1+1) - (2\text{A})(1) - i_3 = 0$

$\Rightarrow i_3 = 3i_2 - 2$

Mesh 3: $i_3(1+1+1+1) - i_2(1) - 2\text{A}(1) + 2 = 0$

$\Rightarrow 4i_3 - i_2 = 0$

Solving $\Rightarrow i_2 = \frac{8}{11}\text{A} \Rightarrow \boxed{i_o = -\frac{8}{11}\text{A}}$

4.32

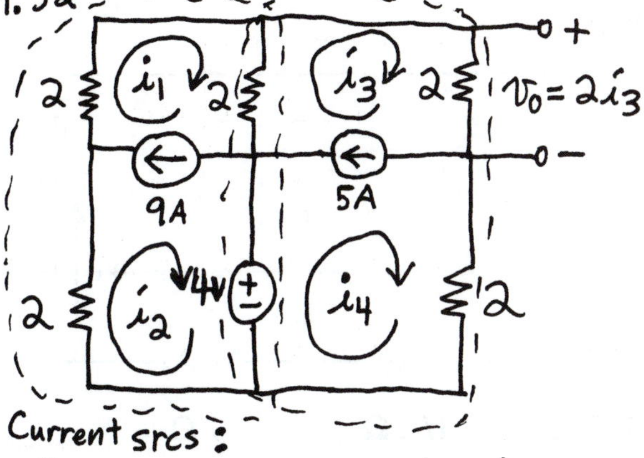

Current srcs:

$9A = i_1 - i_2$, $5A = i_3 - i_4$

Supermeshes (dashed lines):

$$i_2(2) + i_1(2+2) - i_3(2) + 4 = 0$$

$$\Rightarrow 2i_1 + i_2 - i_3 = -2$$

$$-2i_1 + i_3(2+2) + 2i_4 - 4 = 0$$

$$\Rightarrow 2i_3 + i_4 - i_1 = 2$$

Solving... $\Rightarrow i_3 = 3.5A$

$\therefore$ $\boxed{v_0 = 7.0 \text{ V}}$

4.33

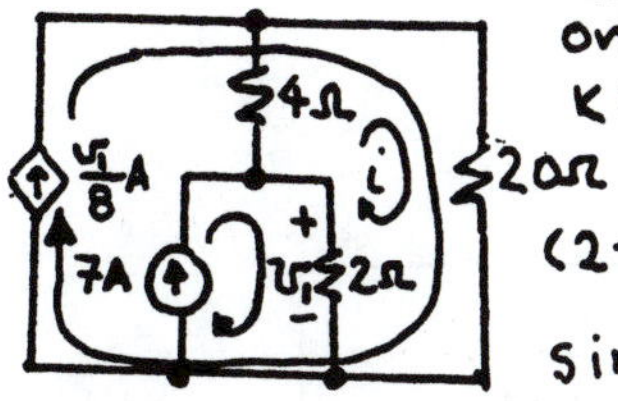

For the mesh with only resistors, KVL yeilds.

$(2+4+20)i - 7(2) + 20(\frac{v_1}{8}) = 0$

since $v_1 = 2(7-i)$

or $i = \frac{14-v_1}{2}$ then $\frac{26}{2}(14-v_1) + \frac{20}{8}v_1 = 14$

or $v_1 = \underline{16V}$.

4.34

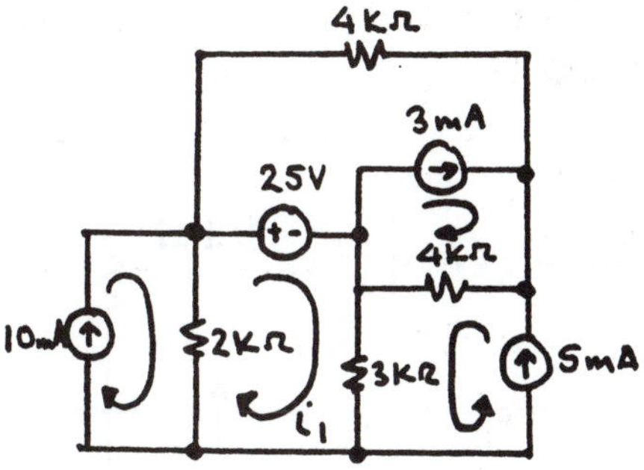

i_1 in mA , KVL for the mesh of i_1

$2(i_1 - 10) + 3(i_1 + 5) = -25$ then $i_1 = -4mA$

$i_{3k\Omega} = i_1 + 5mA = 1mA$,

$P_{3k\Omega} = i_{3k\Omega}^2 R = (3k\Omega)(1mA)^2 = \underline{3mW}$

4.35

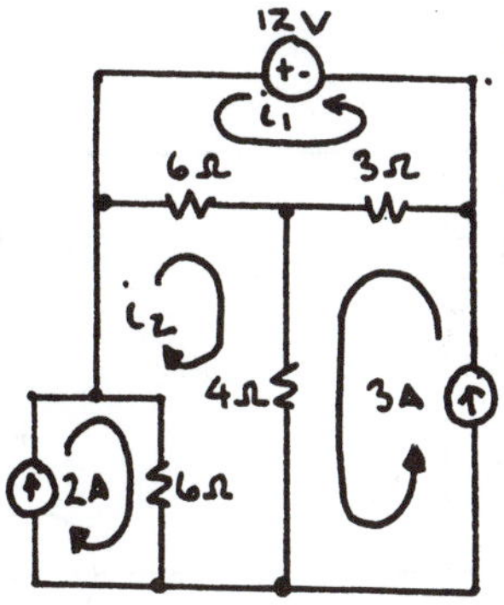

using mesh

$i_1(6+3) + i_2 6 - 3(3) = 12$

or $i_1 9 + i_2 6 = 21$

$i_2(6+4+6) + i_1 6 + 4(3) - 6(2) = 0$

or $i_1 6 + i_2 16 = 0$

$i_1 = \begin{vmatrix} 21 & 6 \\ 0 & 16 \end{vmatrix} \Big/ \begin{vmatrix} 9 & 6 \\ 6 & 16 \end{vmatrix}$

$= \frac{336}{108} = 3.111A$

$i_{3\Omega} = i_1 - 3A = \underline{1/9 A}$ to the right.

4.36

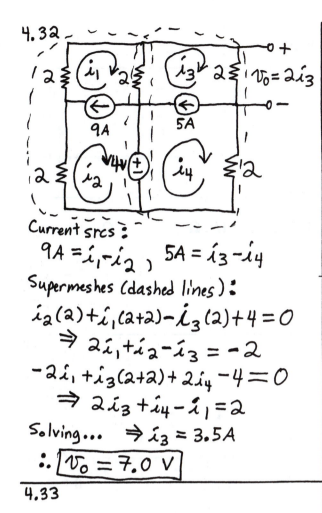

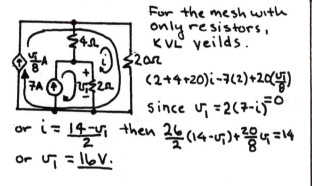

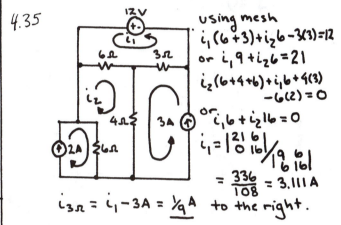

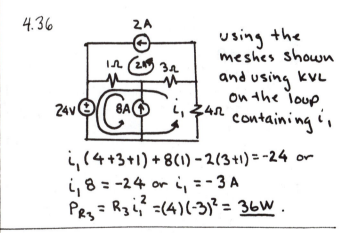

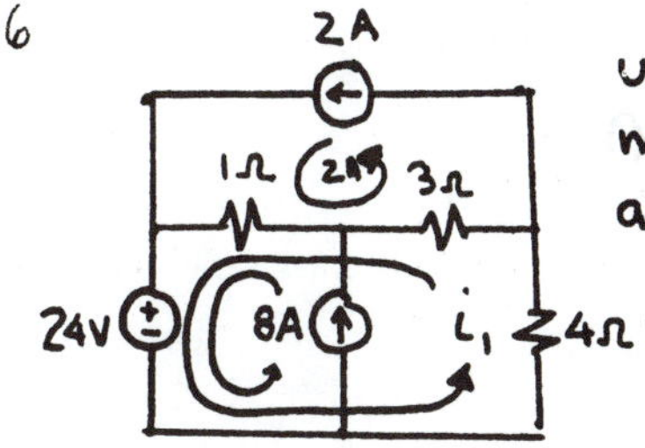

using the meshes shown and using KVL on the loop containing i_1

$i_1(4+3+1) + 8(1) - 2(3+1) = -24$ or

$i_1 8 = -24$ or $i_1 = -3A$

$P_{R_3} = R_3 i_1^2 = (4)(-3)^2 = \underline{36W}$.

4.37 Let i be the mesh current in the all resistor loop then by KVL in that loop.

$i(6+4+2) - 4(2) - 7(4) = 0$ or

$12i = 36$ $\therefore i = \underline{3A}$.

4.38

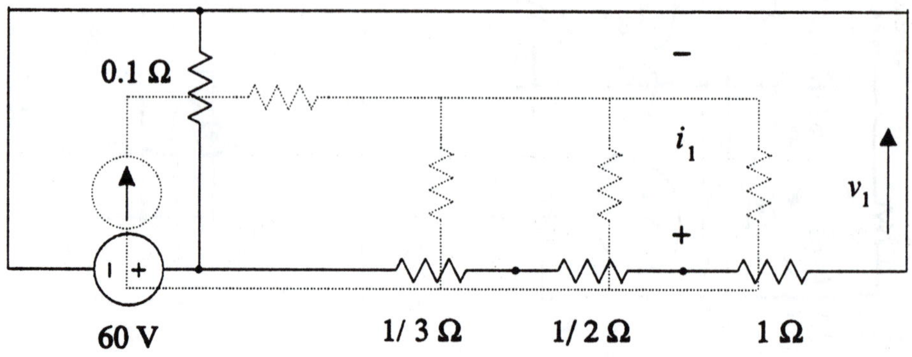
0.1 Ω
−
i_1
+
v_1
60 V
1/ 3 Ω
1/ 2 Ω
1 Ω

4.39

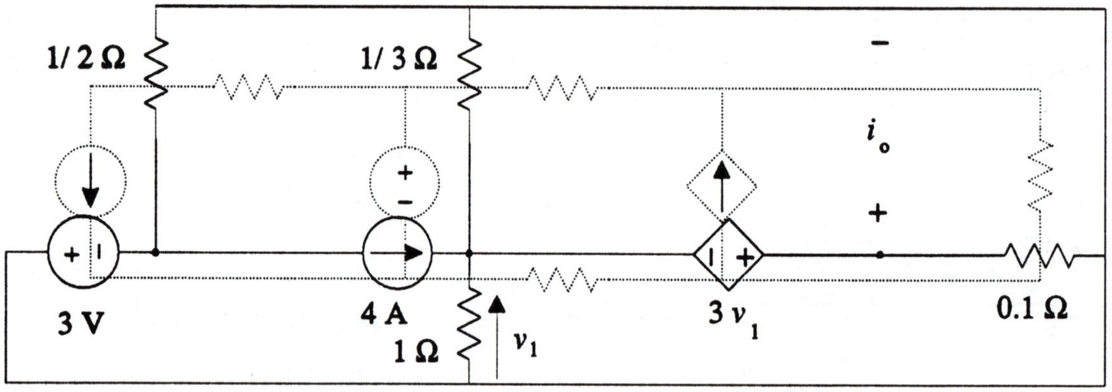
1/ 2 Ω
1/ 3 Ω
−
i_o
+
3 V
4 A
1 Ω
v_1
$3 v_1$
0.1 Ω

4.40

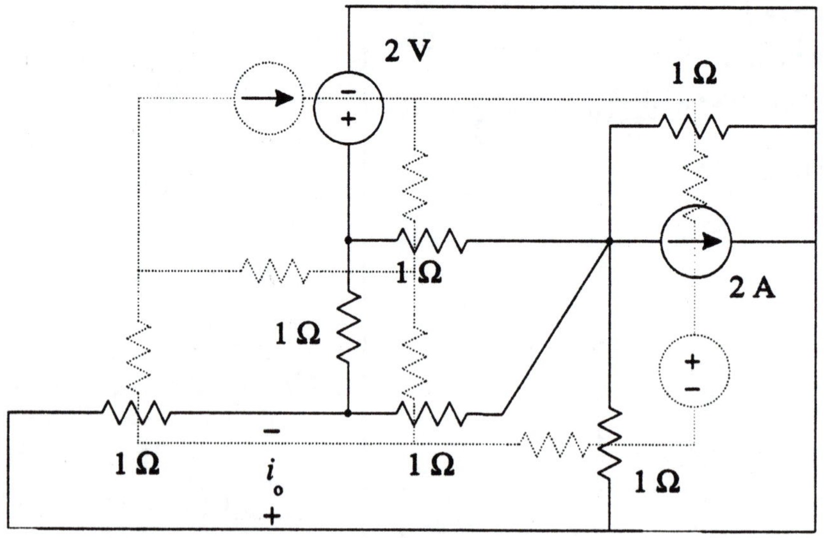
2 V
1 Ω
1 Ω
2 A
1 Ω
1 Ω
−
i_o
+
1 Ω
1 Ω

4.41

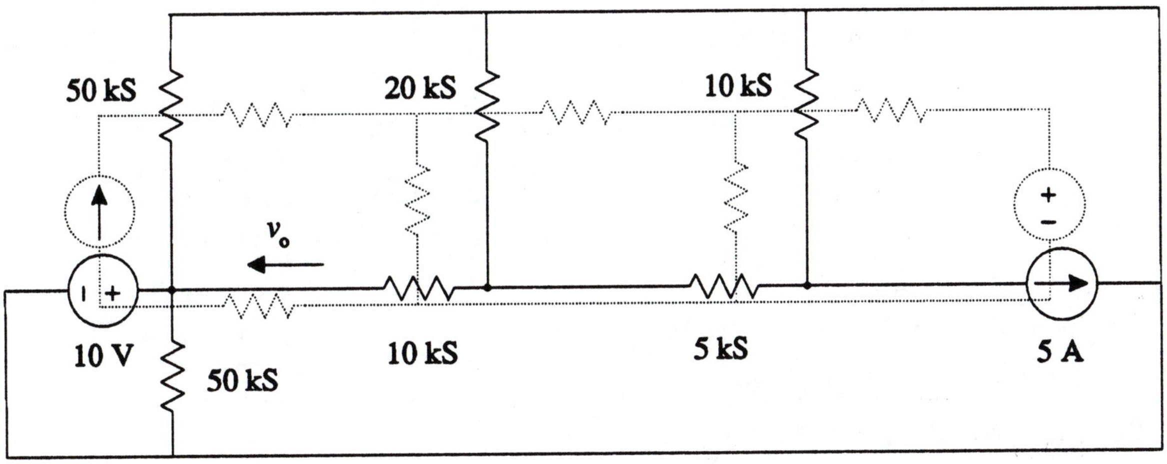
50 kS
20 kS
10 kS
v_o
10 V
50 kS
10 kS
5 kS
5 A

4.42

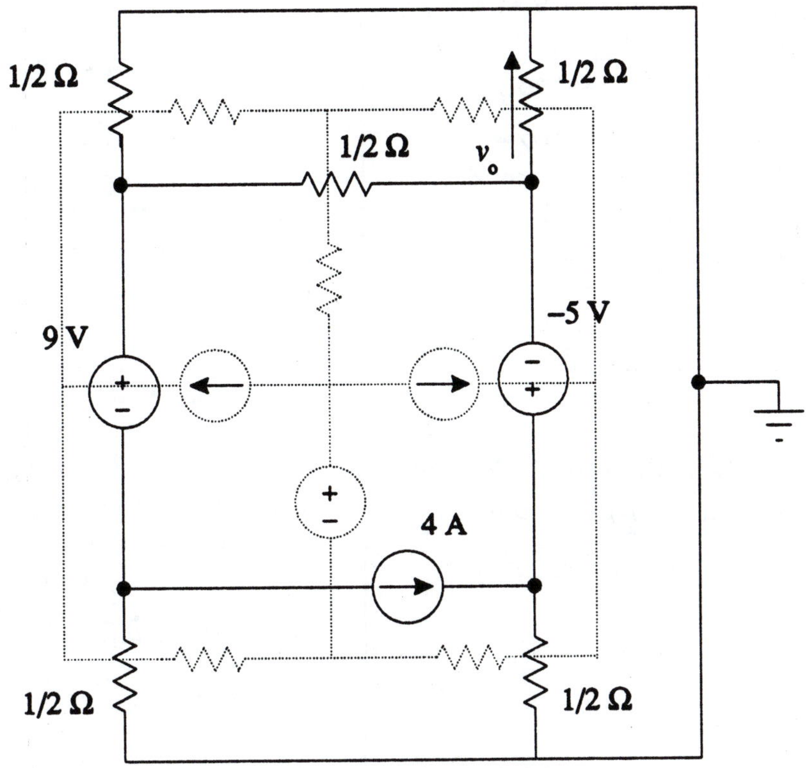
1/2 Ω
1/2 Ω
1/2 Ω
v_o
9 V
−5 V
4 A
1/2 Ω
1/2 Ω

4.43 v_o = node voltage at top of 3-Ω resistor. Voltages between input terminals of op-amp is zero. Node analysis at (+) terminal

$$\frac{3\cos 4t}{2\,k\Omega} + \frac{3\cos 4t - v_o}{8\,k\Omega} = 0$$

$v_o = 15\cos 4t$ V, $i = v_o/3\,k\Omega$

$i =$ **5 cos 4t mA**

4.44 Using Node analysis

$v_1\left(\frac{1}{2}+1\right) = 6\cos 2t$ or $v_1 = 4\cos 2t$

$v_1\left(\frac{1}{2}+\frac{1}{2}\right) - \frac{v}{2} = \frac{6}{2}\cos 2t$ or

$2v_1 - v = 6\cos 2t$, $v =$ **$2\cos 2t$ V**

4.45 Using node analysis at v_3 & 0V

$-v_1\left(\frac{1}{4}\right) - v_2(2) + v_3\left(\frac{1}{4}+\frac{1}{2}+1+2\right) = 0$ or

$-8v_2 + 15v_3 = v_1 = 8\cos 1000t$,

$-v_2\left(\frac{1}{2}\right) - v_3(1) + 0 = 0$ or $-v_2 - 2v_3 = 0$

Adding 15/2 times the second

$-15.5\,v_2 = 8\cos 1000t$ ∴

$v_2 =$ **$0.516\cos 1000t$ V**

4.46 By KVL, $v_1 = v_g = 8$V, and

$v_1\left(1+\frac{1}{2}\right) - v_3\left(\frac{1}{2}\right) = 0$ or

$v_3 = 3v_1 = 3(8) =$ **24V**.

4.47 Let v_1 be the output of the first op amp and v_2 be the output of the second. The voltage at the noninverting input of the voltage follower is also v_2. KCL at inverting input of first op amp yeilds (currents in mA)

$\frac{v_g}{1} + \frac{v_1}{10} + \frac{v_2}{5}$ or $v_1 + 2v_2 = -(10)(5)$

KCL at noninverting input of voltage follower yields.

$v_2\left(1+\frac{1}{2}\right) - \frac{v_1}{2} = 0$ or $-v_1 + 3v_2 = 0$

∴ $5v_2 = -50$ or $v_2 = -10$V

$i_{5k\Omega} = \frac{v_2}{5} =$ **-2mA** to the left.

4.48 Let v_3 be the node voltage of the op amp input terminals (the input voltage is zero.) KCL at these inputs yeilds

non inverting: $v_3\left(\frac{1}{1}+\frac{1}{3.3}\right) - \frac{1}{1}v_2 = 0$

inverting: $v_3\left(\frac{1}{1.2}+\frac{1}{4.7}\right) - \frac{1}{1.2}v_1 - \frac{1}{4.7}v_0 = 0$

$v_3 = \frac{3.3}{1+3.3}v_2 = 1.535\cos 377t$ V

$v_0 = -\frac{4.7}{1.2}v_1 + \frac{1.2+4.7}{(1.2)}v_3$

$=$ **$-15.67 + 7.547\cos 377t$ V**

4.49 Let v_o be the op amp output voltage. Then

$v_o = -8/2\,(12\cos 2t) = -48\cos 2t$ V

By voltage division

$12i = \frac{(12)(6)/(12+6)}{4+4}v_o$, $i = \frac{v_o}{(2)(12)}$

∴ **$i = -2\cos 2t$ A**

Problem 4.50:

```
SPICE CODE FOR PROB 4.50 (PROB 4.2)
* ELEMENT STATEMENTS
I1 0 1 DC 60
R1 1 2 10OHM
R2 2 0 3OHM
R3 2 0 2OHM
R4 2 0 1OHM
* CONTROL STATEMENTS
.PRINT DC V(2)
* ANALYSIS TYPE DEFAULTS TO DC
.END
```

SPICE Output:

```
*****************************************************************************
 NODE   VOLTAGE

(    2)   32.7270
```

Problem 4.51:

```
SPICE CODE FOR PROB 4.51 (PROB 4.8)
* ELEMENT STATEMENTS
I1 2 0 DC 5
V2 0 3 DC 4
I2 0 4 DC 9
R1 1 2 2OHM
R2 2 3 2OHM
R3 3 4 2OHM
R4 4 1 2OHM
R5 1 0 2OHM
* CONTROL STATEMENTS
.PRINT DC V(1,2)
* ANALYSIS TYPE DEFAULTS TO DC
.END
```

SPICE Output:

```
*****************************************************************************
NODE   VOLTAGE      NODE   VOLTAGE      NODE   VOLTAGE      NODE   VOLTAGE

(    1)    0.0000  (    2)   -7.0000  (    3)   -4.0000  (    4)    7.0000
```

Thus, $v_o = v_1 - v_2 = +7.0$ V.

Problem 4.52:

```
SPICE CODE FOR PROB 4.52 (PROB 4.20)
* ELEMENT STATEMENTS
V1 1 0 DC 18
R1 2 1 2OHM
R2 2 4 4OHM
VDUMMY 4 0 0V
R3 2 3 1OHM
R4 3 0 5OHM
F1 2 3 VDUMMY 5
* CONTROL STATEMENTS
.PRINT DC I(VDUMMY)
* ANALYSIS TYPE DEFAULTS TO DC
.END
```

SPICE Output:

```
*******************************************************************************
VOLTAGE SOURCE CURRENTS
    NAME          CURRENT

    VDUMMY        2.000E+00
```

Problem 4.53:

```
SPICE CODE FOR PROB 4.53 (PROB 4.24)
* ELEMENT STATEMENTS
VG 4 0 DC 4
R1 4 1 0.5
R2 1 2 2
R3 1 3 4
R4 3 0 1
R5 2 0 1
G1 2 3 1 3 5
* CONTROL STATEMENTS
.PRINT DC V(2)
* ANALYSIS TYPE DEFAULTS TO DC
.END
```

SPICE Output:

```
*******************************************************************************
   NODE   VOLTAGE

(    2)    -.6061
```

Problem 4.54:

```
SPICE CODE FOR PROB 4.54 (PROB 4.44)
* ELEMENT STATEMENTS:
* SINCE WE ONLY HAVE RESISTORS,
* ONLY THE AMPLITUDE WILL BE CHANGED.
* THEREFORE USE DC ANALYSIS.
VG 1 0 DC 6
R1 1 2 2OHM
RF 2 4 2OHM
R2 1 3 1OHM
RA 3 0 2OHM
XOP 3 2 4 OPAMP
* CONTROL STATEMENTS
.PRINT DC V(4)
* ANALYSIS TYPE DEFAULTS TO DC
* ******* OPAMP SUBCIRCUIT ********
* INPUTS: NONINVERT (1), INVERT (2)
* OUTPUT: (3)
* **********************************
.SUBCKT OPAMP 1 2 3
RI 1 2 1MEG
E1 4 0 1 2 1E5
RO 4 3 30OHM
.ENDS OPAMP
.END
```

SPICE Output:

```
*******************************************************************************
 NODE   VOLTAGE

(    4)    2.0006
```

Problem 4.55:

```
SPICE CODE FOR PROB 4.55 (PROB 4.47)
* ELEMENT STATEMENTS:
* SINCE WE ONLY HAVE RESISTORS,
* ONLY THE AMPLITUDE WILL BE CHANGED.
* THEREFORE USE DC ANALYSIS.
VG 1 0 DC 5
R1 1 2 1KOHM
RF 2 3 10KOHM
R2 2 6 5KOHM
VDUMMY 5 6 0VOLTS
R3 3 4 2KOHM
R4 4 0 1KOHM
ROUT 5 0 4KOHM
XOP1 0 2 3 OPAMP
XOP2 4 5 5 OPAMP
* CONTROL STATEMENTS
.PRINT DC I(VDUMMY)
* ANALYSIS TYPE DEFAULTS TO DC
* ******* OPAMP SUBCIRCUIT ********
* INPUTS: NONINVERT (1), INVERT (2)
* OUTPUT: (3)
* *********************************
.SUBCKT OPAMP 1 2 3
RI 1 2 1MEG
E1 4 0 1 2 1E5
RO 4 3 30OHM
.ENDS OPAMP
.END
```

SPICE Output:

```
******************************************************************************
VOLTAGE SOURCE CURRENTS
    NAME         CURRENT

    VDUMMY       -2.000E-03
```

Chapter 5

Energy Storage Elements

5.1 Capacitors

5.1 A constant current of 10μA is charging a 5μF capacitor. If the capacitor was previously uncharged, find the charge and voltage on it after 20s.

5.2 A 10 μF capacitor has a voltage $v(t) = f(t)$ V, as shown. Find the current i flowing through the capacitor at $t = -7, -3, 1, 3,$ and 7 ms.

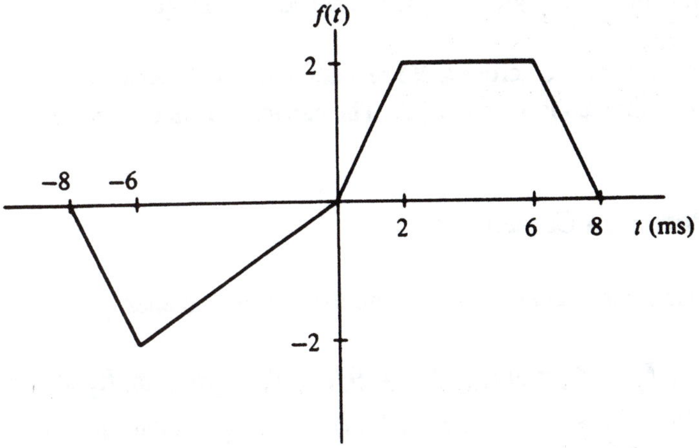

PROBLEM 5.2

5.3 If $f(t)$ in Prob. 5.2 is the current in mA flowing in a 1μF capacitor, find its voltage at $t = -6, 0, 2,$ and 8 ms.

5.4 A 0.1μF capacitor has a current of $3\cos(2000t)$ mA. Find its voltage $v(t)$ if $v(0) = -15$ V.

5.5 Find the charge residing on each plate of a 2μF capacitor that is charged to 50V. If the same charge resides on a 1μF capacitor, what is the voltage?

5.6 Determine the voltage required to store 100μC on a 2μF capacitor. What time will be required for a constant current of 50mA to deliver this charge?

5.7 The voltage across a $10\mu F$ capacitor is as shown. Find the current i at $t = 5ms$, and the power p at $t = 20ms$.

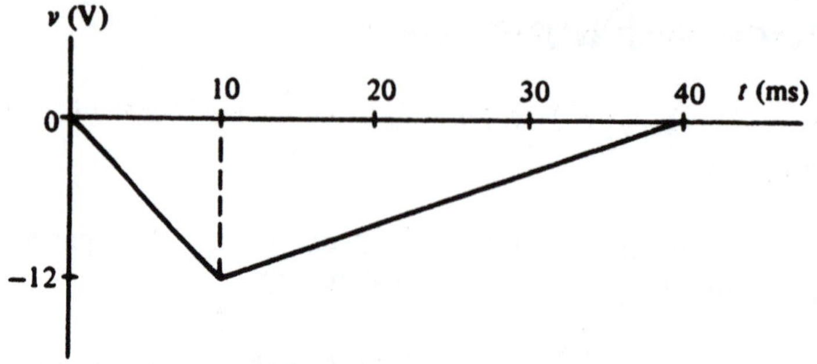

PROBLEM 5.7

5.8 How long will it take for a constant 25mA current to deliver a charge of $100\mu C$ to a $1\mu F$ capacitor? What will then be the capacitor voltage?

5.9 Find the voltage for $t > 0$ across a 2Ω resistor and a 0.5F capacitor in series if the current entering their positive terminals is $4\cos(2t)$ A. The capacitor is uncharged at $t = 0$.

5.2 Energy Storage in Capacitors

5.10 A $10\mu F$ capacitor is charged to 20V. Find the charge and energy.

5.11 Let $C = ⅓$F, $R_1 = R_2 = 3\Omega$, and $V = 9$ V. If the current in R_2 at $t = 0^-$ is 1A directed downward, find at $t = 0^-$ and at $t = 0^+$ (a) the charge on the capacitor, (b) the current in R_1 directed to the right, and (c) the current in C directed downward.

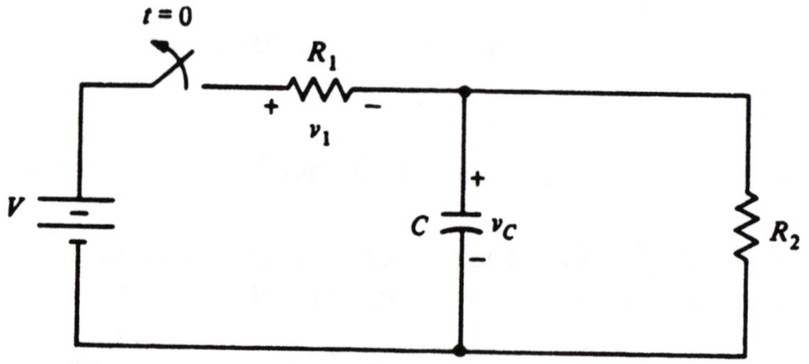

PROBLEM 5.11

5.12 The current entering the positive terminal of a 0.25F capacitor is $i = 6\cos(2t)$ A. If the initial capacitor voltage (at $t = 0$) is 4V, find the energy stored in the capacitor at $t = 3\pi/4$ s.

5.13 The current in a 0.5 F capacitor is $i = 6t$ A and the voltage at $t = 0$ is $v(0) = 2$ V. Find the energy stored in the capacitor at $t = 1$ s.

5.14 Find the work required to charge a 0.1μF capacitor to 200V.

5.15 A voltage of $4e^{-t}$ V appears across a parallel combination of a 1Ω resistor and a 0.25F capacitor. Find the power absorbed by the combination.

5.16 Find $i_1(0^-)$, $i_1(0^+)$, $i_2(0^-)$, and $i_2(0^+)$ if the switch is opened at $t = 0$, and $v_1(0^-) = 18\,\text{V}$ and $v_2(0^-) = 6\,\text{V}$.

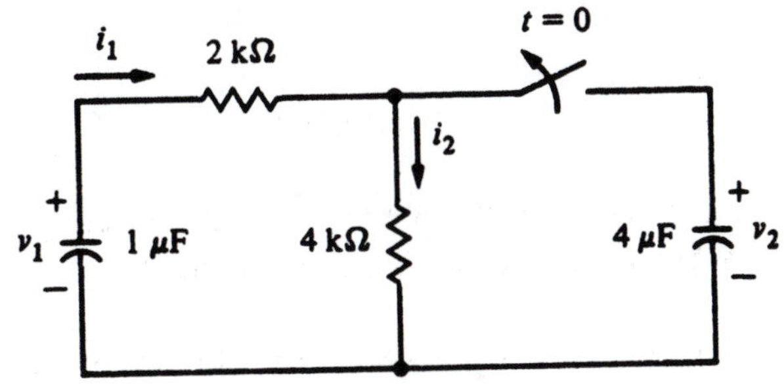

PROBLEM 5.16

5.3 Series and Parallel Capacitors

5.17 Find the equivalent capacitance and the initial voltage.

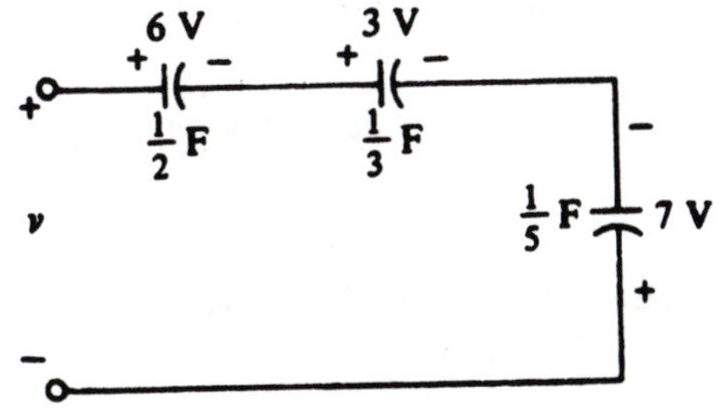

PROBLEM 5.17

5.18 Find the equivalent capacitance.

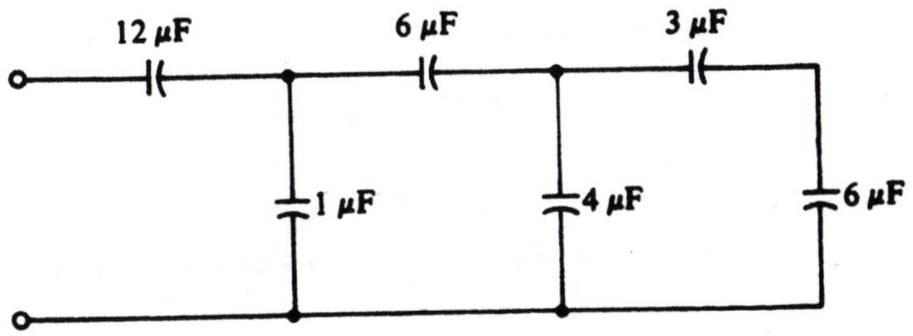

PROBLEM 5.18

5.19 The capacitances shown are all in μF. Find the equivalent capacitance at terminals *a*-*b*.

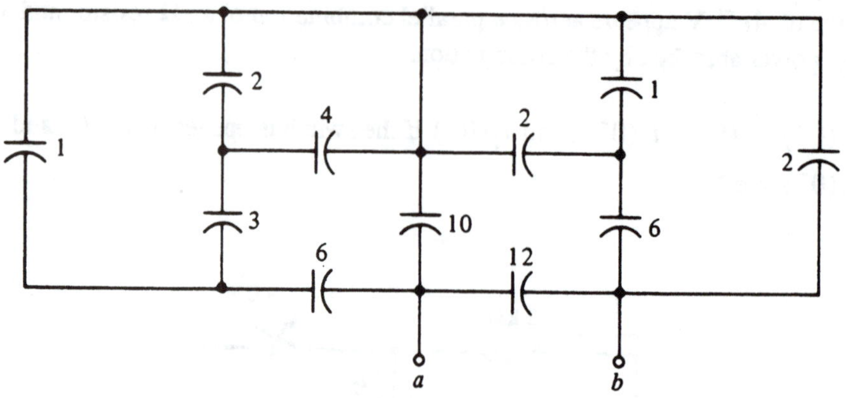

PROBLEM 5.19

5.4 Inductors

5.20 If a 600mH inductor has $i(t) = f(t)$ mA, where $f(t)$ is given in Prob. 5.2, find its voltage at $t = -3$, 1, 3, and 7ms.

5.21 If a 1mH inductor has $v(t) = f(t)$ mV, where $f(t)$ is given in Prob. 5.2, find its current at $t = -6$, 0, and 8ms if the current at –8ms is 0.

5.22 A 100mH inductor has a terminal current of $100\sin(100t)$ mA. Find $v(t)$.

5.23 Find the current for $t > 0$ in a 100mH inductor having a terminal voltage of $10e^{-100t}$ V if $i(0) = 0$ A.

5.24 Find the flux linkage of a 50mH inductor at 10ms, 40ms, and 80ms if the current is $f(t)$ mA, as shown.

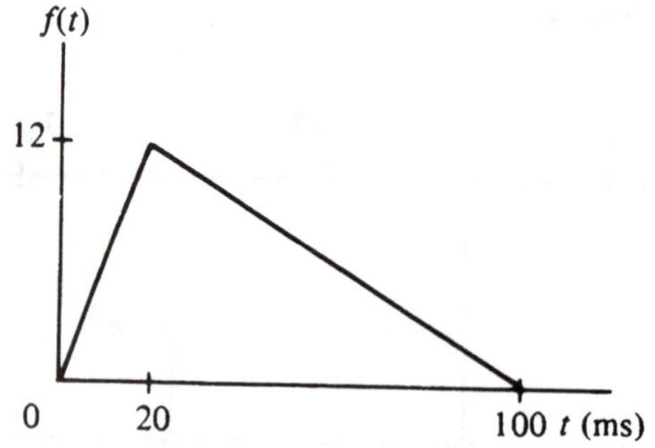

PROBLEM 5.24

5.25 Find the voltage across a 10H inductor at $t = 10$ ms, 40 ms, and 80 ms if its current is $f(t)$ mA, given in Prob. 5.24.

5.26 Find the current i through a 120H inductor at $t = 20$ ms, 60 ms, and 100 ms if its voltage is $f(t)$ V, given in Prob. 5.24 and $i(0) = 2$ mA.

5.27 Find the terminal voltage of a 100mH inductor if its current is (a) 5A, (b) $100t$ A, (c) $10\sin(200t)$ A, and (d) $10e^{-50t}$ A.

5.28 Find the current $i(t)$ of a 10mH inductor with a voltage of $2\sin(1000t)$ V, if $i(0) = 0$.

5.5 Energy Storage in Inductors

5.29 A 5mH inductor has a current of 200mA. Find the flux linkage and the energy.

5.30 Let $I = 3$ A, $R_1 = 6\Omega$, $R_2 = 12\Omega$, $L = 2$H, and $i_1(0^-) = 2$ A. If the switch is open at $t = 0^-$, find $i_L(0^-)$, $i_L(0^+)$, $i_1(0^+)$, and $di_L(0^+)/dt$.

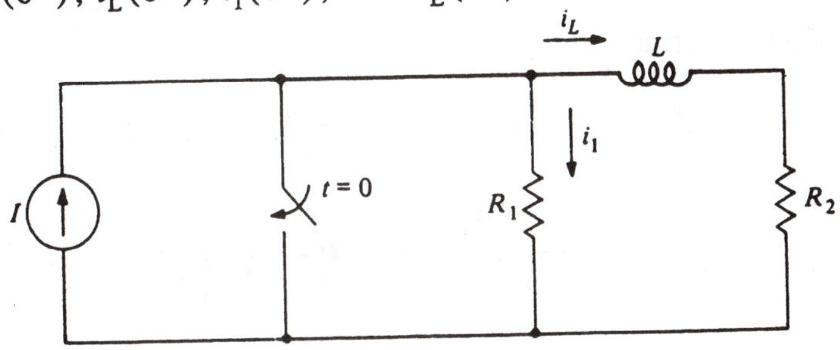

PROBLEM 5.30

5.31 The voltage of a 0.5H inductor is $v = 6t$ V and the initial current is $i(0) = 2$ A. Find the energy stored in the inductor at $t = 1$ s.

5.32 The voltage of a 0.25H inductor is $v = 6\cos(2t)$ V and the current at $t = 0$ is 0. Find the energy stored in the inductor at $3\pi/4$ s and the power delivered to the inductor at $\pi/8$ s.

5.33 Find the work required to establish a current of 60mA in a 100mH inductor.

5.34 If $v(0^-) = 9\,\text{V}$, $i(0^-) = 1\,\text{A}$, and the switch is opened at $t = 0$, find $i(0^+)$ and $di(0^+)/dt$.

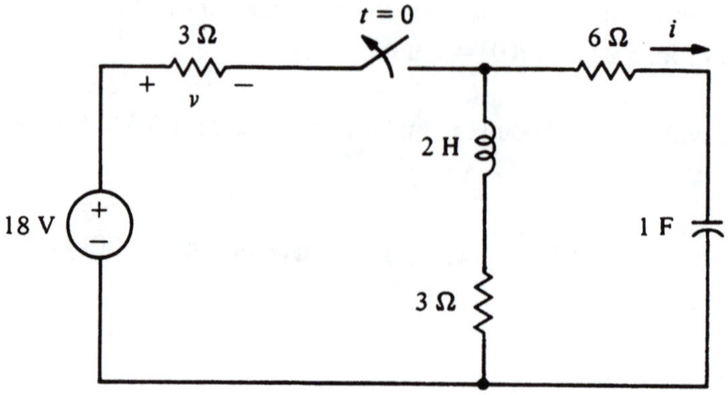

PROBLEM 5.34

5.6 Series and Parallel Inductors

5.35 Find the equivalent inductance.

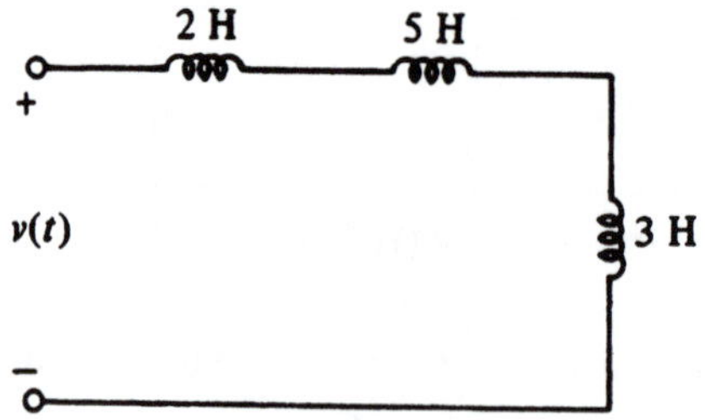

PROBLEM 5.35

5.36 Find the equivalent inductance.

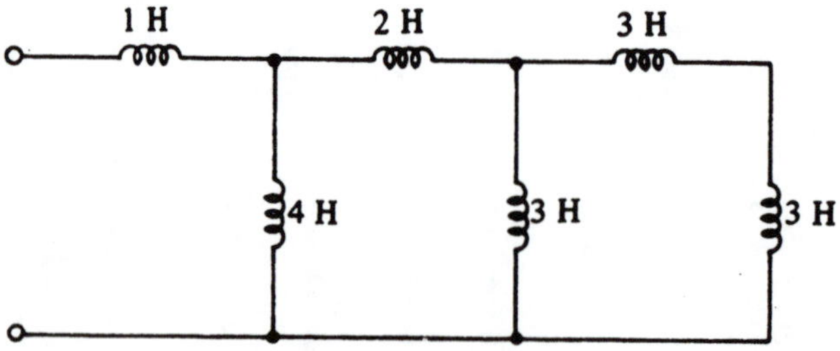

PROBLEM 5.36

5.37 Determine L_{eq}.

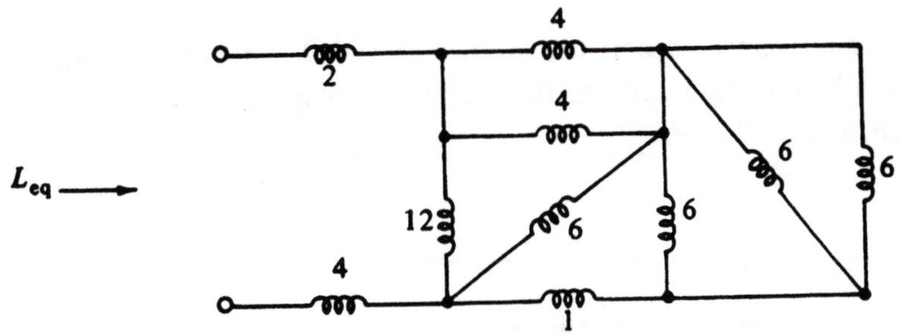

PROBLEM 5.37

5.7 DC Steady State

5.38 Find v_o.

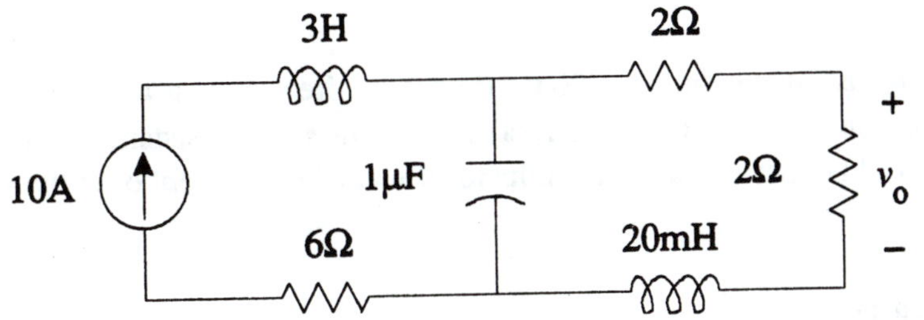

PROBLEM 5.38

5.39 Find v_o and i_o.

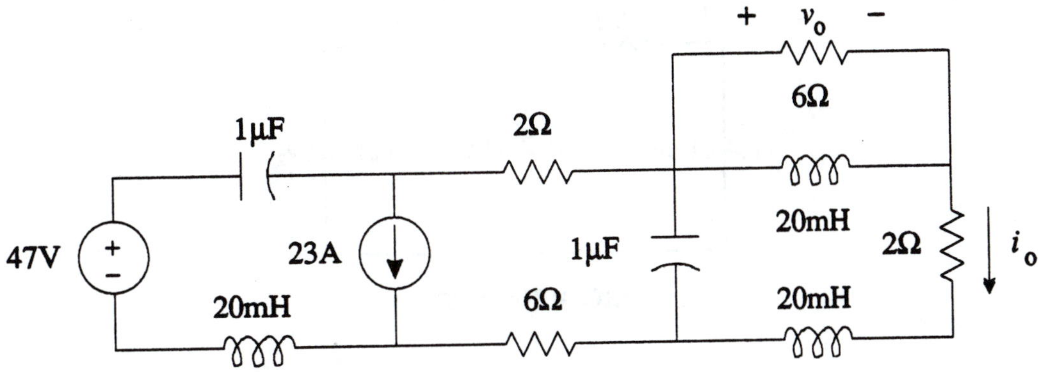

PROBLEM 5.39

5.8 Practical Capacitors and Inductors

5.40 Mylar capacitors have a resistance-capacitance product of $10^5\,\Omega$. Find the equivalent parallel resistor in the following figure for the following mylar capacitors: (a) 33pF, (b) 0.001μF, and (c) 2.7μF.

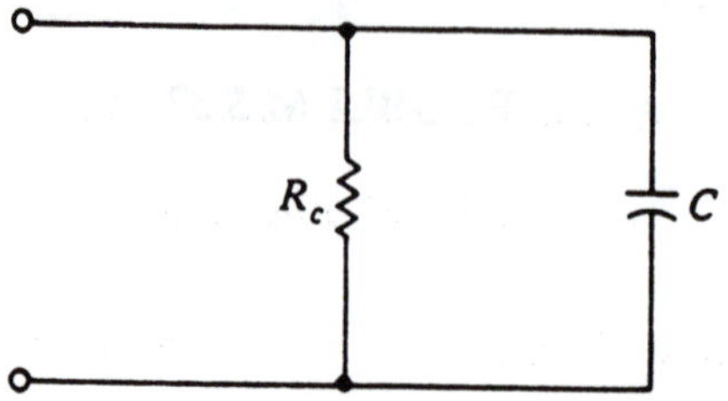

PROBLEM 5.40

5.41 An inductor is created by winding 50,000 turns of thin wire around a 1cm x 1cm square form. If the wire has a resistance of 2.0 Ω/meter, (a) what is the series resistance of the inductor and (b) what power will be dissipated by the inductor if it carries a current of 100mA dc?

5.9 Singular Circuits

5.42 Find $v_1(0^+)$, $v_2(0^+)$, $i(0^+)$, and $i(0^-)$ if the switch is closed at $t = 0$, and $v_1(0^-) = 14$ V and $v_2(0^-) = 6$ V.

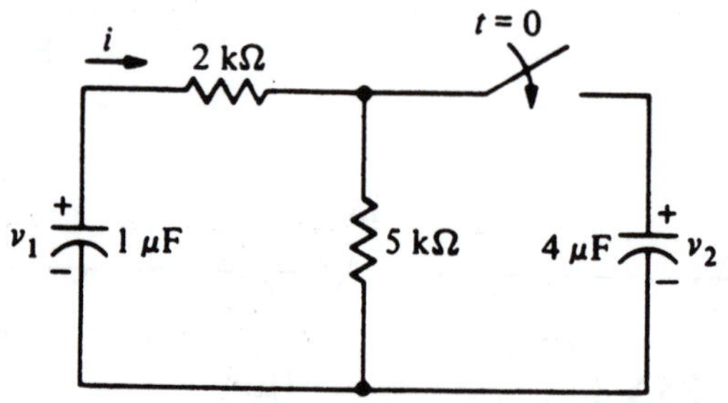

PROBLEM 5.42

5.43 If $v(0^-) = 6$ V and the switch is opened at $t = 0$, find $i_1(0^-)$, $i_2(0^-)$, $i_1(0^+)$, and $i_2(0^+)$.

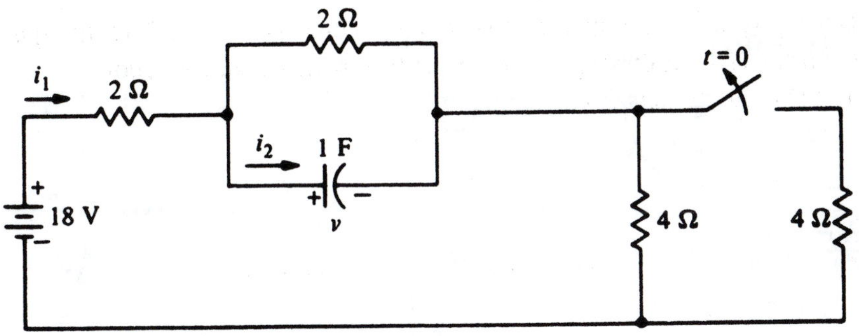

PROBLEM 5.43

5.44 If $i_1(0^-) = 1$ A and the switch is opened at $t = 0$, find $i_1(0^+)$, $i_2(0^-)$, $i_2(0^+)$, $v(0^-)$, and $v(0^+)$.

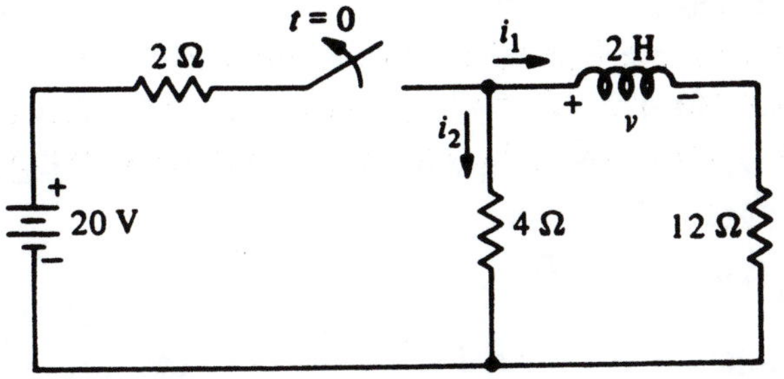

PROBLEM 5.44

5.45 If $v_1(0^-) = 9$ V, $i_1(0^-) = 3$ A, and the switch is opened at $t = 0$, find $v_1(0^+)$, $i_1(0^+)$, $i_2(0^-)$, $i_2(0^+)$, $v_2(0^-)$, and $v_2(0^+)$.

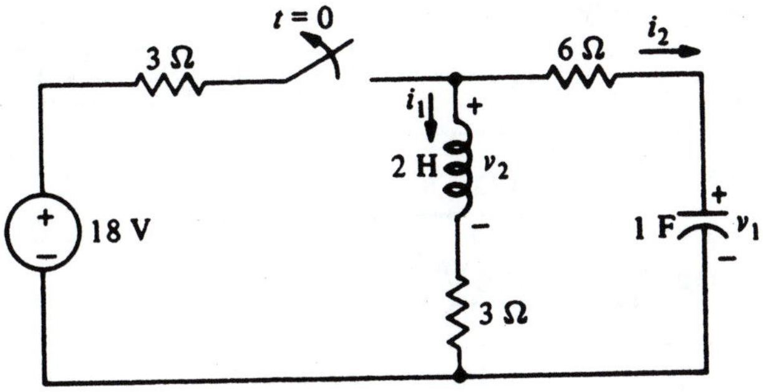

PROBLEM 5.45

CHAPTER 5 SOLUTIONS

5.1 $q = \int_0^{20} i\,dt + q(0) = \int_0^{20} 10^{-5} dt$

$v = q/C = \frac{200\times10^{-6}}{5\times10^{-6}} = 40V$. $= 10^{-5}(20) = 200\mu C$

5.2 $i = C\frac{dv}{dt}$; $\frac{dv}{dt}$ is the slope of graph.

$t = -7ms: i = (10^{-5})\frac{0-2}{[-6-(-8)]10^{-3}} = -10mA$

$t = -3ms: i = (10^{-5})\frac{0+2}{[0+6]10^{-3}} = 3.33mA$

$t = 1ms: i = (10^{-5})\frac{2-0}{[2-0]10^{-3}} = 10mA$

$t = 3ms: i = (10^{-5})(0) = 0mA$

$t = 7ms: i = (10^{-5})\frac{0-2}{[8-6]10^{-3}} = -10mA$.

5.3 $v = \frac{1}{C}\int_{-\infty}^{t} i\,dt + v(t_0)$;

area under the graph is the integral and $v(t_0) = 0$ at less than $-8ms$.

$v(-6ms) = \frac{1}{10^{-6}}\left[\frac{1}{2}(2ms)(-2mA)\right] + 0 = -2V$.

$v(0) = \frac{1}{10^{-6}}\left[\frac{1}{2}(6ms)(2mA)\right] - 2 = -8V$.

$v(2ms) = \frac{1}{10^{-6}}\left[\frac{1}{2}(2ms)(2mA)\right] - 8 = -6V$.

$v(8ms) = \frac{10^{-6}}{10^{-6}}\left[(2)(4) + \frac{1}{2}(2)(2)\right] - 6 = 4V$.

5.4 $v = \frac{1}{C}\int_0^t i\,dt + v(0)$

$= \frac{1}{10^{-7}}\int_0^t [3\cos 2000t]\times10^{-3}dt - 15$

$= 15\sin(2000t) - 15\ V$

5.5 $q = Cv = (2\times10^{-6})(50) = 100\mu C$ on each plate ($\pm$)

$v = q/C = 100\times10^{-6}/10^{-6} = 100V$.

5.6 $v = q/C = 100\times10^{-6}/2\times10^{-6} = 50V$

$q = \int_{t_0}^{t_1} i\,dt = i(t_1 - t_0)$

$\therefore (t_1 - t_0) =$ time to deliver charge

$= q/i = \frac{100\times10^{-6}}{50\times10^{-3}} = 2ms$

5.7 $i = C\frac{dv}{dt}$; $\frac{dv}{dt}$ is the slope of graph

$i(5ms) = (10^{-5})\left[\frac{-12-0}{(10-0)10^{-3}}\right] = -12mA$.

$i(20ms) = (10^{-5})\left[\frac{0+12}{(40-10)10^{-3}}\right] = 4mA$

$v(20ms) = \frac{2}{3}(12) = 8V$; $p = vi = 32mW$

5.8 $q = it$ or $t = \frac{100\times10^{-6}}{25\times10^{-3}} = 4ms$

5.9 $v = Ri + \frac{1}{C}\int_0^t i\,dt + v(0)$, $v(0) = 0$

$v = 2(4\cos 2t) + \frac{1}{0.5}\int_0^t 4\cos 2t\,dt$

$= 8\cos 2t + 4\sin 2t\ V$

5.10 $q = Cv = (10^{-5})(20) = 200\mu C$

$W = \frac{1}{2}Cv^2 = \frac{10^{-5}}{2}(20)^2 = 2mJ$.

5.11 $v_C(0-) = IR_2 = 1(3) = 3V = v_C(0+)$

(a) $q(0^-) = q(0^+) = Cv(0^-) = \frac{1}{3}(3) = 1C$

(b) $i_{R_1}(0-) = \frac{v_1(0^-)}{R_1} = \frac{V - v_C(0-)}{R_1} = \frac{9-3}{3} = 2A$

$i_{R_1}(0+) = 0$ since the circuit is open

(c) $i_C(0-) = i_{R_1}(0-) - \frac{v_C(0-)}{R_2} = 2 - \frac{3}{3} = 1A$

$i_C(0+) = i_{R_1}(0+) - \frac{v_C(0+)}{R_2} = 0 - \frac{3}{3} = -1A$

5.12 $v = \frac{1}{C}\int_0^{3\pi/4} i\,dt + v(0)$

$= \frac{1}{0.25}\int_0^{3\pi/4} 6\cos 2t\,dt + 4 = -12 + 4 = -8V$

$W = \frac{1}{2}Cv^2 = \frac{1}{2}(.25)(-8)^2 = 8J$

5.13 $v = \frac{1}{C}\int_0^1 i\,dt + v(0)$

$= \frac{1}{0.5}\int_0^1 6t\,dt + 2 = 6 + 2 = 8V$

$W = \frac{1}{2}Cv^2 = \frac{1}{2}(0.5)(8)^2 = 16J$

5.14 $W = \frac{1}{2}Cv^2 = \frac{1}{2}(10^{-7})(200)^2 = 2mJ$

5.15

$$i = \frac{v}{R} + C\frac{dv}{dt} = \frac{4e^{-t}}{1} + \frac{1}{4}\frac{d}{dt}(4e^{-t})$$
$$= 3e^{-t}\ \text{A}$$
$$p = vi = (4e^{-t})(3e^{-t}) = \underline{12e^{-t}\ \text{W}}.$$

5.16

$v_1(0^+) = v_1(0^-) = 18\text{V}$;

$v_2(0^+) = v_2(0^-) = 6\text{V}$;

$$i_2(0^-) = \frac{v_2(0^-)}{4\text{k}\Omega} = \underline{1.5\text{mA}}$$

$$i_1(0^-) = \frac{v_1(0^-) - v_2(0^-)}{2\text{k}\Omega} = \underline{6\text{mA}}$$

At $t = 0+$, the switch is open and $i_1(0+) = i_2(0+)$.

$$\therefore\ i_1(0^+) = i_2(0^+) = \frac{v_1(0^+)}{(2+4)\text{k}\Omega} = \underline{3\text{mA}}$$

5.17

$$\frac{1}{C_s} = \frac{1}{\ } + \frac{1}{1/3} + \frac{1}{1/5} \text{ or } C_s = \underline{.1\ \text{F}}$$

$$v(t_0) = 6 + 3 - 7 = \underline{2\text{V}}$$

5.18

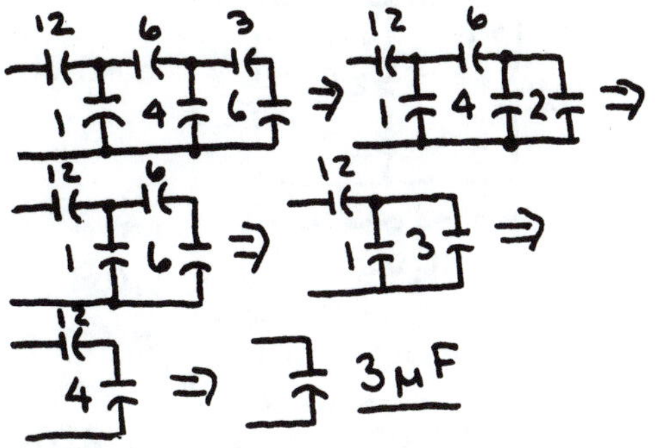

5.19

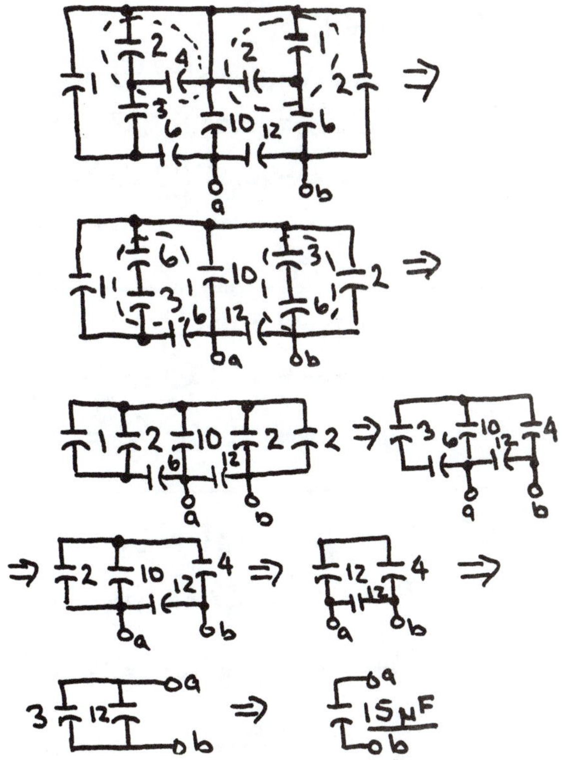

5.20

$v = L\frac{di}{dt}$; slope of graph equals $\frac{di}{dt}$

$$v(-3\text{ms}) = (600\times10^{-3})\frac{(0+2)}{(0+6)} = \underline{0.2\text{V}}$$

$$v(1\text{ms}) = (600\times10^{-3})\left[\frac{2-0}{2-0}\right] = \underline{0.6\text{V}}$$

$$v(3\text{ms}) = (600\times10^{-3})(0) = \underline{0\ \text{V}}$$

$$v(7\text{ms}) = (600\times10^{-3})\left[\frac{0-2}{8-6}\right] = \underline{-0.6\text{V}}$$

5.21

$$i = \frac{1}{L}\int_{-\infty}^{t} v\,dt + i_0,$$

$$i(-6) = \frac{1}{10^{-3}}\left[-\frac{1}{2}(2\text{mV})(2\text{ms})\right] + 0 = \underline{-2\text{mA}}$$

$$i(0) = \frac{1}{10^{-3}}\left[-\frac{1}{2}(2\text{mV})(6\text{ms})\right] - 2\text{mA} = \underline{-8\text{mA}}$$

$$i(8) = \frac{1}{10^{-3}}\left[(2\text{mV})(6\text{ms})\right] - 8\text{mA} = \underline{4\text{mA}}$$

5.22

$$v = L\,di/dt$$
$$v = (0.1)\frac{d}{dt}(100\sin 100t)$$
$$= \underline{\cos 100t\ \text{V}}.$$

5.23

$$i = \frac{1}{L}\int_0^t v\,dt + i_0$$
$$= \frac{1}{0.1}\int_0^t 10e^{-100t}\,dt + 0$$
$$= \underline{1 - e^{-100t}\ \text{A}}$$

5.24 $\lambda = Li$,

$\lambda(10ms) = (50\times10^{-3})(\frac{12}{2}mA) = \underline{300\mu Wb}$

$\lambda(40ms) = (0.05)[\frac{3}{4}(12mA)] = \underline{450\mu Wb}$

$\lambda(80ms) = (0.05)(\frac{12}{4}mA) = \underline{150\mu Wb}$

5.25 $v = L\,di/dt$, di/dt equals slope,

$v(10ms) = (10)\frac{12}{20} = \underline{6V}$

$v(40ms) = (10)\frac{(-12)}{(100-20)} = \underline{-1.5V}$

$v(80ms) = (10)(-\frac{12}{80}) = \underline{-1.5V}$

5.26 $i = \frac{1}{L}\int_0^t v\,dt + i_0$

$i(20ms) = \frac{1}{120}\left[\frac{1}{2}(20ms)12\right] + 2mA = \underline{3mA}$

$i(60ms) = \frac{1}{120}\left[\frac{1}{2}(12+\frac{12}{2})(40ms)\right] + 3mA$

$= \underline{6mA}$

$i(100ms) = \frac{1}{120}\left[\frac{1}{2}(12)(80ms)\right] + 3mA$

$= \underline{7mA}$

5.27 $v = L\,di/dt$

(a) $v = (0.1)\,d/dt(5) = \underline{0V}$.

(b) $v = (0.1)\,d/dt(100t) = \underline{10V}$.

(c) $v = (0.1)\,d/dt(10\sin 200t)$

$= \underline{200\cos 200t \text{ V}}$

(d) $v = (0.1)\,d/dt(10e^{-50t}) = \underline{-50e^{-50t}}$ V.

5.28 $i = \frac{1}{L}\int_0^t v\,dt + i_0$

$= \frac{1}{10^{-2}}\int_0^t 2\sin 1000t\,dt + 0$

$= \underline{200(1-\cos 1000t)\text{ mA}}$

5.29 $\lambda = Li = (0.005)(0.2) = \underline{1mWb}$

$W = \frac{1}{2}Li^2 = \frac{1}{2}(0.005)(0.2)^2 = \underline{100\mu J}$

5.30 $i_L(0-) = I - i_1(0-) = 3-2 = \underline{1A}$

The inductor current is continuous so that $i_L(0+) = i_L(0-) = \underline{1A}$. $v_{R_1}(0+) = 0$ = voltage across a short ciruit $\therefore i_1(0+) = \frac{v_{R_1}}{R_1} = \underline{0}$.

By KVL, $L\frac{d\,i_L(0+)}{dt} + R_2 i_L(0+) = 0$

$\therefore \frac{d\,i_L(0+)}{dt} = -\frac{R_2 i_L(0+)}{L} = -\frac{12(1)}{2} = \underline{-6\frac{A}{s}}$

5.31 $i = \frac{1}{L}\int_0^t v\,dt + i_0$

$= \frac{1}{0.5}\int_0^1 6t\,dt + 2 = 8A$

$W = \frac{1}{2}Li^2 = \frac{1}{2}(0.5)(8)^2 = \underline{64J}$

5.32 $i(3\pi/4) = \frac{1}{.25}\int_0^{3\pi/4} 6\cos 2t\,dt + 0$

$= 12(\sin\frac{3\pi}{2}) = -12A$

$W(3\pi/4) = \frac{1}{2}Li^2 = \frac{1}{2}(.25)(12)^2 = \underline{18J}$

$i(\frac{\pi}{8}) = 12(\sin\frac{\pi}{4}) = \frac{12}{\sqrt{2}}A$

$v(\frac{\pi}{8}) = 6\cos\pi/4 = \frac{6}{\sqrt{2}}V$

$p(\frac{\pi}{8}) = vi = (\frac{12}{\sqrt{2}})(\frac{6}{\sqrt{2}}) = \underline{36W}$

5.33

$W = \frac{1}{2}Li^2 = \frac{1}{2}(0.1)(60\times10^{-3})^2 = \underline{180\mu J}$

5.34 At $t = 0^-$, $v(0-) = 9V$, $i(0-) = 1A$

$i_1(0^-) = \frac{9}{3} = 3A$, $i_L(0^-) = i_1(0-) - i(0^-)$
$= 2A$

KVL: $v_c(0-) = 18 - v(0-) - 6i(0-)$
$= 3V$

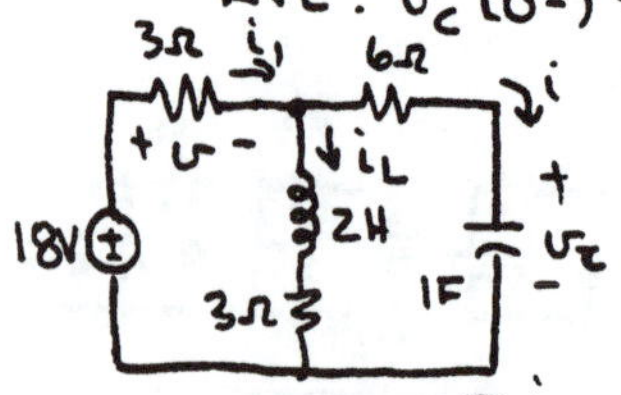

circuit for t=0−

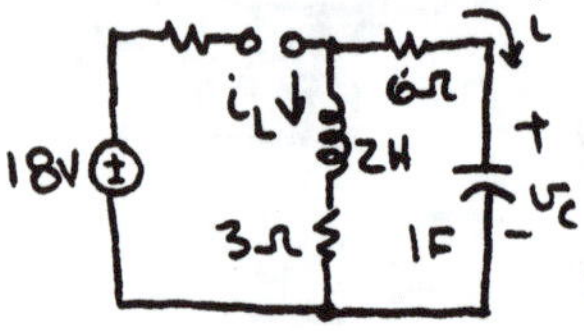

t=0+

At $t = 0+$,

$i_L(0+) = i_L(0^-) = 2A$

$v_c(0+) = v_c(0^-) = 3$

KCL: $i(0+) = -i_L(0+) = \underline{-2A}$

KVL: $2\frac{di(0+)}{dt} + (6+3)i(0+) + v_c(0+) = 0$

$\frac{di(0^+)}{dt} = [-9(-2)-3]/2 = \underline{7.5A/s}$

5.35

$L_s = 2 + 5 + 3 = \underline{10\,H}$

5.36

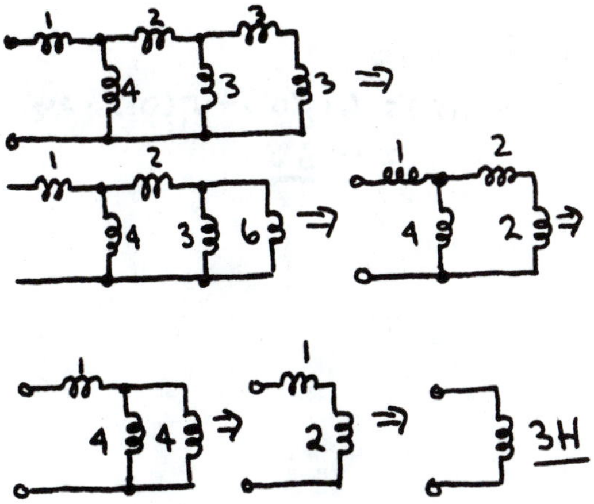

5.37

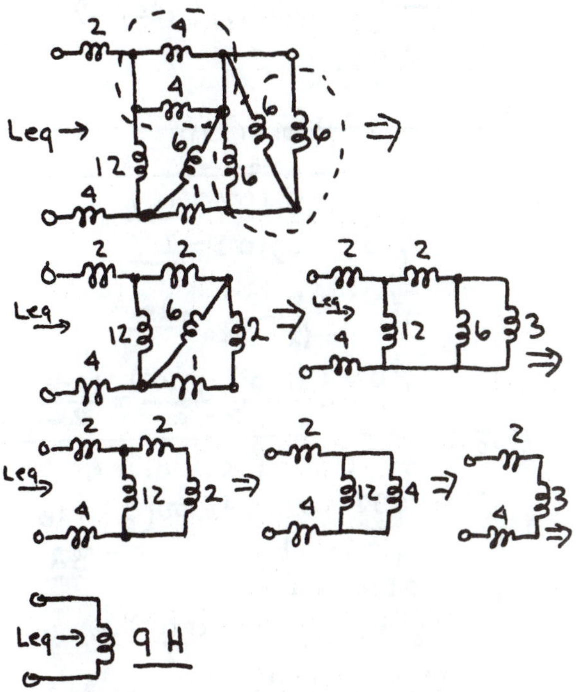

$L_{eq} = \underline{9\,H}$

5.38

DC steady state ⇒ Inductor = short ckt, Capacitor = open ckt

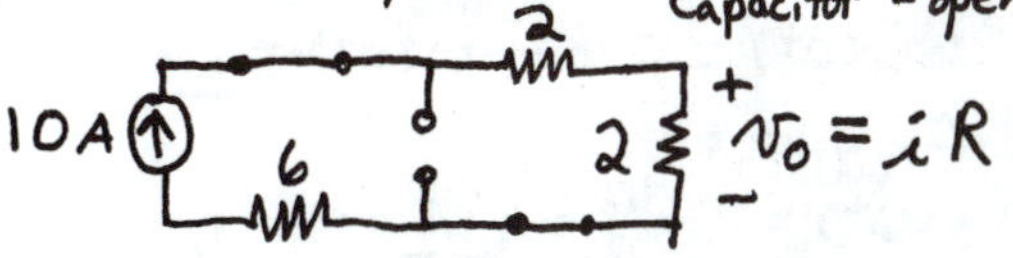

$v_o = iR$

$\therefore\ v_o = (10A)(2\,\Omega) = \underline{20\,V}$

5.39

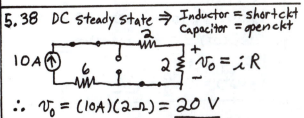

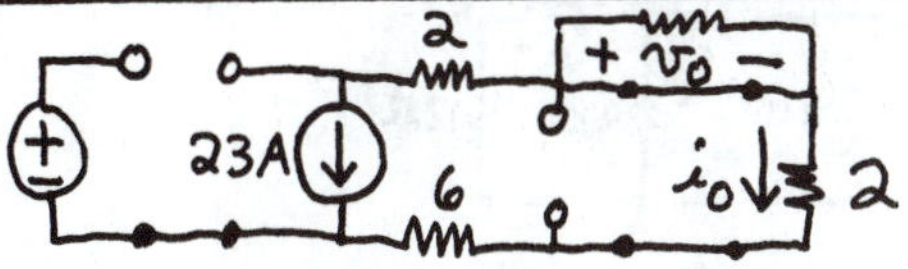

$\underline{v_o = 0\,v}$ due to short ckt

$\underline{i_o = -23A}$ due to current src

5.40

(a) $(33\times10^{-12})R = 10^5$ or

$\underline{R = 3.03\times10^{15}\,\Omega}$

(b) $(.001\times10^{-6})R = 10^5$

$\Rightarrow\ \underline{R = 10^{14}\,\Omega}$

(c) $(2.7\times10^{-6})R = 10^5$

$\Rightarrow\ \underline{R = 37.04\times10^9\,\Omega}$

5.41 Each turn = 4cm

(a) 

$\Rightarrow 4\text{cm} \times 50000 = 200000\,\text{cm} = 2000\,\text{m wire}$

$\Rightarrow R = (2.0\,\Omega/\text{m})(2000\text{m}) = \underline{4\,k\Omega}$

(b) $P = i^2 R = \underline{40\,W}$

5.42 $v_1(0^+) = v_1(0^-) = \underline{14V}$

$v_2(0^+) = v_2(0^-) = \underline{6V}$

$i_1(0^-) = \dfrac{v_1(0^-)}{(2+5)k\Omega} = \underline{2mA}$.

$i_1(0^+) = \dfrac{v_1(0^+) - v_2(0^+)}{2\,K\Omega} = \dfrac{14-6}{2k\Omega} = \underline{4mA}$

5.43 At $t = 0^-$; KVL gives

$i_1(0^-)2 + v(0^-) + i_1(0^-)\left[\frac{4}{2}\right] = 18$ or

$4i_1(0^-) = 12$ or $i_1(0^-) = \underline{3A}$

KCL gives:

$i_2(0^-) = i_1(0^-) - \dfrac{v(0^-)}{2} = 3 - \dfrac{6}{2} = \underline{0A}$

$v(0^-) = v(0^+) = 6V$ the At $t = 0^+$

KVL gives

$i_1(0^+)2 + v(0^+) + i_1(0^+)4 = 18$ or

$6i_1(0^+) = 12$ or $i_1(0^+) = \underline{2A}$

KCL gives

$i_2(0^+) = 2 - \frac{6}{2} = \underline{-1A}$

5.44

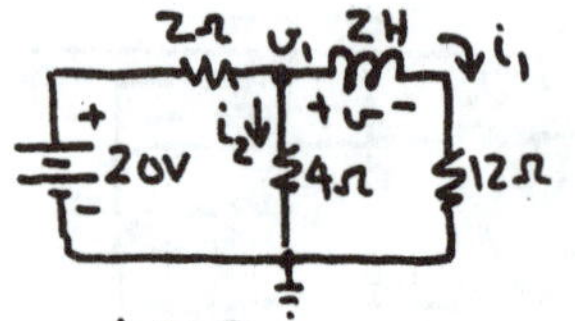

$t = 0^-$

By KCL,

$i_1(0^-) + i_2(0^-) + \dfrac{v_1(0^-) - 20}{2} = 0$ or

$2i_2(0^-) + v_1(0^-) = 18$, since

$v_1(0^-) = i_2(0^-)4$ then

$6i_2(0^-) = 18$ or $i_2(0^-) = \underline{3A}$

$v(0^-) = v_1(0^-) - i_1(0^-)12 = 12 - 12 = \underline{0V}$

At $t = 0^+$; $i_1(0^+) = i_1(0^-) = \underline{1A}$,

$i_2(0^+) = -i_1(0^+) = \underline{-1A}$

$v(0^+) = i_2(0^+)(12+4) = \underline{-16V}$

5.45

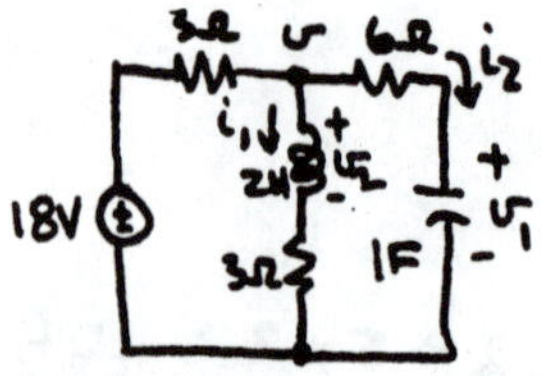

$t = 0^-$

At $t = 0^-$,
$v_1(0^-) = 9V$,
$i_1(0^-) = 3A$,
By KCL,

$\dfrac{v(0^-) - 18}{3} + i_1(0^-) + \dfrac{v(0^-) - v_1(0^-)}{6} = 0$ or

$3v(0^-) = 27$ or $v(0^-) = 9V$

$i_2(0^-) = \dfrac{v(0^-) - v_1(0^-)}{6} = \dfrac{9-9}{6} = \underline{0A}$

$v_2(0^-) = v(0^-) - i_1(0^-)3 = 9 - 9 = \underline{0V}$

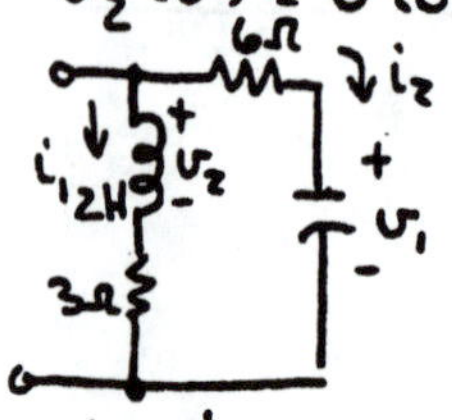

$t = 0^+$

At $t = 0^+$,

$i_1(0^+) = i_1(0^-) = \underline{3A}$

$v_1(0^+) = v_1(0^-) = \underline{9V}$

$i_2(0^+) = -i_1(0^+) = \underline{-3A}$

By KVL,

$V_2(0^+) = 6i_2(0^+) + v_1(0^+) + 3i_2(0^+) = \underline{-18V}$.

Chapter 6

First-Order Circuits

6.1 Simple RC and RL Circuits without Sources

6.1 The capacitor is charged to a voltage of 100V prior to the closing of the switch. For $t > 0$, find (a) $v(t)$, (b) $i(t)$, (c) $w_c(t)$, and (d) the time at which $v(t) = 50$ V.

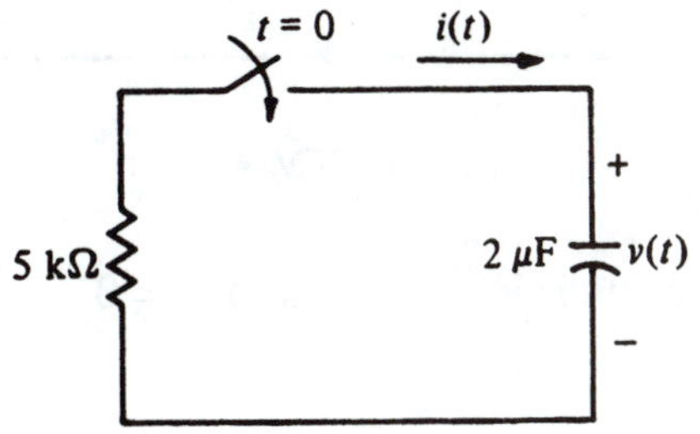

PROBLEM 6.1

6.2 The switch in the network opens at $t = 1$ s, at which time $v = 10$ V. For $t > 1$, find $v(t)$ and $w_c(t)$.

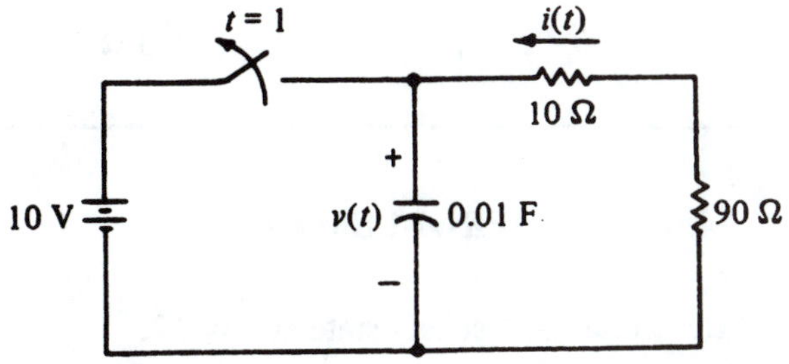

PROBLEM 6.2

6.3 Find $v(t)$ for $t > 0$ if $i(0) = 2$ A.

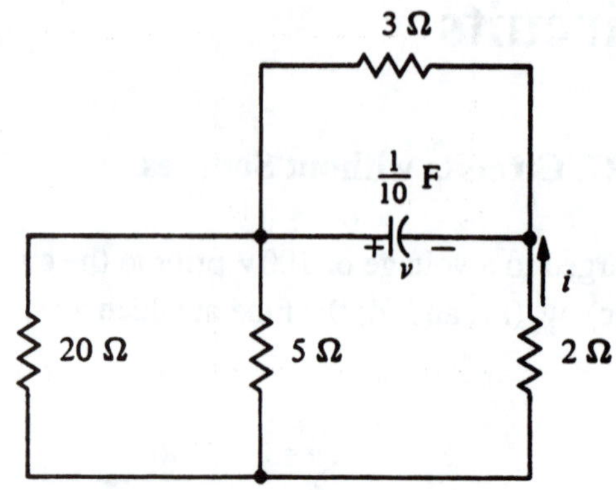

PROBLEM 6.3

6.4 Find $v(t)$ for $t > 0$ if the circuit is in steady state at $t = 0^-$.

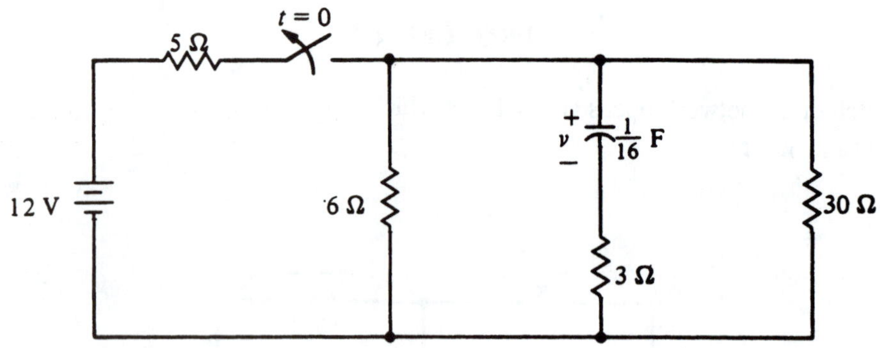

PROBLEM 6.4

6.5 Find $v(t)$ for $t > 0$ if the circuit is in steady state at $t = 0^-$.

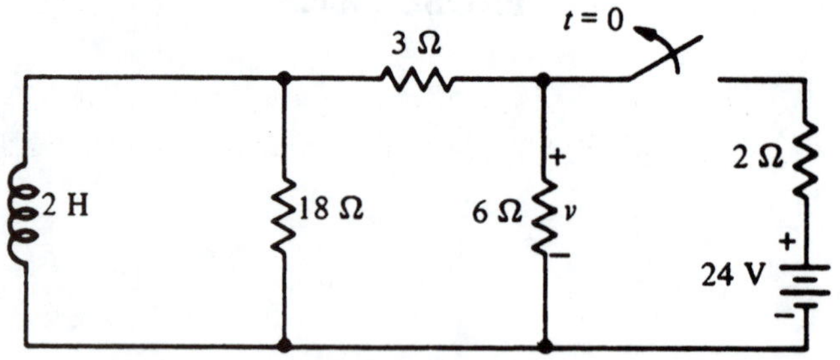

PROBLEM 6.5

6.6 Find $i(t)$ for $t > 0$ if the circuit is in steady state at $t = 0^-$.

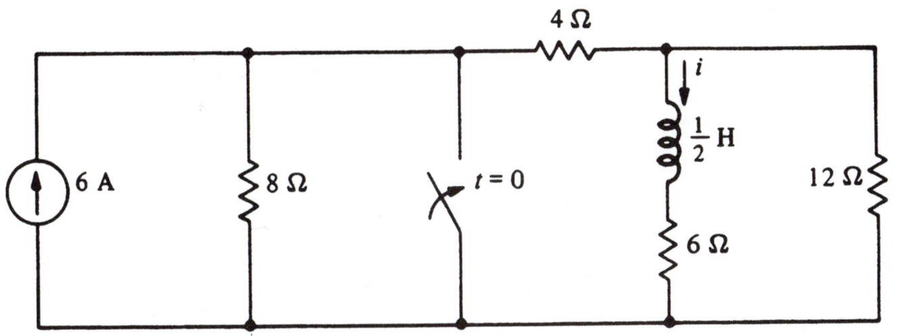

PROBLEM 6.6

6.7 In a source-free *RL* circuit with $R = 10\text{k}\Omega$ and $L = 50\text{mH}$, find the current in the circuit for $t > 0$ if $i(0) = 83$ A. At what time does the current decay to 100mA?

6.8 Find $v_o(t)$ for $t > 0$ if the circuit is in steady state until $t = 0$ when the switch is opened.

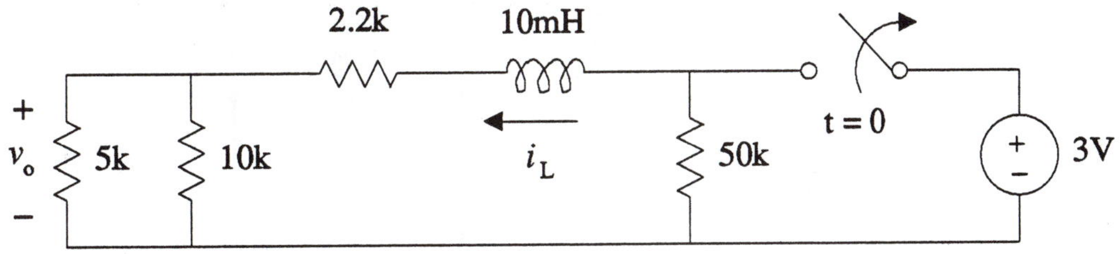

PROBLEM 6.8

6.2 Time Constants

6.9 In a series *RC* circuit, determine (a) τ for $R = 2\text{k}\Omega$ and $C = 5\text{mF}$, (b) *C* for $R = 2\text{k}\Omega$ and $\tau = 10\text{ms}$, and (c) *R* for $v(t)$ to half every 2ms on a 0.1μF capacitor.

6.10 A series *RC* circuit consists of a 2kΩ and 0.1μF capacitor. It is desired to increase the current in the network by a factor of 3 without changing the capacitor voltage. Find the necessary values of *R* and *C*.

6.11 Find $i(t)$ for $t > 0$ if the circuit is in steady state at $t = 0^-$.

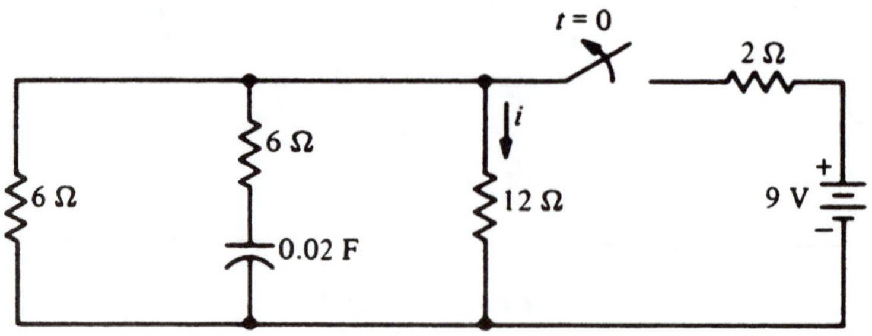

PROBLEM 6.11

6.12 Find $v(t)$ for $t > 0$ if the circuit is in steady state at $t = 0^-$.

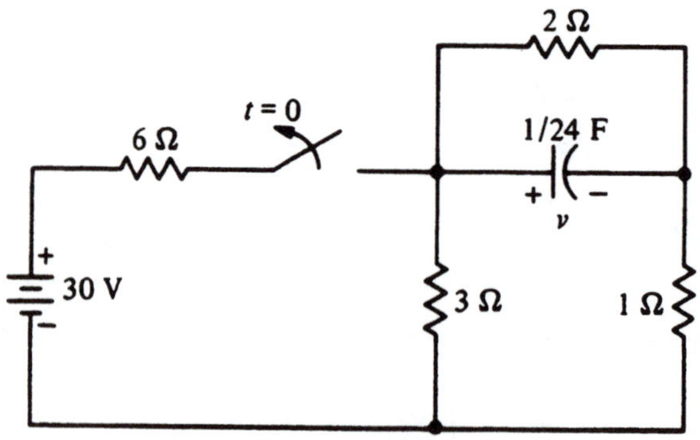

PROBLEM 6.12

6.13 Find $v(t)$ for $t > 0$ if the circuit is in steady state at $t = 0^-$.

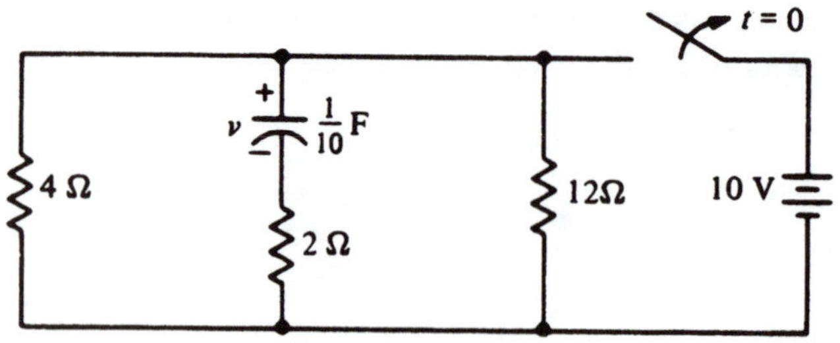

PROBLEM 6.13

6.14 Find $i(t)$ for $t > 0$ if the circuit is in steady state at $t = 0^-$.

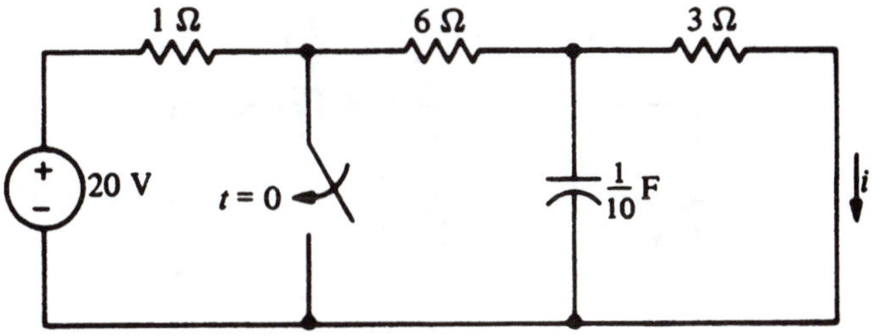

PROBLEM 6.14

6.15 Find $v(t)$ for $t > 0$ if $v(0) = 4$ V.

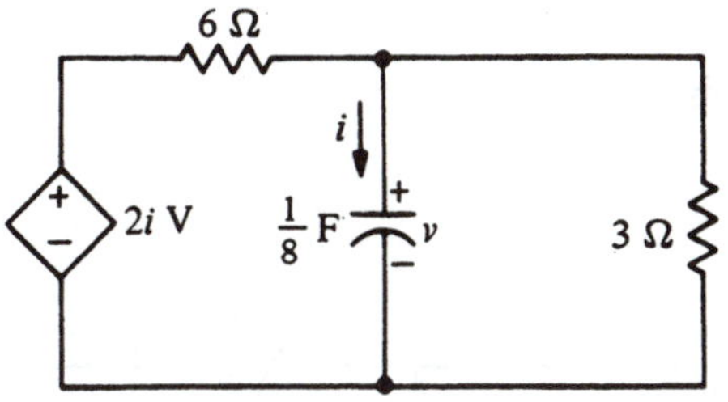

PROBLEM 6.15

6.16 In a series *RL* circuit, determine (a) τ for $R = 1\text{k}\Omega$ and $L = 15\text{mH}$, (b) *L* for $R = 10\text{k}\Omega$ and $\tau = 40\text{ms}$, and (c) *R* for the stored energy in a 5mH inductor to halve every 4ms.

6.17 A series *RL* circuit has a time constant of 5ms with an inductance of 4H. It is desired to halve the inductor voltage without changing the current response. Find the new values of inductance and resistance required.

6.18 The circuit shown is in a DC steady-state condition at $t = 0^-$. Find v for $t > 0$.

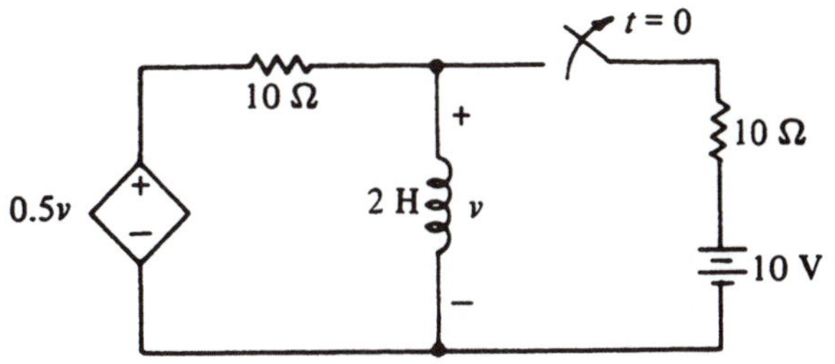

PROBLEM 6.18

6.19 Find i for $t > 0$ if $i(0) = 2$ A.

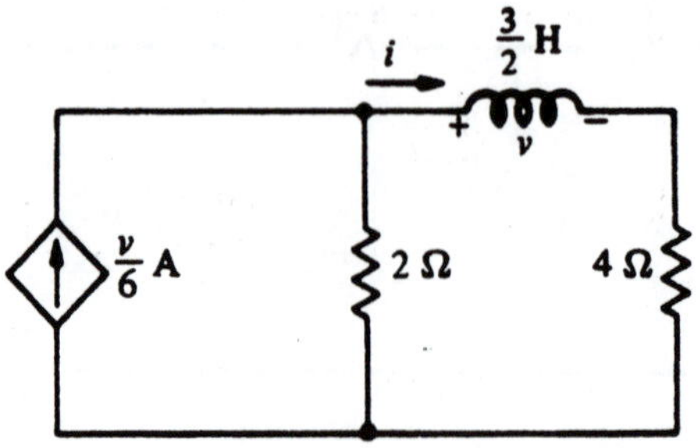

PROBLEM 6.19

6.3 General First-Order Circuits without Sources

6.20 Find the time constant of this circuit.

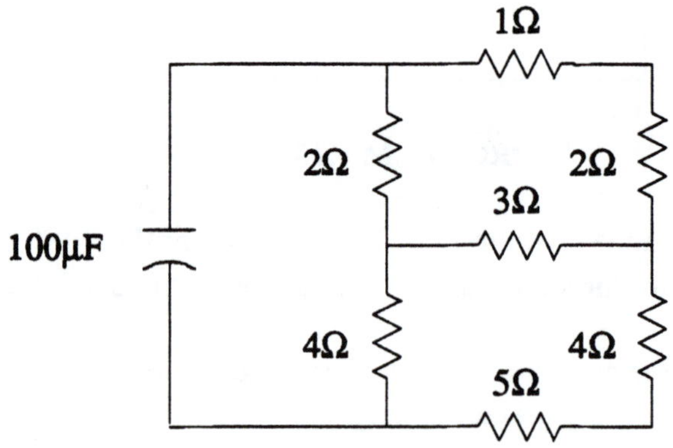

PROBLEM 6.20

6.21 Find the time constant of this circuit.

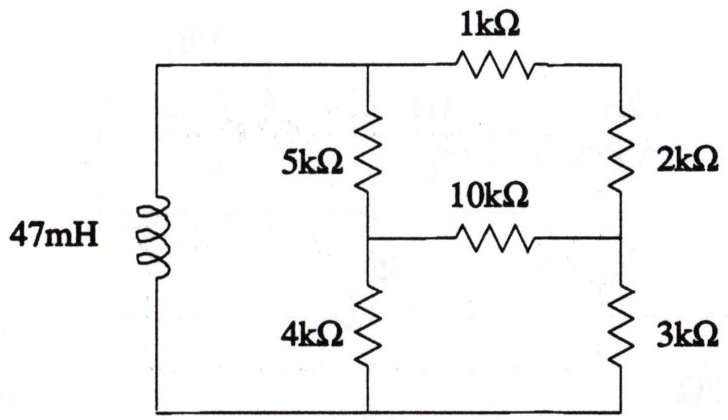

PROBLEM 6.21

6.22 Find the time constant of this circuit.

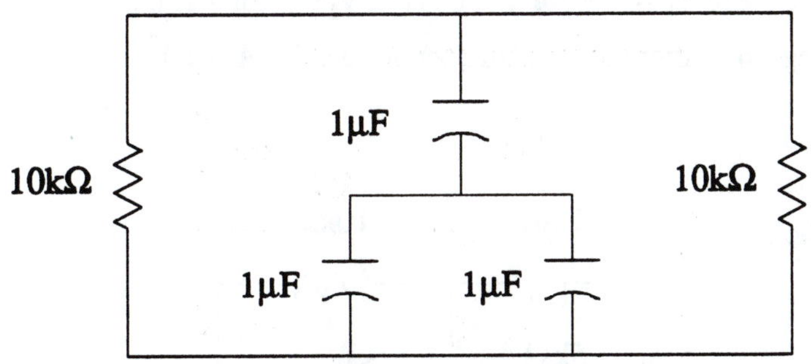

PROBLEM 6.22

6.23 Find the time constant of this circuit.

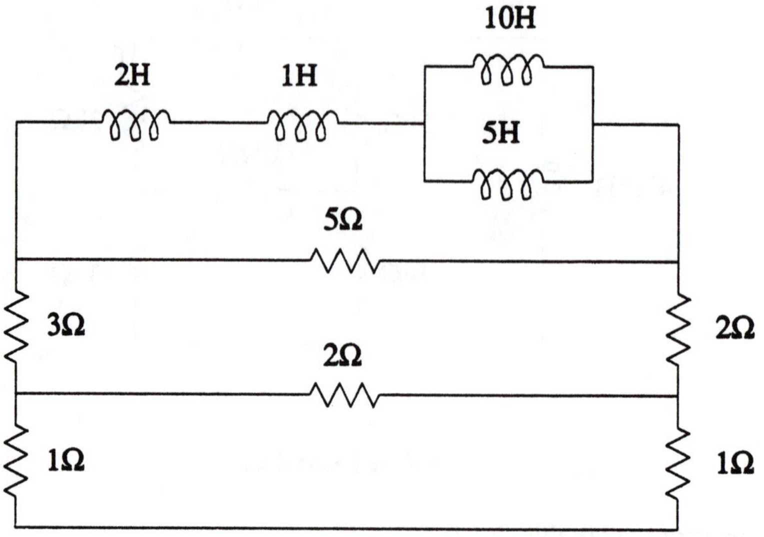

PROBLEM 6.23

6.24 If $v_1 = 1\text{V}$ for $t < 0$ and $v_1 = 0\text{V}$ for $t \geq 0$, find $v_2(t)$ for $t \geq 0$. Use the ideal voltage amplifier model with open loop gain $A = 10^5$.

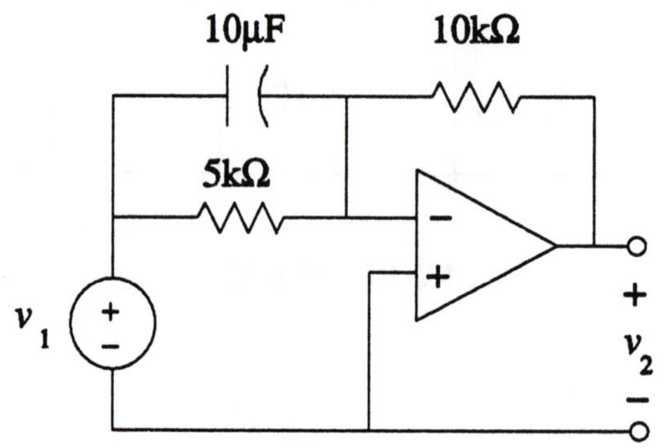

PROBLEM 6.24

6.25 If $v_1 = 20\text{V}$ for $t < 0$, $v_1 = 0\text{V}$ for $t \geq 0$, $v_2 = 10\text{V}$ for $t < 0$, and $v_2 = 0\text{V}$ for $t \geq 0$, find $v_3(t)$ for $t \geq 0$.

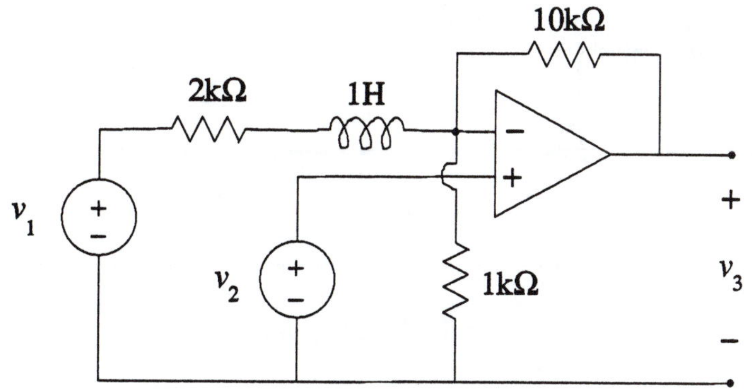

PROBLEM 6.25

6.4 Circuits with DC Sources

6.26 Find $v(t)$ for $t > 0$ if $v(0^-) = 5\,\text{V}$.

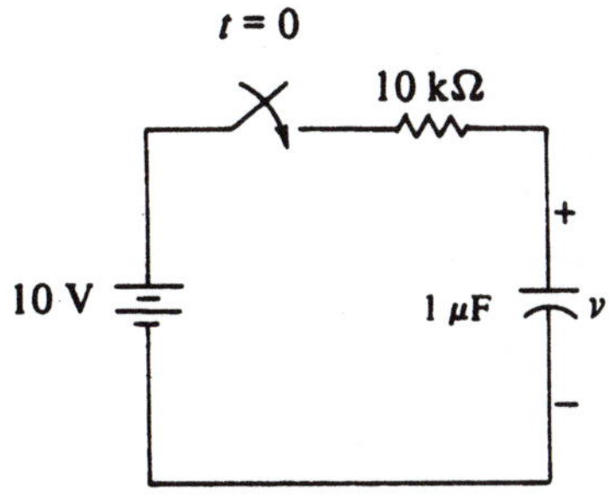

PROBLEM 6.26

6.27 The circuit is in steady state at $t = 0^-$. Find i for $t > 0$ if the switch is moved from position 1 to position 2 at $t = 0$.

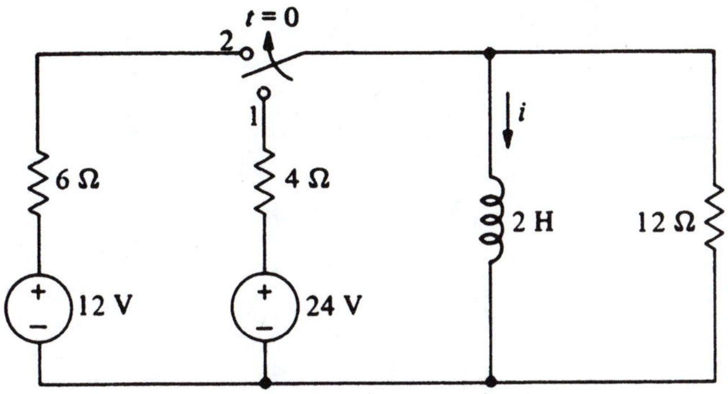

PROBLEM 6.27

6.28 Find the current in the inductor for $t > 0$ if the circuit is in steady state at $t = 0^-$.

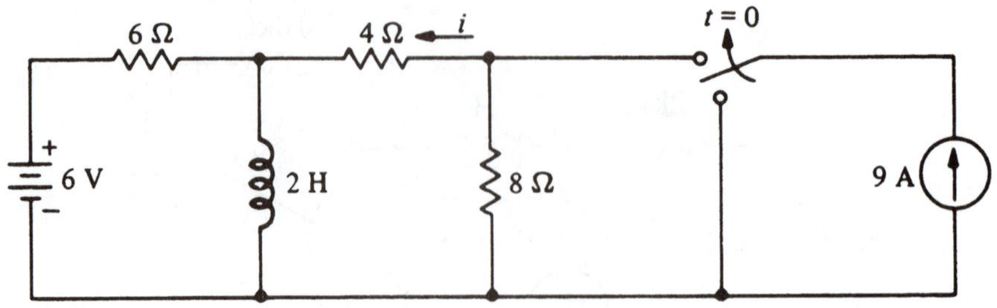

PROBLEM 6.28

6.29 Find i for $t > 0$ if the circuit is in steady state at $t = 0^-$.

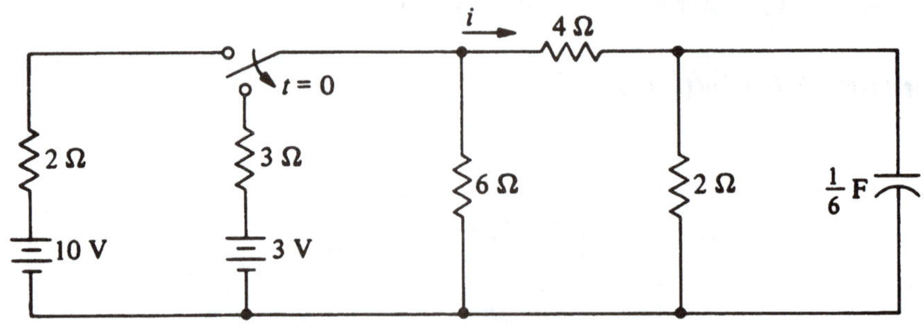

PROBLEM 6.29

6.30 Find i for $t > 0$ if the circuit is in steady state at $t = 0^-$.

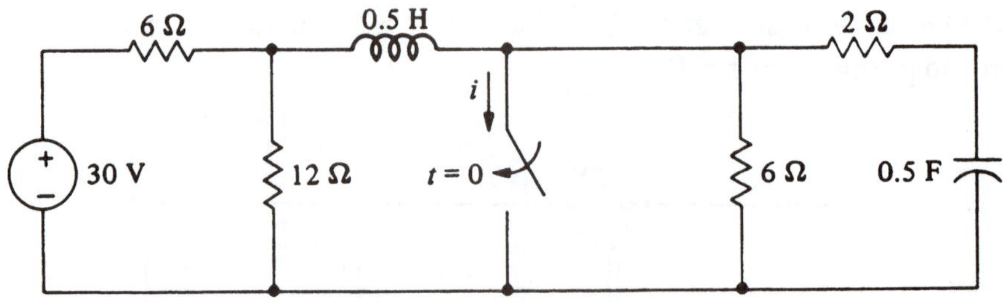

PROBLEM 6.30

6.31 Find $v_o(t)$ for $t > 0$.

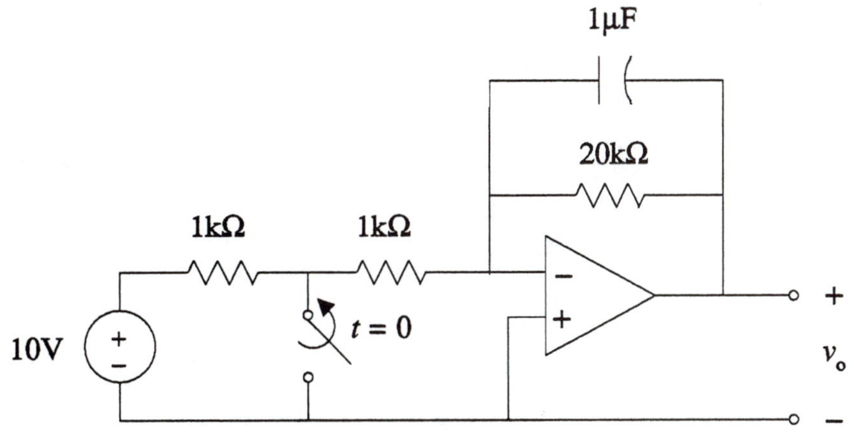

PROBLEM 6.31

6.32 Find $v_o(t)$ for $t \geq 0$ if $i(0) = -1\,\text{mA}$.

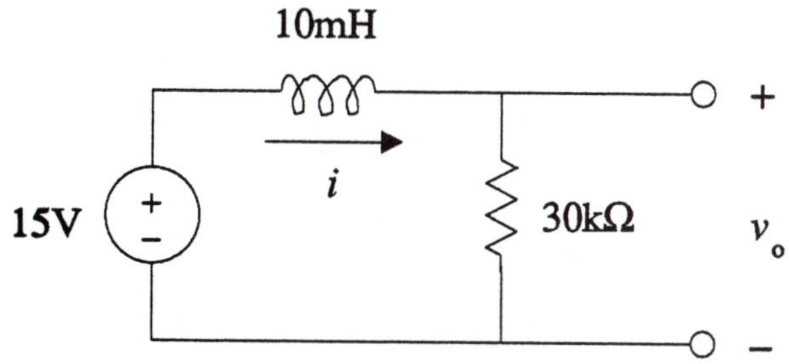

PROBLEM 6.32

6.5 Superposition in First-Order Circuits

6.33 Using superposition, find v_o for $t > 0$ if $i_L(0) = 3$ A.

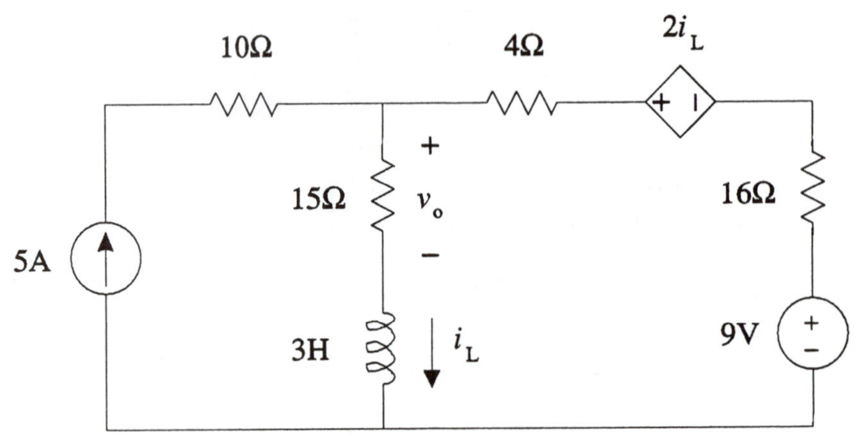

PROBLEM 6.33

6.34 Using superposition, find v_c for $t > 0$ if $v_c(0) = 20$ V.

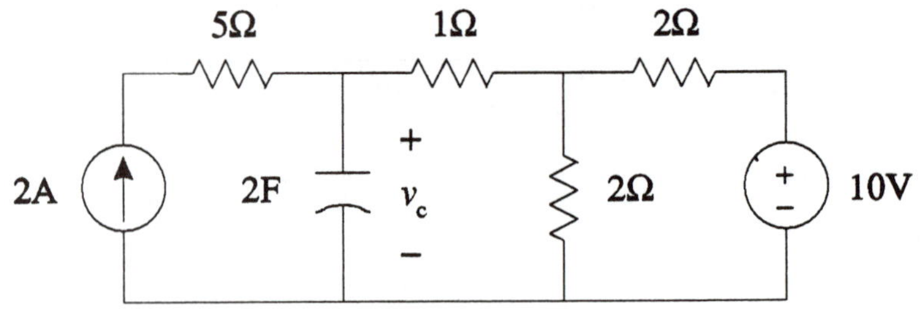

PROBLEM 6.34

6.35 Using superposition, find v_o for $t > 0$ if $i_L(0) = 1$ A.

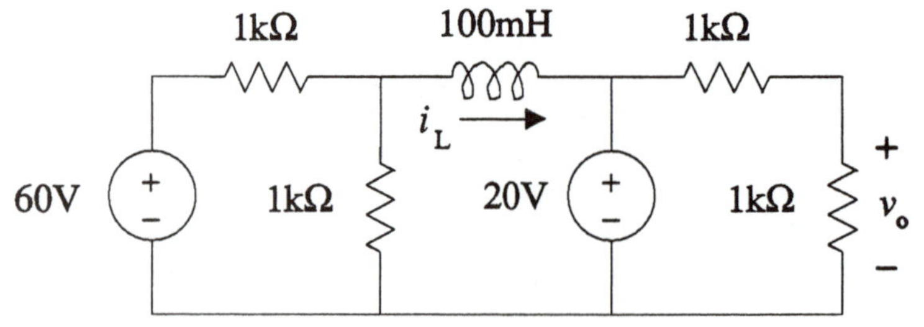

PROBLEM 6.35

6.36 Using superposition, find v_o for $t > 0$ if $v_o(0) = 0$.

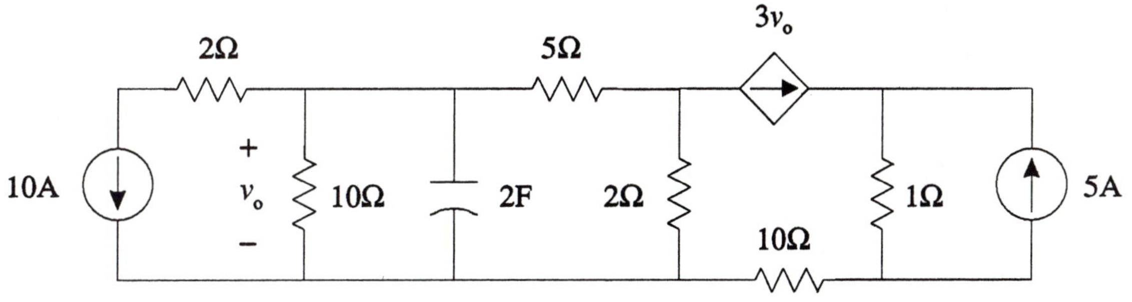

PROBLEM 6.36

6.37 Use superposition to find v_3 (for $t > 0$) in terms of v_1 and v_2 assuming that there is no initial energy storage.

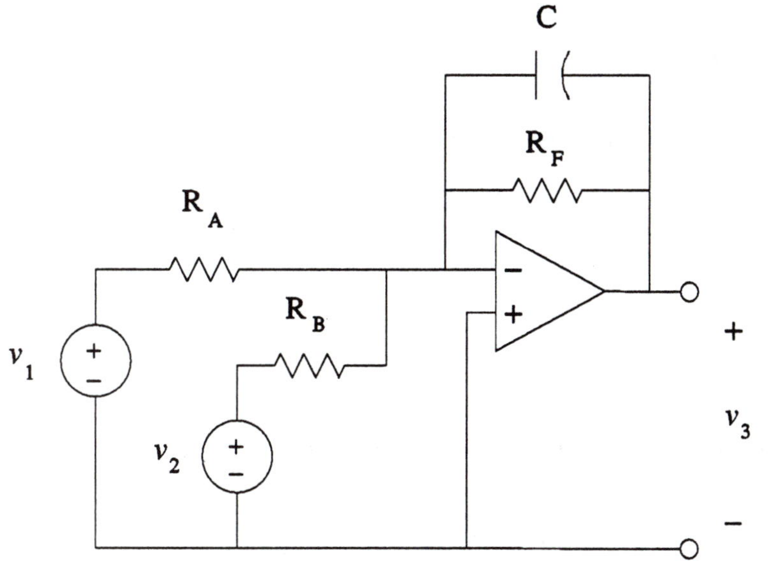

PROBLEM 6.37

6.6 Unit Step Function

6.38 Using unit step functions, write an expression for the current $i(t)$ which satisfies:

(a) $i(t) = 0\text{A}, \ t < 0$
$= 5\text{A}, \ t > 0.$

(b) $i(t) = -1\text{A}, \ t < -20\text{ms}$
$= 2\text{A}, \ -20\text{ms} < t < 40\text{ms}$
$= 3\text{A}, \ 40\text{ms} < t.$

(c) $i(t) = 6\text{A}, \ t < 10\text{s}$
$= -6\text{A}, \ t > 10\text{s}.$

6.39 Sketch the voltage given by $v(t) = 3u(t-3) + tu(t) - 3u(t+3)\,\text{V}$.

6.40 Express $v(t)$ in terms of unit step functions:

$v(t) = 0\text{V}, \ t < -10$
$= -10\text{V}, \ -10 < t < 0$
$= 20\text{V}, \ 0 < t < 10$
$= 15\text{V}, \ 10 < t.$

6.41 Express $f(t)$ in terms of scaled and shifted unit step functions.

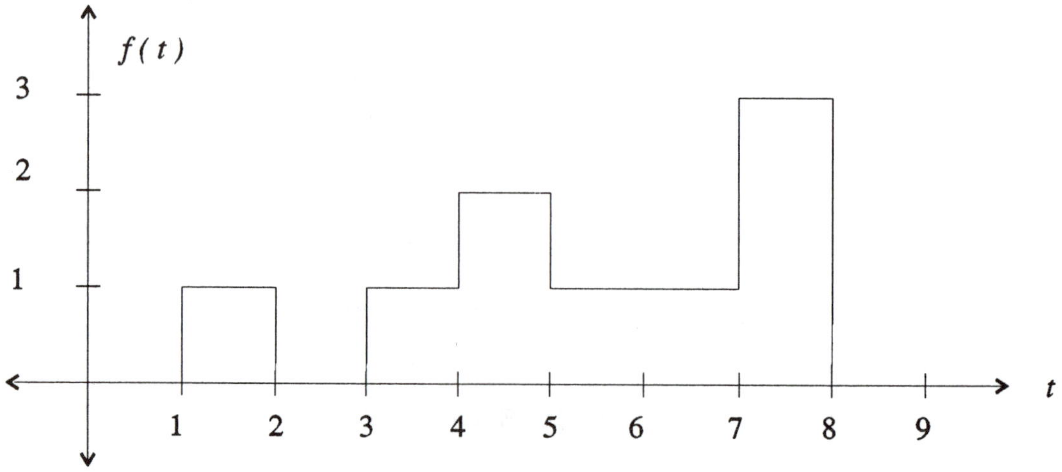

PROBLEM 6.41

6.7 Step and Pulse Responses

6.42 Find the response $v_1(t)$ to the voltage step $v(t) = 10u(t)$ V.

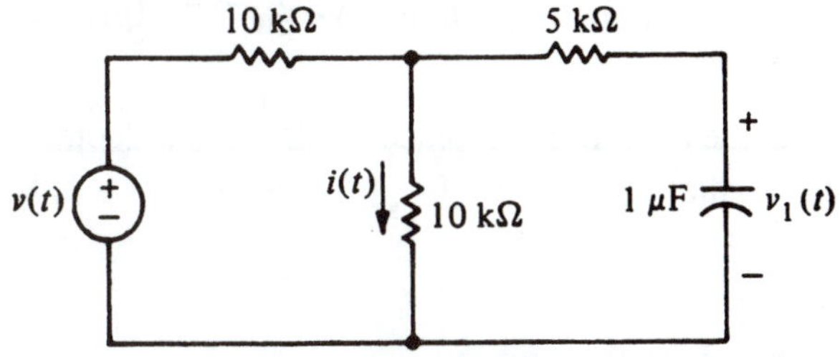

PROBLEM 6.42

6.43 Find v if $v_g = 3e^{-3(t+4)}u(t+4)$ V and there is no initial stored energy.

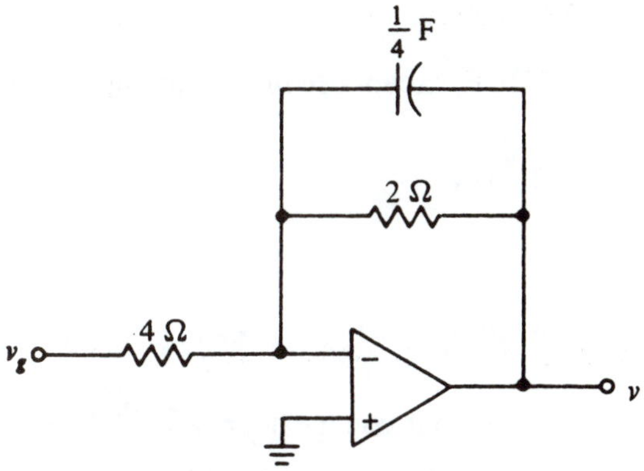

PROBLEM 6.43

6.44 Find the step response i (note that this means $v_g = u(t)$ V).

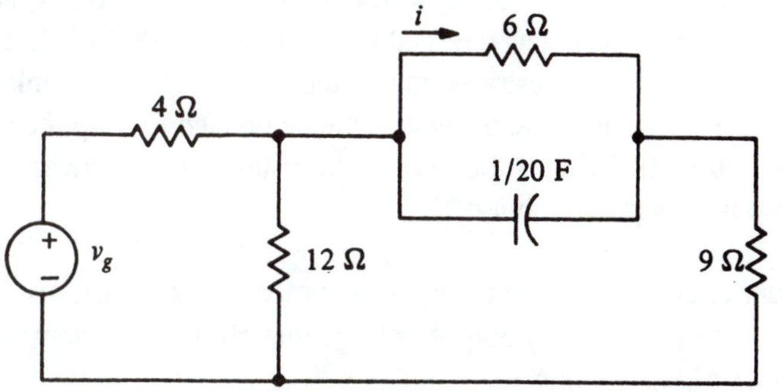

PROBLEM 6.44

6.45 (a) Solve Prob. 6.42 if the capacitor is replaced with a 0.2H inductor.
(b) Find the response to the circuit in (a) if $v(t) = 10[u(t) - u(t-2)]$V.

6.46 (a) Solve Prob. 6.43 if the capacitor is replaced with a 0.25H inductor.
(b) Find the response to the circuit in (a) if $v_g(t) = 3e^{-3(t+4)}[u(t+4) - u(t)]$V.

6.47 (a) Solve Prob. 6.44 if the capacitor is replaced with a 1H inductor.
(b) Find the response to the circuit in (a) if $v_g(t) = u(t) - u(t-1)$ V.

6.8 SPICE and the Transient Response

6.48 Repeat Prob. 6.1 (a), (b), and (d) using SPICE.

6.49 Repeat Prob. 6.30 using SPICE.

6.50 Repeat Prob. 6.45(a) using SPICE.

6.51 Repeat Prob. 6.45(b) using SPICE, except use instead an input voltage:

$$v(t) = 10[u(t) - u(t - 1\times 10^{-5})]\text{V}.$$

6.9 Design of First-Order Circuits

6.52 Design a first-order circuit with output voltage $v_o = 25e^{-3t}$ V for $t > 0$ out of DC sources (for charging prior to $t = 0$), ideal switches, resistors, and a single: (a) capacitor, (b) inductor.

Hint: The circuit should first charge the energy storage element for an appropriate period of time to reach an output voltage $v_o = 25$ V at which time a switch (or switches) are thrown and the energy storage element is discharged with an appropriate time constant.

6.53 A DC power supply contains a bank of capacitors (connected in parallel) equivalent to a 2F capacitance. This 2F capacitance gets charged to a voltage of 1000V DC during operation. When the power supply is shut off, it is desirable to discharge the capacitor bank for safety reasons with a discharge resistance. Find the value of the resistance needed to drop the capacitor bank voltage down to below 10V after 1s. What is the maximum instantaneous power absorbed by this resistor (at the instant the power supply is shut off)?

6.54 Design a first-order circuit using an uncharged inductor, DC sources, resistors, and switches that has an input voltage $5u(t)$ V and an output voltage that climbs exponentially to 4.75V in 1s. What is the time constant of your circuit?

6.55 Using the inverting integrator circuit introduced in Section 6.7 of the text, design a circuit that when given an input of $v_1 = 5u(t)$ V produces an output $v_2 = -100t\, u(t)$ V.

CHAPTER 6 SOLUTIONS

6.1 (a) $v(t) = v(0)e^{-t/RC}$, where
$RC = (5\times10^3)(2\times10^{-6}) = 10^{-2}$ s
$v(0) = 100$V, $v(t) = \underline{100\, e^{-100t}\, V}$

(b) $i(t) = -v(t)/5k\Omega = \underline{20e^{-100t}\, mA}$

(c) $w_c(t) = \frac{1}{2}Cv^2 = \frac{1}{2}(2\mu F)\,v^2(t)$
$= \underline{10\, e^{-200t}\, mJ}$

(d) $50 = 100\, e^{-100t}$, solve for t,
$t = \dfrac{\ln(50/100)}{-100} = \underline{6.93\ \ ms}$

6.2
$v(t) = v(1)\, e^{-(t-1)/RC}$, where
$RC = (0.01)(10+90) = 1$ s, $v(1) = 10$V,
$\underline{v(t) = 10e^{1-t}\, V}$
$w_c(t) = \frac{1}{2}Cv^2 = \frac{1}{2}(0.01)(10e^{1-t})^2$
$= \underline{0.5\, e^{2(1-t)}\, J}$

6.3
Replacing the 20Ω and 5Ω parallel combination by 4Ω, we see that
$v(0) = (4+2)\,i(0) = 12$V.
R_{eq} = resistance seen by capacitor
$= (4+2)(3)/4+2+3 = 2\Omega$.
$1/\tau = 1/R_{eq}C = \dfrac{1}{2(1/10)} = 5\ s^{-1}$
$\therefore\ v = \underline{12\, e^{-5t}\, V}$.

6.4 At $t=0^-$ the capacitor is open-circuited and it's voltage is the voltage across the 6Ω and 30Ω parallel combination. By voltage div.
$v(0^+) = v(0^-) = \dfrac{30(6)/30+6}{5+5}(12) = 6V$
For $t>0$ the resistance seen by the capacitor is
$R_{eq} = 3 + \dfrac{30(6)}{30+6} = 8\Omega$; $\dfrac{1}{\tau} = \dfrac{1}{R_{eq}C} = \dfrac{1}{8(\frac{1}{16})} = 2\,s^{-1}$
$\therefore\ v = \underline{6e^{-2t}\, V}$.

6.5 i_L = inductor current downward.
$v_L(0^-) = 0$ (short circuit). For the circuit shown, voltage division gives,

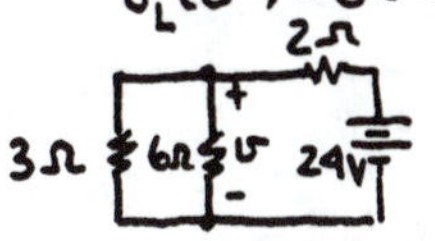

$t = 0^-$ $\qquad v(0^-) = \dfrac{\frac{(3)(6)}{3+6}}{2+2}(24) = 12V$
the $i_L(0^-) = \dfrac{v(0^-)}{3} = 4A$, for $t>0$
$R_{eq} = \dfrac{18(3+6)}{18+3+6} = 6\Omega$ (seen by inductor)
$\dfrac{1}{\tau} = \dfrac{R_{eq}}{L} = \dfrac{6}{2} = 3s^{-1}$ $\therefore\ i_L(0^+) = 4e^{-3t}A$
$v_L(0^+) = L\dfrac{d}{dt} = 2(4)(-3)e^{-3t} = -24e^{-3t}V$
By voltage division,
$v(0^+) = \dfrac{6}{6+3}v_L = \underline{-16e^{-3t}\, V}$

6.6 At $t=0^-$, the inductor is a short circuit. Thus for the circuit shown current division gives

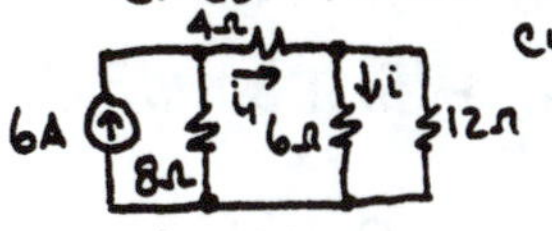

$t=0^-$ $\qquad i_1(0^-) = \dfrac{8\ (6)}{4+8+\frac{6(12)}{6+12}} = 3A$;
$i(0^-) = \dfrac{12}{6+12}(i_1) = 2A$; For $t>0$
$R_{eq} = 6 + \dfrac{12(4)}{4+12} = 9\Omega$ (seen by inductor)
$\dfrac{1}{\tau} = \dfrac{R_{eq}}{L} = \dfrac{9}{1/2} = 18s^{-1}$
$\therefore\ i = \underline{2\, e^{-18t}\, A}$

6.7 KVL: $iR + L\dfrac{di}{dt} = 0$
$\Rightarrow (50\times10^{-3})\dfrac{di}{dt} + 10000i = 0$
$\Rightarrow \dfrac{di}{dt} + (2\times10^5)i = 0$
Trial natural Soln: $i(t) = Ke^{-(2\times10^5)t}$
Match initial conditions:
$i(0) = Ke^0 = K = 83A$
$\therefore\ \underline{i(t) = 83\, e^{-(2\times10^5)t}\, A}$ for $t \geq 0$.
$100mA = 0.1A = 83e^{-(2\times10^5)t}$
$\Rightarrow e^{-2\times10^5 t} = 1.205\times10^{-3}$
$\Rightarrow -2\times10^5 t = \ln(1.205\times10^{-3}) = -6.72$
$\therefore$ At $\underline{t = 3.36\times10^{-5}s = 33.6\mu s}$, $i = 0.1A$

6.8 Circuit at $t=0-$ (DC steady-state)

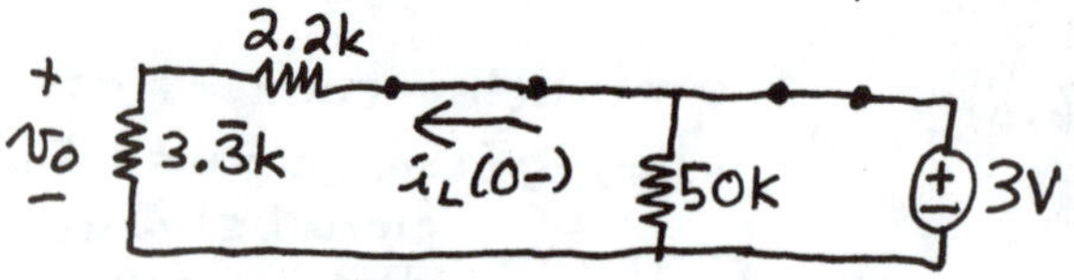

$$i_L(0-) = \frac{3\text{ V}}{2.2k + 3.\bar{3}k} = 0.542\text{ mA}$$

Circuit for $t>0$:

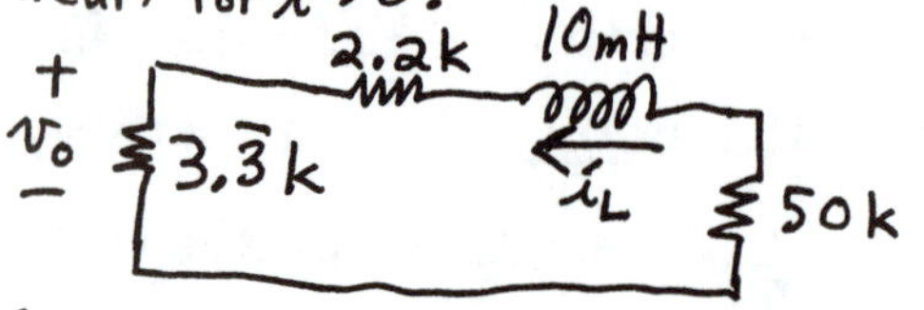

$$i_L(0+) = i_L(0-) = 0.542\text{mA}$$

$$R_{eq} = 2.2k + 3.\bar{3}k + 50k = 55.5\bar{3}\text{ k}$$

$$\text{KVL: } (10\times10^{-3})\frac{di}{dt} + i(55.5\bar{3}\times10^{3}) = 0$$

$$\Rightarrow \frac{di}{dt} + (5.55\bar{3}\times10^{6})i = 0$$

$$\Rightarrow i(t) = Ke^{-(5.55\bar{3}\times10^{6})t}$$

$$\text{I.C. } i(0+) = Ke^{0} = K = 0.542\text{mA}$$

$$\Rightarrow i(t) = 0.542e^{-(5.55\bar{3}\times10^{6})t}\text{ mA}$$

$$v_0(t) = (3.\bar{3}\text{ k})\,i(t)$$

$$\therefore\ \underline{v_0(t) = 1.81e^{-(5.55\bar{3}\times10^{6})t}\text{ V}}$$

6.9

(a) $\tau = RC = (2\times10^{3})(5\times10^{-3}) = \underline{10\text{ s}}$

(b) $C = \tau/R = (10\times10^{-3})/(10\times10^{3}) = \underline{1\mu\text{F}}$

(c) $v(t_0) = V_0$; $v(t_0 + 2\times10^{-3}) = V_0e^{-2\times10^{-3}/\tau}$

$\therefore\ \tau = RC = \frac{-2\times10^{-3}}{\ln 0.5} \Rightarrow R = \frac{-2\times10^{-3}}{(0.1\times10^{-6})\ln 0.5} = \underline{28.85\text{ k}\Omega}$

6.10

Since $i(t) = \frac{V_0}{R}e^{-(t-t_0)/\tau}$, we must have R replaced by R/3. Therefore $\tau = RC = (\frac{R}{3})(3C)$

$\therefore\ R_{new} = \frac{R}{3} = \underline{666.7\Omega}$; $C_{new} = 3C = \underline{0.3\mu\text{F}}$

6.11

At $t=0^-$, since $i_c(0^-) = 0$, the capacitor voltage is that across the parallel combination of 12 Ω and 6 Ω, equivalently 4 Ω. By voltage division

$$v_c(0^-) = \frac{4}{4+2}(9) = 6\text{V}$$

The resistance seen by the capacitor is $R_{eq} = \frac{6(12)}{6+12} + 6 = 10\Omega$

$\tau = R_{eq}C = 10(0.02) = 0.2\text{ s}$

$\therefore\ v_c = 6e^{-5t}\text{V}$; $i_c = C\frac{dv}{dt} = (0.02)(-30e^{-5t}) = -0.6e^{-5t}\text{A}$

By current division,

$i = -\frac{6}{6+12}i_c = \underline{0.2e^{-5t}\text{A}}$

6.12

At $t=0^-$, the capacitor is open-circuited and its voltage is the voltage across the 2Ω resistor.

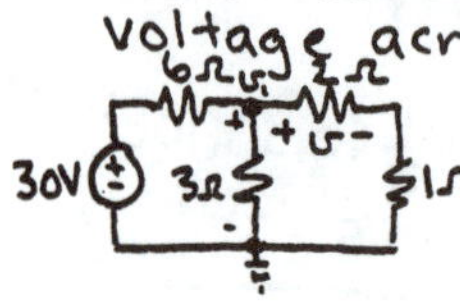

By $v_1 = \frac{\frac{(2+1)3}{3+3}}{6+3/2}(30) = 6\text{V}$

$v(0^-) = \frac{2}{1+2}(6) = 4\text{V}$; for $t>0$,

R_{eq} = resistance seen by capacitor

$= \frac{(3+1)2}{3+1+2} = 1.33\Omega$

$\frac{1}{\tau} = \frac{1}{R_{eq}C} = \frac{1}{(\frac{1}{24})(1.33)} = 18\text{ s}^{-1}$

$\therefore\ v = \underline{4e^{-18t}\text{ V}}$

6.13

At $t=0^-$, since $i_c(0^-) = 0$, the capacitor voltage is that of 10V source. The resistance seen by the capacitor $R_{eq} = \frac{(4)(12)}{4+12} + 2 = 5\Omega$

$\frac{1}{\tau} = \frac{1}{R_{eq}C} = \frac{1}{(5)(1/10)} = 2\text{ s}^{-1}$

$\therefore\ v = \underline{10e^{-2t}\text{V}}$

6.14

At $t=0^-$ the capacitor is open-circuited. By voltage division $v_c(0^-) = \frac{3}{1+6+3}(20) = 6\text{V}$. For $t>0$

R_{eq} = resistance seen by capacitor

$= \frac{(6)(3)}{6+3} = 2\Omega$

$\frac{1}{\tau} = \frac{1}{R_{eq}C} = \frac{1}{2(\frac{1}{10})} = 5\text{ s}^{-1}$

$\therefore\ v_c = 6e^{-5t}\text{V}$; $i = \frac{v_c}{3} = \underline{2e^{-5t}\text{A}}$

6.15 KCL gives $\frac{1}{8}\frac{dv}{dt} + \frac{v}{3} + \frac{v-2i}{6} = 0$;
$i = \frac{1}{8}\frac{dv}{dt}$ ∴ $\frac{dv}{dt} + 6v = 0$ and
$\frac{dv}{v} = -6dt$, $\ln v = -6t + \ln K$
or $v = Ke^{-6t}$, $v(0) = 4 = K$,
∴ $v = \underline{4e^{-6t}}$ V

6.16 (a) $\tau = L/R = \frac{15\times10^{-3}}{10^3} = \underline{15\,\mu s}$

(b) $L = R\tau = (10^4)(40\times10^{-3}) = \underline{400\,H}$

(c) $w(t_0 + 4\times10^{-3}) = \frac{1}{2}w(t_0)$

$= \frac{1}{2}\left[\frac{1}{2}LI_0^2 e^{-2Rt_0/L}\right]$

∴ $\frac{1}{2}LI_0^2 e^{-2R(t_0+4\times10^{-3})/L} = \left(\frac{1}{2}\right)\left[\frac{1}{2}LI_0^2 e^{-2Rt_0/L}\right]$

Then $e^{-2R(4\times10^{-3})/L} = \frac{1}{2}$ and

$-2R(4\times10^{-3})/5\times10^{-3} = \ln 0.5 = -\ln 2$

$R = \frac{5}{8}\ln 2 = \underline{0.433\,\Omega}$.

6.17 $R = L/\tau = 4/5\times10^{-3} = 800\Omega$

$v = RI_0 e^{-(t-t_0)/\tau}$

If R is replaced by $\frac{R}{2} = 400\Omega$, the voltage is halved. Then

$\tau = \frac{L}{R} = \frac{L/2}{R/2} \Rightarrow L_{new} = \frac{L}{2} = \underline{2\,H}$.

6.18 i_L = inductor current downward.

$v(0^-) = 0$ (short circuit)

$i_L(0^-) = \frac{10V}{10\Omega} = 1A$, For $t>0$, KVL,

$v + 10i_L - \frac{v}{2} = 0$; $v = 2\frac{di_L}{dt}$. Then

$\frac{di_L}{dt} + 10i_L = 0$ ∴ $i_L(t) = i_L(0+)e^{-10t}$

$= e^{-10t}A$

Then $v = 2\frac{d}{dt}(e^{-10t}) = \underline{-20e^{-10t}V}$

6.19 KVL around the right mesh gives

$2(i - \frac{v}{6}) + v + 4i = 0$; $v = L\frac{di}{dt}$ then

$\frac{di}{dt} + 6i = 0$ $\frac{di}{i} = -6dt$.

∴ $i = Ke^{-6t} = i(0)e^{-6t} = \underline{2e^{-6t}\,A}$

6.20 Simplify the resistor network:

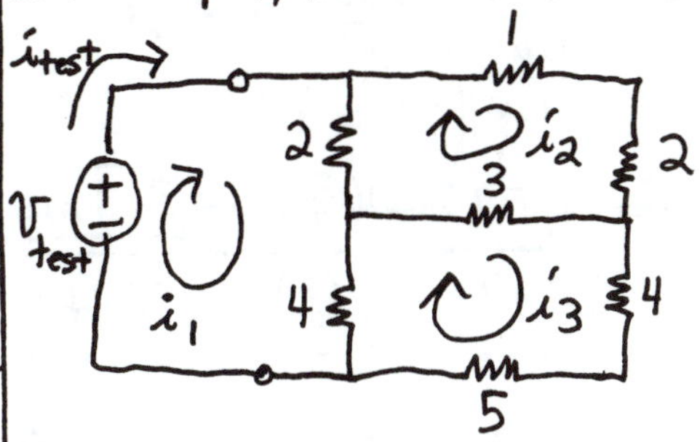

Mesh 1: $i_1 = i_{test}$

Mesh 2: $i_2(1+2+3+2) - i_1(2) - i_3(3) = 0$

$\Rightarrow 8i_2 - 2i_{test} = 3i_3$

Mesh 3: $i_3(3+4+5+4) - i_1(4) - i_2(3) = 0$

$\Rightarrow 16i_3 - 4i_{test} = 3i_2$

Solving $\Rightarrow i_2 = \frac{132}{357}i_{test}$, $i_3 = \frac{38}{119}i_{test}$

KVL around outside loop:

$v_{test} = (1+2)i_2 + (4+5)i_3$

$= \frac{474}{119}i_{test} = 3.98\,i_{test}$

∴ $R_{eq} = 3.98\,\Omega$

$\Rightarrow \tau = R_{eq}C = (3.98\Omega)(100\mu F)$

$= 3.98\times10^{-4}s = \underline{0.398ms}$

6.21 Simplify the resistor network:

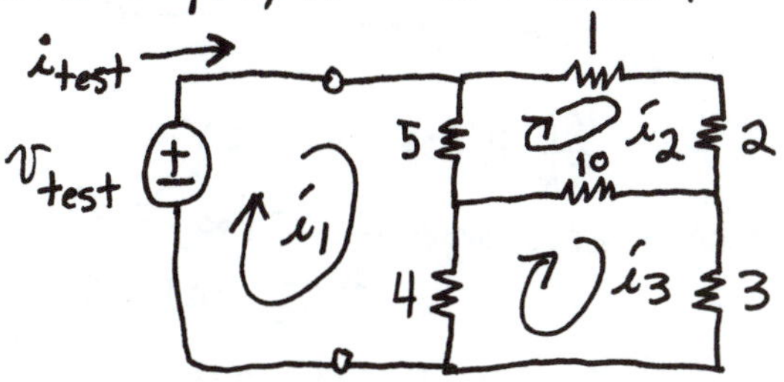

Mesh 1: $i_1 = i_{test}$

Outside Mesh: $v_{test} = 3i_2 + 3i_3$

Mesh 2: $18i_2 - 5i_{test} - 10i_3 = 0$

Mesh 3: $17i_3 - 4i_{test} - 10i_2 = 0$

Solving $\Rightarrow v_{test} = (3.597)i_{test}$

$\Rightarrow R_{eq} = 3.597\Omega$ ∴ $\tau = L/R_{eq} = \underline{13.1ms}$

6.22 $C_{eq} = \frac{(1\mu F)(2\mu F)}{1\mu F + 2\mu F} = \frac{2}{3}\mu F$

$R_{eq} = 10k \| 10k = 5k\Omega$

$\therefore \tau = R_{eq}C_{eq} = (5k\Omega)(\frac{2}{3}\mu F) = \underline{3.\bar{3}\ ms}$

6.23 $L_{eq} = 2 + 1 + \frac{(10)(5)}{10+5} = 6.\bar{3}\ H$

Resistors:

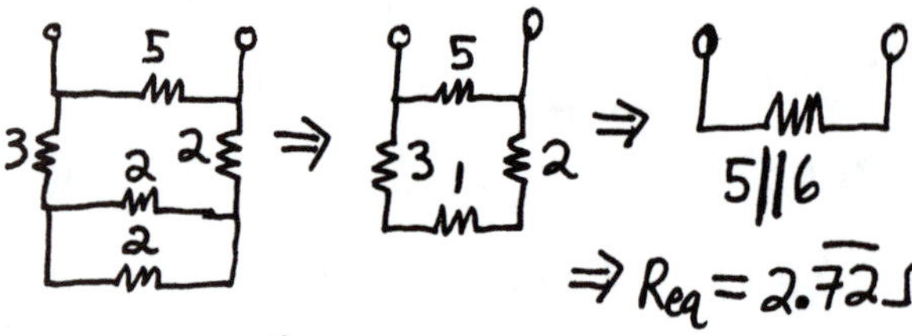

$\Rightarrow R_{eq} = 2.\overline{72}\ \Omega$

$\therefore \tau = \frac{L_{eq}}{R_{eq}} = \frac{6.\bar{3}}{2.\overline{72}} = \underline{2.3\bar{2}\ s}$

6.24 DC steady-state at $t = 0-$:

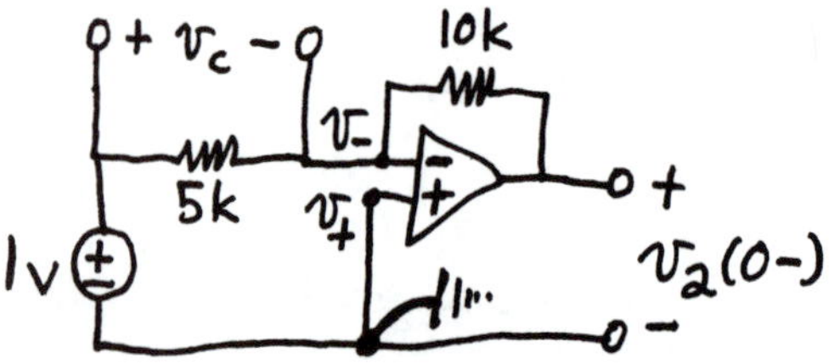

Virtual short: $v_- \simeq v_+ = 0v$

$v_c(0-) = v_1 - 0v = v_1 = 1v = v_c(0+)$

For $t > 0$:

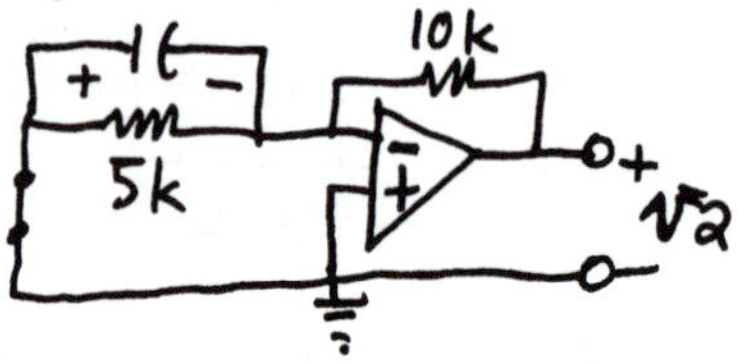

At $t = 0+$:

$v_2 = -\frac{10k}{5k} v_c(0+) = -2V$

Capacitor discharges $\Rightarrow v_c(\infty) = 0$

$\Rightarrow v_2(\infty) = 0$

First order $\Rightarrow v_2(t) = v_2(0+)e^{-t/\tau}$

$\Rightarrow v_2 = -2e^{-t/\tau}, \quad \tau = R_{eq}C.$

Find Req using ideal op amp model:

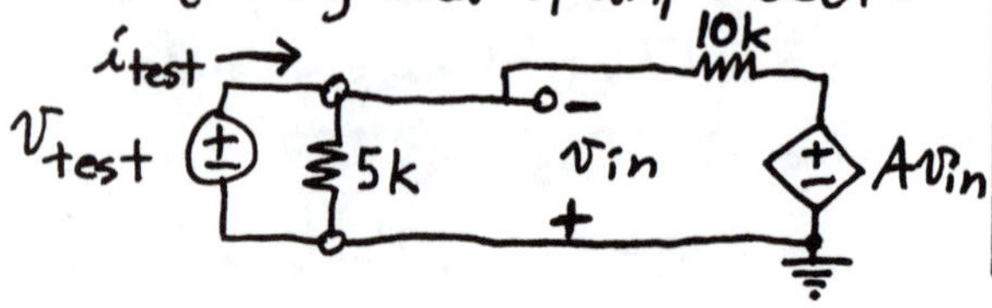

KCL: $i_{test} + \frac{v_{in}}{5k} + \frac{(A+1)v_{in}}{10k} = 0$

$v_{in} = -v_{test}$

$\Rightarrow i_{test} = v_{test}\left(\frac{1}{5k} + \frac{A+1}{10k}\right) = v_{test}(3+A)$

$\Rightarrow R_{eq} = \frac{1}{3+A} = \frac{1}{3+10^5} \simeq 10^{-5}$

$\therefore \tau = R_{eq}C = (10^{-5})(10^{-6}) = 10^{-11}s$

$\Rightarrow v_2(t) = -2e^{-(10^{11})t}\ V$

6.25 DC steady-state at $t = 0-$:

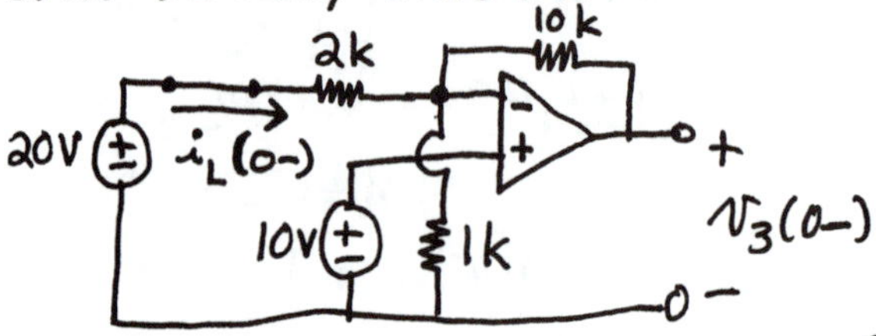

Virtual short: $v_- \simeq v_+ = 10V \Rightarrow i_L(0-) = \frac{20-10}{2k}$

$\Rightarrow i_L(0-) = i_L(0+) = 0.005A$

For $t > 0$, inductor discharges.

At $t = \infty$:

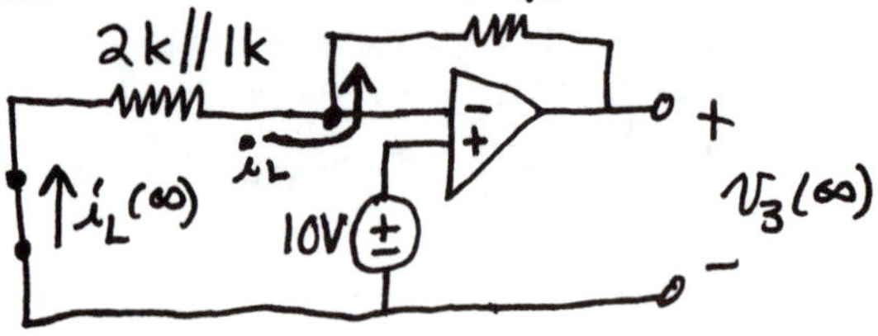

Virtual short: $v_- \simeq v_+ = 10v$

Virtual open: $i_L = \frac{v_- - v_3}{10k} = \frac{-v_-}{0.67k}$

$\Rightarrow \frac{10V}{10k} + \frac{10V}{0.67k} = \frac{v_3}{10k}$

$\Rightarrow (10V)(1 + 15) = 160V = v_3(\infty)$

At $t = 0+$: From above, $i_L(0+) = .005A$

KCL at v_-: $0.005A = \frac{v_-}{1k} + \frac{v_- - v_3(0+)}{10k}$

$= \frac{10}{1k} + \frac{10 - v_3(0+)}{10k}$

$\Rightarrow v_3(0+) = (0.005)(10k) + 10(10) + 10$

$\Rightarrow v_3(0+) = 60V$

6.25 (cont'd) Find R_{eq} since $\tau = L/R_{eq}$

Using ideal voltage amplifier model:

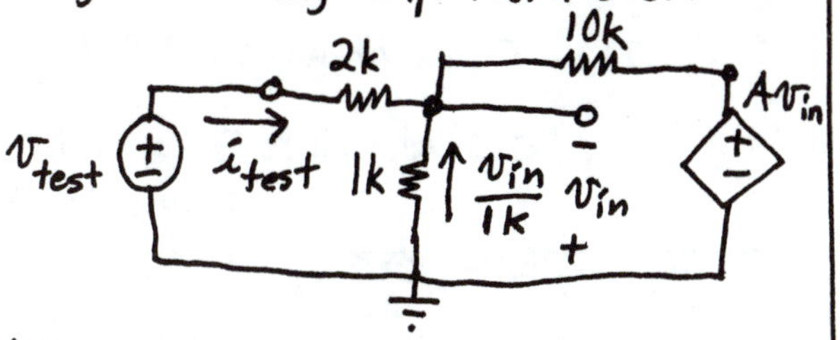

KVL: $v_{test} + v_{in} = 2000\, i_{test}$

$$\Rightarrow v_{in} = 2000\, i_{test} - v_{test}$$

KVL: $A v_{in} + v_{in} + 10000\left(i_{test} + \frac{v_{in}}{1k}\right) = 0$

Solving $\Rightarrow v_{test} = i_{test}\left[\frac{2000A + 32000}{A + 11}\right]$

$$\Rightarrow R_{eq} = \lim_{A\to\infty}\left[\frac{2000A + 32000}{A+11}\right] = 2k$$

Putting it all together:

$$\tau = L/R_{eq} = \frac{1H}{2k} = 5\times 10^{-4}\,s$$

$v_3(0+) = +60\,V,\ v_3(\infty) = +160\,V$

Try $v_3(t) = Ke^{-t/\tau} + A$

Initial cond: $v_3(0+) = K + A = 60$

Final cond: $v_3(\infty) = A = 160$

$$\therefore\ \boxed{v_3(t) = 160 - 100e^{-(2\times 10^3)t}}$$

6.26

capacitor current downward $= i_c$

$i_c = 10^{-6}\frac{dv}{dt}$. By KVL

$10k\Omega(i_c) + v = 10$ or $\frac{dv}{dt} + 100v = 10^3$

separating the variables

$\frac{dv}{v-10} = -100dt$ or $\ln(v-10) = -100t + \ln K$

$v - 10 = Ke^{-100t}$, $v(0^+) = v(0^-) = 5V$

then $5 - 10 = K = -5$

$\therefore\ v = \underline{10 - 5e^{-100t}\,V}$

6.27 At $t = 0^-$, the inductor is a short circuit. thus $i(0^-) = 24/4 = 6A$. For $t > 0$, KCL gives for v_L

$\frac{v_L - 12}{6} + i + \frac{v_L}{12} = 0$ or $3v_L + 12i = 24$

setting $v_L = 2\frac{di}{dt}$ gives $\frac{di}{dt} + 2i = 4$.

separation of variables yeilds

$\frac{di}{i-2} + 2dt = 0$ and $\ln(i-2) = -2t + \ln K$

$\therefore\ i - 2 = Ke^{-2t}$, since $i(0^-) = i(0^+)$

$i(0^+) - 2 = K \Rightarrow K = 6 - 2 = 4$.

then $i = \underline{4e^{-2t} + 2\,A}$ for $t > 0$

6.28 At $t = 0^-$, the inductor voltage is zero, thus $i_L(0^-) = \frac{6V}{6\Omega} = 1A$, down. For $t > 0$, for the circuit shown

$R_{eq} = \frac{6(4+8)}{6+4+8} = 4\Omega$ (seen by inductor)

$\frac{1}{\tau} = \frac{R_{eq}}{L} = \frac{4}{2} = 2\ s^{-1}$;

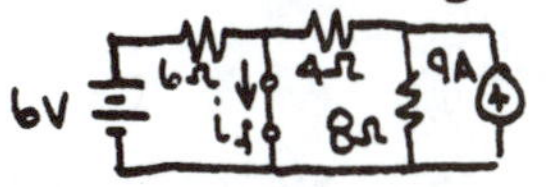

$i_f = \frac{6}{6} + \frac{8}{4+8}(9) = 7A$, since there are essentially two circuits. Therefore

$i_L = i_n + i_f = Ke^{-R_{eq}t/L} + 7$

$i_L(0^+) = i_L(0^-) = 1 = K + 7 \Rightarrow K = -6$

$\therefore\ i_L = \underline{7 - 6e^{-2t}}\,A$.

6.29 At $t = 0^-$, the capacitor is open-circuited. By voltage division

$v_c(0^-) = \frac{2}{2+4}(v_{6\Omega})$, where $v_{6\Omega}$ is the voltage across the 6Ω resistor

Since $v_{6\Omega} = 3v_c(0^-)$, KVL at the top of 6Ω resistor yeilds.

$3v_c(0^-)\left(\frac{1}{6} + \frac{1}{6} + \frac{1}{2}\right) = \frac{10}{2} \Rightarrow v_c(0^-) = 2V$

Now $i = i_n + i_f$, where i_f is the dc steady-state value and i_n is an exponential with $\tau = R_{eq}C$ with the source dead, the resistance seen by the capacitor is

$R_{eq} = \frac{2[4 + (6)(3)/(6+3)]}{2 + 4 + 2} = \frac{3}{2}\Omega$

$\tau = (\frac{3}{2})(\frac{1}{6}) = \frac{1}{4}\ s^{-1}$

with the circuit in dc steady-state, the capacitor is open, and KCL

$v_{6\Omega}\left(\frac{1}{6} + \frac{1}{6} + \frac{1}{3}\right) = \frac{3}{3} \Rightarrow v_{6\Omega} = \frac{3}{2}$

$\therefore\ i_f = \frac{3}{2}/(4+2) = \frac{1}{4}A$

$i = \frac{1}{4} + Ke^{-4t}$

6.29 (cont'd)

To determine K we need $i(0^+)$:

$v_{6\Omega}(0^+) = v_c(0^+) + 4i(0^+) = 4i(0^+) + 2$

KCL at top of 6Ω resistor yeilds

$$\frac{v_{6\Omega}(0^+)-3}{3} + \frac{v_{6\Omega}(0^+)}{6} + i(0^+) = 0 \text{ or}$$

$$\frac{4i(0^+)+2-3}{3} + \frac{4i(0^+)+2}{6} + i(0^+) = 0$$

$\therefore i(0^+) = 0 = \frac{1}{4} + k \Rightarrow k = -\frac{1}{4}$

$i = \frac{1}{4}(1 - e^{-4t})\,A$

6.30

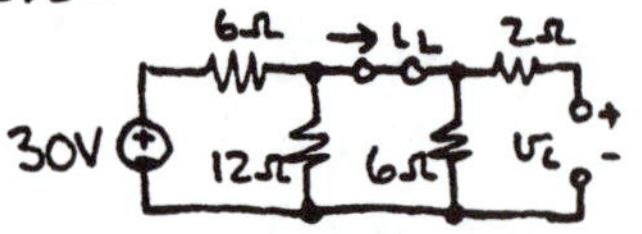

At $t = 0^-$, by voltage division

$$v_c(0^-) = \frac{6(12)/6+12}{6+4}(30) = 12V, \text{ then}$$

$i_L(0^-) = \frac{v_c(0^-)}{6} = 2A$. At $t = 0^+$, the closing of the switch seperates the circuit into two independent circuits due to the short-circuit.

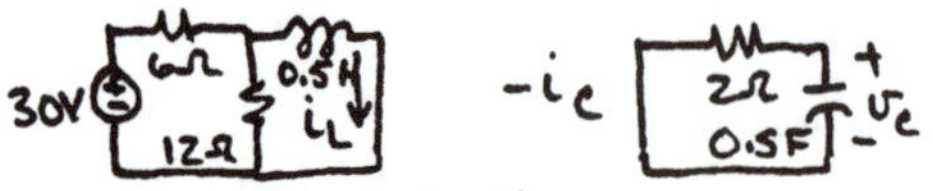

For $t > 0$, $i = i_L - i_c$ as shown above In the capacitor circuit.

$$-i_c = -0.5\frac{dv_c}{dt} = -0.5\,v_c(0^+)\frac{d}{dt}\left(e^{-\frac{t}{RC}}\right)$$

$$= (0.5)(12)\left(\frac{1}{2(0.5)}\right)e^{-t} = 6e^{-t}A$$

In the inductor circuit $i_L = i_{Ln} + i_{Lf}$ where $i_{Lf} = 30/6 = 5A$ (inductor is a short-circuit), with the source dead the inductor sees the resistance $R_{eq} = \frac{6(12)}{6+12} = 4\Omega$. Then $\frac{R_{eq}}{L} = \frac{4}{1/2} = 8$ and $i_L = 5 + Ke^{-8t}$;

$i_L(0^+) = i_L(0^-) = 2 = 5 + K \Rightarrow K = -3$

and $i_L = 5 - 3e^{-8t}A$.

$\therefore i = 5 - 3e^{-8t} + 6e^{-t}A \quad (i_L - i_c = i)$

6.31 Since input voltage is zero for $t<0$,

$v_c(0-) = v_c(0+) = 0$ uncharged cap.

$\Rightarrow v_o(0+) = 0$ by virtual short principle.

At $t = \infty$, DC steady-state $\Rightarrow$ capacitor is opencircuit.

$$\Rightarrow v_o(\infty) = -\frac{20k}{2k}(10V) = -100V$$

First order $\Rightarrow v_o(t) = Ke^{-t/\tau} + A$

$v_o(0) = K + A = 0 \Rightarrow K = -A$

$v_o(\infty) = A = -100$

$\Rightarrow v_o(t) = -100(1 - e^{-t/\tau})\,V$

Using actual circuit methods:

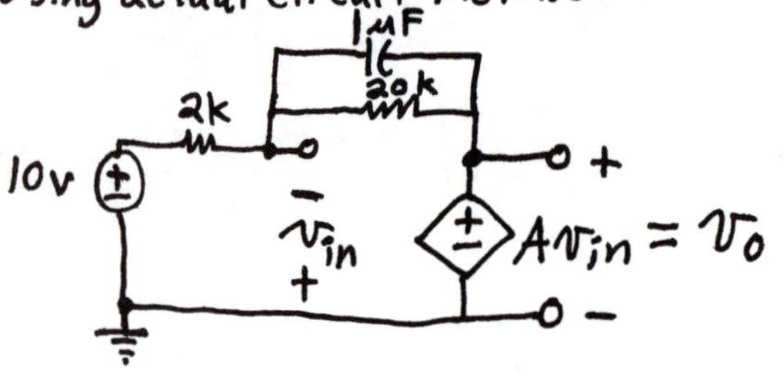

$$\text{KCL: } \frac{10-(-v_{in})}{2k} = \frac{-v_{in} - Av_{in}}{20k} + 1\mu\frac{dv_c}{dt}$$

$$\Rightarrow \frac{10}{2k} + \frac{v_{in}}{2k} = \frac{-(A+1)v_{in}}{20k} + (1\mu)\frac{d}{dt}\left[-(A+1)v_{in}\right]$$

$$\Rightarrow \frac{dv_{in}}{dt} + 50\frac{(A+11)}{(A+1)}v_{in} = \frac{-5000}{(A+1)}$$

$$\lim_{A\to\infty}: \frac{dv_{in}}{dt} + 50\,v_{in} = 0$$

$$\Rightarrow \tau = \frac{1}{50} \quad (\text{or } R_{eq} = 20k\Omega)$$

From above: $\boxed{v_o(t) = -100(1 - e^{-50t})V}$

6.32 $v_o(0) = (30k\Omega)(i(0)) = -30\text{ V}$

At $t=\infty$, ckt at DC steady state

$\Rightarrow v(\infty) = 15\text{ V}.$

$\tau = L/R = 10mH/30k = 3.\overline{3}\times10^{-7}\text{ s}$

$\therefore v_o(t) = A + Be^{-t/\tau}$

Match initial cond: $v_o(0) = A+B = -30$

Match final: $v_o(\infty) = A = 15$

$\therefore$ $v_o(t) = 15 - 45e^{-(3\times10^6)t}\text{ V}$

6.33 ① Response due to initial conditions: kill independent sources.

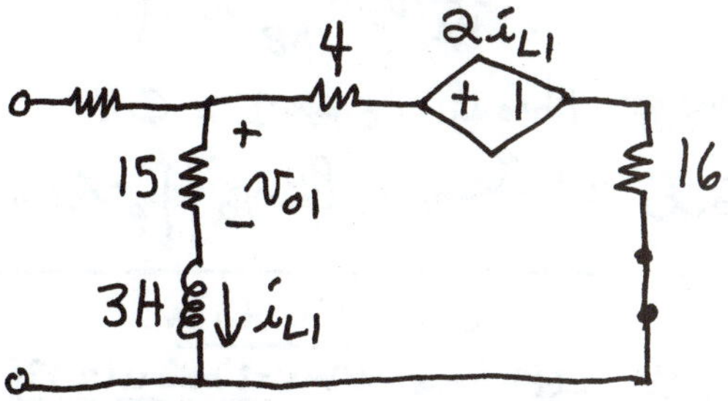

KVL: $i_{L1}(16+4+15) + 3\frac{di_{L1}}{dt} - 2i_{L1} = 0$

$\Rightarrow \frac{di_{L1}}{dt} + 11i_{L1} = 0$

$i_{L1}(t) = K_1e^{-11t}$. $K_1 = i_{L1}(0) = 3$.

$\Rightarrow v_{o1}(t) = 15i_{L1} = 45e^{-11t}$

② Response due to independent sources: Set I.C.'s to zero.

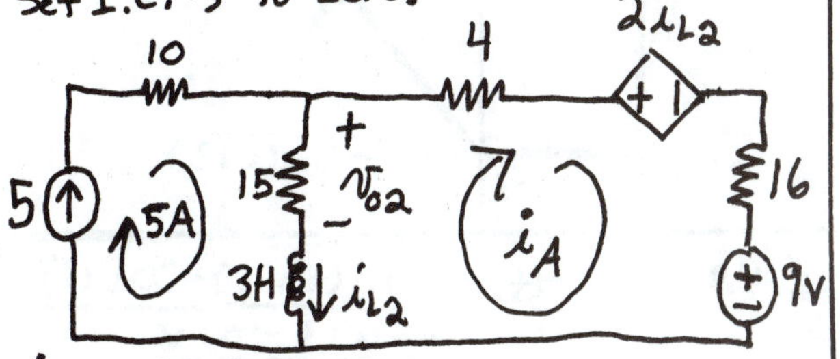

$i_{L2} = 5 - i_A$

Mesh i_A: $15(i_A-5) + i_A(4+16) + 2i_{L2} + 9 + 3\frac{d[i_A-5]}{dt} = 0$

Simplify $\Rightarrow \frac{di_{L2}}{dt} + 11i_{L2} = 36.\overline{3}$

$i_{L2}(0) = 0$

$i_{L2}(t) = Ke^{-11t} + A \Rightarrow i_{L2}(0) = K + A = 0$

$\Rightarrow i_{L2}(t) = K(1-e^{-11t})\text{ A}$

Subsititude:

$\frac{d}{dt}[K(1-e^{-11t})] + 11K - 11Ke^{-11t} = 36.\overline{3}$

$\Rightarrow 11Ke^{-11t} + 11K - 11Ke^{-11t} = 36.\overline{3}$

$\Rightarrow K = 3.3\overline{03} \Rightarrow i_{L2}(t) = 3.3\overline{03}(1-e^{-11t})$

$\Rightarrow v_{o2} = 15i_{L2}(t) = 49.5\overline{45}(1-e^{-11t})$

$\therefore v_o = v_{o1} + v_{o2} = 45e^{-11t} + 49.5\overline{45}(1-e^{-11t})$

$= [49.5\overline{45} - 4.5\overline{45}e^{-11t}]\text{ V}$

6.34 ① I.C. Response:

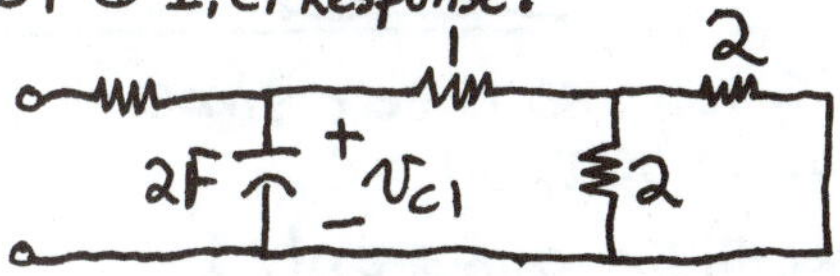

$\Rightarrow$

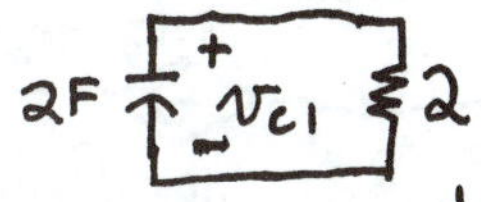

$\tau = RC = 4\text{s}$

$v_{c1}(0) = 20\text{ V}$

$\Rightarrow v_{c1}(t) = 20e^{-t/4}\text{ V}$

② Source response:

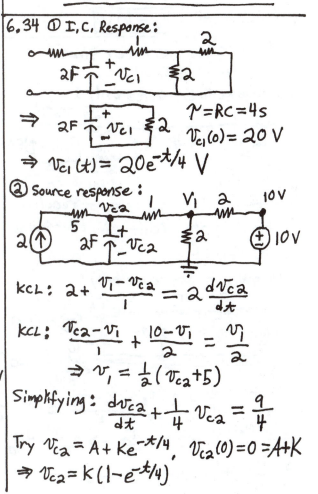

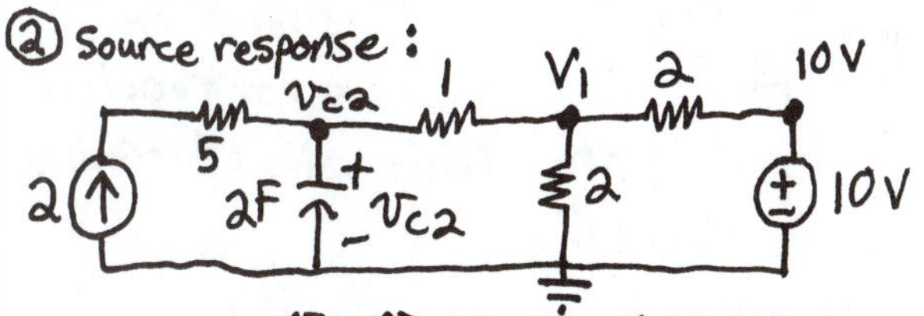

KCL: $2 + \frac{v_1 - v_{c2}}{1} = 2\frac{dv_{c2}}{dt}$

KCL: $\frac{v_{c2}-v_1}{1} + \frac{10-v_1}{2} = \frac{v_1}{2}$

$\Rightarrow v_1 = \frac{1}{2}(v_{c2}+5)$

Simplifying: $\frac{dv_{c2}}{dt} + \frac{1}{4}v_{c2} = \frac{9}{4}$

Try $v_{c2} = A + Ke^{-t/4}$, $v_{c2}(0) = 0 = A+K$

$\Rightarrow v_{c2} = K(1-e^{-t/4})$

6.34 (cont'd)

Substituting $\Rightarrow \frac{9}{4} = \frac{d}{dt}\left[K(1-e^{-t/4})\right] + \frac{K}{4}(1-e^{-t/4})$

$\Rightarrow K = 9$

$\therefore v_c = v_{c1} + v_{c2} = 20e^{-t/4} + 9 - 9e^{-t/4}$

$= 9 + 11e^{-t/4}$ V

6.35 ① I.C. response:

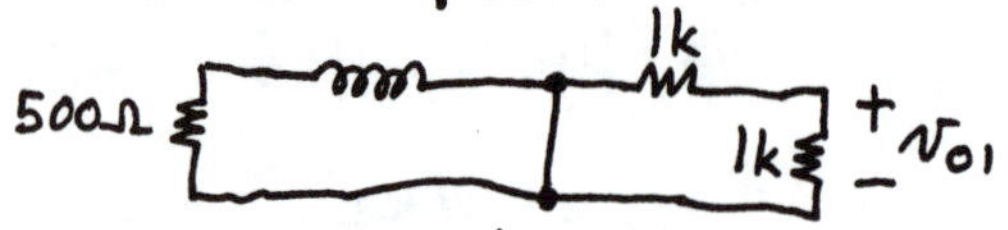

$v_{01} = 0$ due to short ckt

② Source response:

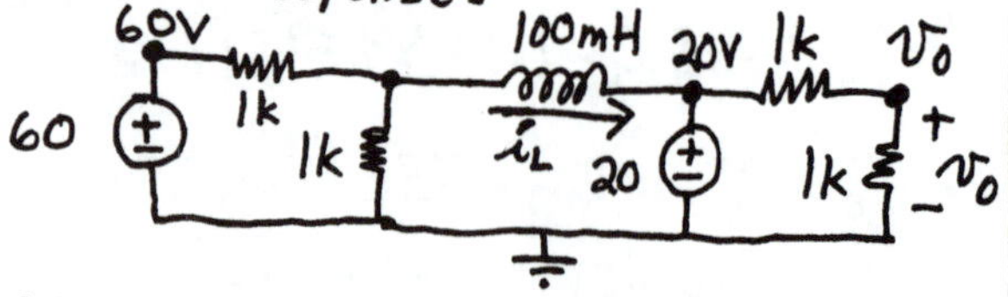

Since 20V src is fixed, by voltage division $\Rightarrow$ $\boxed{v_o = 10\,V}$

6.36 ① I.C. response is zero since $v_o(0) = v_c(0) = 0$.

② Response to 5A source only:

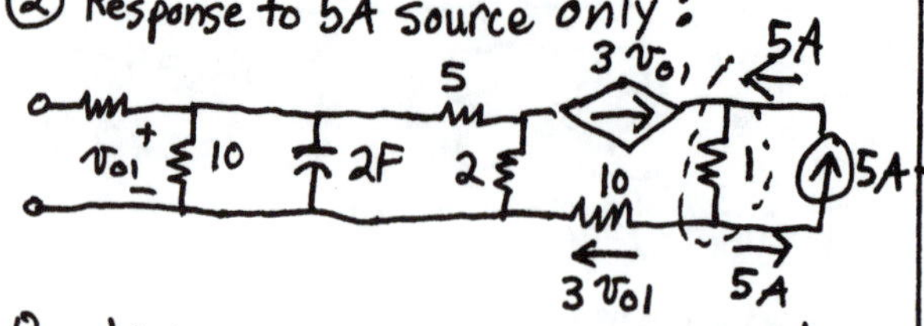

By KCL, 5A source does not reach left part of the circuit. No other sources. $\Rightarrow v_{01} = 0$.

③ Response to 10A source only:

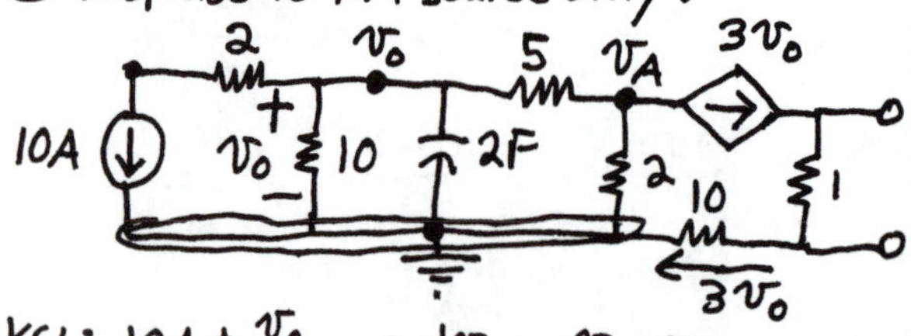

KCL: $10A + \frac{v_o}{10} + 2\frac{dv_o}{dt} + \frac{v_o - v_A}{5} = 0$

KCL: $\frac{v_o - v_A}{5} = 3v_o + \frac{v_A}{2} \Rightarrow v_A = -4v_o$

$\Rightarrow \frac{dv_o}{dt} + 0.55\,v_o = -5$

Try $v_o(t) = A + Ke^{-0.55t}$

$v_o(0) = 0 = A + K \Rightarrow v_o(t) = K(1 - e^{-0.55t})$

Substitute $\Rightarrow K\frac{d}{dt}\left[1 - e^{-0.55t}\right] + 0.55K(1 - e^{-.55t}) = -5$

$\Rightarrow K = -9.\overline{09}$

$\therefore v_o(t) = 9.\overline{09}\,(e^{-.55t} - 1)$ V

6.37 Initially, $v_c = 0 \Rightarrow v_3(0) = 0$.

At $t = \infty$, DC steadystate $\Rightarrow$ C open ckt.

Inverting summer: $v_3(\infty) = -\frac{R_F}{R_A}v_1 - \frac{R_F}{R_B}v_2$

$\therefore v_3(t) = K(1 - e^{-t/\tau})$

$= \left[-\frac{R_F}{R_A}v_1 - \frac{R_F}{R_B}v_2\right](1 - e^{-t/\tau})$

C charges through $R_F \Rightarrow \tau = CR_F$

$\therefore v_3(t) = \left[\frac{R_F}{R_A}v_1 + \frac{R_F}{R_B}v_2\right]\left[e^{-t/CR_F} - 1\right]$

6.38 (a) $i(t) = 5u(t)$ A.

(b) $i(t) = -u(-20ms - t) + 3u(t + 20ms) + u(t - 40ms)$ A.

(c) $i(t) = 6u(10 - t) - 12u(t - 10)$ μA.

6.39

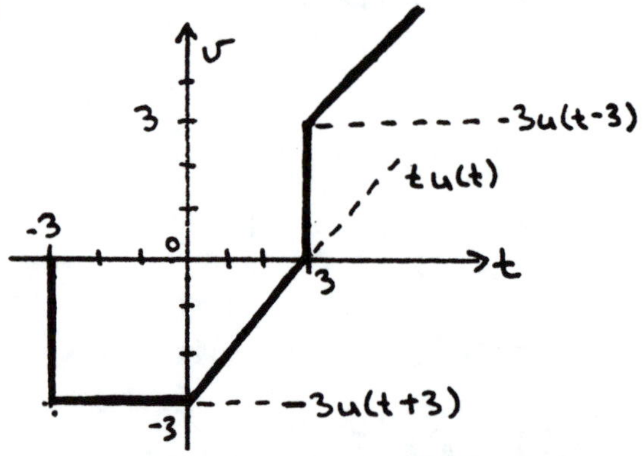

6.40 $v(t) = -10u(t + 10) + 30u(t) - 5u(t - 10)$ V

6.41 $f(t) = u(t-1) - u(t-2) + u(t-3) + u(t-4) - u(t-5) + 2u(t-7) - 3u(t-8)$

6.42 For $t<0$, $v_1(t)=0$. For $t>0$, $v_1 = v_n + v_f$. With the voltage source shorted, the capacitor sees the resistance

$Req = 5 + \frac{10}{2} = 10k\Omega$

$\therefore v_n = Ae^{-t/R_{eq}C} = Ae^{-100t}$

v_f = dc steadystate $= \frac{10}{2} = 5V$

since $v_1(0^-) = v_1(0^+) = 0$

$v_1 = Ae^{-100t} + 5$

$v_1(0) = A + 5 = 0 \Rightarrow A = -5$

$v_1(t) = (5 - 5e^{-100t})u(t)\ V$

6.43 KCL gives at the inverting op amp terminals,

$\frac{1}{4}\frac{dv}{dt} + \frac{v}{2} + \frac{v_g}{4} = 0$ or

$\frac{dv}{dt} + 2v = -3e^{-3(t+4)}u(t+4)$

For $t > -4$,

$ve^{2t} = A + \int(-3e^{-3(t+4)})(e^{2t})dt$

$= A + 3e^{-(t+12)}$

$v = Ae^{-2t} + 3e^{-3(t+4)}$

$v(-4) = Ae^{8} + 3 = 0 \Rightarrow A = -3e^{-8}$

$\therefore v = [3e^{-3(t+4)} - 3e^{-2(t+4)}]u(t+4)$ V.

6.44 For $t<0$, $v_g = 0$ and therefore the capacitor voltage is zero. Hence $i(0+)=0$. With the source dead the resistance seen by the capacitor is

$$Req = \frac{6[9 + \frac{4(12)}{4+12}]}{6+9+3} = 4\Omega$$

$ReqC = 4(\frac{1}{20}) = \frac{1}{5}s$

Now $i = i_n + i_f = A_1e^{-5t} + i_f$, where i_f is the dc steady-state current. With the capacitor open circuited the voltage across the 12Ω resistor is $(6+9)i_f = 15 i_f$, By voltage div.

$$15 i_f = \frac{15(12)/(15+12)}{4 + \frac{(15)(12)}{(15+12)}}(1) \Rightarrow i_f = \frac{1}{24}A$$

6.44 (cont'd)

$\therefore i = A_1e^{-5t} + 1/24,\ t>0$

$i(0^+) = 0 = A_1 + 1/24 \Rightarrow A_1 = -1/24$

$\therefore i = \frac{1}{24}(1 - e^{-5t})u(t)\ A$

6.45 (a) Circuit for $t>0$:

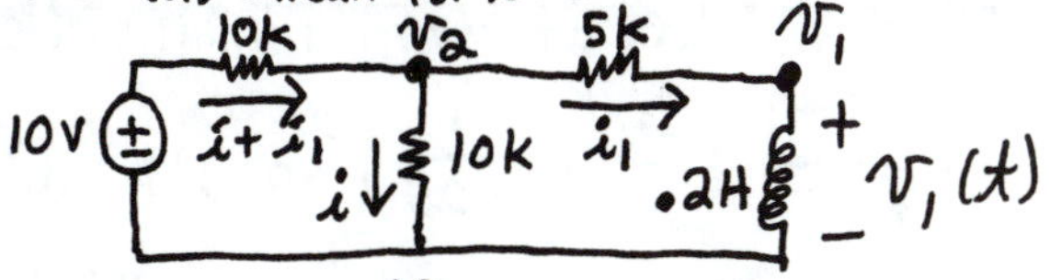

$v_1(t) = 0.2\frac{di_1}{dt}$ $\qquad i_1(0) = 0A$.

KVL: $(5k)i_1 + 0.2\frac{di_1}{dt} - (10k)i = 0$

KVL: $(10k)(i+i_1) + (10k)i = 10$

$\Rightarrow i = 5\times10^{-4} - \frac{1}{2}i_1$

$\Rightarrow \frac{di_1}{dt} + (50k)i_1 = 25$

Try $i_1(t) = K(1 - e^{-50000t})$

$\Rightarrow \frac{d}{dt}[K - Ke^{-50000t}] + (50k)[K(1-e^{-50000t})]$
$= 25$

$\Rightarrow K = 5\times10^{-4}$

$\Rightarrow i_1(t) = (5\times10^{-4})[1 - e^{-50000t}]$

$\Rightarrow v_1(t) = 0.2[25e^{-50000t}]$
$= \boxed{5e^{-50000t}\ V}$

(b) For $0 \le t < 2s$, just like (a)

$\Rightarrow v_1(0 \le t < 2) = 5e^{-50000t}\ V$

At $t = (2-)s \Rightarrow i_1 = (5\times10^{-4})[1 - e^{-50000(2)}]$
$\simeq 5\times10^{-4}A = i_1(2+)$

For $t > 2s$, source-free with $\tau = 1/50000$

$\Rightarrow i_1(t>2) = (5\times10^{-4})e^{-50000(t-2)}$

$v_1(t>2) = [0.2]\frac{di_1(t>2)}{dt} = -5e^{-50000(t-2)}$

$\therefore \boxed{v_1(t) = 5e^{-50000t}[u(t) - u(t-2)] - 5e^{-50000(t-2)}u(t-2)}$

6.46 (a) For $t < -4s$, $v = 0$

Initially (at $t = -4s$): $i_L = 0$

$\Rightarrow$ inverting amp $\Rightarrow v = -\frac{2}{4}\left[3e^{-3(-4+4)}\right]$

$= -\frac{3}{2}e^0 = -\frac{3}{2}$ V

At $t = \infty$, DC steady-state and source = 0 V

$\Rightarrow v(\infty) = 0$

For $t > -4$: KCL at inverting input yields

$$\frac{3e^{-3(t+4)}}{4} + \frac{v}{2} + \frac{1}{0.25}\int_{-4}^{t} v(\tau)d\tau = 0$$

$$\Rightarrow \frac{1}{2}v + 4\int_{-4}^{t} v(\tau)d\tau = -\frac{3}{4}e^{-3(t+4)}$$

$$\frac{d}{dt} \Rightarrow \frac{dv}{dt} + 8v = \frac{9}{2}e^{-3(t+4)}$$

$$v(-4s) = -\frac{3}{2}\text{ V}$$

Try $v(t) = \underbrace{K_1 e^{-3(t+4)}}_{\text{forced soln.}} + \underbrace{K_2 e^{-(t+4)/\tau}}_{\text{natural decay}} + \underbrace{A}_{\text{constant}}$

(for $t > -4$)

Where $\tau = \frac{L}{R} = \frac{0.25\text{H}}{2\Omega} = \frac{1}{8}$

Substitute:

$$\frac{d}{dt}\left[K_1 e^{-3(t+4)} + K_2 e^{-8(t+4)} + A\right] + 8K_1 e^{-3(t+4)}$$

$$+ 8K_2 e^{-8(t+4)} + 8A = \frac{9}{2}e^{-3(t+4)}$$

$$\Rightarrow 5K_1 e^{-3(t+4)} + 8A = \frac{9}{2}e^{-3(t+4)}$$

At $t = -4$: $5K_1 + 8A = \frac{9}{2} \Rightarrow K_1 = \frac{9}{10} - \frac{8}{5}A$

$$v(-4) = -\frac{3}{2} = \left[\frac{9}{10} - \frac{8}{5}A\right]e^0 + K_2 e^0 + A$$

$$\Rightarrow A = 4 + \frac{5}{3}K_2$$

$v(\infty) = 0 = K_1 e^{-\infty} + K_2 e^{-\infty} + A \Rightarrow$

$\underline{A = 0} \Rightarrow K_1 = \frac{9}{10}$, $K_2 = -\frac{12}{5}$

$$\therefore \boxed{v(t) = \left[\frac{9}{10}e^{-3(t+4)} - \frac{12}{5}e^{-8(t+4)}\right]u(t+4)}$$

(b) $v_g = 3e^{-3(t+4)}\left[u(t+4) - u(t)\right]$

For $t < 0$, same answer as part (a).

At $t = 0$, source shut off.

For $t > 0$, v decays to zero with

$\tau = \frac{L}{R} = \frac{0.25}{2} = \frac{1}{8} \Rightarrow v(t>0) = v(0+)e^{-8t}$

To find $v(0+)$, need to first find $i_L(0-) = i_L(0+)$

For $-4s < t < 0$:

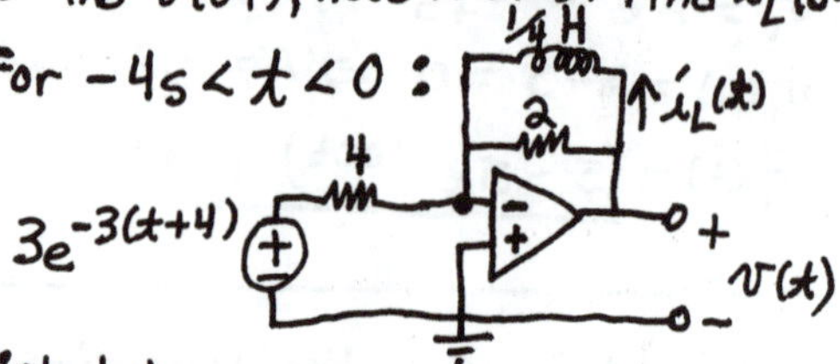

Virtual short: $v_- \simeq v_+ = 0$ V

$$\Rightarrow i_L(t) = \frac{1}{(1/4)}\int_{-4}^{t} v(\tau)d\tau$$

$$\Rightarrow i_L(t) = 4\left[-\frac{3}{10}e^{-3(t+4)} + \frac{3}{10}e^{-8(t+4)}\right]$$

$$\Rightarrow i_L(0-) = i_L(0+) = 4\left[-.3e^{-12} + .3e^{-32}\right]$$

$$= -7.373 \times 10^{-6}\text{ A}$$

For $t > 0$, $i_L(t) = (-7.373 \times 10^{-6})e^{-8t}$

$$\Rightarrow v(t) = \frac{1}{4}\frac{di_L}{dt} = 1.47 \times 10^{-5}e^{-8t}$$

$$\Rightarrow v(0+) = 1.47 \times 10^{-5}\text{ V}$$

$$\therefore v(t) = \left[\frac{9}{10}e^{-3(t+4)} - \frac{12}{5}e^{-8(t+4)}\right] \cdot \left[u(t+4) - u(t)\right] + (1.47 \times 10^{-5})e^{-8t}u(t)$$

6.47 (a) For $t>0$:

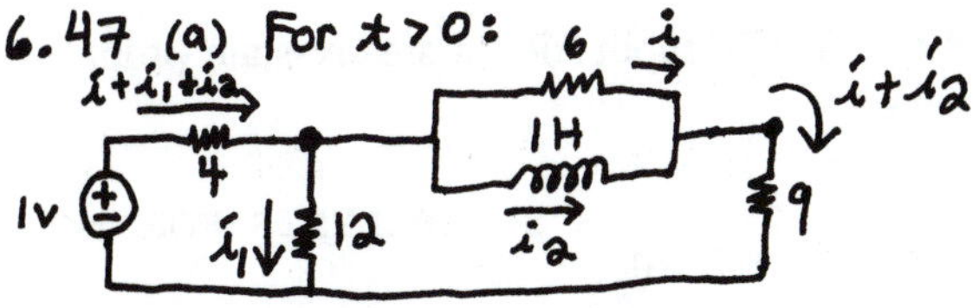

KVL left: $1v = 4(i+i_1+i_2)+12i_1$

KVL: $6i+9(i+i_2)=12i_1$

KVL: $6i=(1)\frac{di_2}{dt}$

Simplifying: $\frac{di_2}{dt}+4i_2=\frac{1}{4}$, $i_2(0)=0$

$\Rightarrow \tau = 1/4$

At $t=\infty$, DC steady state:

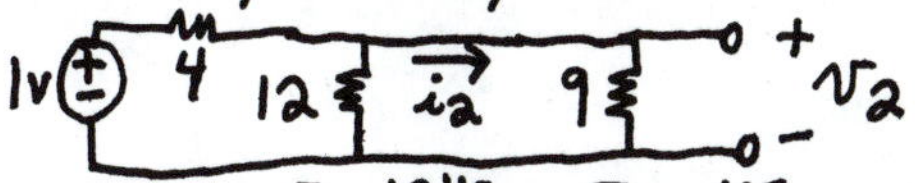

$v_2=(1v)\left[\frac{12\|9}{4+(12\|9)}\right]=\frac{108}{192}=0.5625$

$\Rightarrow i_2(\infty)=0.0625=1/16$

$\Rightarrow i_2(t)=\frac{1}{16}\left[1-e^{-4t}\right]$

At $t=0+$, $i_2(0+)=0$

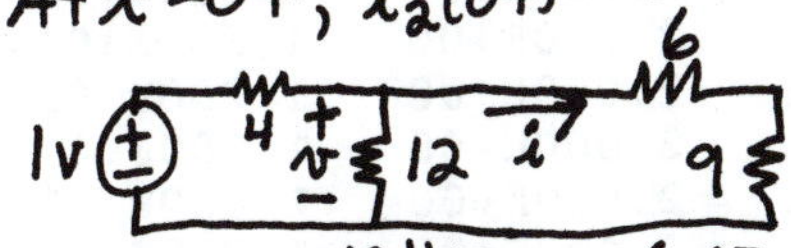

$v=(1V)\frac{12\|15}{4+12\|15}=\frac{6.67}{10.67}=0.625$

$\Rightarrow i(0+)=\frac{0.625}{15\Omega}=0.04167\text{ A}$

$i(\infty)=0\text{ A}$, $\tau=1/4$

$\therefore$ $i(t)=0.04167e^{-4t}u(t)$
$=\frac{1}{24}e^{-4t}u(t)$

(b) For $t<1s$, same as part (a)

$i=\frac{1}{24}e^{-4t}u(t)$

At $t=1s$, $i=\frac{1}{24}e^{-4(1)}=7.6315\times10^{-4}$

Source is shut off:

$i(t>1s)=(7.6315\times10^{-4})e^{-4(t-1)}$

$\therefore$ For all t,

$i(t)=\frac{1}{24}e^{-4t}\left[u(t)-u(t-1)\right]$
$+(7.6315\times10^{-4})e^{-4(t-1)}u(t-1)$

6.48 SPICE input file:

```
PROBLEM 6.48 - TRANSIENT ANALYSIS
VDUMMY 1 2 DC 0
R 0 1 5KOHM
C 2 0 2UF IC=100
.TRAN 1.1E-4 7.15E-3 UIC
.PRINT TRAN I(VDUMMY) V(2)
.END
```

SPICE output:

```
*************************************

TIME         I(VDUMMY)     V(2)

0.000E+00    -2.000E-02    1.000E+02
1.100E-04    -1.978E-02    9.891E+01
2.200E-04    -1.957E-02    9.783E+01
3.300E-04    -1.935E-02    9.675E+01
4.400E-04    -1.914E-02    9.570E+01
5.500E-04    -1.893E-02    9.465E+01
6.600E-04    -1.872E-02    9.362E+01
7.700E-04    -1.852E-02    9.259E+01
8.800E-04    -1.832E-02    9.158E+01
9.900E-04    -1.812E-02    9.058E+01
1.100E-03    -1.792E-02    8.959E+01
1.210E-03    -1.772E-02    8.860E+01
1.320E-03    -1.753E-02    8.763E+01
1.430E-03    -1.734E-02    8.668E+01
1.540E-03    -1.715E-02    8.573E+01
1.650E-03    -1.696E-02    8.479E+01
1.760E-03    -1.677E-02    8.386E+01
1.870E-03    -1.659E-02    8.295E+01
1.980E-03    -1.641E-02    8.204E+01
.
.
.
5.610E-03    -1.141E-02    5.706E+01
5.720E-03    -1.129E-02    5.644E+01
5.830E-03    -1.116E-02    5.582E+01
5.940E-03    -1.104E-02    5.521E+01
6.050E-03    -1.092E-02    5.461E+01
6.160E-03    -1.080E-02    5.401E+01
6.270E-03    -1.068E-02    5.342E+01
6.380E-03    -1.057E-02    5.284E+01
6.490E-03    -1.045E-02    5.226E+01
6.600E-03    -1.034E-02    5.169E+01
6.710E-03    -1.022E-02    5.112E+01
6.820E-03    -1.011E-02    5.056E+01
6.930E-03    -1.000E-02    5.001E+01
7.040E-03    -9.892E-03    4.946E+01
7.150E-03    -9.784E-03    4.892E+01
```

6.49 SPICE input file for steady-state (initial) values:

```
PROBLEM 6.49 - S.S. ANALYSIS FOR T<0
VSRC 1 0 DC 30
R1 1 2 6
R2 2 0 12
R3 4 0 6
R4 4 5 2
L 2 3 0.5
C 5 0 0.5
VDUMMY 3 4 0
.TRAN 0.1 1.0
.PRINT TRAN I(VDUMMY) V(5)
.END
```

SPICE output:

```
*************************************

TIME         I(VDUMMY)     V(5)

0.000E+00    2.000E+00     1.200E+01
1.000E-01    2.000E+00     1.200E+01
2.000E-01    2.000E+00     1.200E+01
3.000E-01    2.000E+00     1.200E+01
4.000E-01    2.000E+00     1.200E+01
5.000E-01    2.000E+00     1.200E+01
6.000E-01    2.000E+00     1.200E+01
7.000E-01    2.000E+00     1.200E+01
8.000E-01    2.000E+00     1.200E+01
9.000E-01    2.000E+00     1.200E+01
1.000E+00    2.000E+00     1.200E+01
```

SPICE input file for $t > 0$ using initial conditions from above:

```
PROBLEM 6.49 - TRANS ANALYSIS FOR T>0
VSRC 1 0 DC 30
R1 1 2 6
R2 2 0 12
R3 3 0 6
R4 3 4 2
L 2 3 0.5 IC=2
C 4 0 0.5 IC=12
VDUMMY 3 0 0
.TRAN 5E-2 10.0 UIC
.PRINT TRAN I(VDUMMY)
.END
```

SPICE output:

```
*************************

     TIME            I(VDUMMY)

     0.000E+00       8.001E+00
     5.000E-02       8.682E+00
     1.000E-01       9.048E+00
     1.500E-01       9.230E+00
     2.000E-01       9.263E+00
     2.500E-01       9.297E+00
     3.000E-01       9.182E+00
     3.500E-01       9.047E+00
     4.000E-01       8.913E+00
     4.500E-01       8.778E+00
     5.000E-01       8.617E+00
     5.500E-01       8.452E+00
     6.000E-01       8.288E+00
     6.500E-01       8.124E+00
     7.000E-01       7.981E+00
     7.500E-01       7.841E+00
     8.000E-01       7.700E+00
     8.500E-01       7.560E+00
     9.000E-01       7.442E+00
     9.500E-01       7.326E+00
     1.000E+00       7.210E+00
     1.050E+00       7.095E+00
     1.100E+00       6.998E+00
     1.150E+00       6.903E+00
     .
     .
     .
     7.700E+00       5.003E+00
     7.750E+00       5.003E+00
     7.800E+00       5.002E+00
     7.850E+00       5.002E+00
     7.900E+00       5.002E+00
     7.950E+00       5.002E+00
     8.000E+00       5.002E+00
     .
     .
     .
     9.300E+00       5.001E+00
     9.350E+00       5.001E+00
     9.400E+00       5.000E+00
     9.450E+00       5.000E+00
     9.500E+00       5.000E+00
     9.550E+00       5.000E+00
     9.600E+00       5.000E+00
     9.650E+00       5.000E+00
     9.700E+00       5.000E+00
     9.750E+00       5.000E+00
     9.800E+00       5.000E+00
     9.850E+00       5.000E+00
     9.900E+00       5.000E+00
     9.950E+00       5.000E+00
     1.000E+01       5.000E+00
```

6.50 SPICE input file:

```
PROBLEM 6.50 - TRANS ANALYSIS FOR T>0
VSRC 1 0 DC 10
R1 1 2 10K
R2 2 0 10K
R3 2 3 5K
L 3 0 0.2
.TRAN 1E-6 2E-4 UIC
.PRINT TRAN V(3)
.END
```

SPICE output:

```
*************************

     TIME            V(3)

     0.000E+00       5.000E+00
     1.000E-06       4.757E+00
     2.000E-06       4.526E+00
     3.000E-06       4.308E+00
     4.000E-06       4.102E+00
     5.000E-06       3.895E+00
     6.000E-06       3.715E+00
     7.000E-06       3.539E+00
     8.000E-06       3.363E+00
     9.000E-06       3.187E+00
     1.000E-05       3.040E+00
     1.100E-05       2.896E+00
     1.200E-05       2.752E+00
     1.300E-05       2.608E+00
     1.400E-05       2.487E+00
     1.500E-05       2.369E+00
     1.600E-05       2.252E+00
     1.700E-05       2.134E+00
     1.800E-05       2.035E+00
     1.900E-05       1.938E+00
     .
     .
     .
     1.900E-04       3.640E-04
     1.910E-04       3.467E-04
     1.920E-04       3.295E-04
     1.930E-04       3.123E-04
     1.940E-04       2.978E-04
     1.950E-04       2.837E-04
     1.960E-04       2.696E-04
     1.970E-04       2.555E-04
     1.980E-04       2.434E-04
     1.990E-04       2.315E-04
     2.000E-04       2.197E-04
```

6.51 SPICE input file:

```
PROBLEM 6.51 - TRANS ANALYSIS FOR T>0
VSRC 1 0 PULSE(0 10 0 1P 1P 1E-5 1E-
4)
R1 1 2 10K
R2 2 0 10K
R3 2 3 5K
L 3 0 0.2
.TRAN 1E-7 1E-4 UIC
.PRINT TRAN V(3)
.END
```

SPICE output:

```
************************

  TIME          V(3)

  0.000E+00     3.802E-09
  1.000E-07     4.975E+00
  2.000E-07     4.950E+00
  3.000E-07     4.926E+00
  4.000E-07     4.901E+00
  5.000E-07     4.877E+00
  6.000E-07     4.852E+00
  7.000E-07     4.828E+00
  8.000E-07     4.804E+00
  9.000E-07     4.781E+00
  1.000E-06     4.757E+00
  1.100E-06     4.733E+00
  1.200E-06     4.709E+00
  1.300E-06     4.686E+00
  1.400E-06     4.663E+00
  1.500E-06     4.640E+00
  1.600E-06     4.617E+00
  1.700E-06     4.595E+00
  1.800E-06     4.572E+00
  1.900E-06     4.549E+00
  2.000E-06     4.526E+00
  2.100E-06     4.503E+00
  2.200E-06     4.480E+00
  .
  .
  .
  8.000E-06     3.354E+00
  8.100E-06     3.336E+00
  8.200E-06     3.319E+00
  8.300E-06     3.302E+00
  8.400E-06     3.284E+00
  8.500E-06     3.267E+00
  8.600E-06     3.250E+00
  8.700E-06     3.233E+00
  8.800E-06     3.216E+00
  8.900E-06     3.199E+00
  9.000E-06     3.182E+00
  9.100E-06     3.165E+00
  9.200E-06     3.148E+00
  9.300E-06     3.131E+00
  9.400E-06     3.114E+00
  9.500E-06     3.097E+00
  9.600E-06     3.080E+00
  9.700E-06     3.063E+00
  9.800E-06     3.046E+00
  9.900E-06     3.029E+00
  1.000E-05     3.012E+00
  1.010E-05    -1.958E+00
  1.020E-05    -1.948E+00
  1.030E-05    -1.938E+00
  1.040E-05    -1.929E+00
  1.050E-05    -1.919E+00
  1.060E-05    -1.909E+00
  1.070E-05    -1.900E+00
  1.080E-05    -1.891E+00
  1.090E-05    -1.881E+00
  1.100E-05    -1.872E+00
  1.110E-05    -1.862E+00
  1.120E-05    -1.853E+00
  1.130E-05    -1.844E+00
  1.140E-05    -1.835E+00
  1.150E-05    -1.826E+00
  1.160E-05    -1.817E+00
  1.170E-05    -1.808E+00
  1.180E-05    -1.799E+00
  1.190E-05    -1.790E+00
  1.200E-05    -1.781E+00
  1.210E-05    -1.772E+00
  1.220E-05    -1.763E+00
  1.230E-05    -1.755E+00
  1.240E-05    -1.746E+00
  1.250E-05    -1.737E+00
  1.260E-05    -1.728E+00
  1.270E-05    -1.719E+00
  1.280E-05    -1.711E+00
  1.290E-05    -1.703E+00
  1.300E-05    -1.695E+00
  .
  .
  .
  7.000E-05    -9.783E-02
  7.010E-05    -9.733E-02
  7.020E-05    -9.684E-02
  7.030E-05    -9.634E-02
  7.040E-05    -9.584E-02
  7.050E-05    -9.535E-02
  7.060E-05    -9.485E-02
  7.070E-05    -9.436E-02
  7.080E-05    -9.391E-02
  .
  .
  .
  9.900E-05    -1.816E-02
  9.910E-05    -1.659E-02
  9.920E-05    -1.501E-02
  9.930E-05    -1.344E-02
  9.940E-05    -1.187E-02
  9.950E-05    -1.029E-02
  9.960E-05    -8.717E-03
  9.970E-05    -7.142E-03
  9.980E-05    -5.568E-03
  9.990E-05    -3.993E-03
  1.000E-04    -2.419E-03
```

6.52 (MANY SOLUTIONS)

(a)

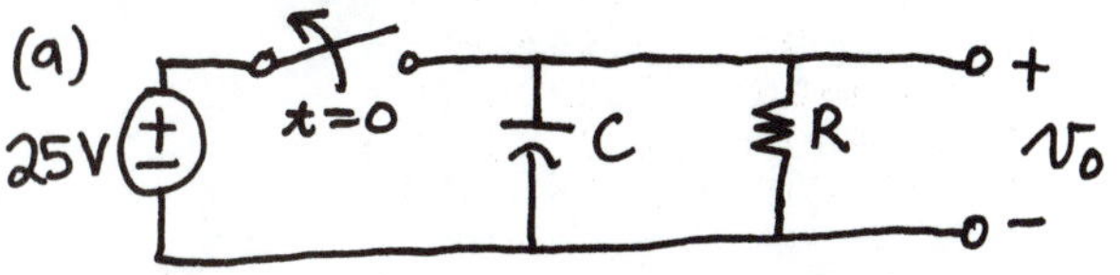

$\tau = RC = \frac{1}{3}$

$\Rightarrow$ Could pick $R = 1\Omega$, $C = \frac{1}{3}F$ for example.

(b)

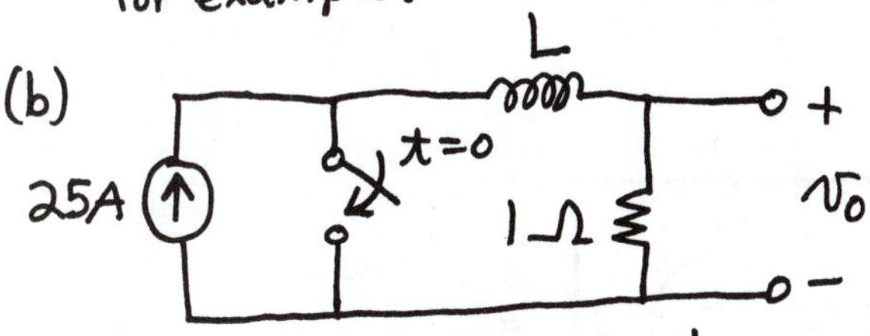

$\tau = \frac{L}{R} = \frac{1}{3} \Rightarrow$ choose $L = \frac{1}{3}H$

6.53

(a)

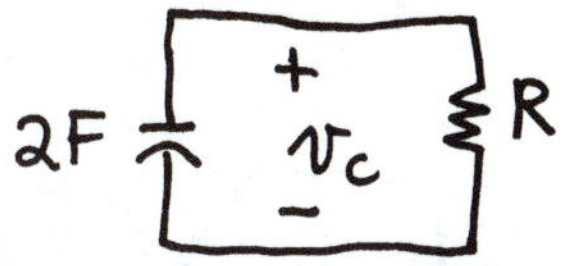

$v_c(0) = 1000\,V \Rightarrow v_c(t) = 1000e^{-t/\tau}\,V$

At $t = 1s$, $v_c(1) = 1000e^{-1/\tau} \leq 10$

$\Rightarrow e^{-1/\tau} \leq 0.01$

$\Rightarrow -1/\tau \leq \ln 0.01$

$\Rightarrow -\frac{1}{R(2F)} \leq -4.60517$

$\Rightarrow R \leq \frac{1}{9.21034}$

$\Rightarrow \boxed{R \leq 0.10857\,\Omega}$

(b) $P = \frac{v_{max}^2}{R} = \frac{(1000)^2}{0.10857}$

$= 9.21 \times 10^6\,W = \boxed{9.21MW}$

6.54 (MANY SOLUTIONS)

(a)

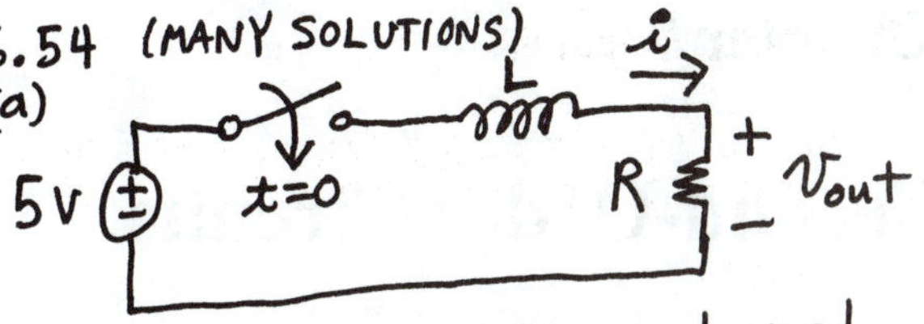

$i(0) = 0$ since inductor uncharged.

$i(t) = K(1 - e^{-t/\tau})$

But $i(\infty) = \frac{5V}{R} \Rightarrow i(t) = \frac{5}{R}(1 - e^{-t/\tau})$

$\Rightarrow i(t) = \frac{5}{R}\left[1 - e^{-\frac{R}{L}t}\right]$

Want $i(1s) = 4.75 = \frac{5}{R}\left[1 - e^{-R/L}\right]$

Choose $\underline{R = 1\Omega} \Rightarrow 5(1 - e^{-1/L}) = 4.75$

$\Rightarrow 0.25 = e^{-1/L} \Rightarrow \ln 0.25 = -\frac{1}{L}$

$\Rightarrow \underline{L = 0.72135H}$

(b) $\tau = \frac{L}{R} = 0.72135s$

6.55

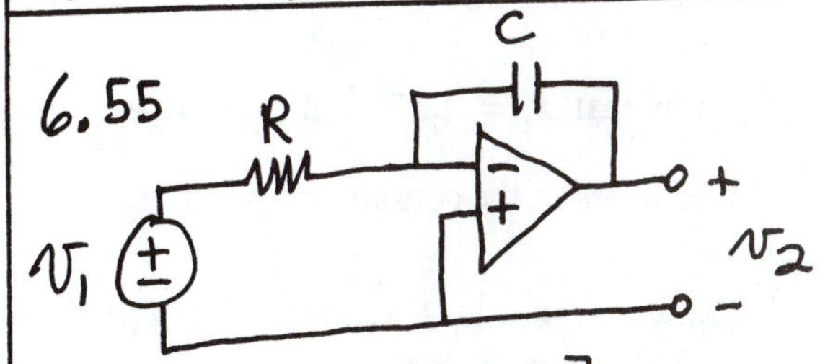

$v_2 = -\frac{1}{RC}t\,[5u(t)]$

$\Rightarrow \frac{5}{RC} = 100 \Rightarrow \frac{1}{RC} = 20$

Choose $\underline{R = 10k}$, $\underline{C = 5\mu F}$ for example.

Chapter 7

Second-Order Circuits

7.2 Second-Order Equations

7.1 Find the differential equation satisfied by the mesh current i_1.

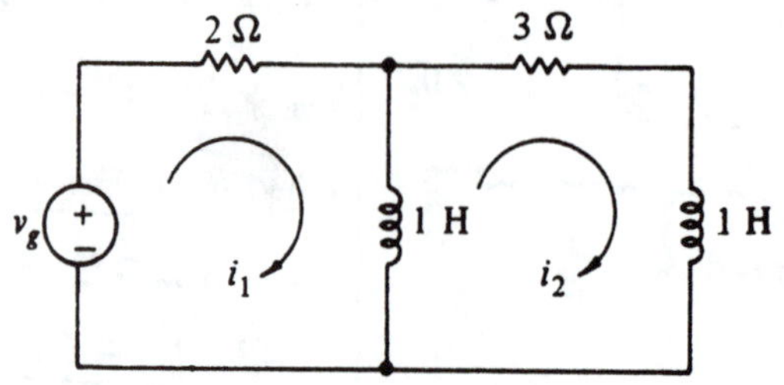

PROBLEM 7.1

7.2 Show that $x_1 = A_1e^{-t}$ and $x_2 = A_2e^{-2t}$ are solutions of $\frac{d^2x}{dt^2} + 3\frac{dx}{dt} + 2x = 0$ regardless of the values of the constants A_1 and A_2.

7.3 Show that $x = x_1 + x_2 = A_1e^{-t} + A_2e^{-2t}$ is also a solution of the second-order differential equation in Prob. 7.2.

7.4 Show that if the right member of the differential equation of Prob. 7.2 is changed from 0 to 6, then $x = A_1e^{-t} + A_2e^{-2t} + 3$ is a solution.

7.3 Natural Response

7.5 Given $\frac{d^2x}{dt^2} + 5\frac{dx}{dt} + 4x = 0$:

(a) Find the characteristic equation.
(b) Find the natural frequency.
(c) Find the damping ratio.
(d) Find the characteristic exponents.
(e) Is the system overdamped, underdamped, or critically damped?
(f) What is the trial form of the solution for $x(t)$.

7.6 Given $\frac{d^2x}{dt^2}+6\frac{dx}{dt}+9x=0$:

(a) Find the characteristic equation.
(b) Find the natural frequency.
(c) Find the damping ratio.
(d) Find the characteristic exponents.
(e) Is the system overdamped, underdamped, or critically damped?
(f) What is the trial form of the solution for $x(t)$.

7.7 Find the characteristic exponents of a circuit described by $\frac{d^2x}{dt^2}+a_1\frac{dx}{dt}+a_0x=0$ if

(a) $a_1=6,\ a_0=8$; (b) $a_1=4,\ a_0=5$; and (c) $a_1=2,\ a_0=1$.

7.8 Find x if $\frac{d^2x}{dt^2}+9x=0$.

7.9 Find i for $t>0$ if $i(0)=4$ A and $v(0)=6$ V.

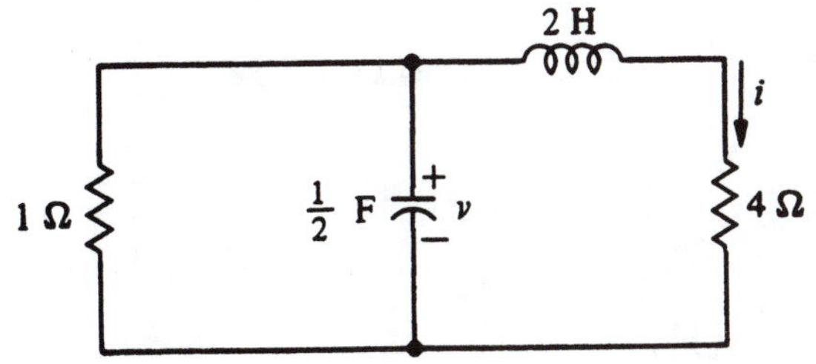

PROBLEM 7.9

7.10 Find i for $t>0$ if the circuit is in steady-state at $t=0^-$.

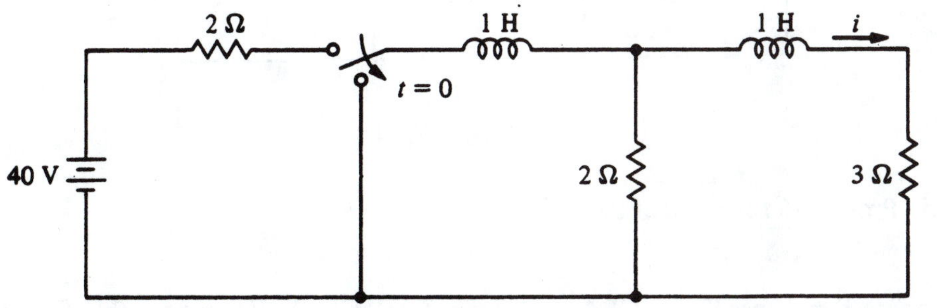

PROBLEM 7.10

7.11 In a source-free parallel *RLC* circuit, $R = 4\text{k}\Omega$ and $C = 0.1\mu\text{F}$. Find L so that the circuit is critically damped.

7.12 In a source-free parallel *RLC* circuit, $R = 4\text{k}\Omega$ and $C = 0.1\mu\text{F}$. Find L so that the circuit is underdamped with $\omega_d = 250$.

7.13 In a source-free parallel *RLC* circuit, $R = 4\text{k}\Omega$ and $C = 0.1\mu\text{F}$. Find L so that the circuit is overdamped with $s_{1,2} = 2000,\ 500$.

7.14 Let $R = 2\Omega$, $L = 4\text{H}$, $v_g = 0$, $v(0) = 4\text{V}$, and $i(0) = 0$. Find i for $t > 0$ if C is 1/8 F.

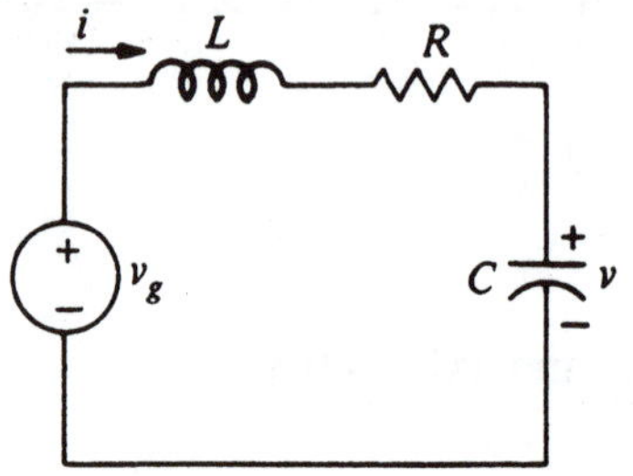

PROBLEM 7.14

7.4 Forced Response

7.15 Find i for $t > 0$ if $L = \frac{8}{3}\text{H}$.

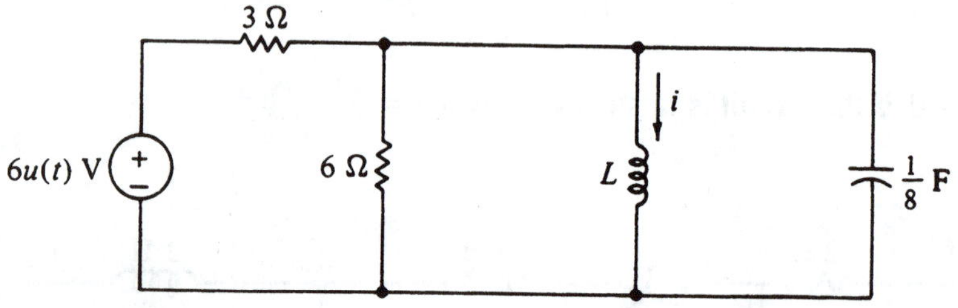

PROBLEM 7.15

7.16 Find i in Prob. 7.15 for $t > 0$ if $L = 2\text{H}$.

7.17 Find the forced response if $\dfrac{d^2x}{dt^2} + 4\dfrac{dx}{dt} + 3x = f(t)$ and (a) $f(t) = 6t^2$, (b) $f(t) = 4e^t$.

7.18 Find the forced response if $\dfrac{d^2x}{dt^2} + 4\dfrac{dx}{dt} + 3x = 2e^{-3t} - e^{-t}$.

7.5 Total Response

7.19 If $x(0) = 8$ and $dx(0)/dt = -1$, find the complete solution in Prob. 7.17.

7.20 Find the complete response if $\frac{d^2x}{dt^2} + 9x = \sin(3t)$ and $x(0) = dx(0)/dt = 0$.

7.21 Find i_1 in Prob. 7.1 for $t > 0$, if $v_g = 12\text{V}$, $i_1(0) = -1\text{A}$, and $i_2(0) = 2\text{A}$.

7.22 Find i for $t > 0$ if $v(0) = 0\text{V}$, $i(0) = 0\text{A}$, $L = 2\text{H}$ and $R = 2\Omega$.

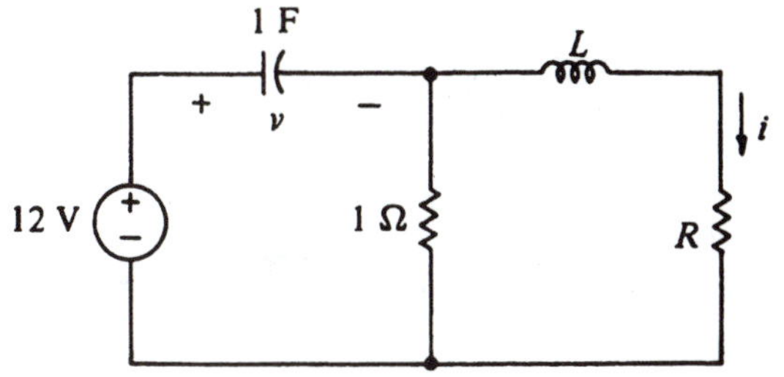

PROBLEM 7.22

7.23 Repeat Prob. 7.22 with $L = 2\text{H}$ and $R = 0\Omega$.

7.24 Find i for $t > 0$ if the circuit is in steady-state at $t = 0^-$.

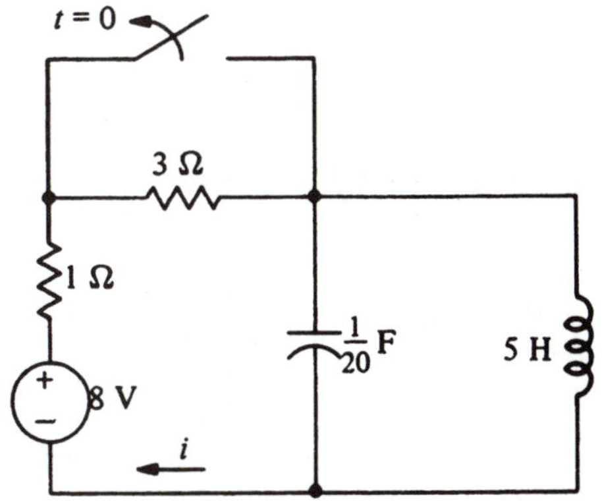

PROBLEM 7.24

7.25 Solve Prob. 7.24 if the 8V source is replaced by a source of $36e^{-4t}$ V with the same polarity, and there is no initial stored energy (at $t = 0$).

7.26 Find v for $t > 0$ if the circuit is in steady-state at $t = 0^-$.

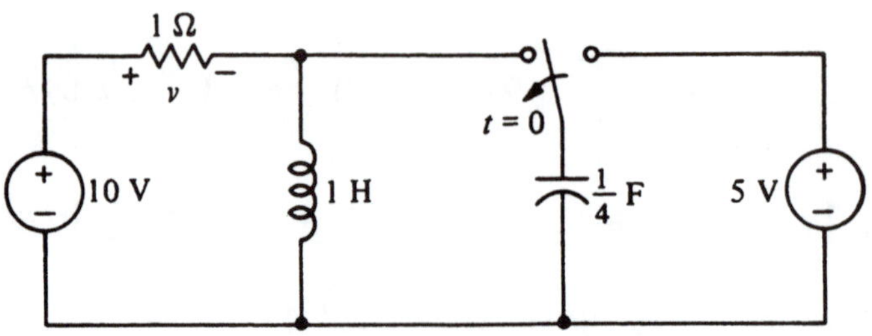

PROBLEM 7.26

7.27 Find v for $t > 0$ if $i_1(0) = -1\text{A}$ and $i_2(0) = 0$.

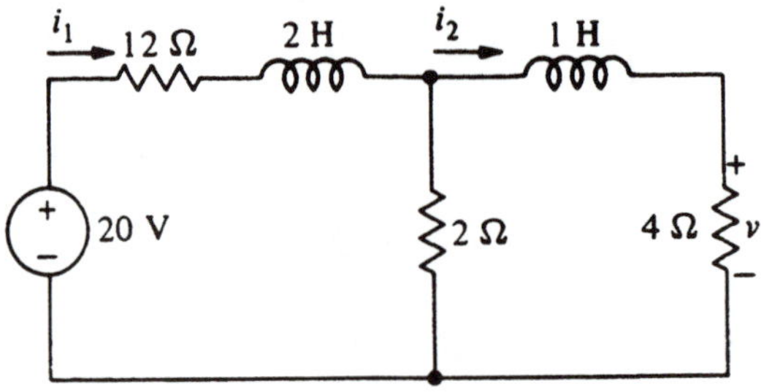

PROBLEM 7.27

7.28 Find v for $t > 0$ if $v(0) = 4\text{V}$ and $i(0) = 3\text{A}$.

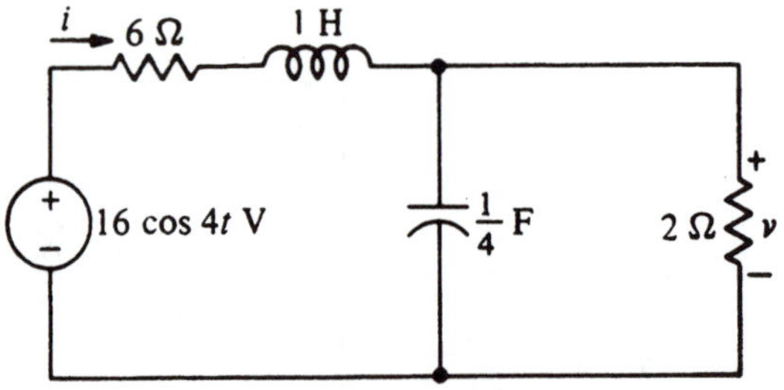

PROBLEM 7.28

7.29 Find v for $t > 0$ if there is no initial stored energy.

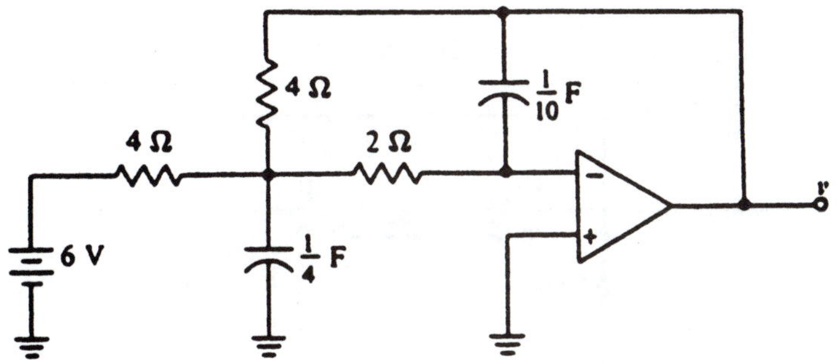

PROBLEM 7.29

7.30 Find i for $t > 0$ if the circuit is in steady-state at $t = 0^-$.

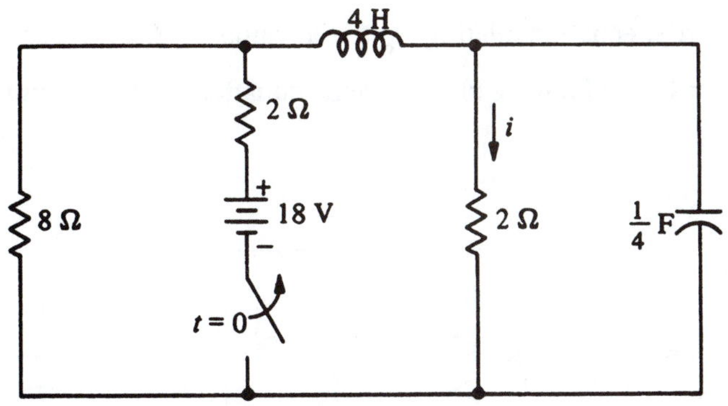

PROBLEM 7.30

7.31 Find v for $t > 0$ if (a) $i_g = 2u(t)$ A and (b) $i_g = 2e^{-t}u(t)$ A.

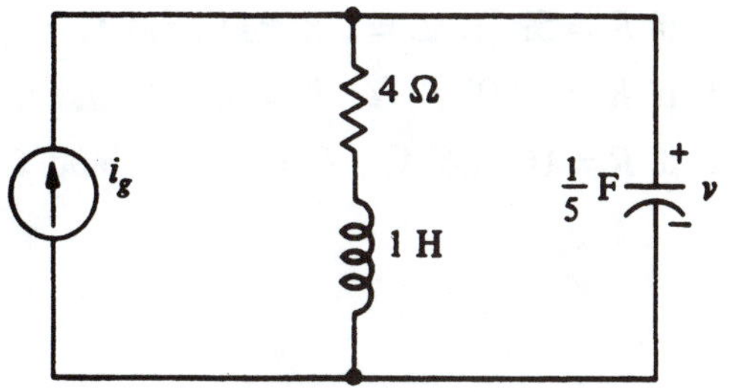

PROBLEM 7.31

7.32 Find v for $t > 0$ if $v(0) = 4\text{V}$ and $i(0) = 2\text{A}$.

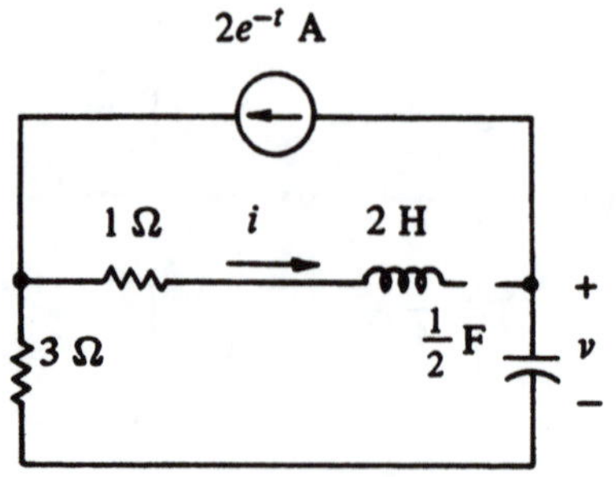

PROBLEM 7.32

7.6 Unit Step Response

7.33 (a) Find a second-order equation relating v_o to the values R, L, and C for $t > 0$ if $i_L(0^-) = v_o(0^-) = 0$. (b) Find the characteristic equation for this circuit.

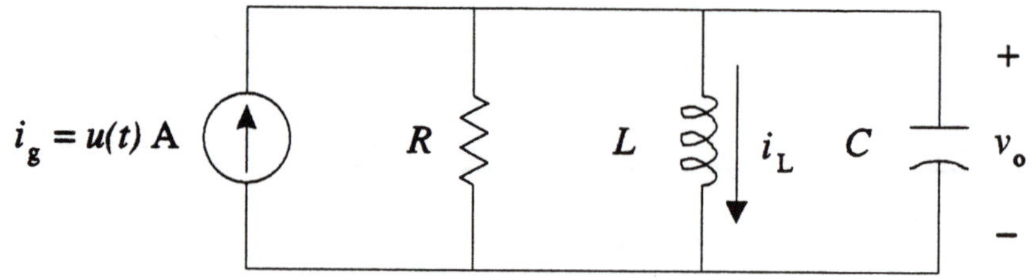

PROBLEM 7.33

7.34 (a) Find $v_o(t)$ for $t > 0$ if $R = 5\text{k}\Omega$, $L = 100\text{mH}$, and $C = 1\mu\text{F}$.

(b) Find $v_o(t)$ for $t > 0$ if $R = 4.8077\text{k}\Omega$, $L = 625\text{H}$, and $C = 1\mu\text{F}$.

(c) Find $v_o(t)$ for $t > 0$ if $R = 16.6667\Omega$, $L = 1/9\ \text{H}$, and $C = 100\mu\text{F}$.

7.35 (a) Find a second-order equation relating i_o to the values R, L, and C for $t > 0$ if $i_o(0^-) = v_c(0^-) = 0$. (b) Find the characteristic equation for this circuit.

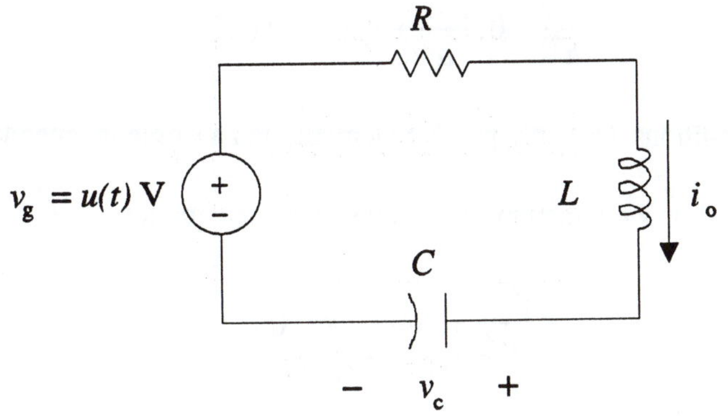

PROBLEM 7.35

7.36 (a) Find $i_o(t)$ for $t > 0$ if $R = 5\text{k}\Omega$, $L = 100\text{mH}$, and $C = 1\mu\text{F}$.
(b) Find $i_o(t)$ for $t > 0$ if $R = 0$, $L = 100\text{mH}$, and $C = 10\mu\text{F}$.
(c) Find $i_o(t)$ for $t > 0$ if $R = 4.8077\text{k}\Omega$, $L = 625\text{H}$, and $C = 1\mu\text{F}$.
(d) Find $i_o(t)$ for $t > 0$ if $R = 16.6667\Omega$, $L = 1/9$ H, and $C = 0.0016\text{F}$.

7.7 Design of Second-Order Circuits

7.37 Design a passive analog computer circuit which solves the differential equation

$$\frac{d^2y}{dt^2} + 0.5\frac{dy}{dt} + y = 2\cos(50t)$$

with zero initial conditions. Use only passive elements and a single independent source.

7.38 Design an active analog computer circuit which solves the equation

$$\frac{d^2y}{dt^2} + 100\frac{dy}{dt} = x(t)$$

where $x(t)$ is an arbitrary input voltage and $y(t)$ is the output of your circuit.

7.39 Design a passive circuit with a step response that includes real exponentials with time constants $\tau = 3\ \mu s$ and $\tau = 90\ \mu s$.

7.40 Using a single op amp, design an inverting integrator circuit which satisfies the equation

$$v_2 = -\int_{-\infty}^{t} v_1(\tau)\, d\tau .$$

CHAPTER 7 SOLUTIONS

7.1 KVL yields

(1) $\frac{di_1}{dt}+2i_1-\frac{di_2}{dt}=v_g$; (2) $-\frac{di_1}{dt}+2\frac{di_2}{dt}+3i_2=0$

Adding 2 times the first equation

(3) $\frac{di_1}{dt}+4i_1+3i_2=2v_g$, or

$\frac{d^2i_1}{dt^2}+4\frac{di_1}{dt}+3\frac{di_2}{dt}=2\frac{dv_g}{dt}$

Substituting for $\frac{di_2}{dt}$ from (1) yields

(4) $\frac{d^2i_1}{dt^2}+7\frac{di_1}{dt}+6i_1=2\frac{dv_g}{dt}+3v_g$

7.2 $x_1=A_1e^{-t}$, $x_2=A_2e^{-2t}$

$\frac{d^2x_1}{dt^2}+3\frac{dx_1}{dt}+2x_1=A_1(-1)^2e^{-t}+A_13(-1)e^{-t}+2A_1e^{-t}=0$

$\frac{d^2x_2}{dt^2}+3\frac{dx_2}{dt}+2x_2=A_2(-2)^2e^{-2t}+3(-2)A_2e^{-2t}+2A_2e^{-2t}=0$

7.3 $\frac{d^2(x_1+x_2)}{dt^2}+3\frac{d(x_1+x_2)}{dt}+2(x_1+x_2)=$

$A_1(-1)^2e^{-t}+A_2(-2)^2e^{-2t}+3(A_1(-1)e^{-t}+A_2(-2)e^{-2t})+$

$2(A_1e^{-t}+A_2e^{-2t})=0$

7.4 $\frac{d^2x}{dt^2}+3\frac{dx}{dt}+2x=(A_1(-1)^2e^{-t}+A_2(-2)^2e^{-2t})$

$+3(A_1(-1)e^{-t}+A_2(-2)e^{-2t})$

$+2(A_1e^{-t}+A_2e^{-2t}+3)=6$

7.5

(a) $s^2+5s+4=0$

(b) $\omega_0=\sqrt{4}=2$

(c) $\zeta=\frac{5}{2(2)}=\frac{5}{4}$

(d) $s_{1,2}=\left(-\frac{5}{4}\pm\sqrt{(\frac{5}{4})^2-1}\right)(2)=-1,-4$

or $(s+1)(s+4)=0\Rightarrow s_{1,2}=-1,-4$

(e) $\zeta=\frac{5}{4}>1\Rightarrow$ overdamped

(f) $x(t)=K_1e^{-t}+K_2e^{-4t}$

7.6 (a) $s^2+6s+9=0$

(b) $\omega_0=\sqrt{9}=3$

(c) $\zeta=\frac{6}{2(3)}=1$

(d) $s_{1,2}=\left(-\frac{6}{6}\pm\sqrt{1^2-1}\right)3=-3,-3$

or $(s+3)(s+3)=0\Rightarrow s_{1,2}=-3,-3$

(e) $\zeta=1\Rightarrow$ Critically damped

(f) $x(t)=K_1e^{-3t}+K_2te^{-3t}$

7.7 $s^2+a_1s+a_0=0$

(a) $s^2+6s+8=0$ or $(s+2)(s+4)=0$

$s_{1,2}=-2,-4$

(b) $s^2+4s+5=0$ or $(s+2)^2+1=0$

$s_{1,2}=-2\pm j$

(c) $s^2+2s+1=0$ or $(s+1)^2=0$

$s_{1,2}=-1,-1$

7.8 characteristic equation:

$s^2+9=0$; $s_{1,2}=\pm j3$

$\therefore x=A_1\cos 3t+A_2\sin 3t$

7.9 Let i and i_1 be the clockwise mesh currents. KVL for left mesh:

$i_1+2\int_0^t(i_1-i)dt+6=0$ or

$\frac{di_1}{dt}+2i_1-2i=0$, KVL for outer loop:

$2\frac{di}{dt}+4i+i_1=0\Rightarrow i_1=-2\frac{di}{dt}-4i$

Eliminating i_1: $\frac{d^2i}{dt^2}+4\frac{di}{dt}+5i=0$

$s^2+4s+5=0\Rightarrow s_{1,2}=-2\pm j1$

$i=e^{-2t}(A_1\cos t+A_2\sin t)$; $i(0)=4=A_1$

KVL for right mesh at $t=0$:

$v(0)=6=2\frac{di(0)}{dt}+4i(0)$

$\therefore\frac{di(0)}{dt}=-5=A_2-2A_1\Rightarrow A_2=3$

$i=e^{-2t}(4\cos t+3\sin t)$ A

7.10

Let i_1 = current in left 1-H inductor to right. At $t=0^-$ the inductors are short circuits.

$\therefore i_1(0^-) = \frac{40}{2+\frac{2(3)}{5}} = \frac{25}{2}A$;

$i(0^-) = \frac{2}{2+3} i_1(0^-) = 5A$

For $t>0$ the mesh equations are

(1) $\frac{di_1}{dt} + 2(i_1 - i) = 0$; (2) $\frac{di}{dt} + 3i + 2(i - i_1) = 0$

From (2) $i_1 = \frac{1}{2}\frac{di}{dt} + \frac{5}{2}i$. Substitution into (1): $\frac{d^2i}{dt^2} + 7\frac{di}{dt} + 6i = 0 \Rightarrow$

$s^2 + 7s + 6 = 0 \Rightarrow s_{1,2} = -1, -6$

$\therefore i = A_1e^{-t} + A_2e^{-6t}$

$i(0^+) = A_1 + A_2$; From(2) $\frac{di(0^+)}{dt} = 2i_1(0^+) - 5i(0^+) = 0$

$\frac{di(0^+)}{dt} = -A_1 - 6A_2 = 0$; $\therefore A_1 = 6, A_2 = -1$

$\therefore i = \underline{6e^{-t} - e^{-6t}}$ A

7.11 $L = 4R^2C = 4(4000)^2(10^{-7}) = \underline{6.4H}$

7.12 $\sqrt{\frac{1}{LC} - (\frac{1}{2RC})^2} = \omega_d$

$L = \frac{1}{(10^{-7})[(661)^2 + [\frac{1}{(2\times4\times10^3)(10^{-7})}]^2]} = \underline{5H}$

7.13 $s_{1,2} = -\frac{1}{2RC} \pm \sqrt{(\frac{1}{2RC})^2 - \frac{1}{LC}}$

$2000 = -1250 + \sqrt{(1250)^2 - \frac{10^7}{L}} \Rightarrow L = \underline{10H}$

7.14 The differential equation for the circuit is

$L\frac{di}{dt} + Ri + \frac{1}{C}\int_0^t i\,dt + V_0 = v_g$

For $t=0^+$, $L\frac{di(0^+)}{dt} + Ri(0^+) + V_0 = v_g(0^+) = 0$

$\frac{di(0^+)}{dt} = -\frac{2}{4}(0) - \frac{4}{4} = -1$ A/s

$s_{1,2} = -\frac{R}{2L} \pm \sqrt{(\frac{R}{2L})^2 - \frac{1}{LC}} = -\frac{2}{2(4)} \pm \sqrt{(\frac{1}{4})^2 - \frac{8}{4}}$

$= -\frac{1}{4} \pm j1.392$

$\therefore i = e^{-t/4}(A_1\cos 1.39t + A_2\sin 1.39t)$

$i(0^+) = 0 = A_1$; $\frac{di(0^+)}{dt} = A_2 1.39 = -1 \Rightarrow$

$A_2 = .7184$; $i = \underline{.718e^{-t/4}\sin 1.39t}$ A

7.15

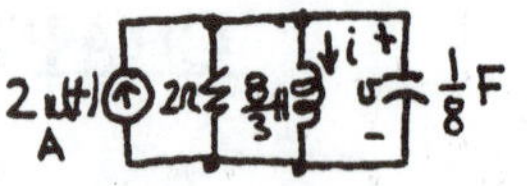

Using the norton circuit of the first 3 elements, we obtain the parallel RLC circuit. The differential equation is

$\frac{d^2i}{dt^2} + \frac{1}{CR}\frac{di}{dt} + \frac{1}{LC}i = \frac{i_g}{LC}$

For $t=0^+$; $i(0^+) = i(0^-) = 0$,

$\frac{di(0^+)}{dt} = \frac{v(0^+)}{L} = \frac{v(0^-)}{L} = 0$,

$s_{1,2} = -\frac{1}{2RC} \pm \sqrt{(\frac{1}{2RC})^2 - \frac{1}{LC}}$, $\frac{1}{2RC} = 2$

$= -2 \pm \sqrt{4-3} = -1, -3$

$i_f = i_g = 2A$

$i = A_1e^{-t} + A_2e^{-3t} + 2$

$i(0^+) = 0 = A_1 + A_2 + 2$

$\frac{di(0^+)}{dt} = 0 = -A_1 - 3A_2$, Adding,

$-2A_2 = -2 \Rightarrow A_2 = 1$, $A_1 = -3$

$\therefore i = \underline{e^{-3t} - 3e^{-t} + 2 \text{ A}}$ $(t>0)$

7.16 Using the differential equation of Prob. 9.31

$s_{1,2} = -2 \pm \sqrt{4 - \frac{1}{2(\frac{1}{8})}} = -2$

$i_f = i_g = 2A$;

$i = (A_1 + A_2t)e^{-2t} + 2$

$i(0^+) = 0 = A_1 + 2 \Rightarrow A_1 = -2$

$\frac{di(0^+)}{dt} = 0 = -2A_1 + A_2 \Rightarrow A_2 = -4$

$\therefore i = \underline{2 - (2+4t)e^{-2t} \text{ A}}$ $(t>0)$

7.17 (a) $x_f = At^2 + Bt + C$:

$\frac{dx_f}{dt} = 2At + B$; $\frac{d^2x_f}{dt^2} = 2A$;

$\frac{d^2x_f}{dt^2} + 4\frac{dx}{dt} + 3x = 6t^2$

$(2A) + 4(2At + B) + 3(At^2 + Bt + C) =$

$3At^2 + (8A + 3B)t + (2A + 4B + 3C) = 6t^2$

$3A = 6 \Rightarrow A = 2$; $8A + 3B = 0 \Rightarrow B = -\frac{16}{3}$

$2A + 4B + 3C = 0 \Rightarrow C = \frac{52}{9}$

$\therefore x_f = \underline{2t^2 - \frac{16}{3}t + \frac{52}{9}}$

(b) $x_f = Ae^t$: $\frac{dx_f}{dt} = \frac{d^2x_f}{dt^2} = Ae^t$;

$Ae^t + 4(Ae^t) + 3(Ae^t) = 8Ae^t = 4e^t$

$8A = 4$; $A = \frac{1}{2}$ $\therefore x_f = \underline{\frac{1}{2}e^t}$

7.18

Since $s_{1,2} = -1, -3$ and $f(t) = 2e^{-3t} - e^{-t}$, try

$x_f = Ate^{-3t} + Bte^{-t}$

$\frac{dx_f}{dt} = -3Ate^{-3t} + Ae^{-3t} - Bte^{-t} + Be^{-t}$

$\frac{d^2x_f}{dt^2} = 9Ate^{-3t} - 6Ae^{-3t} + Bte^{-t} - 2Be^{-t}$

$\therefore (9Ate^{-3t} - 6Ae^{-3t} + Bte^{-t} - 2Be^{-t}) + 4(-3Ate^{-3t} + Ae^{-3t} - Bte^{-t} + Be^{-t}) + 3(Ate^{-3t} + Bte^{-t}) =$

$-2Ae^{-3t} + 2Be^{-t} = 2e^{-3t} - e^{-t}$

Equating coefficients

$e^{-3t}: -2A = 2 \Rightarrow A = -1$

$e^{-t}: 2B = -1 \Rightarrow B = -\frac{1}{2}$

$\therefore x_f = \underline{-t(e^{-3t} + \frac{1}{2}e^{-t})}$

7.19

The characteristic equation is $s^2 + 4s + 3 = 0$; $s_{1,2} = -1, -3$ and $x_n = A_1e^{-t} + A_2e^{-3t}$

(a) $x = x_n + x_f$; using x_f of Prob. 7.17

$= A_1e^{-t} + A_2e^{-3t} + 2t^2 - \frac{16}{3}t + \frac{52}{9}$

$x(0) = 8 = A_1 + A_2 + \frac{52}{9}$ $\quad A_1 + A_2 = \frac{20}{9}$

$\frac{dx(0)}{dt} = -1 = -A_1 - 3A_2 - \frac{16}{3}$ $\quad -A_1 - 3A_2 = \frac{13}{3}$

Adding, $-2A_2 = \frac{59}{9} \Rightarrow A_2 = \frac{-59}{18}$, $A_1 = \frac{11}{2}$;

$\therefore x = \underline{\frac{11}{2}e^{-t} - \frac{59}{18}e^{-3t} + 2t^2 - \frac{16}{3}t + \frac{52}{9}}$.

(b) $x = A_1e^{-t} + A_2e^{-3t} + \frac{1}{2}e^t$

$x(0) = A_1 + A_2 + \frac{1}{2} = 8$ $\quad A_1 + A_2 = \frac{15}{2}$

$\frac{dx(0)}{dt} = -A_1 - 3A_2 + \frac{1}{2} = -1$ $\quad -A_1 - 3A_2 = -\frac{3}{2}$

Adding, $-2A_2 = 6 \Rightarrow A_2 = -3$, $A_1 = \frac{21}{2}$

$\therefore x = \underline{21/2\, e^{-t} - 3e^{-3t} + \frac{1}{2}e^t}$

7.20

Since $s_{1,2} = \pm j3$ and $f(t) = \sin 3t$

try $x_f = t(A\cos 3t + B\sin 3t)$

$\frac{dx_f}{dt} = t(-3A\sin 3t + 3B\cos 3t) + A\cos 3t + B\sin 3t$,

$\frac{d^2x_f}{dt^2} = t(-9A\cos 3t - 9B\sin 3t) - 6A\sin 3t + 6B\cos 3t$,

$\frac{d^2x_f}{dt^2} + 9x_f = -6A\sin 3t + 6B\cos 3t = \sin 3t$

Equating coefficients

$A = -1/6$, $B = 0$; $x_f = -\frac{1}{6}t\cos 3t$

$x = x_n + x_f$

$x = A_1\cos 3t + A_2\sin 3t - \frac{1}{6}t\cos 3t$

$x(0) = 0 = A_1$

$\frac{dx}{dt} = -3A_1\sin 3t + 3A_2\cos 3t - \frac{1}{6}\cos 3t - \frac{t}{2}\sin 3t$

$\frac{dx(0)}{dt} = 3A_2 - \frac{1}{6} = 0 \Rightarrow A_2 = \frac{1}{18}$

$\therefore \underline{x = \frac{1}{18}\sin 3t - \frac{1}{6}t\cos 3t}$.

7.21

$\frac{di_1}{dt} = 2v_g - 3i_2 - 4i_1$, then

$\frac{di_1(0)}{dt} = 2(12) - 3(2) - 4(-1) = 22$

Using equation (4) of Prob. 7.1 to find

i_{1f}; $\frac{d^2i_{1f}}{dt^2} + 7\frac{di_{1f}}{dt} + 6i_{1f} = 3v_g + 2\frac{dv_g}{dt}$

since $\frac{di_{1f}}{dt} = \frac{d^2i_{1f}}{dt^2} = \frac{dv_g}{dt} = 0$,

$6i_{1f} = 3v_g \Rightarrow i_{1f} = 6$ A.

The characteristic equation of (4) is $s^2 + 7s + 6 = 0$; $s_{1,2} = -1, -6$

$i_{1n} = A_1e^{-t} + Ae^{-6t}$, the

$i_1 = i_{1n} + i_{1f} = A_1e^{-t} + A_2e^{-6t} + 6$

$i_1(0) = A_1 + A_2 + 6 = -1$ or $A_1 + A_2 = -7$

$\frac{di_1(0)}{dt} = -A_1 - 6A_2 = 22$; Adding,

$-5A_2 = 15 \Rightarrow A_2 = -3$, $A_1 = -4$

$\therefore i_1 = \underline{-4e^{-t} - 3e^{-6t} + 6}$ A.

7.22 Since the capacitor is an open circuit in dc steady-state, $i_f = 0$, and $i(0^+) = 0$. Let i_1 and i be the clockwise mesh currents. KVL gives

$$\int_0^t i_1 dt + i_1 - i = 12 \quad \text{or} \quad \frac{di}{dt} = \frac{di_1}{dt} + i_1$$

$$L\frac{di}{dt} + Ri + i - i_1 = 0 \Rightarrow i_1 = L\frac{di}{dt} + (R+1)i$$

$$\therefore L\frac{di}{dt} + (R+1)i + \frac{d}{dt}\left[L\frac{di}{dt} + (R+1)i\right] - \frac{di}{dt} = 0$$

$$L\frac{d^2i}{dt^2} + (L+R)\frac{di}{dt} + (R+1)i = 0$$

KVL around the outside loop at $t = 0^+$ gives $-12 + v(0^+) + L\frac{di(0^+)}{dt} + Ri(0^+) = 0$

$$\therefore \frac{di(0^+)}{dt} = \frac{12}{L}$$

for $L = 2H$, $R = 2\Omega$

The characteristic equation is $s^2 + 2s + 3/2 = 0$; $s_{1,2} = -1 \pm j\,1/\sqrt{2}$

$$i = e^{-t}(A_1 \cos t/\sqrt{2} + A_2 \sin t/\sqrt{2})$$

$i(0) = A_1 = 0$, then setting $A_1 = 0$

$$\frac{di(0)}{dt} = \frac{A_2}{\sqrt{2}} = \frac{12}{2} \Rightarrow A_2 = 6\sqrt{2}$$

$$\therefore i = \underline{e^{-t}(6\sqrt{2}) \sin t/\sqrt{2}} \text{ A.}$$

7.23 From Prob. 7.22 the characteristic equation is $s^2 + s + 1/2 = 0$; $s_{1,2} = -1/2 \pm j\,1/2$

$$i = e^{-t/2}(A_1 \cos t/2 + A_2 \sin t/2)$$

$i(0) = A_1 = 0$, then setting $A_1 = 0$

$$\frac{di(0)}{dt} = \frac{A_2}{2} = \frac{12}{2} \Rightarrow A_2 = 12$$

$$\therefore i = \underline{12 e^{-t/2} \sin t/2} \text{ A}$$

7.24 At $t = 0^-$, the capacitor is an open circuit; the 3Ω resistor and the inductor are shorted.

$$\therefore i_L(0^-) = \frac{8}{1} = 8\text{A}, \quad v_c(0^-) = 0$$

For $t > 0$, $v_{cf} = 0$, KCL gives

$$(1)\ \frac{v_c - 8}{4} + \frac{1}{20}\frac{dv_c}{dt} + \frac{1}{5}\int_0^t v_c dt + i_L(0^-) = 0$$

or $\frac{d^2v_c}{dt^2} + 5\frac{dv_c}{dt} + 4 = 0$

7.24 (cont'd)

From equation (1) $\frac{dv_c(0)}{dt} = -120$

the characteristic equation is $s^2 + 5s + 4 = 0$; $s_{1,2} = -1, -4$

$$v_c = A_1 e^{-t} + A_2 e^{-4t}$$

$$v_c(0) = A_1 + A_2 = 0$$

$\frac{dv_c(0)}{dt} = -A_1 - 4A_2 = -120$, Adding,

$-3A_2 = -120 \Rightarrow A_2 = 40$, $A_1 = -40$

$$\therefore v_c = 40(e^{-4t} - e^{-t}) \text{ V}$$

$$i = \frac{8 - v_c}{4} = \underline{2 + 10(e^{-t} - e^{-4t}) \text{ A}}$$

7.25 At $t > 0$, KCL gives

$$(1)\ \frac{v_c - 36e^{-4t}}{4} + \frac{1}{20}\frac{dv_c}{dt} + \frac{1}{5}\int_0^t v_c dt = 0$$

, $i_L(0^+) = 0$, then

$$(2)\ \frac{d^2v_c}{dt^2} + 5\frac{dv_c}{dt} + 4v_c = -720e^{-4t}$$

Try $v_{cf} = Ate^{-4t}$

$$\frac{dv_{cf}}{dt} = -4Ate^{-4t} + Ae^{-4t}$$

$$\frac{d^2v_{cf}}{dt^2} = 16Ate^{-4t} - 8Ae^{-4t}$$

Substituting into (2) gives

$$(16Ate^{-4t} - 8Ae^{-4t}) + 5(-4Ate^{-4t} + Ae^{-4t}) + 4(Ate^{-4t}) = -3Ae^{-4t} = -720e^{-4t}$$

therefore $A = \frac{720}{3} = 240$ and

$v_{cf} = 240te^{-4t}$ and

$$v_c = A_1 e^{-t} + A_2 e^{-4t} + 240te^{-4t}$$

$$v_c(0) = 0; \quad \frac{dv_c(0)}{dt} = -5v_c(0) + 180 = 180$$

$$v_c(0) = A_1 + A_2 = 0$$

$$\frac{dv_c(0)}{dt} = -A_1 - 4A_2 + 240 = 180 \text{ or } -A_1 - 4A_2 = -60$$

Adding, $-3A_2 = -60 \Rightarrow A_2 = 20$, $A_1 = -20$

$$v_c = 20e^{-4t} - 20e^{-t} + 240te^{-4t}$$

$$i = \frac{36e^{-4t} - v_c}{4} = \underline{5e^{-t} + 4e^{-4t} - 60te^{-4t} \text{ A}}$$

7.26

At $t=0^-$, $i_L(0^-) = \frac{10}{1} = 10A$, $v_c(0^-) = 5V$; KCL gives

$$\frac{v_c - 10}{1} + \frac{1}{4}\frac{dv_c}{dt} + \int_0^t v_c\,dt + i_L(0) = 0 \text{ or}$$

$$\frac{d^2 v_c}{dt^2} + 4\frac{dv_c}{dt} + 4 = 0\,;\quad s_{1,2} = -2, -2$$

$$v_c = (A_1 + A_2 t)e^{-2t}\,;$$

$$\frac{dv_c(0)}{dt} = -4v_c(0) = -20 = -2A_1 + A_2$$

$$v_c(0) = A_1 = 5\,;\ A_2 = -10$$

$$\therefore v_c = (5 - 10t)e^{-2t}$$

$$v = 10 - v_c = \underline{10 - (5 - 10t)e^{-2t}}\ V$$

7.27

for $t > 0$, KVL gives

$$12i_1 + 2\frac{di_1}{dt} + 2(i_1 - i_2) = 20$$

$$\frac{di_2}{dt} + 4i_2 + 2(i_2 - i_1) = 0 \text{ or}$$

$i_1 = \frac{1}{2}\frac{di_2}{dt} + 3i_2$, Eliminate i_1 from first equation and

$$\frac{d^2 i_2}{dt^2} + 13\frac{di_2}{dt} + 40 i_2 = 20\,,$$

$$\frac{di_2(0)}{dt} = -6i_2(0) + 2i(0) = -2\,,\quad i_2(0) = 0$$

Try $i_{2f} = A$ then

$40 i_{2f} = 20 \Rightarrow i_{2f} = 1/2$

The characteristic equation is

$s^2 + 13s + 40 = 0\,;\ s_{1,2} = -5, -8$

$$i_2 = A_1 e^{-5t} + A_2 e^{-8t} + 1/2$$

$$i_2(0) = A_1 + A_2 + 1/2 = 0$$

$\frac{di_2(0)}{dt} = -5A_1 - 8A_2 = -2$, Adding times the first gives $3A_1 = -6 \Rightarrow A_1 = -2$, $A_2 = 3/2 \therefore i_2 = -2e^{-5t} + 3/2 e^{-8t} + 1/2$,

$$v = 4i = \underline{6e^{-8t} - 8e^{-5t} + 2}\ V.$$

7.28

KVL around the left mesh yields

$$6i + \frac{di}{dt} + v = 16\cos 4t$$

KCL at the top node of capacitor

$$i = \frac{1}{4}\frac{dv}{dt} + \frac{v}{2}$$

Eliminating i results in

$$\frac{d^2 v}{dt^2} + 8\frac{dv}{dt} + 16v = 64\cos 4t$$

$$s^2 + 8s + 16 = 0 \Rightarrow v_n = (A_1 + A_2 t)e^{-4t}$$

$v_f = A\cos 4t + B\sin 4t$ results in

$$32(-A\sin 4t + B\cos 4t) = 64\cos 4t$$

$$\therefore A = 0\,,\ B = 2$$

$$v = (A_1 + A_2 t)e^{-4t} + 2\sin 4t$$

$$v(0^+) = 4 = A_1$$

$$\frac{1}{4}\frac{dv(0^+)}{dt} = \frac{-v(0)}{2} + i(0^+) = 3 - \frac{4}{2} = 1$$

$$\frac{dv(0)}{dt} = 4 = -4A_1 + A_2 + 8 \Rightarrow A_2 = 12$$

$$v = \underline{(4 + 12t)e^{-4t} + 2\sin 4t}\ V.$$

7.29 Let v_1 be the nodal voltage at the top of 1/4 capacitor, then KCL gives

$$v_1\left(\frac{1}{4} + \frac{1}{4} + \frac{1}{2}\right) + \frac{1}{4}\frac{dv_1}{dt} - \frac{v}{4} - \frac{6}{4} = 0 \text{ or}$$

$$4v_1 + \frac{dv_1}{dt} - v = 6$$

node analysis at inverting opamp terminal gives.

$$\frac{v_1}{2} + \frac{1}{10}\frac{dv}{dt} = 0 \Rightarrow v_1 = -\frac{1}{5}\frac{dv}{dt}$$

substituting for v_1 in first equation

$$\frac{d^2 v}{dt^2} + 4\frac{dv}{dt} + 5v = -30$$

since $\frac{d^2 v_f}{dt^2} = \frac{dv_f}{dt} = 0$, $5v_f = -30$;

$$v_f = -6V$$

the characteristic equation is

$$s^2 + 4s + 5 = 0\,;\ s_1 = -2 \pm j$$

$$v = e^{-2t}(A_1\cos t + A_2 \sin t) - 6$$

$$\frac{dv(0)}{dt} = -5v_1(0) = 0$$

$$v(0) = 0 = A_1 - 6 \Rightarrow A_1 = 6$$

$$\frac{dv(0)}{dt} = -2A_1 + A_2 = 0 \Rightarrow A_2 = 12$$

$$\therefore v = \underline{e^{-2t}(6\cos t + 12\sin t) - 6}\ V$$

7.30

At $t=0^-$, the capacitor is open-circuited and the inductor is a short circuit. By current division and Ohm's law

$$i(0^-) = \frac{18}{2+\frac{2(8)}{10}} \cdot \frac{8}{2+8} = 4A; \; v_c(0^-) = 4(2) = 8V$$

Let i_L be the inductor current to the right and i_c be the capacitor current downward. For $t>0$, $v_c = 2i$, $i_c = \frac{1}{4}\frac{dv_c}{dt} = \frac{1}{2}\frac{di}{dt}$ and $i_L = i + i_c = i + \frac{1}{2}\frac{di}{dt}$. KVL around the left mesh yields

$8i_L + 4\frac{di_L}{dt} + 2i = 0$ or

$8(i+\frac{1}{2}\frac{di}{dt}) + 4\frac{d}{dt}(i+\frac{1}{2}\frac{di}{dt}) + 2i = 0$ or

$\frac{d^2i}{dt^2} + 4\frac{di}{dt} + 5i = 0 \Rightarrow s^2+4s+5=0 \Rightarrow$

$s_{1,2} = -2 \pm j$, $i = e^{-2t}(A_1\cos t + A_2 \sin t)$

$i(0^+) = \frac{1}{2}v_c(0^+) = \frac{1}{2}v_c(0^-) = \frac{8}{2} = 4 = A_1$

$\frac{di(0+)}{dt} = 2[i_L(0+) - i(0+)] = 2i(0^-) - v_c(0^+)$

$= 2i(0^-) - v_c(0+) = 2(4) - 8 = 0$

$= A_2 - 2A_1 \Rightarrow A_2 = 8$

$\therefore i = e^{-2t}(4\cos t + 8\sin t)$ A

7.31

v_L = inductor voltage, positive top.

KCL gives (1) $\frac{1}{5}\frac{dv}{dt} + \frac{v - v_L}{4} = i_g$

(2) $-\frac{v-v_L}{4} + \int_0^t v_L\,dt = 0$

Add (1) and (2), differentiate and substitute for v_L from (1):

$$\frac{d^2v}{dt^2} + 4\frac{dv}{dt} + 5v = 5\frac{di_g}{dt} + 20i_g$$

At $t=0^-$, $v(0^-) = i_L(0^-) = 0$.

Therefore $v(0+) = 0$ and $v_L(0+) = 0$ by KVL. From (1) $\frac{dv(0^+)}{dt} = 5i_g(0^+)$

7.31 (cont'd)

(a) $i_g = 2u(t)$A. For $t>0$, $\frac{di_g}{dt} = 0$, $i_g = 2$ A:

$\frac{d^2v}{dt^2} + 4\frac{dv}{dt} + 5v = 40 \Rightarrow s_{1,2} = -2 \pm j$

$v = e^{-2t}(A_1\cos t + A_2\sin t) + 8$

$v(0^+) = 0 = A_1 + 8$; $\frac{dv(0^+)}{dt} = -2A_1 + A_2 = 10$

$A_1 = -8$; $A_2 = -6 \Rightarrow$

$v = 8 - e^{-2t}(8\cos t + 6\sin t)$ V

(b) $i_g = 2e^{-t}u(t)$: For $t>0$; $i_g = 2e^{-t}$ and $\frac{di_g}{dt} = -2e^{-t}$. Therefore

$\frac{d^2v}{dt^2} + 4\frac{dv}{dt} + 5v = 30e^{-t}$

Try $v_f = Ae^{-t}$. Then

$A(1-4+5)e^{-t} = 30e^{-t} \Rightarrow A = 15$.

$v = e^{-2t}(A_1\cos t + A_2\sin t) + 15e^{-t}$

$v(0) = 0 = A_1 + 15 \Rightarrow A_1 = -15$.

$\frac{dv(0)}{dt} = 5(2) = -2A_1 + A_2 - 15 \Rightarrow A_2 = -5$

$\therefore v = 15e^{-t} - e^{-2t}(15\cos t + 5\sin t)$ V ($t>0$)

7.32

KVL gives: $i - i_g = \frac{1}{2}\frac{dv}{dt}$

$i + 2\frac{di}{dt} + v - 3(i_g - i) = 0$;

$4(i_g + \frac{1}{2}\frac{dv}{dt}) + 2\frac{d}{dt}(i_g + \frac{1}{2}\frac{dv}{dt}) + v = 3i_g$

or $\frac{d^2v}{dt^2} + 2\frac{dv}{dt} + v = -i_g - 2\frac{di_g}{dt} = 2e^{-t}$

$s^2 + 2s + 1 = 0 \Rightarrow s_{1,2} = -1, -1$; try

$v_f = At^2e^{-t}$; $\therefore 2Ae^{-t} = 2e^{-t} \Rightarrow A = 1$

$v = (A_1 + A_2t)e^{-t} + t^2e^{-t}$

$= (A_1 + A_2t + t^2)e^{-t}$

$v(0) = 4 = A_1$: $\frac{dv(0)}{dt} = 2[i(0) - i_g(0)]$

$= 2(2-2) = 0$

$\frac{dv(0)}{dt} = 0 = -A_1 + A_2 \Rightarrow A_2 = 4$

$v = (4 + 4t + t^2)e^{-t}$ V

7.33 (a) KCL: $1A = C\frac{dv_o}{dt} + \frac{v_o}{R} + \frac{1}{L}\int_0^t v_o(\tau)d\tau$

Differentiate: $0 = C\frac{d^2v_o}{dt^2} + \frac{1}{R}\frac{dv_o}{dt} + \frac{1}{L}v_o(t)$

$$\Rightarrow \boxed{\frac{d^2v_o}{dt^2} + \left(\frac{1}{RC}\right)\frac{dv_o}{dt} + \left(\frac{1}{LC}\right)v_o = 0}$$

(b) $s^2 + \frac{1}{RC}s + \frac{1}{LC} = 0$

7.34 (a) $\frac{1}{RC} = \frac{1}{(5k)(1\mu F)} = 200$

$\frac{1}{LC} = \frac{1}{(100mH)(1\mu F)} = 1\times10^7$

$\Rightarrow s^2 + 200s + 1\times10^7 = 0$

$\Rightarrow s_{1,2} = \frac{-200 \pm \sqrt{(200)^2 - 4\times10^7}}{2}$

$s_{1,2} = -100 \pm j\,3160.696 \Rightarrow$ Underdamped

$v_o(t) = B_1e^{-100t}\cos(3160.7t) + B_2e^{-100t}\sin(3160.7t)$

Match I.C.'s: $v_o(0+) = 0 = B_1e^0\cos(0) + B_2e^0\sin 0$

$\Rightarrow B_1 = 0 \Rightarrow v_o(t) = B_2e^{-100t}\sin(3160.7t)$

Evaluating KCL eqn. at $t = 0+$:

$1 = (1\times10^{-6})\frac{dv_o(0+)}{dt} + \frac{1}{5\times10^3}v_o(0+) + 0$

$\Rightarrow 1\times10^6 = \frac{dv_o(0+)}{dt} + 200\,v_o(0+)$

Substitute:

$1\times10^6 = -100B_2e^0\sin(0) + 3160.7B_2e^0\cos(0) + 200\,e^0\sin(0)$

$\Rightarrow B_2 = 316.386$

$$\therefore \boxed{v_o(t) = 316.386e^{-100t}\sin(3160.7t)}$$

(b) $\frac{1}{RC} = 208$, $\frac{1}{LC} = 1600$

$\Rightarrow s^2 + 208s + 1600 = 0$

$\Rightarrow (s+8)(s+200) = 0$

$\Rightarrow s_{1,2} = -8, -200 \Rightarrow$ Overdamped

$v_o(t) = K_1e^{-8t} + K_2e^{-200t}$

I.C.'s: $v_o(0+) = 0 = K_1 + K_2 \Rightarrow K_2 = -K_1$

$\Rightarrow v_o(t) = K_1(e^{-8t} - e^{-200t})$

KCL eqn. at $t = 0+$:

$1 = (1\times10^{-6})\frac{dv_o(0+)}{dt} + \frac{1}{4.8077k}v_o(0+) + 0$

$\Rightarrow 1\times10^6 = \frac{dv_o(0+)}{dt} + 208\,v_o(0+)$

Substitute $\Rightarrow 1\times10^6 = -8K_1e^{-8(0)} + 200K_1e^{-200(0)} + 208K_1e^0 - 208K_1e^0$

$\Rightarrow K_1 = 5208.\overline{3}$

$$\therefore \boxed{v_o(t) = 5208.\overline{3}\left[e^{-8t} - e^{-200t}\right]V}$$

(c) $\frac{1}{RC} = 600$, $\frac{1}{LC} = 90000 \Rightarrow s^2 + 600s + 90000 = 0$

$\Rightarrow (s+300)(s+300) = 0 \Rightarrow s_{1,2} = -300, -300$

$\Rightarrow$ Critically damped. $v_o(t) = K_1e^{-300t} + K_2te^{-300t}$

I.C.'s: $v_o(0+) = 0 = K_1 + K_2 \Rightarrow K_2 = -K_1$

$\Rightarrow v_o(t) = K_1[e^{-300t}$ [redacted] $te^{-300t}]$

KCL eqn. at $t = 0+$:

$1 = (100\times10^{-6})\frac{dv_o(0+)}{dt} + \frac{1}{16.6667}v_o(0+) + 0$

$\Rightarrow (1\times10^4) = \frac{dv_o(0+)}{dt} + 600\,v_o(0+)$

Subst: $1\times10^4 = K_1\left[-300e^0 + 300te^0 - e^0\right]_{t=0} + 600\left[e^0 - (0)e^0\right]$

$\Rightarrow 1\times10^4 = 299K_1 \Rightarrow K_1 = 33.445$

$$\therefore \boxed{v_o(t) = 33.445\left[e^{-300t} - te^{-300t}\right]}$$

7.35 (a)

KVL: $1V = Ri_o + L\frac{di_o}{dt} + \frac{1}{C}\int_0^t i_o(\tau)d\tau$

Differentiate: $L\frac{d^2i_o}{dt^2} + R\frac{di_o}{dt} + \frac{1}{C}i_o = 0$

$\Rightarrow \frac{d^2i_o}{dt^2} + \frac{R}{L}\frac{di_o}{dt} + \frac{1}{LC}i_o = 0$

(b) $s^2 + \frac{R}{L}s + \frac{1}{LC} = 0$

7.36(a) $\frac{R}{L} = 5\times10^4$, $\frac{1}{LC} = 1\times10^7$

$\Rightarrow s^2 + (5\times10^4)s + (1\times10^7) = 0$

$\Rightarrow s_{1,2} = -200.8, -49800 \Rightarrow$ Overdamped

$i_o(t) = K_1e^{-200.8t} + K_2e^{-49800t}$

I.C.'s: $i_o(0+) = K_1 + K_2 = 0$

$\Rightarrow i_o(t) = K_1(e^{-200.8t} - e^{-49800t})$

KVL eqn: at $t = 0+$

$\frac{1}{L} = \frac{di_o}{dt}\Big|_{t=0+} + \frac{R}{L}i_o(0+) + 0$

$\Rightarrow 10 = K_1[-200.8e^0 + 49800e^0] + (5\times10^4)K_1(e^0 - e^0)$

$\Rightarrow K_1 = 2.016\times10^{-4}$

$\therefore$ $i_o(t) = (2.016\times10^{-4})[e^{-200.8t} - e^{-49800t}]$

(b) $\frac{R}{L} = 0$, $\frac{1}{LC} = 1\times10^6 \Rightarrow s^2 + (1\times10^6) = 0$

$\Rightarrow s_{1,2} = j1000, j1000$

$\Rightarrow$ Undamped.

$i_o(t) = K_1\cos 1000t + K_2\sin 1000t$

I.C.: $i_o(0+) = 0 = K_1(1) + K_2(0)$

$\Rightarrow i_o(t) = K\sin 1000t$

KVL eqn: $\frac{1}{L} = \frac{di_o}{dt}\Big|_{t=0+} + \frac{R}{L}i_o(0+) + 0$

$\Rightarrow 10 = 1000K\cos(0) + 0 \Rightarrow K = 0.01$

$\therefore$ $i_o(t) = 0.01\sin 1000t$ A $= 10\sin 1000t$ mA

(c) $\frac{R}{L} = 7.6923$, $\frac{1}{LC} = 1600$

$\Rightarrow s^2 + 7.6923s + 1600 = 0$

$\Rightarrow s_{1,2} = -3.846 \pm j39.815 \Rightarrow$ Underdamped

$i_o(t) = K_1e^{-3.846t}\cos(39.815t) + K_2e^{-3.846t}\sin(39.815t)$

I.C.'s: $i_o(0+) = K_1 + K_2(0) = K_1 = 0$

$\Rightarrow i_o(t) = K_2e^{-3.846t}\sin(39.815t)$

KVL eqn. at $t = 0+$: $\frac{1}{L} = \frac{di_o(0+)}{dt} + \frac{R}{L}i_o(0+)$

$\Rightarrow .0016 = [K_2(39.815e^{-3.846(0)}\cos(0) + 0] + 7.6923K_2(1)(0)$

$\Rightarrow .0016 = K_2(39.815) \Rightarrow K_2 = 4.02\times10^{-5}$

$\therefore$ $i_o(t) = (4.02\times10^{-5})e^{-3.846t}\sin(39.8t)$ A

(d) $\frac{R}{L} = 150$, $\frac{1}{LC} = 5625$

$\Rightarrow s^2 + 150s + 5625 = 0 \Rightarrow (s+75)^2 = 0$

$\Rightarrow s_{1,2} = -75, -75 \Rightarrow$ Critically damped

$i_o(t) = K_1e^{-75t} + K_2te^{-75t}$

I.C.'s: $i_o(0+) = K_1 + K_2(0) = K_1 = 0$

$\Rightarrow i_o(t) = K_2te^{-75t}$

KVL eqn. at $t = 0+$: $\frac{1}{L} = \frac{di_o(0)}{dt} + \frac{R}{L}i_o(0)$

$\Rightarrow 9 = [-75K_2(0)e^{-75(0)} + K_2e^{-75(0)}] + 0$

$\Rightarrow 9 = K_2$

$\therefore$ $i_o(t) = 9te^{-75t}$ A

7.37 There are many solutions. Here are two of them:

Series RLC:

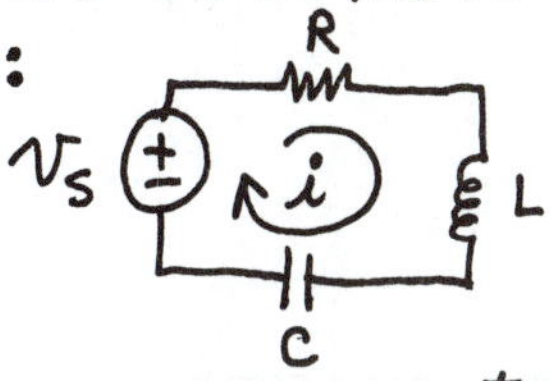

KVL yields $Ri + L\dfrac{di}{dt} + \dfrac{1}{C}\displaystyle\int_{-\infty}^{t} i\,d\tau = v_s(t)$

Differentiating and dividing by L: $\dfrac{d^2i}{dt^2} + \left(\dfrac{R}{L}\right)\dfrac{di}{dt} + \left(\dfrac{1}{LC}\right)i = \left(\dfrac{1}{L}\right)\dfrac{dv_s}{dt}$

$\therefore$ with $\dfrac{R}{L} = 0.5$, such as $R = 1\,\Omega$, $L = 2\,\text{H}$

$\dfrac{1}{LC} = 1 \quad L = 2\,\text{H} \Rightarrow C = \tfrac{1}{2}\,\text{F}$

$\Rightarrow v_s = \dfrac{2}{25}\sin 50t\ \text{V}$

With these circuit parameters and $y \equiv i$, this circuit solves the equation.

Parallel RLC:

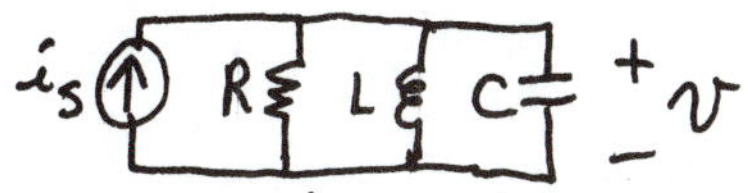

KCL yields $\dfrac{v}{R} + \dfrac{1}{L}\displaystyle\int_{-\infty}^{t} v\,d\tau + C\dfrac{dv}{dt} = i_s$

$\Rightarrow \dfrac{d^2v}{dt^2} + \left(\dfrac{1}{RC}\right)\dfrac{dv}{dt} + \left(\dfrac{1}{LC}\right)v = \left(\dfrac{1}{C}\right)\dfrac{di_s}{dt}$

Thus, with $y \equiv v$, $R = 1\,\Omega$, $C = 2\,\text{F}$, $L = \tfrac{1}{2}\,\text{H}$, and $i_s = \dfrac{2}{25}\sin 50t\ \text{A}$, this also works.

7.38 Using the text example as a model:

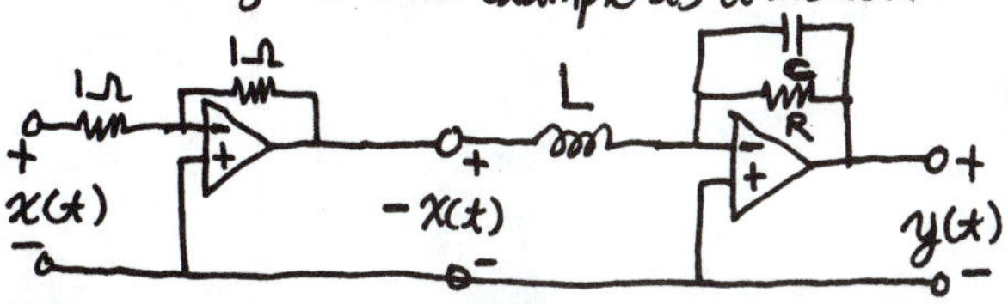

Using the virtual short and open principles, KCL at the inverting terminal yields:

$C\dfrac{dy}{dt} + \dfrac{y}{R} + \dfrac{1}{L}\displaystyle\int -x(\tau)\,d\tau = 0$

$\Rightarrow \dfrac{d^2y}{dt^2} + \left(\dfrac{1}{RC}\right)\dfrac{dy}{dt} = \left(\dfrac{1}{LC}\right)x(t)$

$\Rightarrow \dfrac{1}{RC} = 100,\ \dfrac{1}{LC} = 1$

$\therefore$ We can choose $C = \tfrac{1}{100}\,\text{F}$, $R = 1\,\Omega$, $L = 100\,\text{H}$

7.39

$y = K_1e^{-t/\tau_1} + K_2e^{-t/\tau_2} = K_1e^{-t/(3\times10^{-6})} + K_2e^{-t/(90\times10^{-6})}$

One realization is a series RLC w/current output.

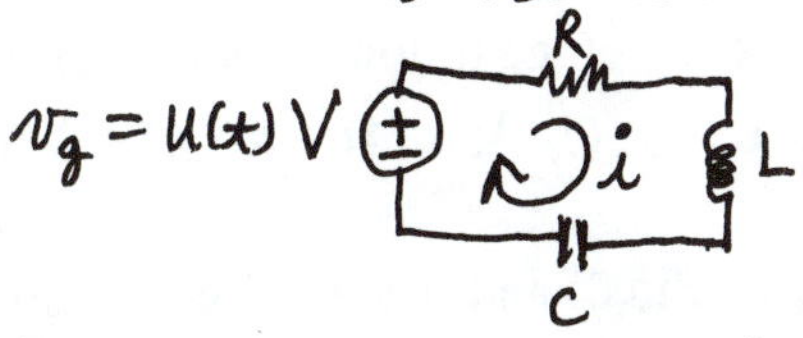

KVL, differentiating, at $t > 0$,

$\Rightarrow \dfrac{d^2i}{dt^2} + \left(\dfrac{R}{L}\right)\dfrac{di}{dt} + \left(\dfrac{1}{LC}\right)i = \dfrac{dv_g}{dt} = 0$

Characteristic eqn: $s^2 + \dfrac{R}{L}s + \dfrac{1}{LC} = 0$

$s_1 = -\dfrac{1}{\tau_1} = -3.\overline{3}\times10^5,\ s_2 = -\dfrac{1}{\tau_2} = -1.\overline{1}\times10^4$

$\Rightarrow (s + 3.\overline{3}\times10^5)(s + 1.\overline{1}\times10^4) = 0$

$\Rightarrow s^2 + (3.\overline{4}\times10^5)s + (3.\overline{7037}\times10^9) = 0$

$\Rightarrow \dfrac{R}{L} = 3.\overline{4}\times10^5,\ \dfrac{1}{LC} = 3.\overline{7037}\times10^9$

Choose $\underline{R = 10\,\text{k}\Omega} \Rightarrow L = \underline{0.0290\,\text{H}}, \underline{C = 9.3\,\text{nF}}$

7.40 Two versions are as follows:

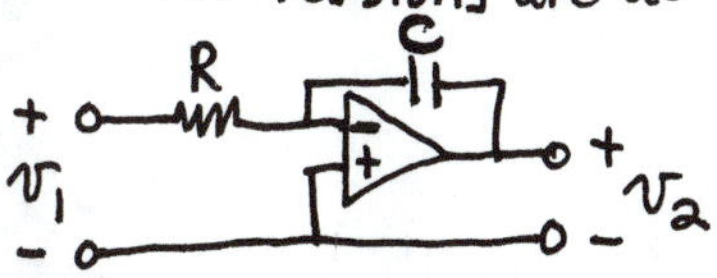

Using virtual short/open with KCL,

$\dfrac{v_1}{R} + C\dfrac{dv_2}{dt} = 0 \Rightarrow v_2 = -\dfrac{1}{RC}\displaystyle\int_{-\infty}^{t} v_1\,d\tau$

Choose $RC = 1$ (Ex: $R = 100\,\text{k}\Omega$, $C = 10\,\mu\text{F}$)

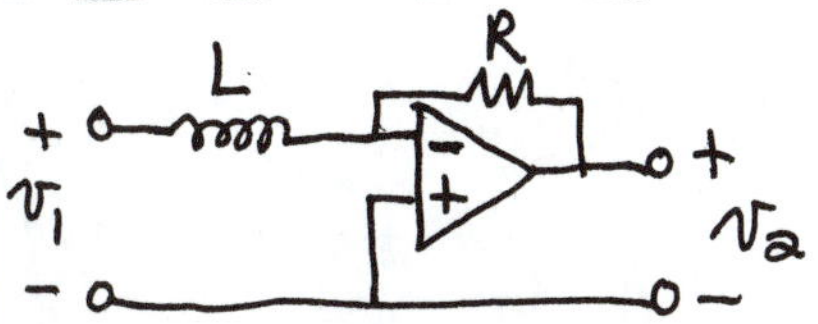

Using virtual short/open with KCL,

$\dfrac{v_2}{R} + \dfrac{1}{L}\displaystyle\int_{-\infty}^{t} v_1\,d\tau = 0 \Rightarrow v_2 = -\dfrac{R}{L}\displaystyle\int_{-\infty}^{t} v_1\,d\tau$

Choose $\dfrac{R}{L} = 1$ (Ex: $R = 1\,\Omega$, $L = 1\,\text{H}$)

Chapter 8

Sinusoidal Sources and Phasors

8.1 Properties of Sinusoids

8.1 Find the period of the following sinusoids: (a) $6\sin(5t+17°)$, (b) $5\sin(6\pi\, t)$, (c) $3\sin(4t+\pi/4)+2\cos(4t+\pi)$.

8.2 Find the amplitude and phase of the sinusoids in Prob. 8.1.

8.3 Find the frequency of the sinusoids in Prob. 8.1.

8.4 Sketch the sinusoids in Prob. 8.1 and find the time in seconds between 0s and the next zero crossing.

8.5 Given $v_1 = 5\sin(4t)$ V and $v_2 = 20\left[\cos(4t)+\sqrt{3}\sin(4t)\right]$V determine (a) if v_1 leads or lags v_2 and by how much in (b) degrees, (c) radians, and (d) seconds.

8.6 Find i_4 using KCL and the properties of sinusoids (trigonometric identities) if $i_1 = 2\cos(3t)$ A, $i_2 = 25\cos(3t+36.9°)$ A, and $i_3 = 13\sin(3t+157.4°)$ A.

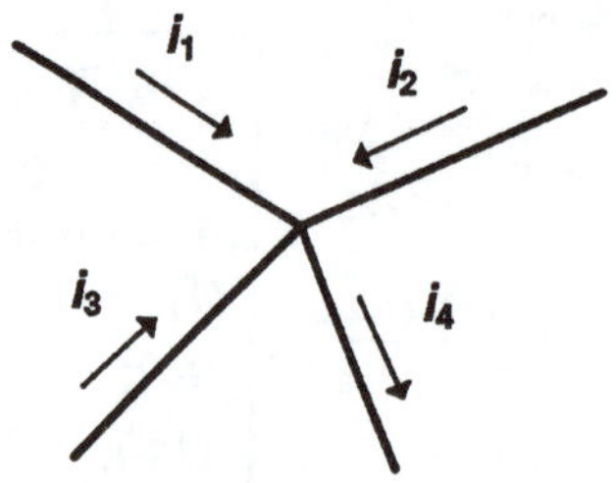

PROBLEM 8.6

8.2 RLC Circuit Example

8.7 Find the forced mesh current, $i(t)$, in the following series RLC circuit.

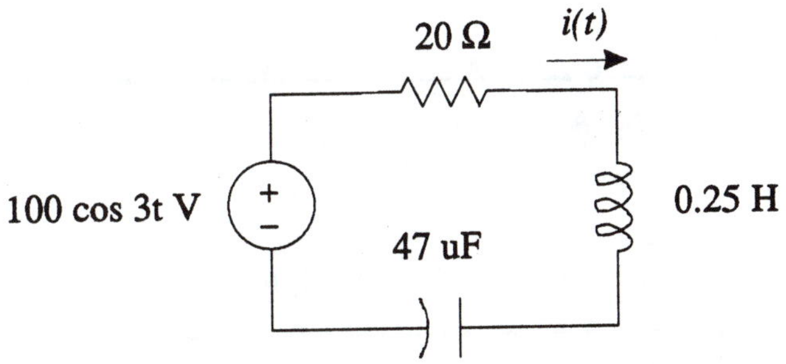

PROBLEM 8.7

8.8 Find the forced voltage $v(t)$ in the following parallel RLC circuit.

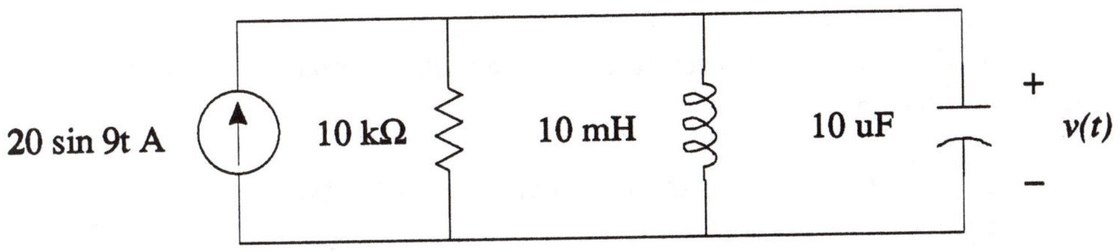

PROBLEM 8.8

8.9 In the following series RL circuit, find the current $i(t)$.

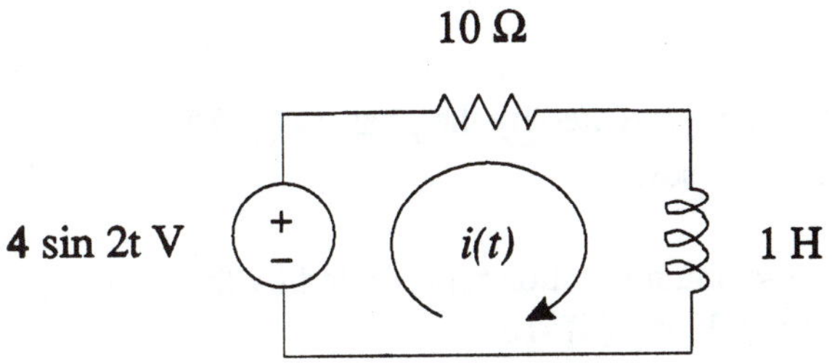

PROBLEM 8.9

8.3 Complex Sources

8.10 A complex voltage input $10e^{j(4t+20°)}$ V into a linear circuit produces a current output $2e^{j(4t-25°)}$ A. Find the output current if the input voltage is (a) $45e^{j(4t+50°)}$ V, (b) $15\sin(4t+60°)$ V, and (c) $25\cos(4t)$ V.

8.11 Find the response v_1 to the source $2e^{j8t}$ A and use the result to find the response v_1 to (a) $2\cos(8t)$ A and (b) $2\sin(8t)$ A.

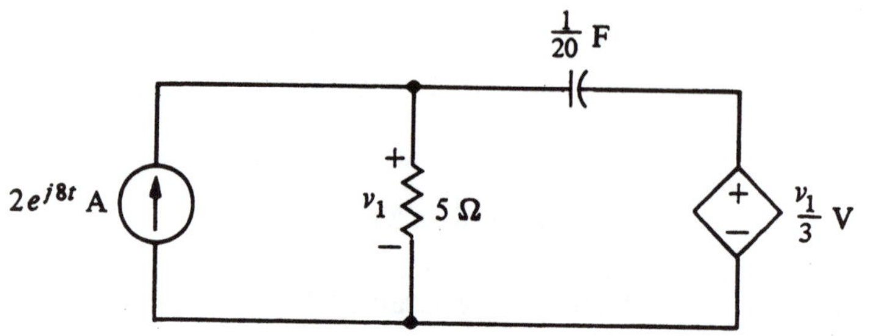

PROBLEM 8.11

8.12 Solve problem 8.7 by replacing the sinusoidal source with a complex source and using the methods illustrated in Section 8.3 of the text.

8.13 Solve problem 8.8 by replacing the sinusoidal source with a complex source and using the methods illustrated in Section 8.3 of the text.

8.14 Solve problem 8.9 by replacing the sinusoidal source with a complex source and using the methods illustrated in Section 8.3 of the text.

8.4 Phasors

8.15 Find the phasor representation of (a) $5\cos(4t+30°)$, (b) $-3\sin(3t)+2\cos(3t)$, and (c) $4\sin(10t-20°)$.

8.16 Find the time-domain function represented by the phasors (a) $15\angle-45°$, (b) 5 + j12, (c) 5 - j12, ω = 5 rad/s for all cases.

8.17 If ω = 5 rad/s, find the time-domain function represented by the phasors (a) -6 - j6, (b) -5 + j12, (c) 4 - j3, and (d) -10.

8.18 Express the source in each of the following problems by the real part of a complex exponential quantity (i.e. $\mathrm{Re}\{Ae^{j(\omega t+\phi)}\}$): (a) Prob. 8.7, (b) Prob. 8.8, (c) Prob. 8.9

8.5 I-V Laws for Phasors

8.19 Using phasors, find the ac steady-state current if $v = 12\cos(1000t + 30°)$ V and $R = 4$ kΩ.

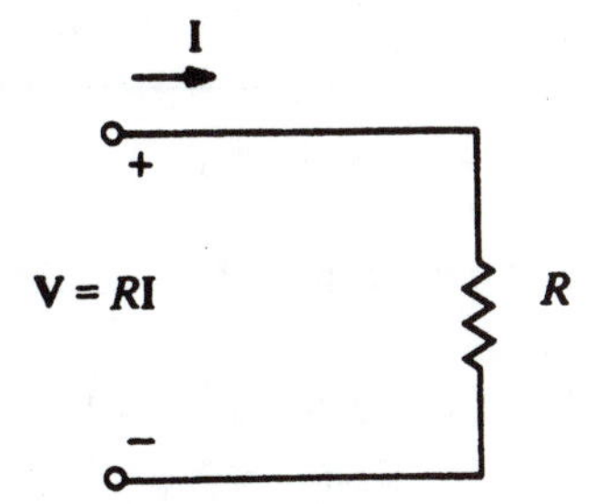

PROBLEM 8.19

8.20 Using phasors, find the ac steady-state current if $v = 10\sin(377t + 15°)$ V, $L = 0.5$ H

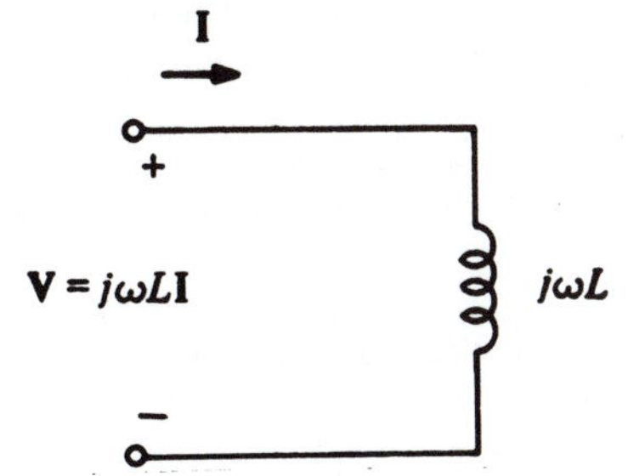

PROBLEM 8.20

8.21 Using phasors, find the ac steady-state current if $v = 12\sin(377t + 15°)$ V, $C = 10$ μF.

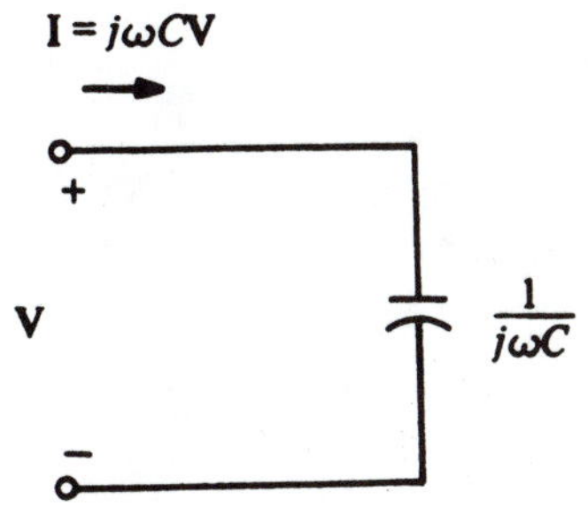

PROBLEM 8.21

8.22 Prove that (a) the current in an inductor lags the voltage across its terminals by 90° and (b) the current in a capacitor leads the voltage across its terminals by 90°, in ac sinusoidal steady state.

8.6 Impedance and Admittance

8.23 Find the conductance and susceptance if **Z** is (a) 12 + j5, (b) 3 - j3, (c) $5\angle 30°$.

8.24 Find the resistance and reactance if **Y** is (a) 12 + j5, (b) 3 - j3, (c) $5\angle 30°$.

8.25 Find the impedance of the circuit shown if the time-domain functions represented by the phasors **V** and **I** are

(a) $v = -15\cos(2t) + 8\sin(2t)$ V, $i = 1.7\cos(2t + 40°)$ A.

(b) $v = \text{Re}\{je^{j\,2t}\}$ V, $i = \text{Re}\{(1+j)\,e^{j\,(2t+30°)}\}$ mA.

(c) $v = aV_m\cos(\omega t + \theta)$ V, $i = V_m\cos(\omega t + \theta - \alpha)$ A.

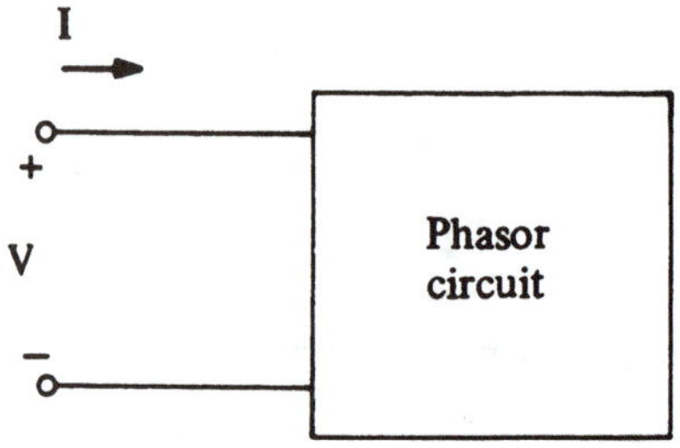

PROBLEM 8.25

8.26 Find the equivalent admittances in Prob. 8.25.

8.7 Kirchhoff's Laws and Impedance Equivalents

8.27 Find both the equivalent impedance and admittance of the following subcircuits.

(a)

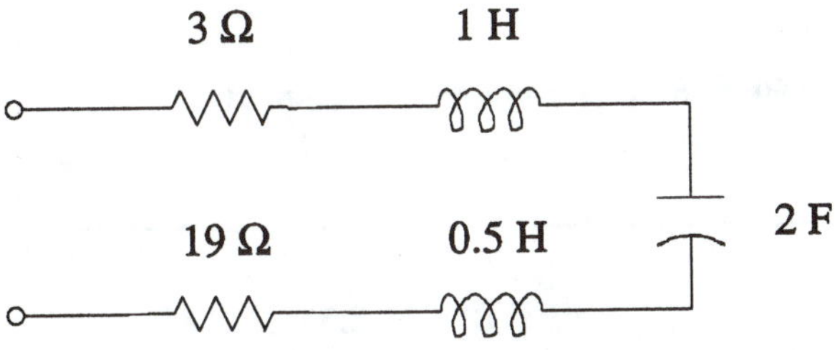

(b)

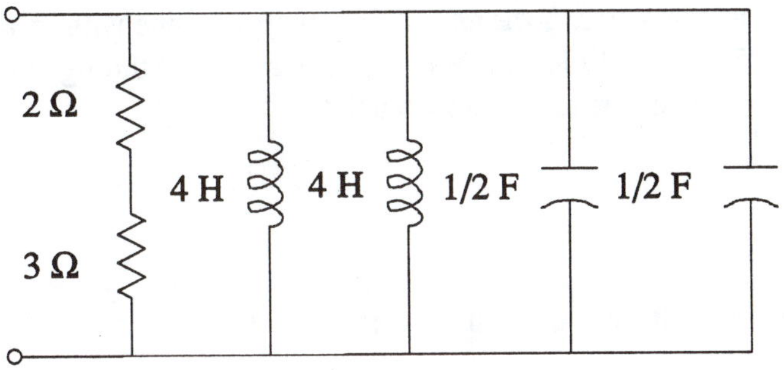

8.28 Design a series RLC subcircuit including a 10H inductor with an equivalent impedance of $\mathbf{Z} = 3 + j8\ \Omega$ when $\omega = 1$ rad/s.

8.29 Design a parallel RLC subcircuit including a 10F capacitor with an equivalent admittance of $\mathbf{Y} = 3 + j8$ S when $\omega = 1$ rad/s.

8.30 A 1 mH inductor is connected in series with a capacitor in order to achieve an equivalent impedance of $\mathbf{Z} = 0$ when $\omega = 1$ rad/s. (a) What is the value of C ? (b) What is the impedance of this LC combination when $\omega = 10$ rad/s ? (c) What is $\mathbf{Z}$ as a function of ω ?

8.31 Find the steady-state voltage v using phasors.

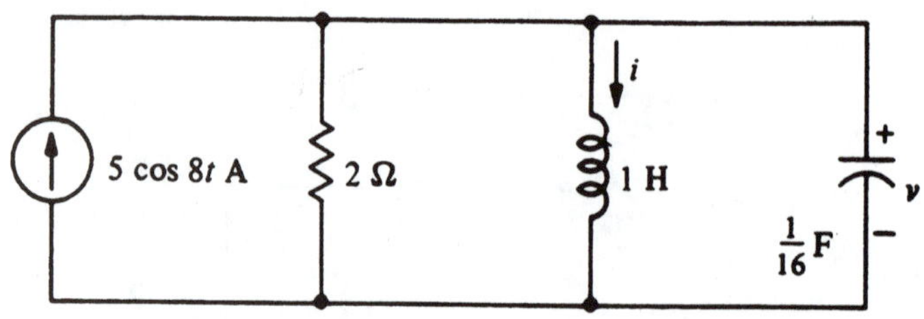

PROBLEM 8.31

8.32 Find the steady-state current i in Prob. 8.31 using phasors and current division.

8.33 Solve 8.7 using phasors, impedance, and Kirchhoff's Laws.

8.34 An audio amplifier outputs a voltage into a speaker which can be modeled as an 8 ohm resistor in series with a 100 mH inductor. If the amplifier outputs a cosine with an amplitude of 3 V and a phase angle of 45° at $f = 159$ Hz, (a) What is the current $i(t)$ through the speaker? (b) Does this current lead or lag the voltage and by how much?

8.8 Phasor Circuits

8.35 Find the steady-state voltage v using the phasor circuit.

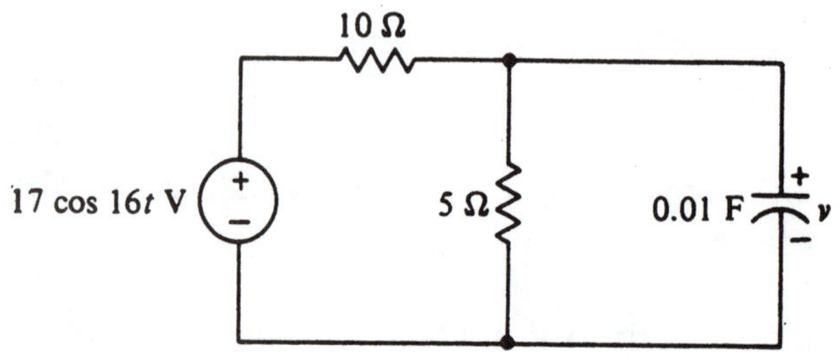

PROBLEM 8.35

8.36 Find the forced voltage v if $i_g = 10\cos(2t)$ A.

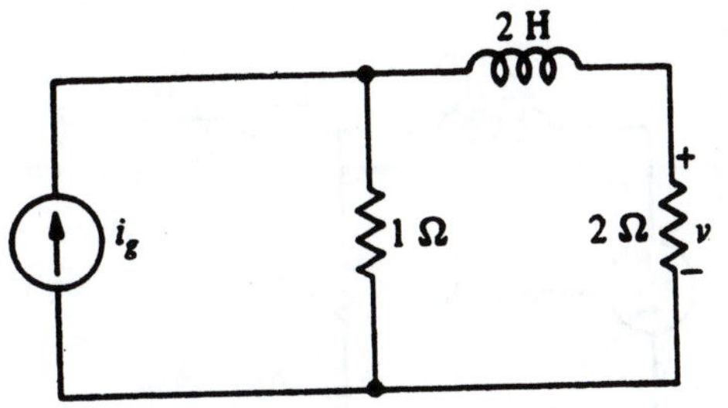

PROBLEM 8.36

8.37 Find the steady-state current i.

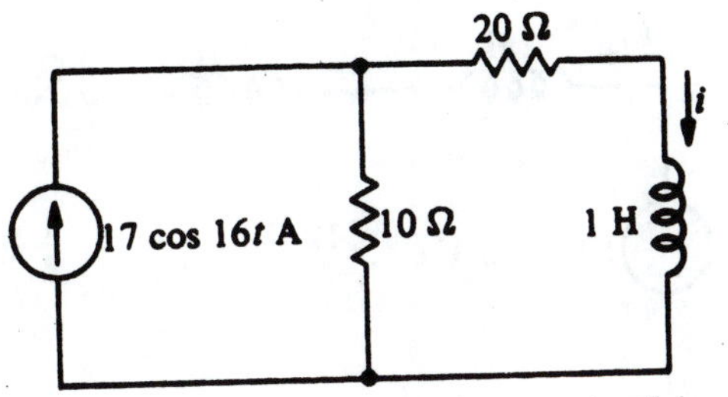

PROBLEM 8.37

8.38 Find the forced current i if $L = 1$H.

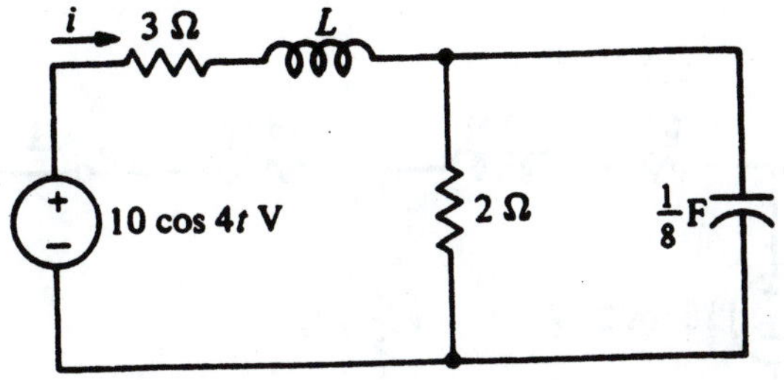

PROBLEM 8.38

8.39 Determine L in Prob. 8.38 so that the impedance seen by the source is real, and for this case find the power delivered by the source at $t = \pi/4$ s.

8.40 Find the forced response i using phasors and current division.

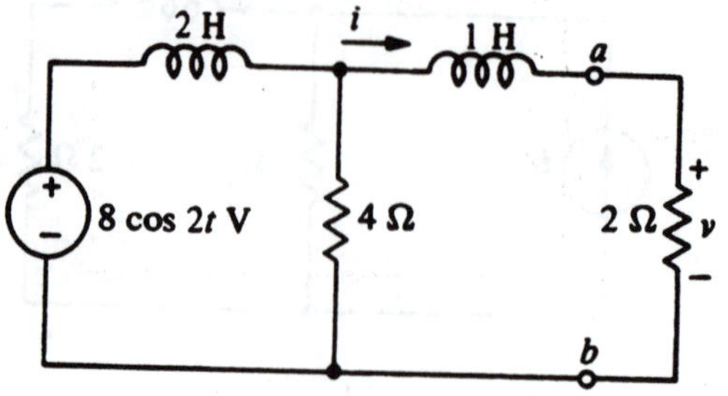

PROBLEM 8.40

8.41 Find the steady-state current i .

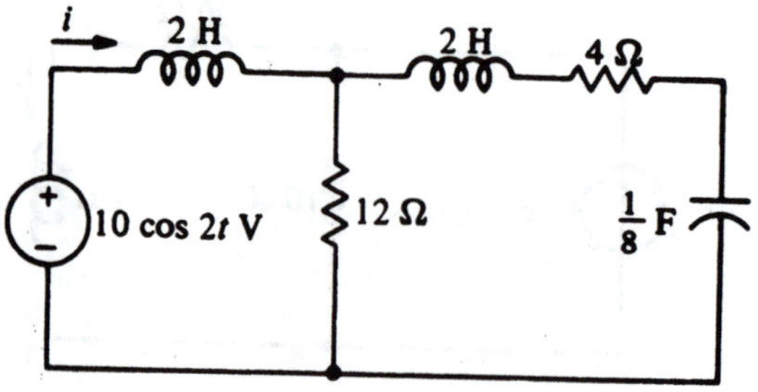

PROBLEM 8.41

8.42 Find the forced response v using the phasor circuit and voltage division.

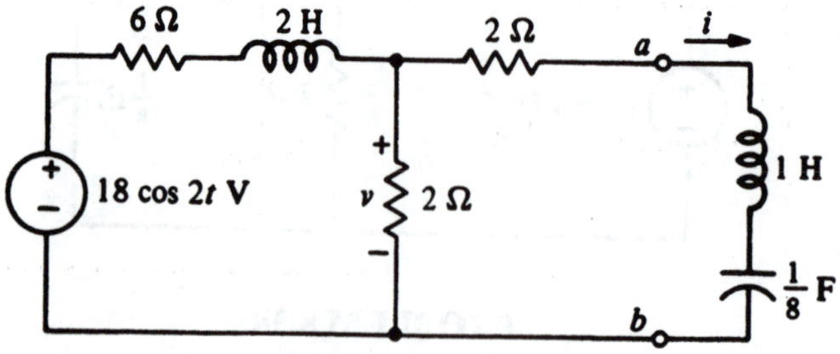

PROBLEM 8.42

8.43 Find the steady-state current i if (a) $\omega = 4$ rad/s and (b) $\omega = 2$ rad/s.

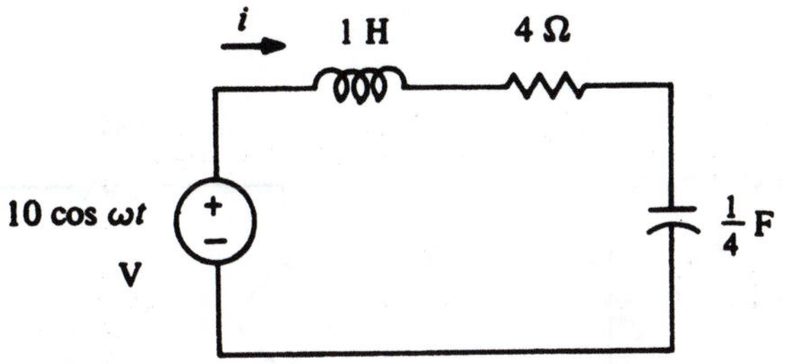

PROBLEM 8.43

8.44 Find the steady-state value of v_x.

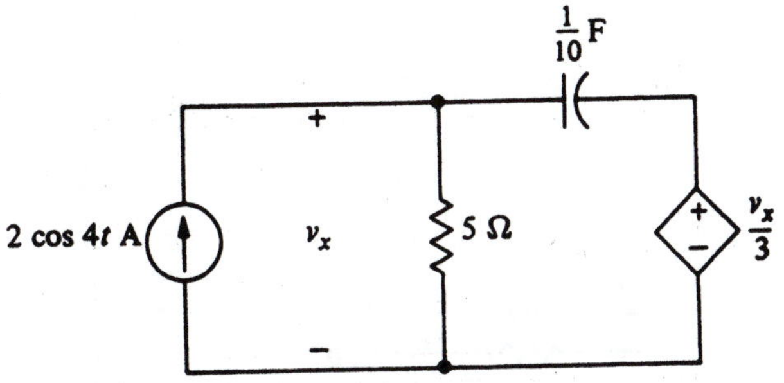

PROBLEM 8.44

8.45 Find the steady-state current i.

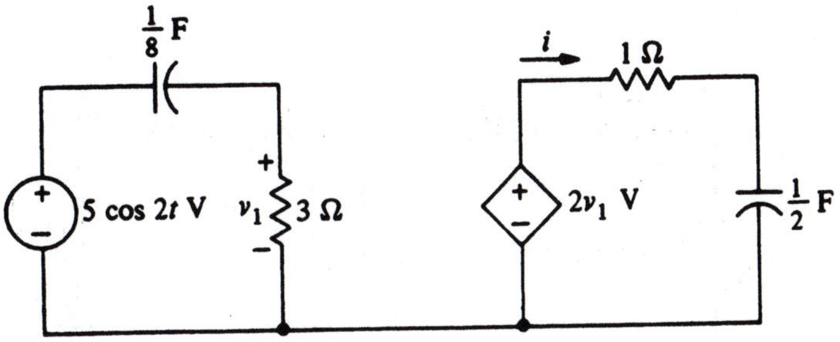

PROBLEM 8.45

8.46 Find the forced response v.

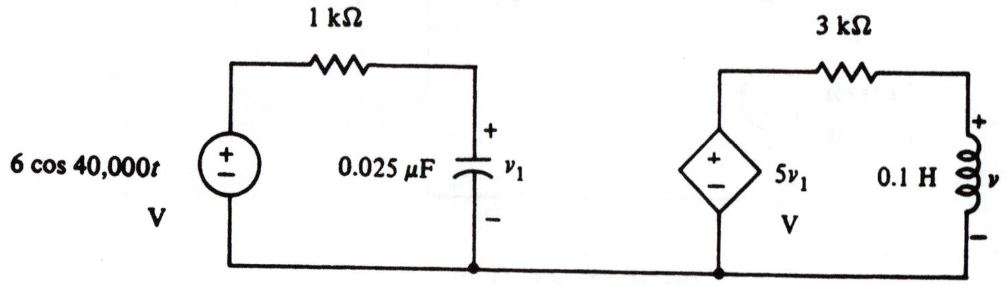

PROBLEM 8.46

8.47 Find the steady-state current i.

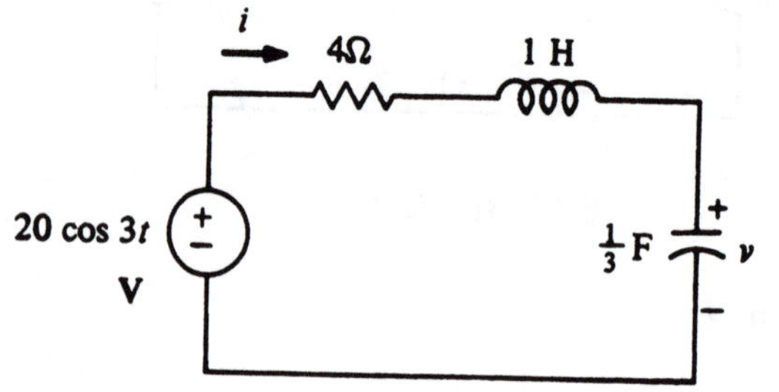

PROBLEM 8.47

CHAPTER 8 SOLUTIONS

8.1 (a) $\omega = 5$ rad/s, $T = 2\pi/\omega = \frac{2}{5}\pi$ s

(b) $\omega = 6\pi$ rad/s, $T = 2\pi/\omega = 1/3$ s

(c) $\omega = 4$ rad/s, $T = \pi/2$ s

8.2 (a) Amplitude = 6, $\phi = 17°$

(b) Amplitude = 5, $\phi = 0°$

(c) $3\sin(4t+\pi/4) + 2\cos(4t+\pi)$

$= 3[\sin 4t \cos\pi/4 + \cos 4t \sin\pi/4]$

$\quad + 2[\cos 4t \cos\pi - \sin 4t \sin\pi]$

$= 2.125 \cos(4t - 86.73°)$

Amplitude = 2.13, $\phi = -86.7°$

8.3 (a) $f = \omega/2\pi = 5/2\pi = 0.80$ Hz

(b) $f = 3.0$ Hz (c) $f = 0.64$ Hz

8.4 (a) $6\sin(5t+17°) \rightarrow$ shift left by 17°

Time in sec. for 17° is $T \times \frac{17°}{360°} = 0.059$ s

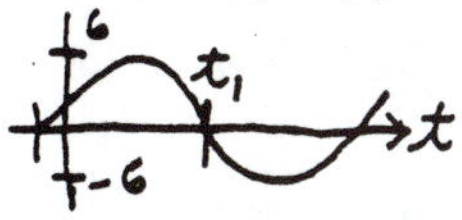

First zero crossing at

$t_1 = \frac{T}{2} - 0.059 = \underline{0.57\,s}$

(b) $5\sin(6\pi t)$

First zero crossing at $\frac{T}{2} = \frac{1}{6}$ s

(c) $2.125\cos(4t - 86.73°) \leftarrow$ Shift right by 86.73°

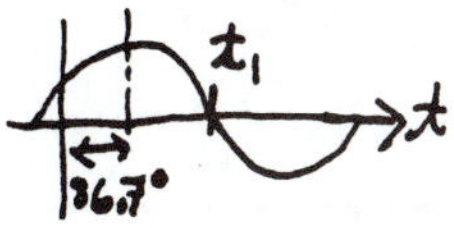

$90° - 86.73° = 3.27°$

First zero crossing at

$t_1 = \frac{T}{2} - \frac{3.27°}{360°} = \underline{0.77\,s}$

8.5 $v_1 = 5\cos(4t - 90°)$

$v_2 = 20\sqrt{1+3}\cos(4t - \tan^{-1}\frac{\sqrt{3}}{1})$

$= 40\cos(4t - 60°)$

(a),(b) v_1 <u>lags</u> v_2 by $90° - 60° = 30°$

(c) $30° \times \frac{\pi \text{ rad}}{180°} = \frac{\pi}{6}$ rad

(d) $\omega = 4 \Rightarrow T = 2\pi/\omega = \frac{\pi}{2}$ s, $\frac{30°}{360°} = \frac{1}{12}$

$\therefore \frac{1}{12} \times \frac{\pi}{2}$ s $= \frac{\pi}{24}$ s $\approx \underline{0.13\,s}$

8.6 By KCL, $i_4 = i_1 + i_2 + i_3$

$i_4 = 25[\cos 3t \cos 36.9° - \sin 3t \sin 36.9°]$

$\quad + 13[\sin 3t \cos 157.4° + \cos 3t \sin 157.4°]$

$\quad + 2\cos 3t = 27\cos 3t - 27\sin 3t$

$i_4 = 27\sqrt{2}\cos(3t + 45°)$ A

8.7 By KVL,

$100\cos 3t = 20i + 0.25\frac{di}{dt} + \frac{1}{47\mu}\int_0^t i(\tau)\,d\tau$

Differentiating and mult. by 4:

$-1200\sin 3t = \frac{d^2 i}{dt^2} + 80\frac{di}{dt} + 8.51\times10^4 i$

Trial forced soln is: $i(t) = A\cos 3t + B\sin 3t$

Plug in and solve $\Rightarrow A = 3.98\times10^{-5}$

$B = -1.41\times10^{-2}$

$\Rightarrow i(t) = (3.98\times10^{-5})\cos 3t + (-1.41\times10^{-2})\sin 3t$

$= 0.0141\cos(3t + 89.8°)$ A

8.8 By KCL at top node,

$20\sin 9t = \frac{v}{10000} + \frac{1}{1\times10^{-2}}\int_{0+}^t v(\tau)\,d\tau + i_L(0+)$

$\quad + (1\times10^{-5})\,dv/dt$

Differentiating and multiply by 1×10^5:

$(1.8\times10^7)\cos 9t = \frac{d^2 v}{dt^2} + 10\frac{dv}{dt} + (1\times10^7)v$

Trial forced soln is: $v(t) = A\cos 9t + B\sin 9t$

Plug in and solve $\Rightarrow A = 1.80$, $B = 1.62\times10^{-5}$

$\Rightarrow v(t) = 1.8\cos 9t + (1.62\times10^{-5})\sin 9t$

$= 1.8\cos(9t - 0.00052°)$ V

8.9 By KVL, $4\sin 2t = \frac{di}{dt} + 10i$

Trial forced soln is $i = A\cos 2t + B\sin 2t$

Plug in and solve $\Rightarrow A = -1/13$, $B = 5/13$

$\Rightarrow i(t) = -\frac{1}{13}\cos 2t + \frac{5}{13}\sin 2t$

$= 0.39\cos(2t - 101.3°)$ A

8.10 The output amplitude is 1/5 that of the input. The phase is 45° less.

(a) $i = \frac{45}{5} e^{j(4t+50°-45°)} = 9e^{j(4t+5°)}$ A

(b) $i = \frac{15}{5} \sin(4t+60°-45°) = 3\sin(4t+15°)$ A

(c) $i = \frac{25}{5} \cos(4t-45°) = 5\cos(4t-45°)$ A

8.11 KCL: $\frac{v_1}{5} + \frac{1}{20}\frac{d}{dt}\left(v_1 - \frac{v_1}{3}\right) = 2e^{j8t}$

$\Rightarrow \frac{dv_1}{dt} + 6v_1 = 60e^{j8t}$. Try $v_1 = Ae^{j8t}$

Plugging in $\Rightarrow A = 6e^{-j53.1°}$

$\therefore v_1 = 6e^{j(8t-53.1°)}$

(a) $v = Re[v_1] = 6\cos(8t-53.1°)$ V

(b) $v = Im[v_1] = 6\sin(8t-53.1°)$ V

8.12 By KVL, $100e^{j3t} = 20i + \frac{1}{4}\frac{di}{dt} + \frac{1}{4\frac{7}{4}}\int_0^t i(\tau)d\tau$

Differentiating and simplifying:

$1200j e^{j3t} = \frac{d^2i}{dt^2} + 80\frac{di}{dt} + (8.51\times10^4)i$

Try $i = Ae^{j3t}$, Plug in $\Rightarrow A = 0.0141e^{j89.8°}$

$i_{forced} = Re[i] = 0.0141\cos(3t+89.8°)$

8.13 $20e^{j9t} = \frac{v}{10000} + \frac{1}{1\times10^{-2}}\int_0^t v(\tau)d\tau + 1\times10^{-5}\frac{dv}{dt}$

Differentiating and simplifying:

$j(1.8\times10^7)e^{j9t} = \frac{d^2v}{dt^2} + 10\frac{dv}{dt} + 10^7 v$

Try $v = Ae^{j9t} \Rightarrow$ Plugging

$A = 1.8e^{j89.99948°}$

$v_{forced} = Im[v] = 1.8\sin(9t+89.99948°)$

$= 1.8\cos(9t-0.00052°)$

8.14 KVL: $4e^{j2t} = \frac{di}{dt} + 10i$

Try $i = Ae^{j2t} \Rightarrow A = 0.39e^{-j11.31°}$

$i_{forced} = Im[i] = 0.39\sin(2t-11.31°)$

$= 0.39\cos(2t-101.31°)$ A

8.15 (a) amplitude = 5, $\phi = 30°$

phasor = $5\angle30°$

(b) $-3\sin3t + 2\cos3t = \sqrt{13}\cos(3t+56.3°)$

phasor = $\sqrt{13}\angle56.3°$

(c) $4\sin(10t-20°) = 4\cos(10t-20°-90°)$

$= 4\cos(10t-110°)$

phasor = $4\angle-110°$

8.16 (a) amplitude = 15, $\phi = -45°$

$v = 15\cos(5t-45°)$

(b) $5+j12 = 13\angle67.38°$

$v = 13\cos(5t+67.38°)$

(c) $5-j12 = 13\angle-67.38°$

$v = 13\cos(5t-67.38°)$

8.17 (a) $-6-j6 = 6\sqrt{2}\angle-135°$

$v = 6\sqrt{2}\cos(5t-135°)$

(b) $-5+j12 = 13\angle112.62°$

$v = 13\cos(5t+112.62°)$

(c) $4-j3 = 5\angle-36.87°$

$v = 5\cos(5t-36.87°)$

(d) $-10 = 10\angle180°$

$v = 10\cos(5t+180°)$

$= -10\cos5t$

8.18 (a) $Re\{100e^{j3t}\}$

(b) $Re\{20e^{j(9t-90°)}\}$

(c) $Re\{4e^{j(2t-90°)}\}$

8.19 $\underline{V} = 12\angle30°$, $R = 4\,k\Omega$

$\underline{I} = \frac{\underline{V}}{R} = \frac{12\angle30°}{4} = 3\angle30°$ mA

$i = 3\cos(1000t+30°)$ mA

8.20 $\underline{V} = 10\angle-75°$

$\underline{I} = \frac{\underline{V}}{j\omega L} = \frac{10\angle-75°}{j(377)(.5)} = 53\angle-165°$

$i = 53\sin(377t-75°)$ A

8.21 $\underline{V} = 12\angle-75°$

$\underline{I} = j\omega C\underline{V} = j(377)(10^{-5})(12\angle-75°)$

$= 45.24\angle15°$ mA

$i = 45.24\sin(377t+105°)$ mA

8.22 (a) $\underline{V} = Z_L \underline{I} = j\omega L \underline{I}$

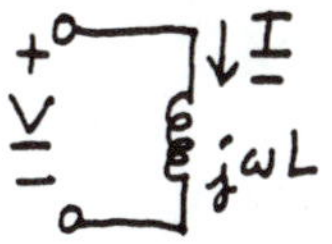

Let $\underline{V} = V_m \angle \theta$, then

$$\underline{I} = \frac{V_m \angle \theta}{\omega L \angle 90^\circ} = \frac{V_m}{\omega L} \angle \theta - 90^\circ$$

$\therefore$ $\underline{I}$ lags $\underline{V}$ by 90°

(b)

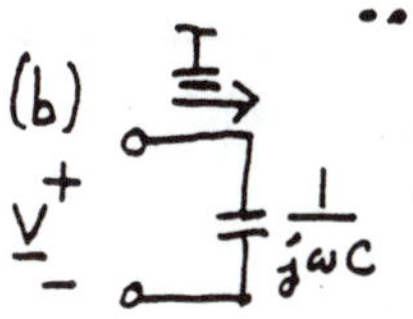

$\underline{V} = Z_C \underline{I} = \frac{1}{j\omega C} \underline{I}$

Let $\underline{V} = V_m \angle \theta$, then

$$\underline{I} = \frac{V_m \angle \theta}{1/(j\omega C)} = \omega C V_m \angle \theta + 90^\circ$$

$\therefore$ $\underline{I}$ leads $\underline{V}$ by 90°

8.23 (a) $\underline{Y} = G + jB = \frac{1}{12 + j5} = .071 - j.03$

(b) $\underline{Y} = G + jB = \frac{1}{3 - j3} = 0.167 + j0.167$

(c) $\underline{Y} = G + jB = \frac{1}{5\angle 30^\circ} = 0.173 - j0.10$

8.24

(a) $\underline{Z} = R + jX = \frac{1}{12 + j5} = .071 - j0.03$

(b) $\underline{Z} = R + jX = \frac{1}{3 - j3} = 0.167 + j0.167$

(c) $\underline{Z} = R + jX = 1/(5\angle 30^\circ) = 0.173 - j0.1$

8.25 (a) $\underline{Z} = \underline{V}/\underline{I} = \frac{-15 - j8}{1.7\angle 40^\circ} = 10\angle 111.9^\circ \, \Omega$

(b) $v = \text{Re}[j(\cos 2t + j \sin 2t)] = -\sin 2t$

$= -\cos(2t - 90^\circ) = \cos(2t + 90^\circ)$ V

$i = \text{Re}\{(1 + j1)[\cos(2t + 30^\circ) + j \sin(2t + 30^\circ)]\}$

$= \cos(2t + 30^\circ) - \sin(2t + 30^\circ)$

$= \sqrt{2} \cos(2t + 75^\circ)$ mA

$$\underline{Z} = \underline{V}/\underline{I} = \frac{1\angle 90^\circ}{\sqrt{2}\angle 75^\circ} = \frac{1}{\sqrt{2}} \angle 15^\circ \, \text{k}\Omega$$

(c) $$\underline{Z} = \frac{a V_m \angle \theta}{V_m \angle \theta - \alpha} = a \angle \alpha \, \Omega$$

8.26 (a) $\underline{Y} = \underline{I}/\underline{V} = \frac{1.7\angle 40^\circ \text{A}}{-15 - j8 \text{ V}} = 0.1 \angle -111.9^\circ$ S

(b) $\underline{Y} = \frac{\underline{I}}{\underline{V}} = \frac{\sqrt{2}\angle 75^\circ \text{mA}}{1\angle 90^\circ \text{ V}} = \sqrt{2} \angle -15^\circ$ mS

(c) $\underline{Y} = \underline{I}/\underline{V} = \frac{V_m \angle \theta - \alpha}{a V_m \angle \theta} = \frac{1}{a} \angle -\alpha$ S

8.27 (a)

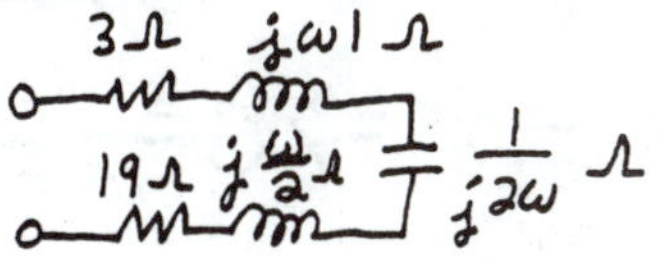

$$\underline{Z}_T = 3 + j\omega - j\frac{1}{2\omega} + j\frac{\omega}{2} + 19$$

$$= 22 + j\left(\omega - \frac{1}{2\omega} + \frac{\omega}{2}\right) \Omega$$

$$\underline{Y}_T = \frac{1}{\underline{Z}_T} = \frac{1}{22 + j(\;)} \cdot \frac{22 - j(\;)}{22 - j(\;)}$$

$$= \frac{22 - j\left(\omega - \frac{1}{2\omega} + \frac{\omega}{2}\right)}{484 + \left(\omega - \frac{1}{2\omega} + \frac{\omega}{2}\right)^2} \text{ S}$$

(b)

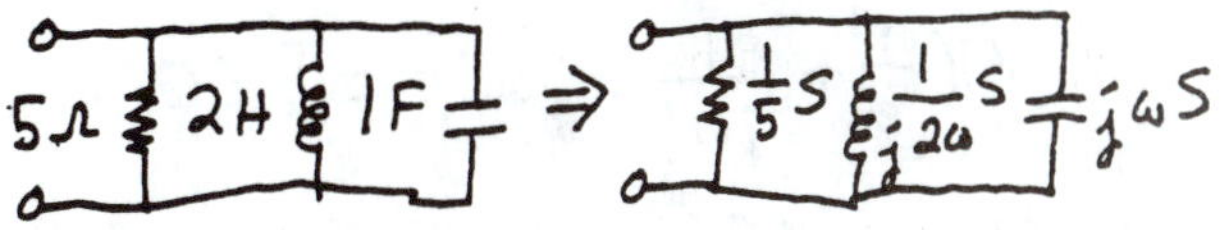

$$\underline{Y}_T = \frac{1}{5}\text{S} + \frac{1}{j2\omega}\text{S} + j\omega\text{S} = \frac{1}{5} + j\left(\omega - \frac{1}{2\omega}\right)\text{S}$$

$$\underline{Z}_T = \frac{1}{\underline{Y}_T} = \frac{\frac{1}{5} - j\left(\omega - 1/2\omega\right)}{1/25 + \left(\omega - 1/2\omega\right)^2} \, \Omega$$

8.28

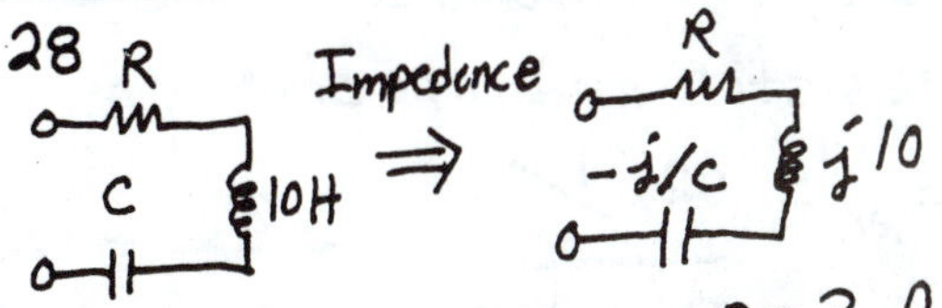

$\underline{Z}_T = R + j10 - j\frac{1}{c} \Rightarrow R = 3\,\Omega$, $C = \frac{1}{2}$ F

8.29

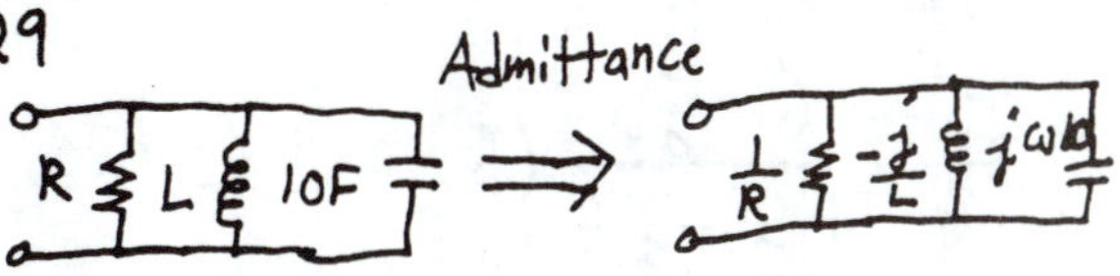

$$\underline{Y}_T = \frac{1}{R} + j10 - j\frac{1}{L}$$

$\Rightarrow R = \frac{1}{3}\,\Omega$, $L = \frac{1}{2}$ H

8.30

(a)

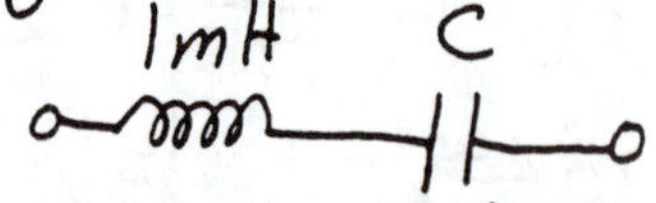

$\Rightarrow$

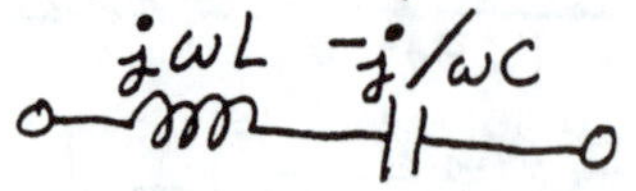

$$\underline{Z}_T = j\left(\omega L - \frac{1}{\omega C}\right) = 0$$

$$\Rightarrow L = \frac{1}{C} \Rightarrow C = \frac{1}{0.001} = 1000\,F$$

(b) $j10(0.001) - \dfrac{j}{10(1000)} = j0.0099\,\Omega$

(c) $\underline{Z} = j\omega L - \dfrac{j}{\omega C} = j\left(.001\omega - \dfrac{.001}{\omega}\right)$

$$= j\left(\frac{\omega - 1/\omega}{1000}\right)\Omega$$

8.31 By KCL at top node,

$$\underline{V}\left(\frac{1}{2} + \frac{1}{j\omega L} + j\omega C\right) = 5\angle 0°\,A$$

$$\underline{V} = \frac{5\angle 0°}{\frac{1}{2} + j\frac{3}{8}} = 8\angle -36.87°$$

$\therefore\ v = 8\cos(8t - 36.87°)\,V$

8.32 The impedance of the 2Ω resistor and capacitor are

$$\underline{Z}_{RC} = \frac{(2)\,1/j\omega C}{2 + \frac{1}{j\omega C}} = \frac{2}{1 + j\omega\frac{2}{16}} = 1 - j$$

By current division,

$$\underline{I} = \frac{(1-j)}{1 - j + j8}(5) = 1\angle -126.9°$$

$\therefore\ i = \cos(8t - 126.9°)\,A$

8.33

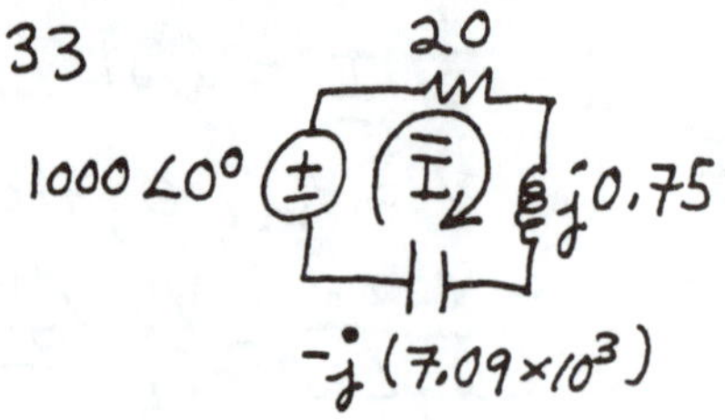

$$\underline{I} = \frac{1000\angle 0°}{20 + j(0.75 - 7.09\times 10^3)}$$

$$= 0.0141\angle 89.84°$$

$\therefore\ i = 0.0141\cos(3t + 89.84°)\,A$

8.34

(a) $\underline{I} = \dfrac{3\angle 45°}{8 + j99.9} = 0.03\angle -40.4°\,A$

$\therefore\ i(t) = 0.03\cos(999t - 40.4°)\,A$

(b) $i(t)$ lags $v(t)$ by $45° - (-40.4°)$
$= 85.4°$

8.35 The impedance of the 5Ω resistor and capacitor is

$$\underline{Z}_{RC} = \frac{(5)\left(\frac{1}{j16(0.01)}\right)}{5 + 1/j16(0.01)}$$

$$= 3.9\angle -38.66°$$

By voltage division,

$$\underline{V} = \frac{\underline{Z}_{RC}}{\underline{Z}_{RC} + 10}(17) = 5\angle -28.07°$$

$\therefore\ v = 5\cos(16t - 28.07°)\,V$

8.36 Let $\underline{I}$ = phasor current of inductor then by current division

$$\underline{I} = \frac{\underline{I}_g}{1+2+j\omega L} = \frac{10}{3+j(2)(2)} = 2\angle -53.13^\circ \text{A}$$

$$\underline{V} = \underline{I}R = (2)(2\angle -53.13^\circ) = 4\angle -53.13 \text{V}$$

$$v = 4\cos(2t - 53.13^\circ)\text{ V}$$

8.37 By current division

$$\underline{I} = \frac{10\,(17)}{10+20+j(16)(1)} = 5\angle -28.07^\circ$$

$$i = 5\cos(16t - 28.07^\circ)\text{ A}$$

8.38 the impedance seen by the source is:

$$\underline{Z} = 3 + j(4)(1) + \frac{2}{1+j(4)(2)(1/8)} = 4+j3 = 5\angle 36.9^\circ$$

$$\underline{I} = \frac{\underline{V}}{\underline{Z}} = \frac{10}{5\angle 36.90} = 2\angle -36.9^\circ \text{A}$$

$$i = 2\cos(4t - 36.9^\circ)\text{ A}$$

8.39

$$Z = 3 + j4L + 1 - j = 4 + j(4L-1)$$

$$\therefore 4L - 1 = 0 \Rightarrow L = 1/4\text{ H}$$

$$\underline{I} = \frac{10}{4} = 2.5\angle 0\text{ A}$$

$$\therefore i = 2.5\cos 4t\text{ A},\quad v = 10\cos 4t$$

$$p = vi = 2.5\cos[4(\tfrac{\pi}{4})] \times 10\cos[4(\pi/4)]$$

$$= 25\text{ W at } t = \pi/4\text{ s}$$

8.40 the impedance seen by the source is.

$$\underline{Z} = j(2)(2) + \frac{4(2+j2)}{4+2+j2} = 1.6 + j4.8 = 5.06\angle 71.6^\circ$$

Let $\underline{I}_1$ be the source current

$$\underline{I}_1 = \frac{8}{5.06\angle 71.6^\circ} = 1.581\angle -71.57^\circ$$

By current division

$$\underline{I} = \frac{4\,\underline{I}_1}{4+2+j2} = \frac{2-j6}{6+j2} = 1\angle -90^\circ$$

$$i = \cos(2t - 90^\circ) = \sin 2t\text{ A}$$

8.41 the impedance seen by the source is

$$\underline{Z} = j(2)(2) + \frac{12\left[4+j4+\frac{1}{j2(1/8)}\right]}{12+4+j4-j4}$$

$$= 3 + j4$$

$$\underline{I} = \underline{V}/\underline{Z} = 10/(3+j4) = 2\angle -53.13^\circ$$

$$i = 2\cos(2t - 53.13^\circ)\text{ A}$$

8.42 By voltage division

$$\underline{V} = \frac{\dfrac{(2)(2+j2-j4)}{2+2-j2}}{6+j4+1.2-j0.4}(18) = 2\sqrt{2}\angle -45^\circ$$

$$v = 2\sqrt{2}\cos(2t - 45^\circ)\text{ V}$$

8.43

(a) $$\underline{I} = \frac{\underline{V}}{\underline{Z}} = \frac{10}{4+j\omega - j4/\omega} = \frac{10}{4+j3} = 2\angle -36.9^\circ$$

$$i = 2\cos(4t - 36.9^\circ)\text{ A}$$

(b) $$\underline{I} = \frac{10}{4+j2-j2} = 2.5\angle 0^\circ$$

$$i = 2.5\cos(2t)\text{ A}$$

8.44 By KCL

$$\underline{V}_x\left(\frac{1}{5} + j\frac{4}{10}\right) - \frac{\underline{V}_x}{3}\left(j\frac{4}{10}\right) = 2 \text{ or}$$

$$\underline{V}_x = \frac{60}{6+j8} = 6\angle -53.13^\circ$$

$$v_x = 6\cos(4t - 53.13^\circ)\text{ V}$$

8.45 By voltage division

$$\underline{V}_1 = \frac{5(3)}{3-j4} = 3\angle 53.13^\circ$$

then,

$$\underline{I} = \frac{2\underline{V}_1}{1+\frac{1}{j\omega C_2}} = \frac{6\angle 53.13^\circ}{1-j(2)(\frac{1}{2})} = 3\sqrt{2}\angle 98.13^\circ$$

$$i = 3\sqrt{2}\cos(2t + 98.13^\circ)\text{ A}$$

8.46 By voltage division

$$\underline{V}_1 = \frac{6}{1+j(40{,}000)(10^3)(0.025\times 10^{-6})}$$

$$= 3\sqrt{2}\angle -45^\circ\text{, and,}$$

$$\underline{V} = \frac{j(40{,}000)(0.1)\,5\underline{V}_1}{3\times 10^3 + j4000} = 12\sqrt{2}\angle -8.13^\circ$$

$$v = 12\sqrt{2}\cos(40{,}000t - 8.13^\circ)\text{ V}$$

8.47

$$\underline{I} = \frac{20}{4+j(3)(1) - j\frac{1}{(3)(1/3)}} = \frac{20}{4+j2} = 4.47\angle -26.57^\circ$$

$$i = 4.47\cos(3t - 26.57^\circ)\text{ A}$$

Chapter 9

AC Steady State Analysis

9.1 Circuit Simplifications

9.1 Find v_1 in the following circuit using voltage division.

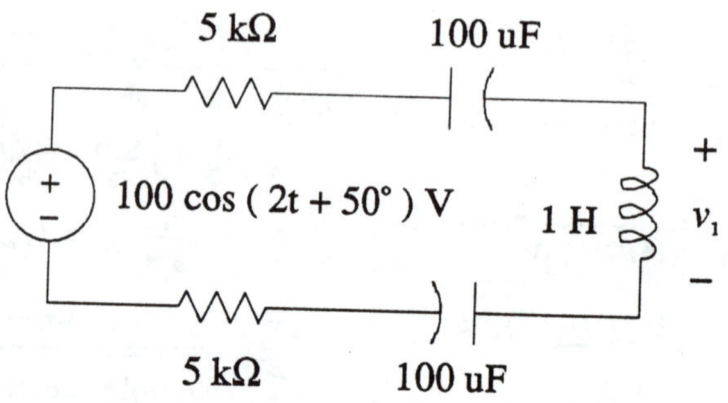

PROBLEM 9.1

9.2 Find v_1 in the following circuit using voltage division.

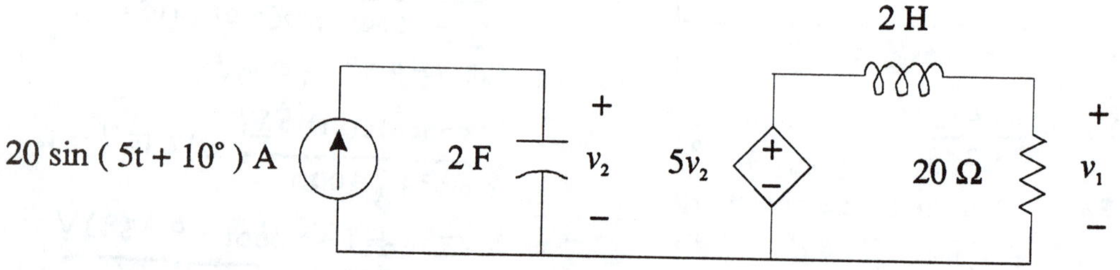

PROBLEM 9.2

9.3 Find i_1 in the following circuit using current division.

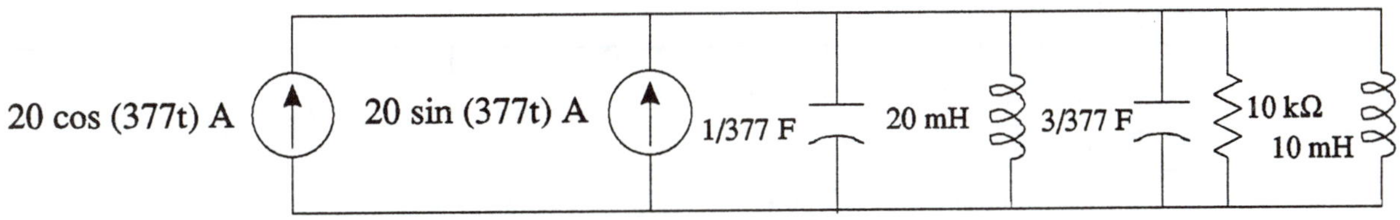

PROBLEM 9.3

9.4 Find i_1 in the following circuit using current division.

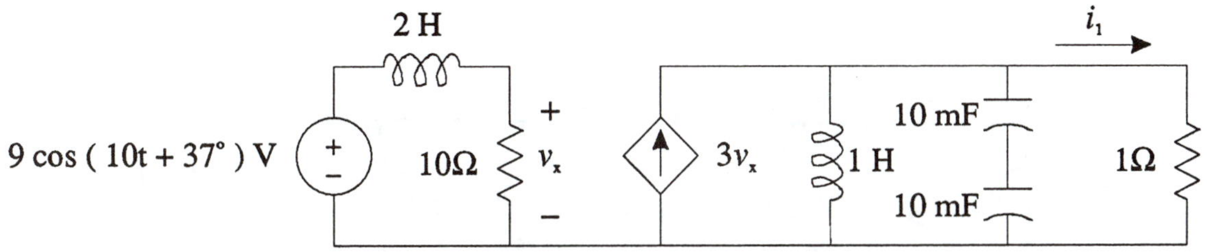

PROBLEM 9.4

9.5 Simplify the following circuit to a single voltage source and impedance and solve for i.

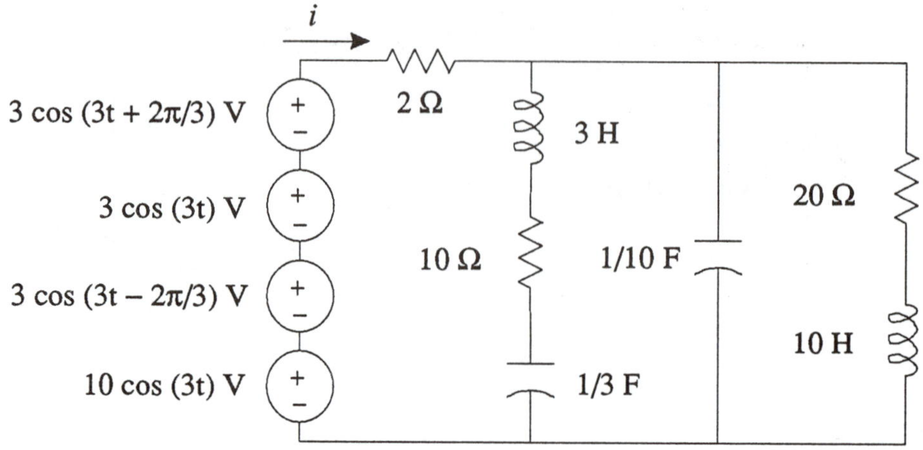

PROBLEM 9.5

9.6 Simplify the following circuit to a single current source and impedance and solve for v.

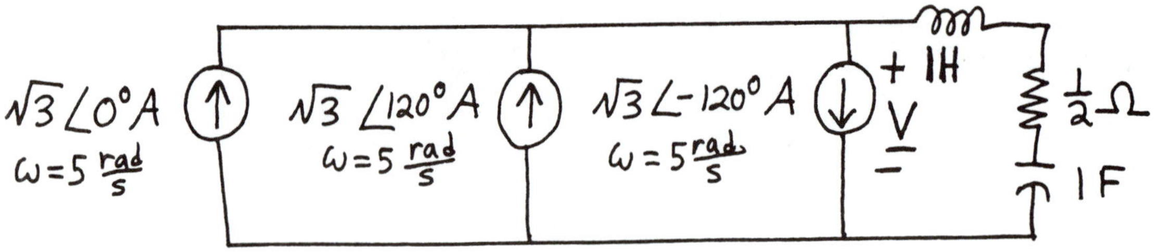

PROBLEM 9.6

9.7 For the phasor circuit corresponding to Prob. 8.40, replace the part to the left of the terminals *a-b* by its Thevenin equivalent and find the steady-state value of v.

9.8 For the phasor circuit corresponding to Prob. 8.42, replace the part to the left of the terminals *a-b* by its Thevenin equivalent and find the steady-state value of i.

9.9 Use the principle or proportionality on the corresponding phasor circuit to find the steady-state value of v.

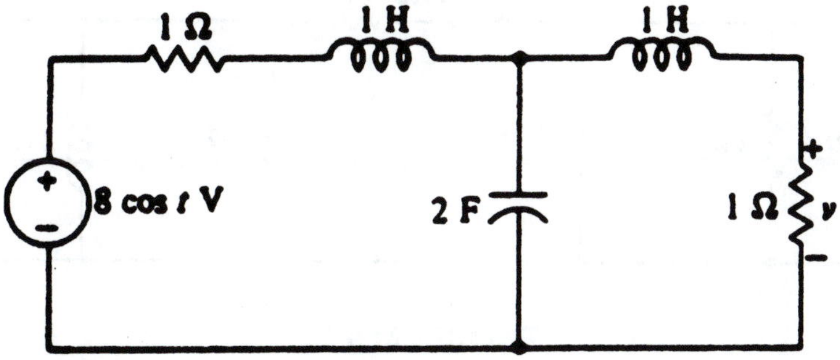

PROBLEM 9.9

9.10 Find the steady-state current i using the principle of proportionality.

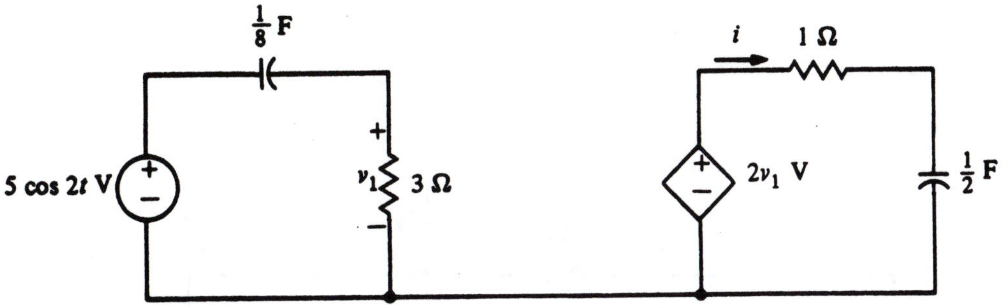

PROBLEM 9.10

9.11 In the corresponding phasor circuit, replace everything except the 1 ohm resistor by its Norton equivalent; use the result to find the steady-state current i.

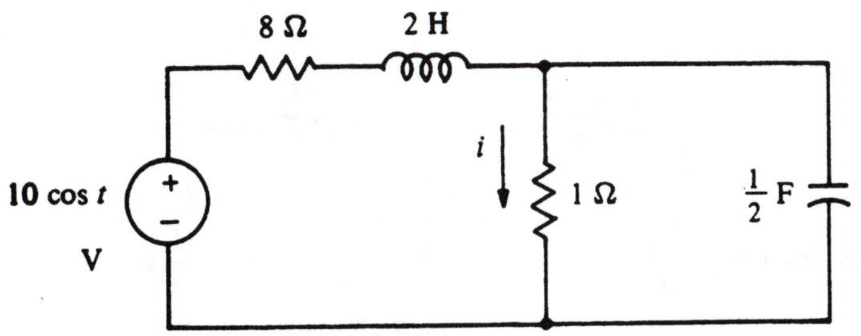

PROBLEM 9.11

9.12 Find the steady-state current i.

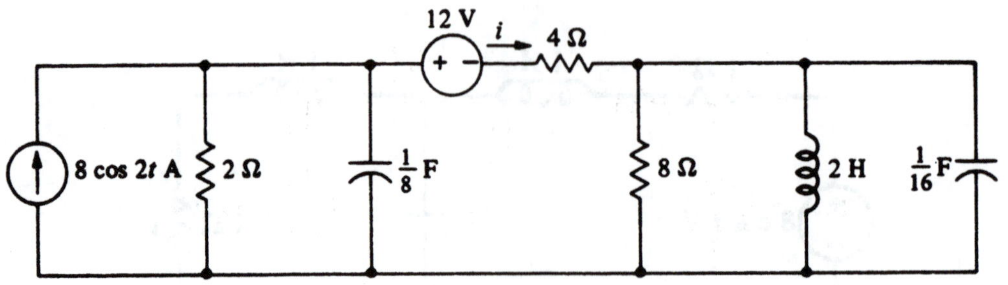

PROBLEM 9.12

9.2 Nodal Analysis

9.13 Find the forced response v using nodal analysis.

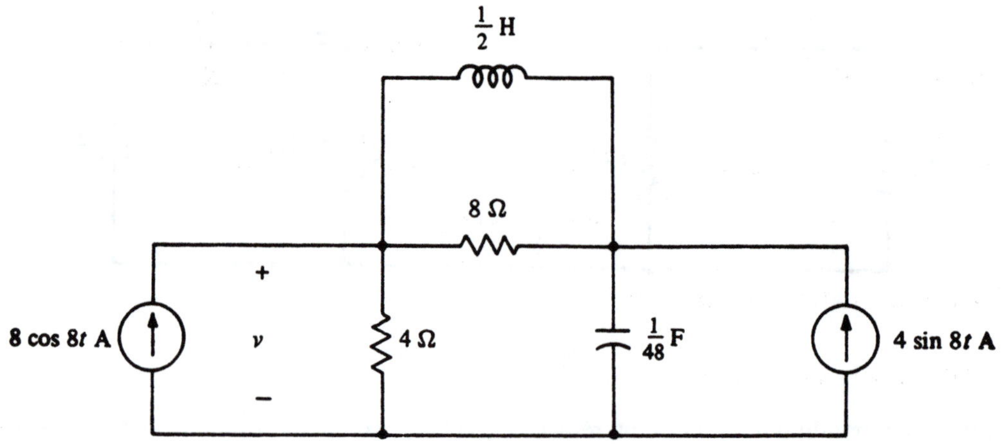

PROBLEM 9.13

9.14 Find the steady-state voltage v using nodal analysis.

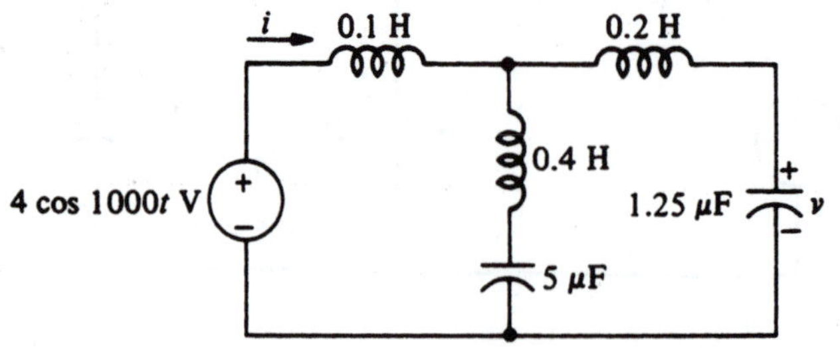

PROBLEM 9.14

9.15 Find the steady-state current i using nodal analysis.

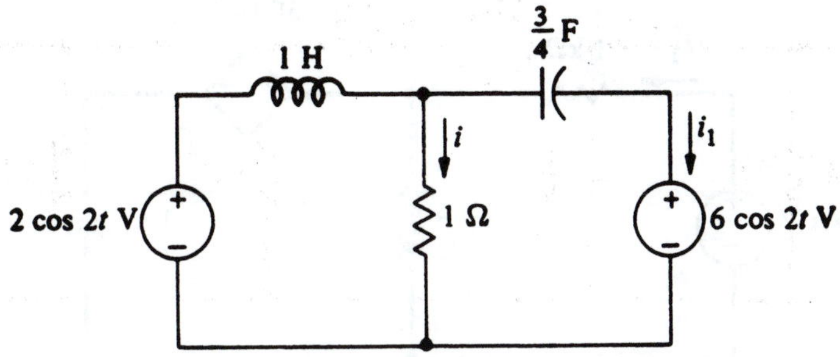

PROBLEM 9.15

9.16 Find the steady-state voltage v using nodal analysis.

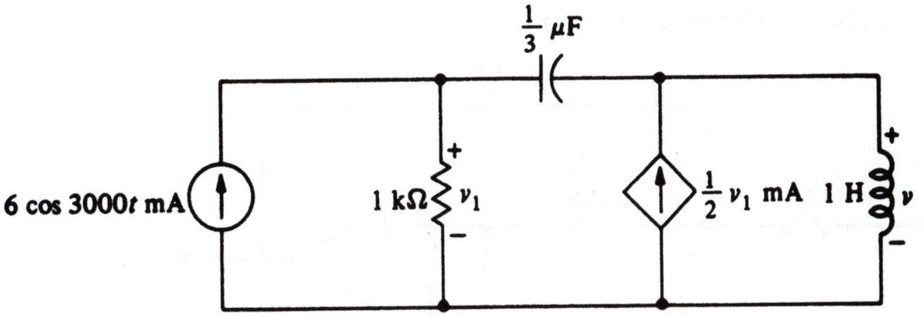

PROBLEM 9.16

9.17 Find the steady-state voltage v_1 in Prob. 9.16.

9.18 Find the steady-state voltage v using nodal analysis.

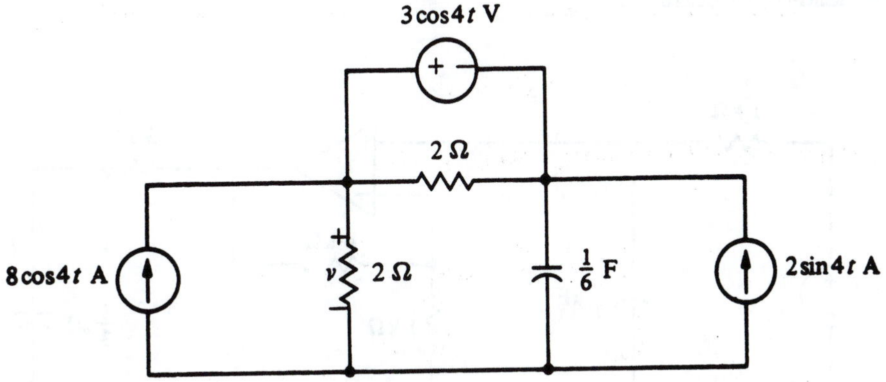

PROBLEM 9.18

9.19 Find the steady-state current i_1 using nodal analysis.

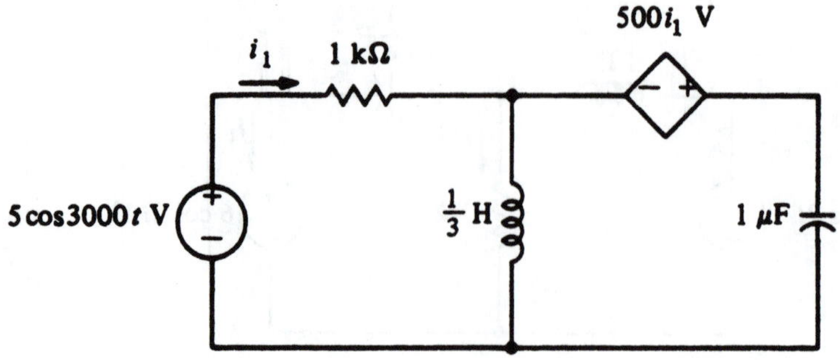

PROBLEM 9.19

9.20 Find the steady-state voltage v if $v_g = 6\cos(5t)$ V.

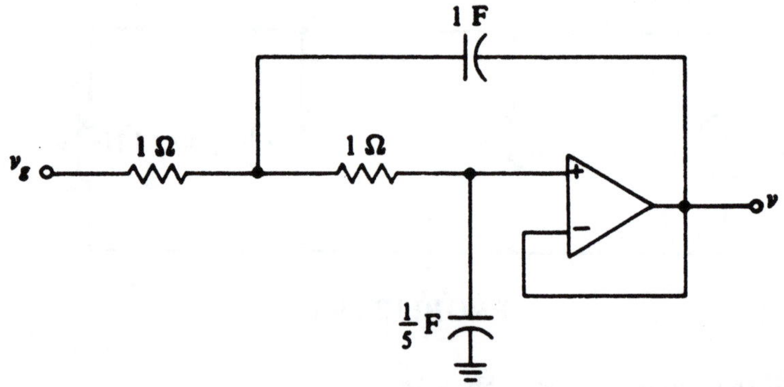

PROBLEM 9.20

9.21 Find the steady-state current i.

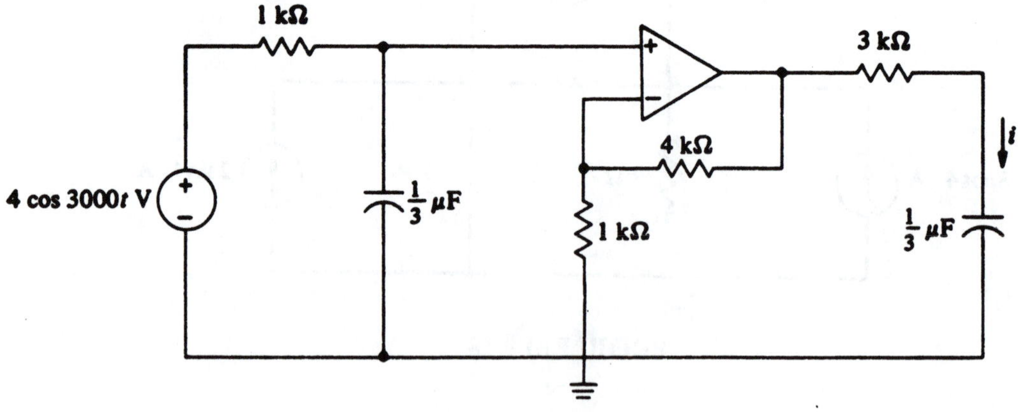

PROBLEM 9.21

9.22 Find the steady-state voltage v if $v_g = 5\cos(2t)$ V.

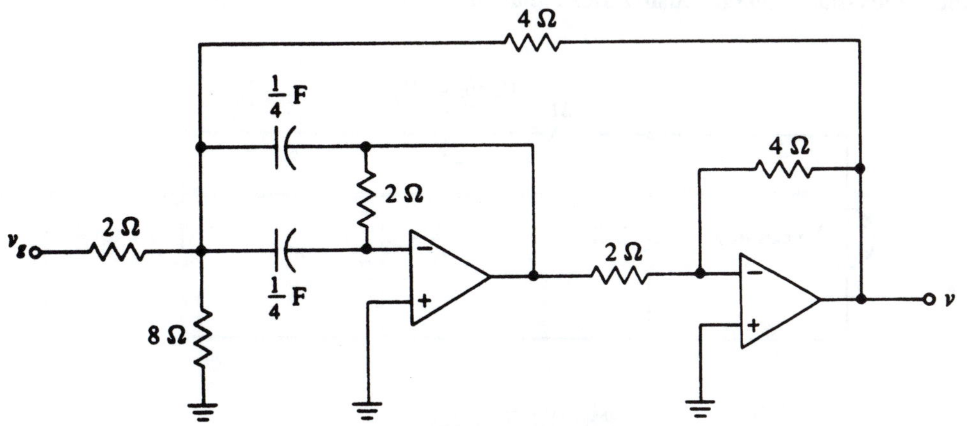

PROBLEM 9.22

9.23 Find the steady-state voltage v if $v_g = 6\cos(t)$ V.

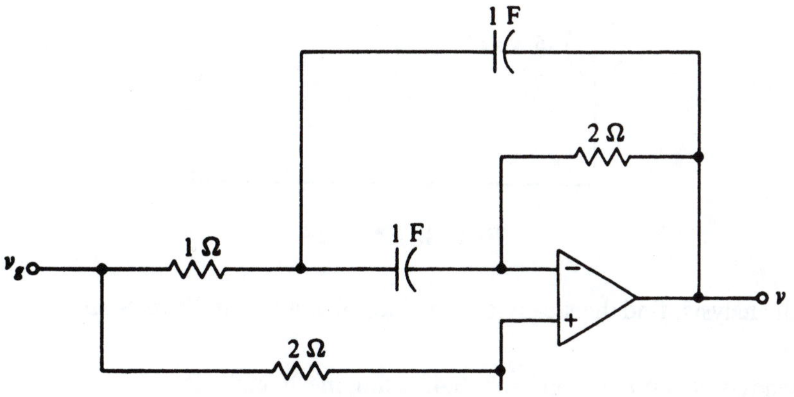

PROBLEM 9.23

9.3 Mesh Analysis

9.24 Find the steady-state voltage v using mesh analysis.

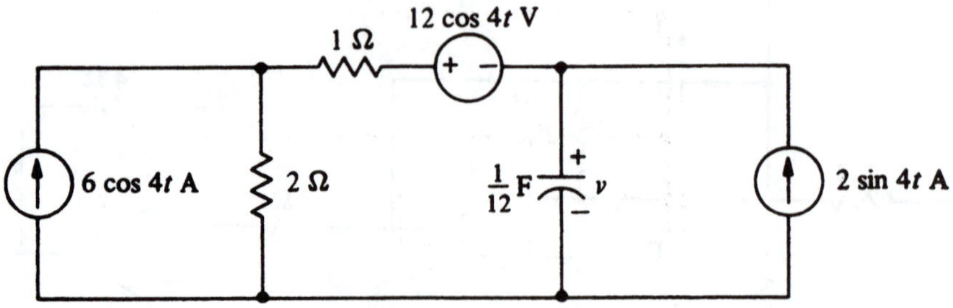

PROBLEM 9.24

9.25 Find the steady-state current i using mesh analysis.

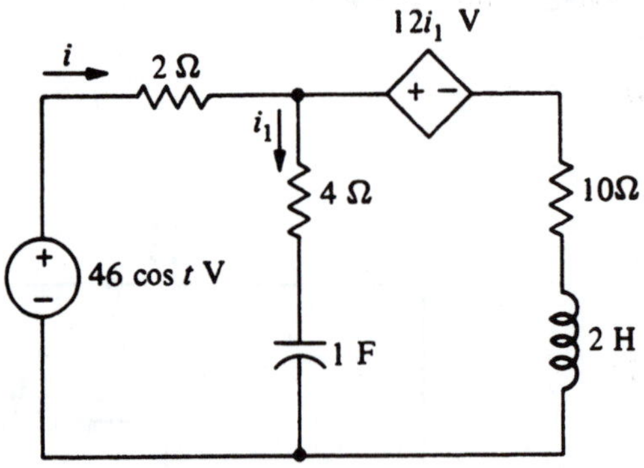

PROBLEM 9.25

9.26 Using mesh analysis, find the steady-state values of i_1 and i in Prob. 9.15.

9.27 Find the steady-state current i in Prob. 8.42 using mesh analysis

9.28 Find the steady-state voltage v using mesh analysis if $i_{g1} = 6\cos(4t)$ A and $i_{g2} = 2\cos(4t)$ A.

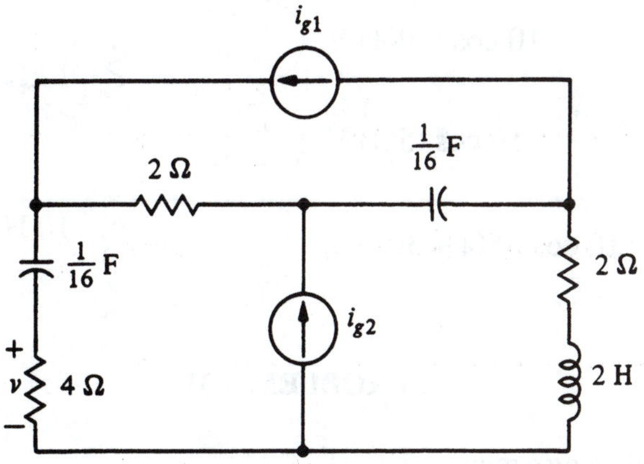

PROBLEM 9.28

9.4 Sources with Different Frequencies

9.29 Find *v(t)* in the following circuit.

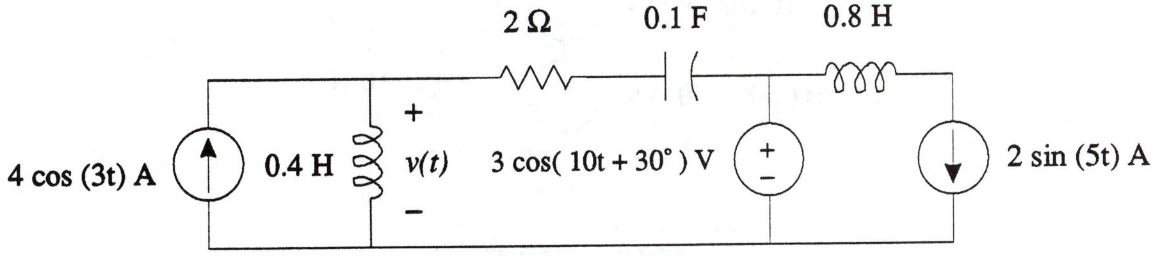

PROBLEM 9.29

9.30 Find $i(t)$ in the following circuit.

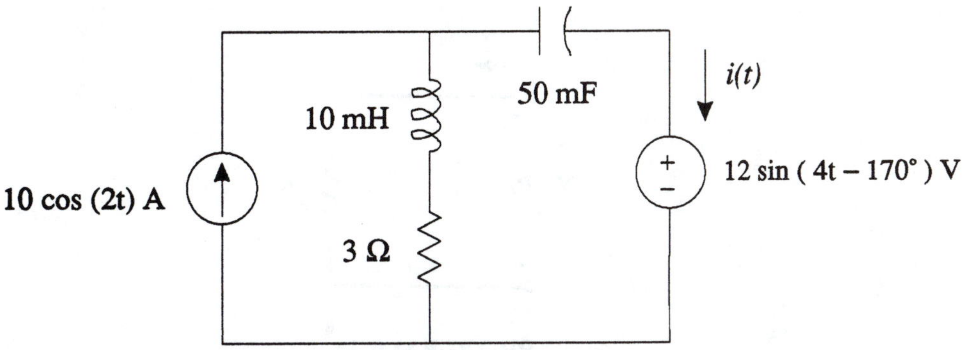

PROBLEM 9.30

9.31 Find v_o in the following circuit.

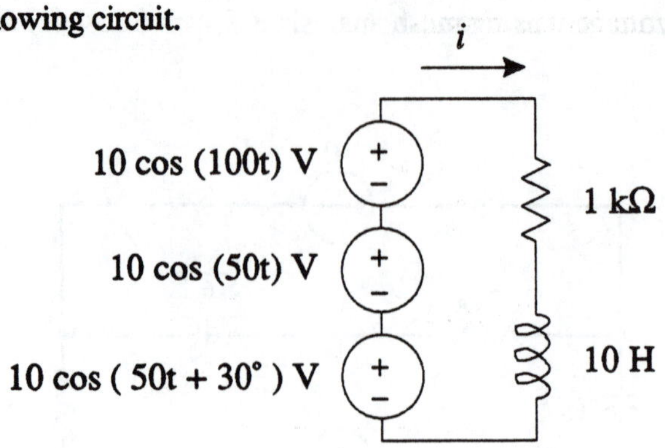

PROBLEM 9.31

9.32 Find $i(t)$ in the following circuit.

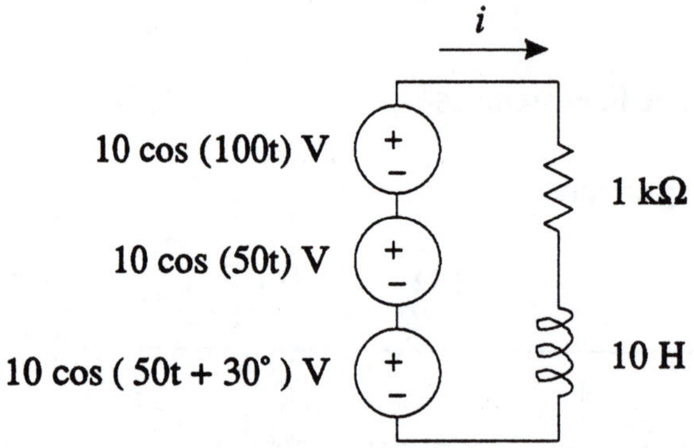

PROBLEM 9.32

9.5 Phasor Diagrams

9.33 Draw the phasor diagram for the circuit.

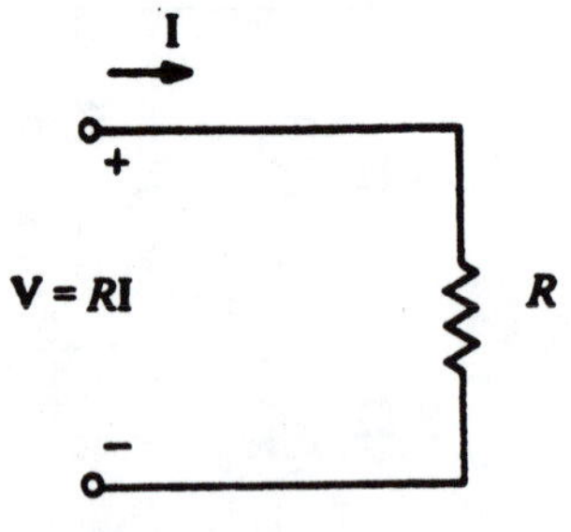

PROBLEM 9.33

9.34 Draw the phasor diagram for the circuit.

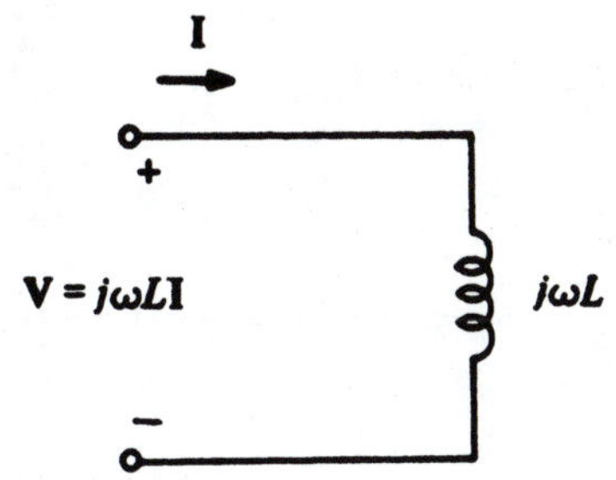

PROBLEM 9.34

9.35 Draw the phasor diagram for the circuit.

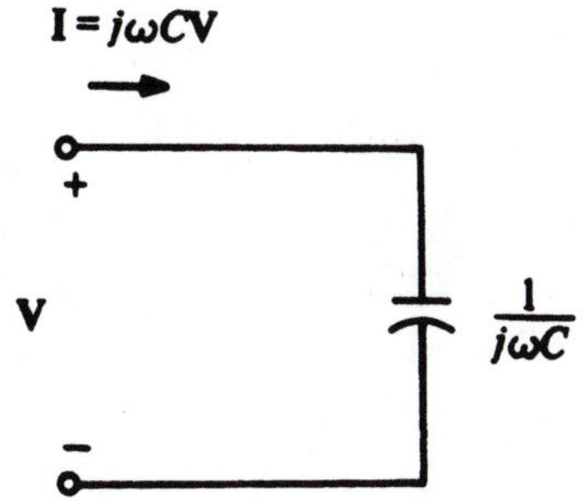

PROBLEM 9.35

9.36 Solve for **V** using a phasor diagram if $R = 2\ \Omega$, $L = 3$H, and $\omega = 1$ rad/s, $\mathbf{I} = 1\angle 0°$ A.

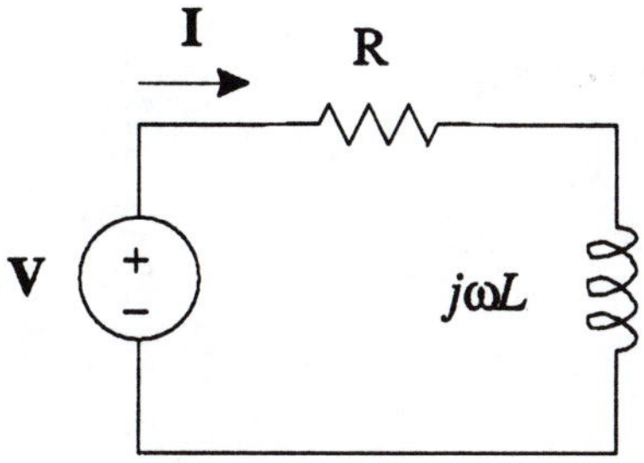

PROBLEM 9.36

9.6 SPICE and AC Steady State

9.37 Solve Prob. 9.2 using SPICE.

9.38 Solve Prob. 9.4 using SPICE.

9.39 Solve Prob. 9.6 using SPICE.

9.40 Solve Prob. 9.12 using SPICE.

9.41 Solve Prob. 9.18 using SPICE.

9.42 Solve Prob. 9.25 using SPICE.

9.43 Solve Prob. 9.28 using SPICE.

9.1 Using phasors and voltage divider,

$$\underline{V}_1 = (100\angle 50° \text{ V})\frac{j2}{10k + j2 - 2\left(j\frac{1}{2(100\mu)}\right)}$$

$$= \frac{(100\angle 50°)(2\angle 90°)}{(10000 - j(9.998\times 10^3))}$$

$$= 0.01414\angle -175°$$

$$\therefore v_1(t) = 0.01414\cos(2t - 175°)\text{ V}$$

9.2 $\underline{V}_2 = \underline{Z}\,\underline{I} = \left(\frac{1}{j10}\right)(20\angle -80°)$

$$= 2\angle -170°$$

Voltage divider: $\underline{V}_1 = 5\underline{V}_2 \cdot \frac{20}{20 + j10}$

$$\Rightarrow \underline{V}_1 = 8.94\angle 163.43°$$

$$\therefore v_1(t) = 8.94\cos(5t + 163.4°)\text{ V}$$

9.3 The total source current is the sum of both parallel sources:

$$\underline{I}_T = 20\angle 0° + 20\angle -90° = 20 - j20$$

Using current divider:

$$\underline{I}_1 = (20 - j20)\frac{j1}{\left(j1 - j\frac{1}{377(20\times 10^{-3})} + j3 + \frac{1}{10000} - j\frac{1}{377(10\times 10^{-3})}\right)}$$

$$= \frac{(28.284\angle -45°)(1\angle 90°)}{(1\times 10^{-4} + j3.602)}$$

$$= 7.852\angle -45°$$

$$\therefore i_1(t) = 7.852\cos(377t - 45°)\text{ A}$$

9.4 Using voltage divider,

$$\underline{V}_x = (9\angle 37°)\frac{10}{10 + j(10)(2)}$$

$$= 4.025\angle -26.435°$$

Using current divider,

$$\underline{I}_1 = 3\underline{V}_x\frac{1}{1 + j(10)(5\times 10^{-3}) - j\frac{1}{10(1)}}$$

$$= \frac{12.075\angle -26.435°}{1 - 0.05j}$$

$$= 12.06\angle -23.57°$$

$$\therefore i_1(t) = 12.06\cos(10t - 23.57°)\text{ A}$$

9.5 Total voltage source = sum

$$\underline{V}_T = 3\angle 120° + 3\angle 0° + 3\angle -120° + 10\angle 0°$$

$$= 10\angle 0°$$

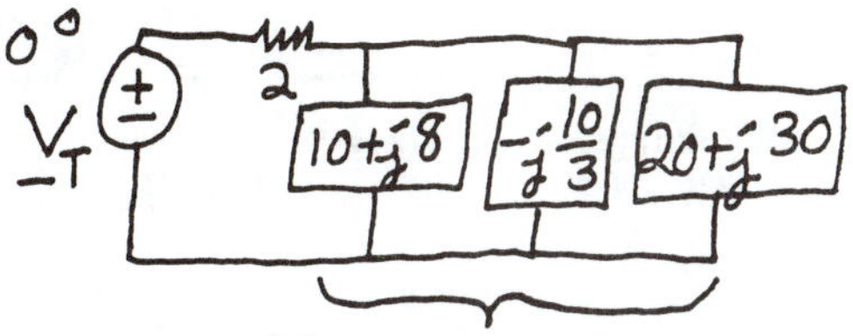

$\underline{Y}_{eq} = \Sigma\underline{Y} = 0.07638 + j0.228$

Total impedance:

$$\underline{Z}_T = 2 + \frac{1}{\underline{Y}_{eq}} = 2 + \frac{1}{.240\angle 71.5°}$$

$$= 2 + 4.17\angle -71.5° = 3.32 - j3.95$$

$$\underline{I} = \frac{\underline{V}_T}{\underline{Z}_T} = \frac{10\angle 0°}{5.16\angle 50°} = 1.94\angle -50°$$

$$\therefore i(t) = 1.94\cos(3t - 50°)\text{ A}$$

9.6 Adding the current sources and series impedances:

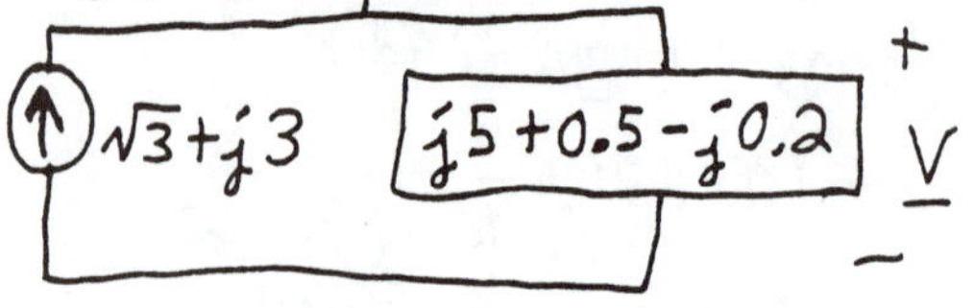

$\underline{V} = \underline{I}\,\underline{Z} = (\sqrt{3}+j3)(0.5+j4.8)$

$= 16.72\angle 144°$

$\therefore\ v(t) = 16.72\cos(5t+144°)$ V

9.7

By voltage division

$\underline{V}_{oc} = (8)\dfrac{4}{4+j4}$

$= \underline{4\sqrt{2}\angle -45°}$ V

$\underline{Z}_{TH} = j2 + \dfrac{4(j4)}{4+j4} = \underline{2+j4}\ \Omega$

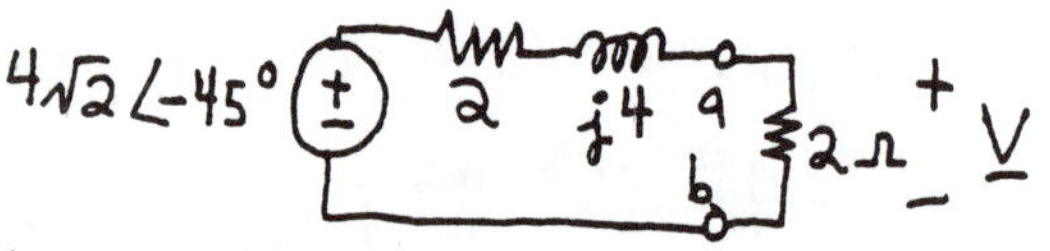

Voltage divider: $\underline{V} = (4\sqrt{2}\angle -45°)\dfrac{2}{2+j4+2}$

$= 2\angle -90°$

$\therefore\ v(t) = 2\sin(2t)$ V

9.8

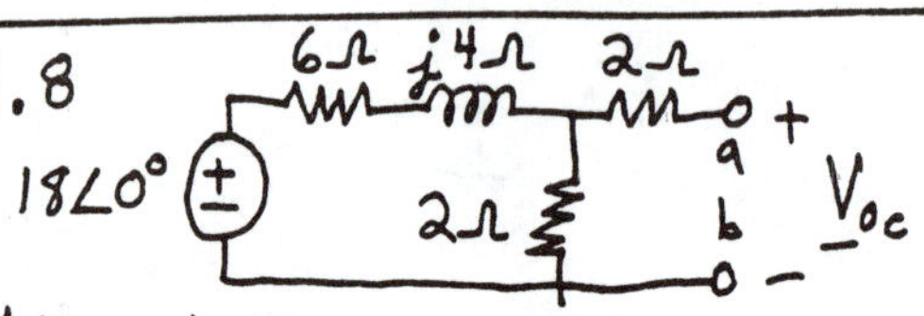

Voltage divider:

$\underline{V}_{oc} = (18)\,2/(6+j4+2) = 3.6-j1.8$

$\underline{Z}_{TH} = 2 + \dfrac{2(6+j4)}{2+6+j4} = 3.6+j0.2$

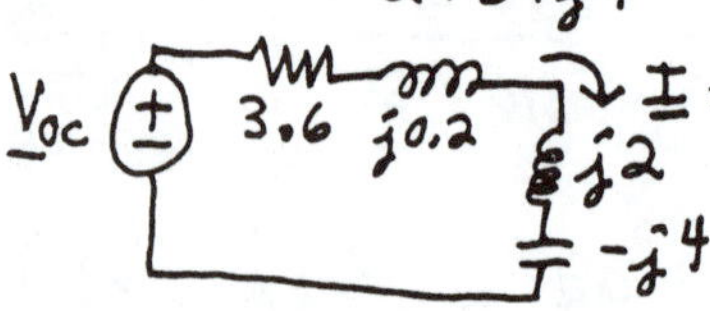

$\underline{I} = \dfrac{\underline{V}_{oc}}{(3.6+j0.2+j2-j4)}$

$\Rightarrow \underline{I} = 1\angle 0°$ A $\quad\therefore\ i(t) = \underline{\cos 2t\ \text{A}}$

9.9

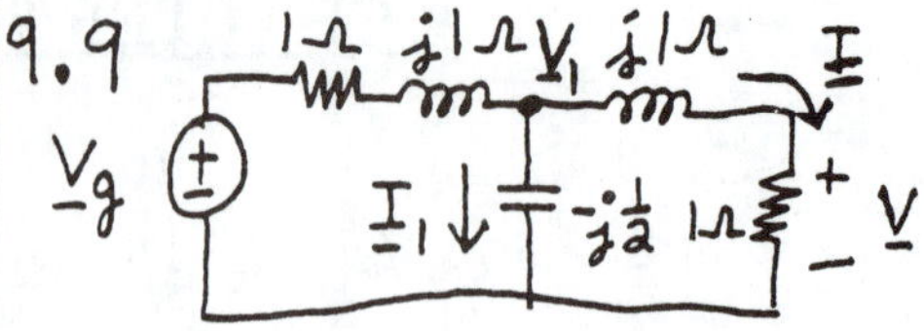

Assume $\underline{V} = 1$ V. Then $\underline{I} = \dfrac{\underline{V}}{1} = 1$A

$\underline{V}_1 = \underline{I}(1+j) = 1+j \Rightarrow \underline{I}_1 = \underline{V}_1(j2) = -2+j2$

$\dfrac{\underline{V}_g - \underline{V}_1}{(1+j)} = \underline{I} + \underline{I}_1 \Rightarrow \underline{V}_g = -2+j2$ V

By proportionality, $\underline{V} = \dfrac{8}{-2+j2} = 2\sqrt{2}\angle -135°$

$\therefore\ v = 2\sqrt{2}\cos(t-135°)$ V

9.10

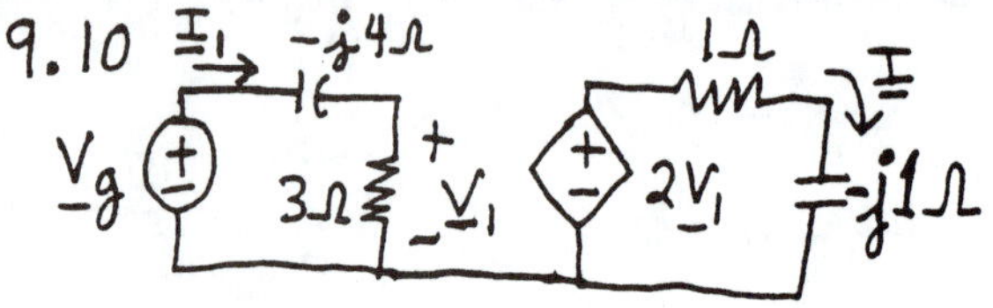

Assume $\underline{I} = 1$A, then $2\underline{V}_1 = (1-j1)\underline{V}$

$\Rightarrow \underline{V}_1 = \frac{1}{2} - j\frac{1}{2}$ V $\Rightarrow \underline{I}_1 = \frac{1}{6} - j\frac{1}{6}$ A

$\Rightarrow \underline{V}_g = (3-j4)(\frac{1}{6} - j\frac{1}{6})$

By proportionality, $\underline{I} = \dfrac{5}{(3-j4)(\frac{1}{6}-j\frac{1}{6})}$

$= 3\sqrt{2}\angle 98.1°$A

$\therefore\ i(t) = 3\sqrt{2}\cos(2t+98.1°)$ A

9.11

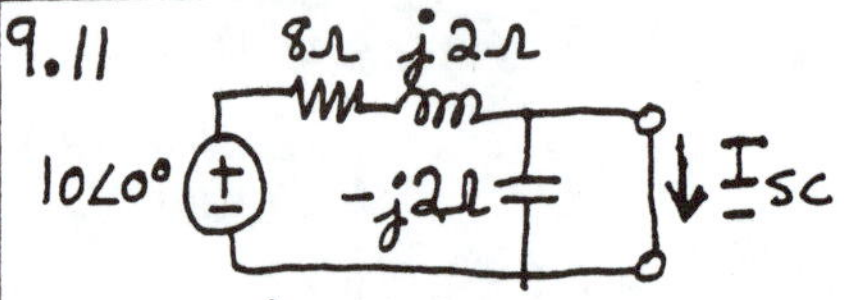

$\underline{Z}_{TH} = \dfrac{(-j2)(8+j2)}{8+j2-j2} = \frac{1}{2} - j2\ \Omega$

$\underline{I}_{sc} = \dfrac{10}{8+j2} = 1.213\underline{\angle -14.04°}$A

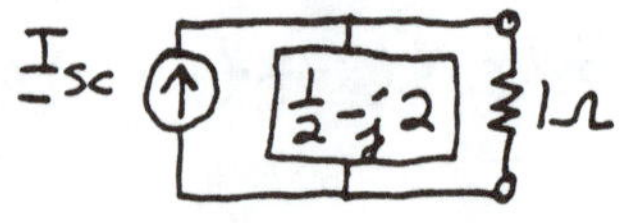

$\underline{I} = \dfrac{(\frac{1}{2}-j2)(1.2\angle 14°)}{1+\frac{1}{2}-j2}$

$= 1\angle -36.87°$

$\therefore\ i = \cos(t-36.87°)$ A

9.12 With the current source dead:

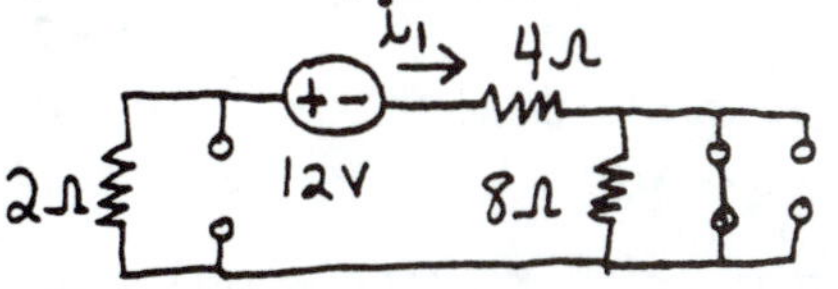

$i_1 = \frac{-12}{4+2} = -2A$

With the voltage source dead:

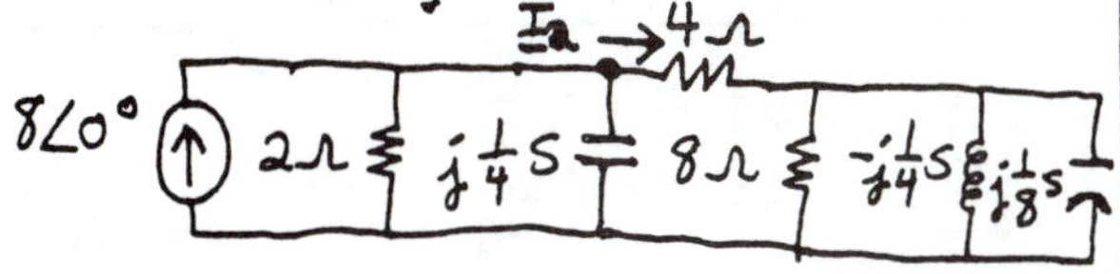

Current divider $I_2 = \dfrac{\frac{1}{\frac{1}{2}+j\frac{1}{4}}(8)}{\frac{1}{\frac{1}{2}+j\frac{1}{4}}+4+\frac{1}{\frac{1}{8}-j\frac{1}{4}+j\frac{1}{8}}}$

$\Rightarrow I_2 = 1-j = \sqrt{2}\angle -45°$

$\therefore i_2 = \sqrt{2}\cos(2t-45°)$ A

$i = i_1 + i_2 = \sqrt{2}\cos(2t-45°) - 2$ A

9.13

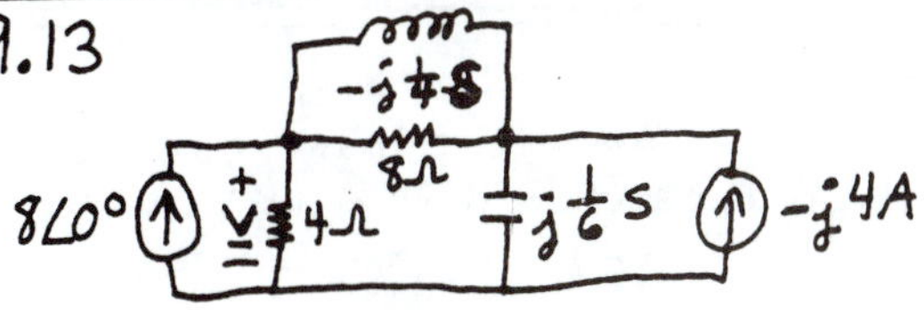

By KCL: $V\left(\frac{1}{4}+\frac{1}{8}-j\frac{1}{4}\right) - V_1\left(\frac{1}{8}-j\frac{1}{4}\right) = 8$

$\Rightarrow V(3-j2) + V_1(-1+j2) = 64$

KCL: $-V\left(\frac{1}{8}-j\frac{1}{4}\right) + V_1\left(\frac{1}{8}+j\frac{1}{6}-j\frac{1}{4}\right) = -j4$

$\Rightarrow V(-3+j6) + V_1(3-j2) = -j96$

$V = \begin{vmatrix} 8 & -1+j2 \\ -j96 & 3-j2 \end{vmatrix} \Big/ \begin{vmatrix} 3-j2 & -1+j2 \\ -3+j6 & 3-j2 \end{vmatrix}$

$= -j16 \Rightarrow v = 16\sin 8t$ V

9.14

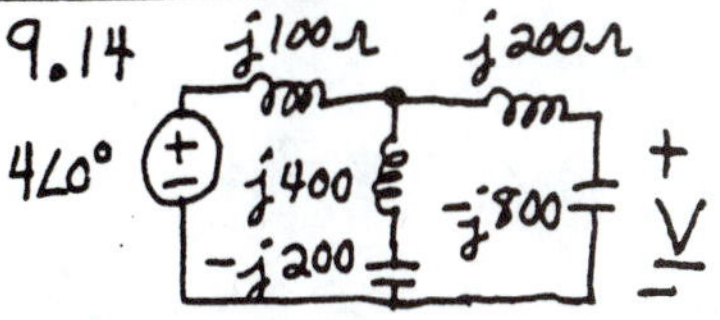

KCL: $V_1\left(\frac{1}{j100}+\frac{1}{j400-j200}+\frac{1}{j200}\right) - V\frac{1}{j200}$

$= 4/j100 \Rightarrow 4V_1 - V = 8$

9.14 (cont.) KCL: $V\left(\frac{1}{j200}-\frac{1}{j800}\right) = V_1\left(\frac{1}{j200}\right)$

$\Rightarrow -4V_1 + 3V = 0$. Adding,

$V = 4 \Rightarrow v = 4\cos(1000t)$ V

9.15 By KCL, $V\left(1+\frac{1}{j2}-\frac{1}{j\,2/3}\right) = \frac{2}{j2} - \frac{6}{j\frac{2}{3}}$

$\Rightarrow V(1+j) = j8 \Rightarrow V = 4\sqrt{2}\angle 45°$

$v = 4\sqrt{2}\cos(2t+45°)$ V

9.16 By KCL using mA

$V_1(1+j) - V(j) = 6$, $V_1\left(-\frac{1}{2}-j\right) + V\left(j\frac{2}{3}\right) = 0$

$V = \begin{vmatrix} 1+j & 6 \\ -\frac{1}{2}-j & 0 \end{vmatrix} \Big/ \begin{vmatrix} 1+j & -j \\ -\frac{1}{2}-j & j2/3 \end{vmatrix}$

$= \dfrac{3+j6}{\frac{1}{3}+j\frac{1}{6}} = 18\angle 36.87°$

$\therefore v = 18\cos(3000t + 36.87°)$ V

9.17 From Prob. 9.16

$V_1 = \dfrac{\begin{vmatrix} 6 & -j \\ 0 & j2/3 \end{vmatrix}}{\frac{1}{3}+j\frac{1}{6}} = 10.73\angle 63.4°$

$\therefore v_1 = 10.73\cos(3000t + 63.4°)$ V

9.18

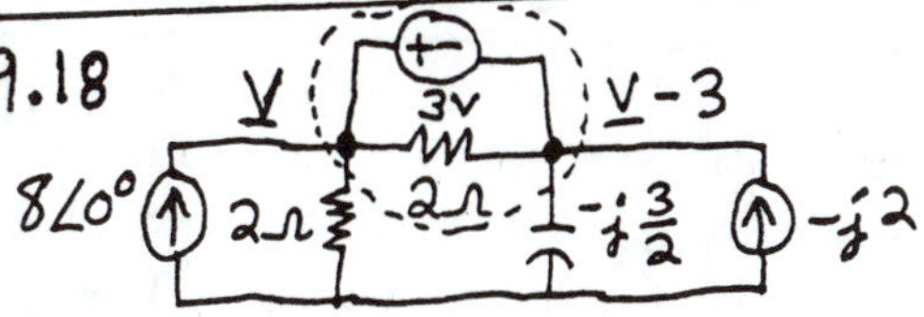

KCL for this generalized node yields:

$\frac{V}{2} + \frac{V-3}{-j3/2} = 8-j2 \Rightarrow V = \frac{48}{3+j4} = 9.6\angle -53.1°$

$\therefore v = 9.6\cos(4t - 53.1°)$ V

9.19 Node voltages are 5, $5-1000I_1$, $5-500I_1$

KCL for supernode containing dependent src:

$-I_1 + \frac{5-1000I_1}{j1000} + (5-500I_1)(j.003) = 0$

$\Rightarrow I_1 = (4\sqrt{5}\times 10^{-3})\angle 63.4°$A

$\therefore i_1 = 4\sqrt{5}\cos(3000t + 63.4°)$ mA

9.20 Let $\underline{V}_1$ be the phasor node voltage at the junction of the 1Ω resistor and the $\underline{V}$ at the input of the op-amp. KCL:

$\underline{V}_1(1+1+j5) - \underline{V}(1+j5) = \underline{V}_g$ or

$\underline{V}_1(2+j5) - \underline{V}(1+j5) = 6$

KCL at the noninverting terminal gives:

$-\underline{V}_1 + \underline{V}(1+j1) = 0 \Rightarrow \underline{V}_1 = \underline{V}(1+j)$

$$\therefore \underline{V} = \frac{6}{(1+j)(2+j5)-(1+j5)}$$

$= 1.34 \angle -153^\circ$

$\Rightarrow v = 1.34 \cos(5t - 153^\circ)$ V

9.21 Let $\underline{V}$ be op-amp output voltage and $\underline{V}_1$ be the voltage at input terminals.

Voltage division: $\underline{V}_1 = \frac{1}{1+j}(4)$

Since $\underline{V} = \left(1+\frac{4}{1}\right)\underline{V}_1 = 5\underline{V}_1$, then

$\underline{V} = 10 - j10$

$$\underline{I} = \frac{\underline{V}}{3000 - j1000} = 4.47 \angle -26.57^\circ \text{ mA}$$

$i = 4.47\cos(3000t - 26.57^\circ)$ mA

9.22 Let v_1 = node voltage at junction of 2Ω and 8Ω resistor, v_2 = output of 1st opamp. Then KCL at v_1 gives.

$\underline{V}_1\left(\frac{1}{2}+\frac{1}{8}+\frac{1}{4}+j\frac{1}{2}+j\frac{1}{2}\right) - \underline{V}_2\left(j\frac{1}{2}\right) - \underline{V}\frac{1}{4} = \frac{5}{2}$

or $\underline{V}_1(7+j8) - \underline{V}_2(j4) - \underline{V}2 = 20$

KCL at inverting terminal of first op amp gives $\underline{V}_1 = j\underline{V}_2$ and KCL at inverting terminal of 2nd op amp gives $\underline{V}_2 = -\underline{V}/2$ therefore $\underline{V}_1 = -j\underline{V}/2$ and

$\underline{V}\left[(7+j8)(-j/2) + j4/2 - 2\right] = 20$ or

$$\underline{V} = \frac{20}{2 - j3/2} = 8\angle 36.87^\circ \text{ V}$$

$v = 8\cos(2t + 36.87^\circ)$ V

9.23 Let v_1 be the voltage at the input terminals of the op amp. Then at the noninverting terminal KCL gives $\underline{V}_1\left(\frac{1}{2}+1\right) = \underline{V}_g\frac{1}{2} \Rightarrow \underline{V}_1 = 2\underline{V}$

KCL at inverting terminal gives

$\underline{V}_1\left(\frac{1}{2}+j\right) - \underline{V}_2(j) - \underline{V}\left(\frac{1}{2}\right) = 0$ or

$\underline{V}_2 j + \underline{V}\frac{1}{2} = 1 + j2$ where v_2 is the node voltage to the right of the 1Ω resistor. KCL at $\underline{V}_2$ yields.

$\underline{V}_2(1+j+j) - \underline{V}_1(j) - \underline{V}(j) = \underline{V}_g$ or

$\underline{V}_2(1+j2) - \underline{V}j = 6 + j2$

$$\underline{V} = \begin{vmatrix} j & 1+j2 \\ 1+j2 & 6+j2 \end{vmatrix} \Big/ \begin{vmatrix} j & \frac{1}{2} \\ 1+j2 & -j \end{vmatrix}$$

$$= \frac{1+j2}{\frac{1}{2} - j} = 2\angle 126.9^\circ \text{ V}$$

$v = 2\cos(t + 126.9^\circ)$ V

9.24 Let $\underline{I}$ be the phasor current in the center mesh. KVL yields

$\underline{I}(1+2-j3) - 6(2) + j3(-j2) + 12 = 0$

$$\underline{I} = \frac{6}{3-j3} = 1+j \text{ A}$$

$\underline{V} = (\underline{I} - j2)(-j3) = 3\sqrt{2}\angle -135^\circ$ V

$v = 3\sqrt{2}\cos(4t - 135^\circ)$ V

9.25 KVL around left and right meshes:

$46 = 2\underline{I} + (\underline{I} - \underline{I}_2)(4-j)$

$12\underline{I} + (10+2j)\underline{I}_2 + (\underline{I}_2 - \underline{I})(4-j) = 0$

$$\Rightarrow \underline{I} = \frac{(-14-j)}{8+j}\underline{I}_2, \quad \underline{I}_2 = \frac{\underline{I}(6-j) - 46}{(4-j)}$$

Solving for $\underline{I}$,

$\Rightarrow \underline{I} = 5.443\angle 9.89^\circ$ A

$\therefore i = 5.443\cos(t + 9.89^\circ)$ A

9.26 KVL around the right mesh and the outside loop yields

$\underline{I} - \underline{I}_1(-j\tfrac{2}{3}) = 6$ or $\underline{I} + \underline{I}_1(j\tfrac{2}{3}) = 6$

$(\underline{I}+\underline{I}_1)(j2) + \underline{I}_1(-j\tfrac{2}{3}) = 2-6$

$\underline{I}(j2) + \underline{I}_1(j\tfrac{4}{3}) = -4$

$$\underline{I} = \begin{vmatrix} 6 & j\tfrac{2}{3} \\ -4 & j\tfrac{4}{3} \end{vmatrix} \Big/ \begin{vmatrix} 1 & j\tfrac{2}{3} \\ j2 & j\tfrac{4}{3} \end{vmatrix}$$

$$= \frac{j32}{4+j4} = 4\sqrt{2}\angle 45^\circ$$

$\therefore\ i = \underline{4\sqrt{2}\cos(2t+45^\circ)\ \text{A}}$

$$\underline{I}_1 = \frac{\begin{vmatrix} 1 & 6 \\ j2 & -4 \end{vmatrix}}{\tfrac{4}{3}+j\tfrac{4}{3}} = \frac{-4-j12}{\tfrac{4}{3}+j\tfrac{4}{3}} = 6.7\angle -153^\circ$$

$i_1 = \underline{6.7\cos(2t-153.4^\circ)\ \text{A}}$

9.27 Let i_1 be the mesh current in the left side (clockwise).

KVL gives:

$\underline{I}_1(6+j4) + (\underline{I}_1-\underline{I})2 = 18$ or

$\underline{I}_1(8+j4) + \underline{I}(-2) = 18$

$\underline{I}(2+j2-j4) + (\underline{I}-\underline{I}_1)2 = 0$ or

$\underline{I}_1(-2) + \underline{I}(4-j2) = 0$

$$\underline{I} = \begin{vmatrix} 8+j4 & 18 \\ -2 & 0 \end{vmatrix} \Big/ \begin{vmatrix} 8+j4 & -2 \\ -2 & 4-j2 \end{vmatrix}$$

$$= \frac{36}{36} = 1;\ \therefore\ i = \underline{\cos 2t\ \text{A}}$$

9.28

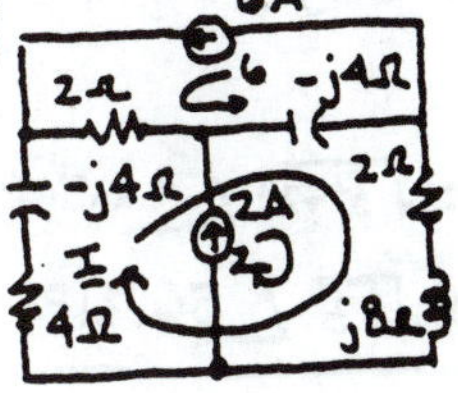

KVL for loop with current $\underline{I}$:

$(2-j4+2+j8+4-j4)\underline{I} + 6(2-j4) + (-j4+2+j8)(2) = 0$

$\underline{I} = 2(-1+j) = 2\sqrt{2}\angle 135^\circ\ \text{A}$

$\underline{V} = -4\underline{I} = 8\sqrt{2}\angle -45^\circ\ \text{V}$

$v = \underline{8\sqrt{2}\cos(4t-45^\circ)\ \text{V}}$

9.29 Using superposition with 3 sources:

(i) Kill right two sources

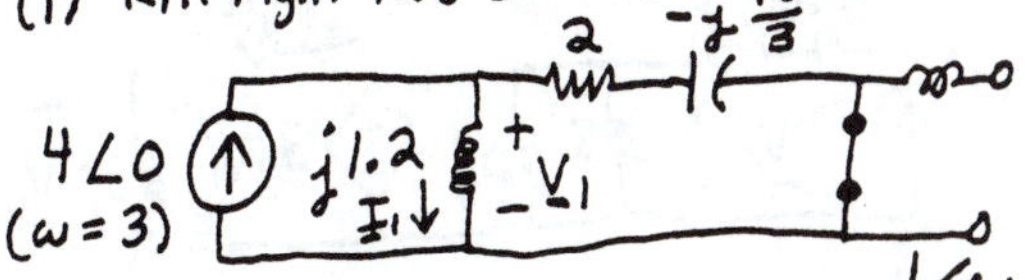

Current divider $\underline{I}_1 = (4\angle 0)\dfrac{1/j1.2}{\dfrac{1}{j1.2} + \dfrac{1}{2-j\frac{10}{3}}}$

$\Rightarrow \underline{I}_1 = 5.32\angle -12.19^\circ\ \text{A}$

$\underline{V}_1 = \underline{I}_1(j1.2) = 6.38\angle 77.81^\circ\ \text{V}$

(ii) Kill two current sources

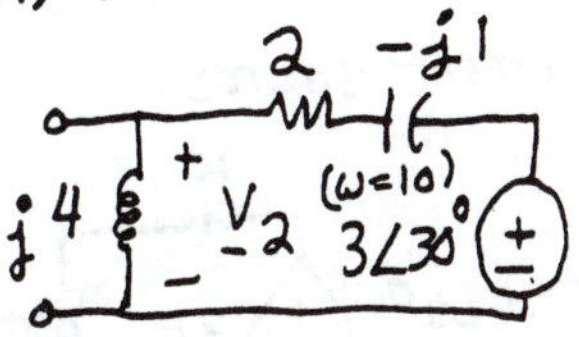

Voltage divider $\underline{V}_2 = (3\angle 30^\circ)\dfrac{4j}{2+3j}$

$\Rightarrow \underline{V}_2 = 3.33\angle 63.69^\circ\ \text{V}$

(iii) Kill two left sources

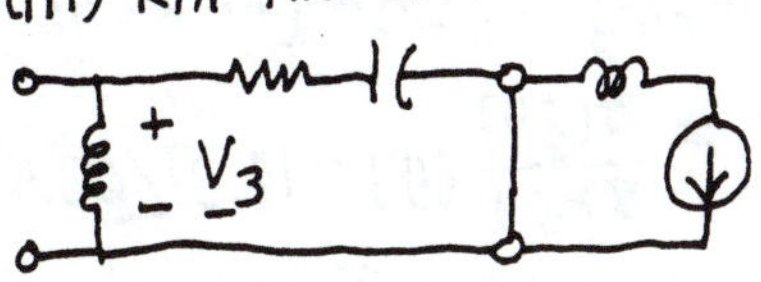

$\underline{V}_3 = 0$ due to short ckt.

$\therefore\ v(t) = v_1 + v_2 + v_3$

$= 6.38\cos(3t+77.8^\circ)$

$\underline{+\ 3.33\cos(10t+63.7^\circ)\text{V}}$

9.30 (i) Superposition − kill current src.

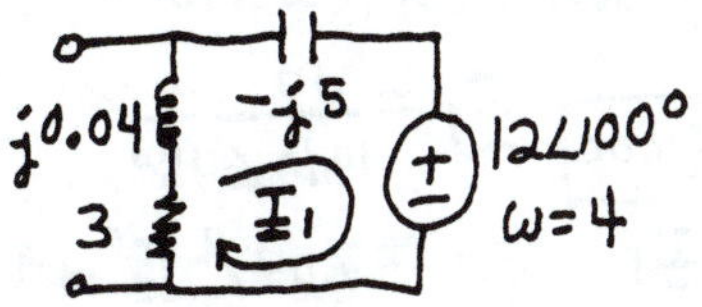

$\underline{I}_1 = \dfrac{-(12\angle 100^\circ)}{3-j4.96}$

$= 2.1\angle -139^\circ\text{A}$

(ii) Kill voltage src.

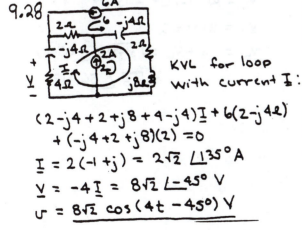

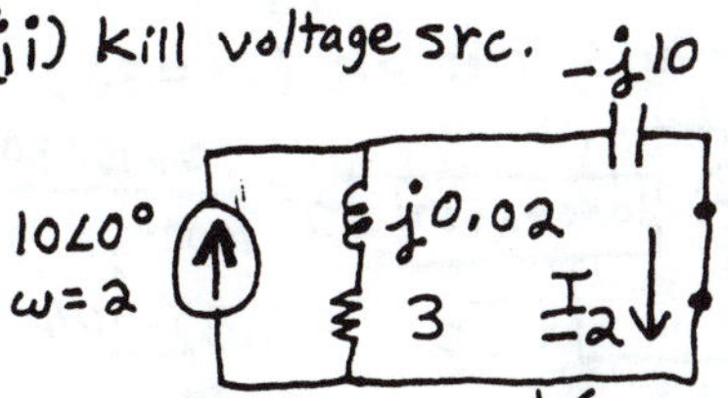

Current divider

$\underline{I}_2 = (10\angle 0^\circ)\dfrac{1/(-j10)}{\dfrac{1}{-j10} + \dfrac{1}{3+j.02}} = 0.29\angle 73.7^\circ\ \text{A}$

$\therefore\ i = i_1 + i_2$

$= 2.1\cos(4t-139^\circ) + 0.29\cos(2t+73.7^\circ)$

9.31 (i) Kill rightmost current source:

Voltage divider $\underline{V}_{01} = 3\underline{V}_{x1} \dfrac{5j}{10+5j}$

$$\underline{V}_{x1} = \underline{Z}_{eq}\underline{I} = \frac{(8)(2j)}{8+2j}(5\angle 30°) = 9.7\angle 106°$$

$\Rightarrow \underline{V}_{01} = 13\angle 169° \Rightarrow v_{01} = 13\cos(t+169°)$

(ii) kill leftmost current source:

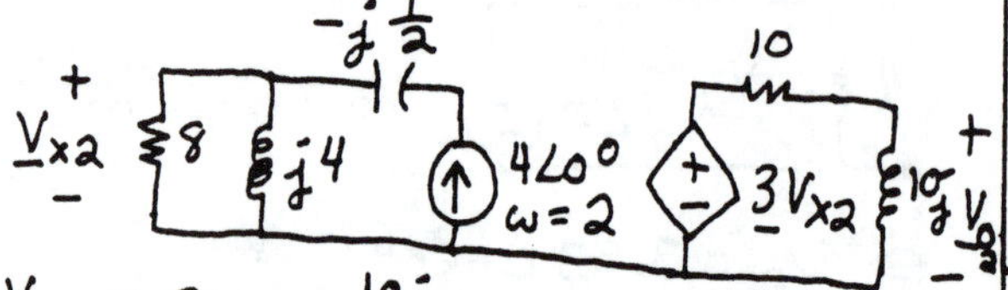

$$\underline{V}_{02} = 3\underline{V}_{x2}\frac{10j}{10+10j} = \underline{V}_{x2}(2.12\angle 45°)$$

$$\underline{V}_{x2} = \underline{Z}_{eq}\underline{I} = \frac{8(j4)}{8+j4}(4) = 14.31\angle 63.4°$$

$\Rightarrow \underline{V}_{02} = 30.3\angle 108° \Rightarrow v_{02} = 30.3\cos(2t+108°)$

$$\therefore v_0 = v_{01} + v_{02}$$
$$= 13\cos(t+169°) + 30.3\cos(2t+108°)\text{ V}$$

9.32 (i) Keep top source on ($\omega = 100$)

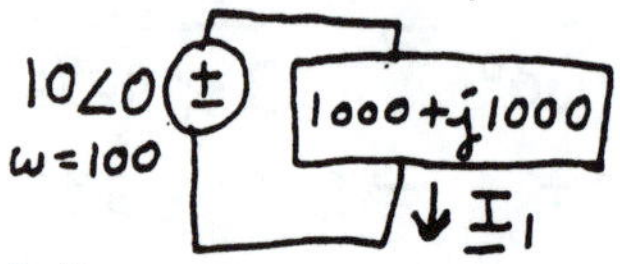

$$\underline{I}_1 = \frac{10}{1414\angle 45°} = 7.07\angle -45°\text{ mA}$$

(ii) Keep two lower sources on ($\omega = 50$)

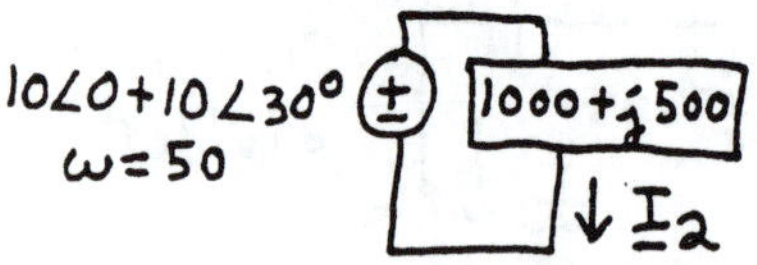

$$\underline{I}_2 = \frac{10+10\angle 30°}{1000+j500} = 17.3\angle -11.6°\text{ mA}$$

$$\therefore i(t) = i_1 + i_2$$
$$= 7.07\cos(100t - 45°)\text{ mA}$$
$$+ 17.3\cos(50t - 11.6°)\text{ mA}$$

9.33

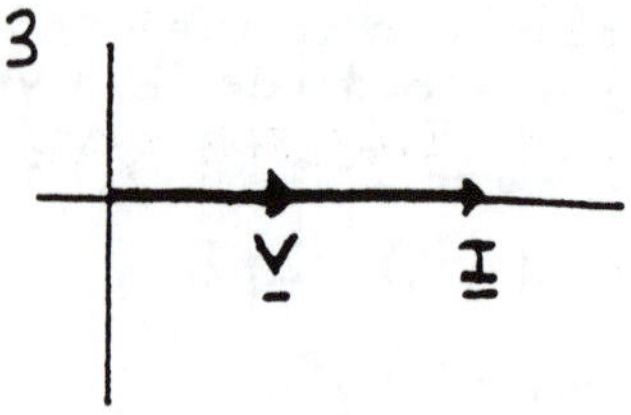

9.34

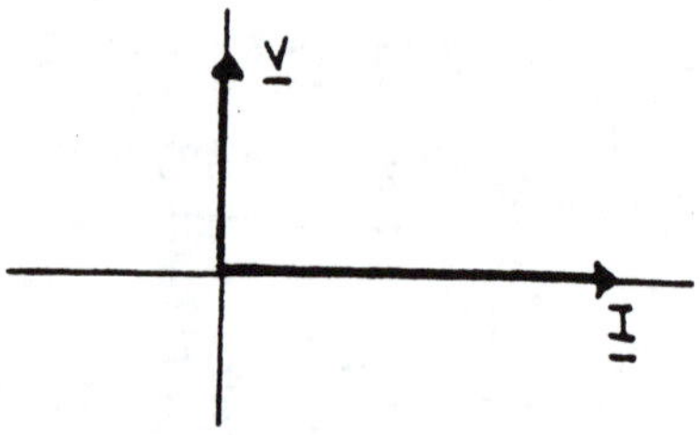

9.35

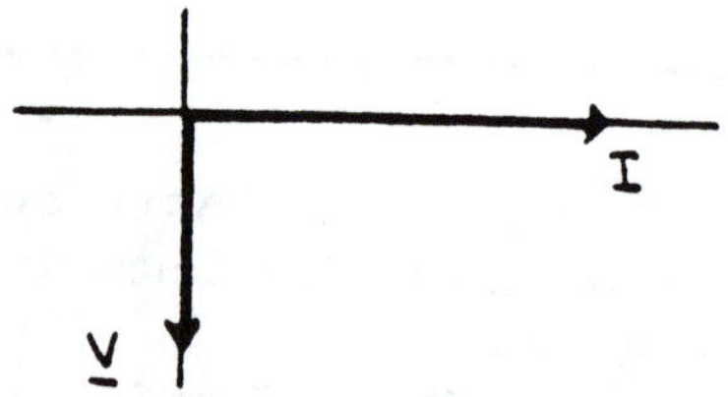

9.36

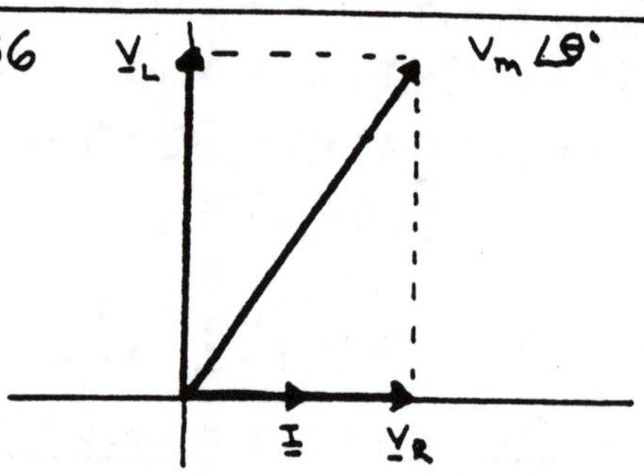

$\underline{V}_L = \omega L|I|\angle 90° = 3\angle 90°$

$\underline{V}_R = R|I| = 2$

$$\underline{V}_m = \underline{V}_R + \underline{V}_L = \sqrt{3^2+2^2}\angle \tan^{-1}\left(\tfrac{3}{2}\right)$$
$$= \sqrt{13}\angle 56.31°$$

9.37 SPICE input file:

```
AC STEADY-STATE SOLUTION FOR PROBLEM 9.2
*DATA STATEMENTS
I1 0 1 AC 20 -80
C1 1 0 2
E1 2 0 1 0 5
L1 2 3 2
R1 3 0 20
*TO ELIMINATE FLOATING NODE IN DC ANALYSIS PUT IN
*HUGE RESISTOR IN PARALLEL WITH THE CAPACITOR
RDUMMY 1 0 1GIG
*CONTROL STATEMENT FOR AC ANALYSIS [F=5/(2*PI) HZ]
.AC LIN 1 0.7957747 0.7957747
*OUTPUT CONTROL STATEMENT FOR V(3)
.PRINT AC VM(3) VP(3)
.END
```

SPICE output

```
FREQ          VM(3)         VP(3)

7.958E-01     8.944E+00     1.634E+02
```

9.38 SPICE input file

```
AC STEADY-STATE SOLUTION FOR PROBLEM 9.4
V1 1 0 AC 9 37
L1 1 2 2
R1 2 0 10
G1 0 3 2 0 3
L2 3 0 1
C1 3 4 10M
C2 4 0 10M
R2 3 5 1
VDUMMY 5 0 0
RHUGE 4 0 1G
.AC LIN 1 1.59155 1.59155
.PRINT AC IM(VDUMMY) IP(VDUMMY)
.END
```

SPICE output

```
FREQ          IM(VDUMMY)    IP(VDUMMY)

1.592E+00     1.206E+01    -2.357E+01
```

9.39 SPICE input file

```
AC STEADY-STATE SOLUTION FOR PROBLEM 9.6
I1 0 1 AC 1.732051 0
I2 0 1 AC 1.732051 120
I3 1 0 AC 1.732051 -120
L1 1 2 1
R1 2 3 0.5
C1 3 0 1
RHUGE 3 0 1G
.AC LIN 1 0.7957747 0.7957747
.PRINT AC VM(1) VP(1)
.END
```

SPICE output

```
FREQ          VM(1)          VP(1)

7.958E-01     1.672E+01      1.440E+02
```

9.40 SPICE input file

```
AC STEADY-STATE SOLUTION FOR PROBLEM 9.12
I1 0 1 AC 8 0
R1 1 0 2
C1 1 0 0.125
V1 1 2 DC 12
R2 2 3 4
R3 3 0 8
L1 3 0 2
C2 3 0 0.0625
.AC LIN 1 0.31831 0.31831
.PRINT AC IM(V1) IP(V1)
.END
```

SPICE output

```
******************************************************************************
****      SMALL SIGNAL BIAS SOLUTION         TEMPERATURE =    27.000 DEG C
******************************************************************************

 NODE   VOLTAGE      NODE   VOLTAGE      NODE   VOLTAGE      NODE   VOLTAGE

(    1)     4.0000  (    2)    -8.0000  (    3)     0.0000

    VOLTAGE SOURCE CURRENTS
    NAME          CURRENT

    V1            -2.000E+00

    TOTAL POWER DISSIPATION    2.40E+01  WATTS

******************************************************************************
****      AC ANALYSIS                        TEMPERATURE =    27.000 DEG C
******************************************************************************

 FREQ          IM(V1)         IP(V1)

 3.183E-01     1.414E+00     -4.500E+01
```

9.41 SPICE input file

```
AC STEADY-STATE SOLUTION FOR PROBLEM 9.18
I1 0 1 AC 8 0
I2 0 2 AC 2 -90
V1 1 2 AC 3 0
R1 1 2 2
R2 1 0 2
C1 2 0 0.16667
.AC LIN 1 0.63662 0.63662
.PRINT AC VM(1) VP(1)
.END
```

SPICE output

```
FREQ         VM(1)        VP(1)
6.366E-01    9.600E+00   -5.313E+01
```

9.42 SPICE input file

```
AC STEADY-STATE SOLUTION FOR PROBLEM 9.25
V1 1 0 AC 46 0
VDUMMY 1 2 AC 0 0
R1 2 3 2
R2 3 4 4
C1 4 0 1
RHUGE 4 0 1G
H1 3 5 VDUMMY 12
R3 5 6 10
L1 6 0 2
.AC LIN 1 0.159155 0.159155
.PRINT AC IM(VDUMMY) IP(VDUMMY)
.END
```

SPICE output

```
FREQ         IM(VDUMMY)   IP(VDUMMY)
1.592E-01    5.443E+00    9.892E+00
```

9.43 SPICE input file

```
AC STEADY-STATE SOLUTION FOR PROBLEM 9.28
I1 0 1 AC 2 0
I2 2 3 AC 6 0
R1 3 1 2
C1 1 2 .0625
R2 2 4 2
L1 4 0 2
C2 3 5 .0625
RHUGE 3 5 1G
R3 5 0 4
.AC LIN 1 0.63662 0.63662
.PRINT AC VM(5) VP(5)
.END
```

SPICE output

```
FREQ         VM(5)        VP(5)
6.366E-01    1.131E+01   -4.500E+01
```

Chapter 10

AC Steady State Power

10.1 Average Power

10.1 The graph below shows $v(t)$ flowing through a 5 kΩ resistor. Find (a) T, the period of $v(t)$, (b) T_P, the period of the instantaneous power, and (c) P, the average power absorbed.

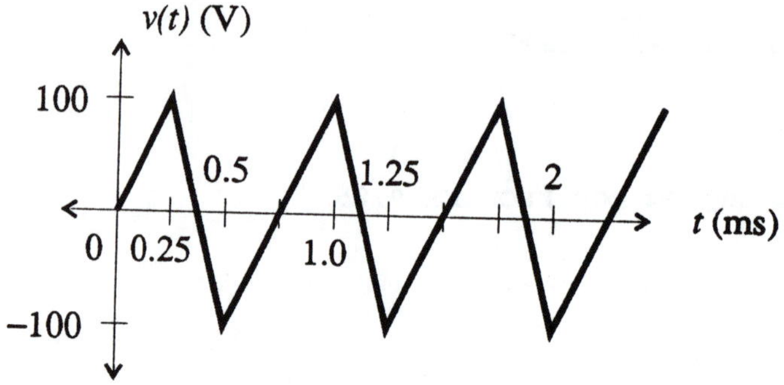

PROBLEM 10.1

10.2 The graph below shows $i(t)$ flowing through a 10 Ω resistor. Find (a) T, the period of $i(t)$, (b) T_P, the period of the instantaneous power, and (c) P, the average power absorbed.

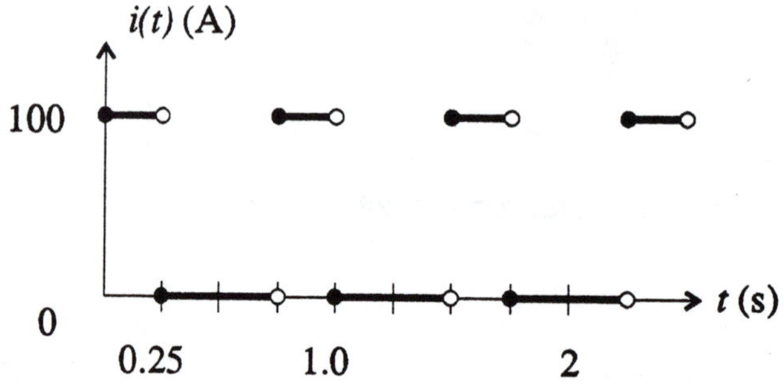

PROBLEM 10.2

10.3 Determine the average power absorbed for instantaneous power $p(t) = RI_m^2[1+\cos(\omega t)]^2$.

10.4 (a) In the circuit below, find the average power P_L absorbed by the load when $\phi = 0$ and the load is a 5 Ω resistor. (b) Find P_L when $\phi = 25°$.

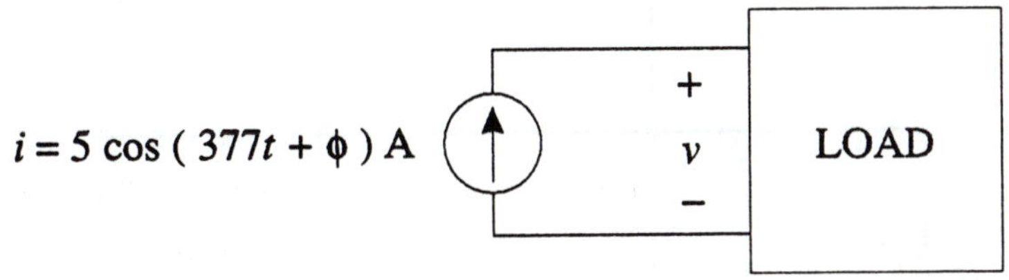

PROBLEM 10.4

10.5 Repeat Prob. 10.4 when the load is a 1 H inductor.

10.6 Repeat Prob. 10.4 when the load is a 1 F capacitor.

10.7 Find the average power absorbed by each resistor, the capacitor, and the source.

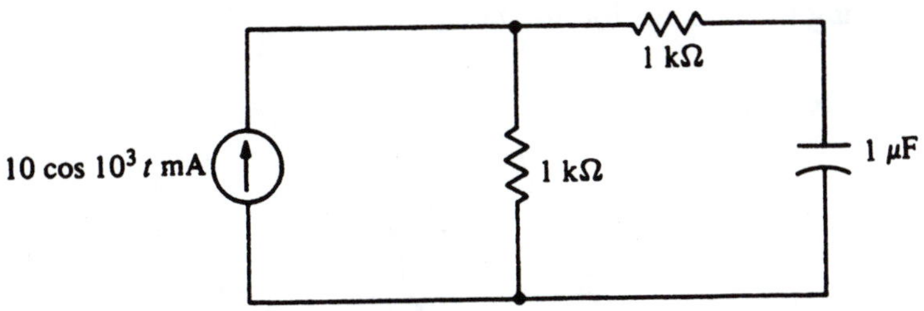

PROBLEM 10.7

10.8 Find the average power absorbed by the 2 Ω resistor.

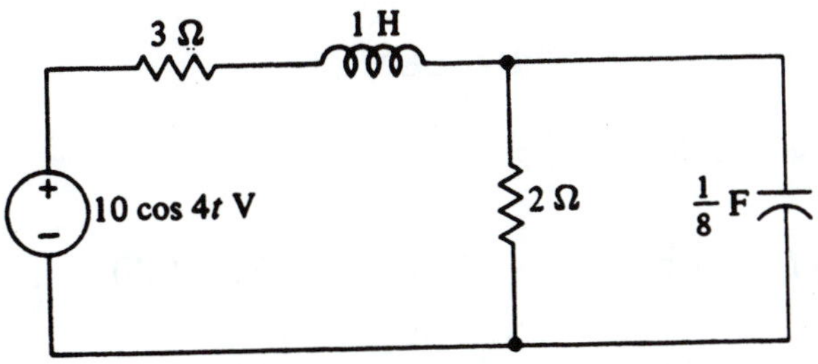

PROBLEM 10.8

10.9 Find the average power delivered by the 1/2 Ω resistor.

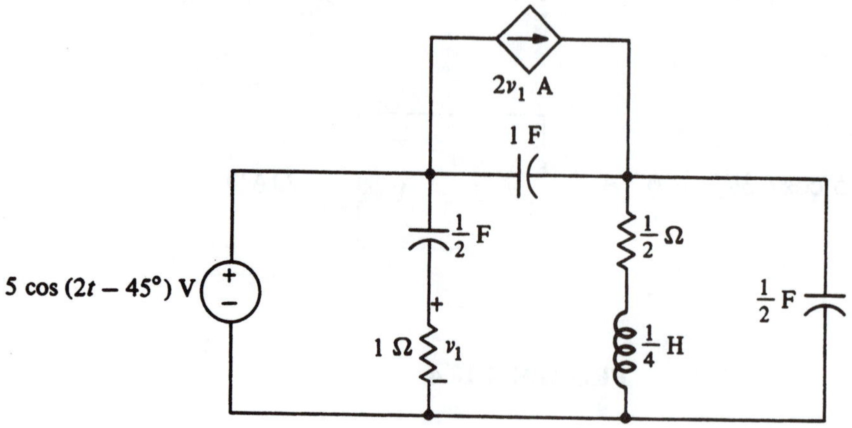

PROBLEM 10.9

10.2 RMS Values

10.10 Repeat Prob. 10.9 using rms phasors for the sources.

10.11 Find I_{rms} in the circuit below.

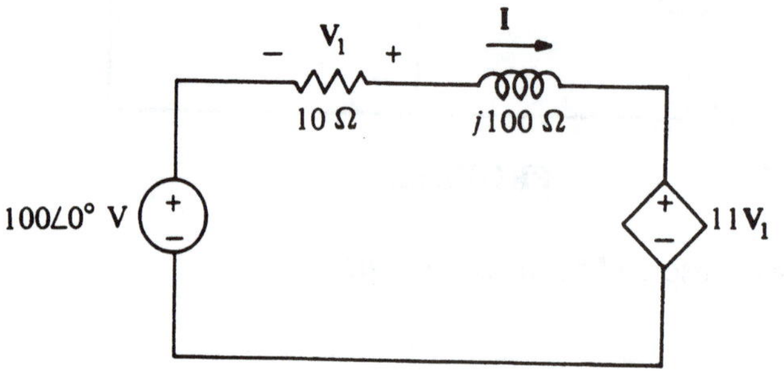

PROBLEM 10.11

10.12 Find the rms value of $i = 5\sin(\omega t) + 8$ A.

10.13 Find the rms value of $i = 6\cos(3\omega t + 30°) + 2\cos(2\omega t + 20°)$ A.

10.14 Find the rms value of $i = 4\cos(\omega t) + 5\cos(\omega t + 15°)$ A.

10.15 Find the rms value of $v = 12 + 5\sqrt{2}\sin t$ V.

10.16 Find the rms value of $v = 9\sqrt{2}\cos(5t) + 3\sqrt{2}\cos(3t) + 2\sqrt{5}\cos(4t - 30°)$ V.

10.17 Find the rms value of $v = \cos(2t) - 6\cos(t) + 3\sqrt{2}\cos(3t)$ V.

10.18 Find the rms value of a periodic current for which one cycle is given by

$$i = \begin{cases} \sqrt{t}\,\text{A}, & 0 < t < 1\text{s} \\ 0\,\text{A}, & 1 < t < 3\text{s}. \end{cases}$$

10.19 Find the rms value of a periodic current for which one cycle is given by

$$i = \begin{cases} t^2\text{A}, & 0 < t < 3\text{s} \\ 0\,\text{A}, & 3 < t < 4\text{s}. \end{cases}$$

10.3 Complex Power

10.20 In the circuit shown below, $\mathbf{Y}_1 = 4 + j3$ S and $\mathbf{Y}_2 = 4 - j3$ S. If $\mathbf{I}_g = 2\sqrt{2}\angle 15°$ A rms, find the complex power absorbed by each load $\mathbf{Z}_1$ and $\mathbf{Z}_2$.

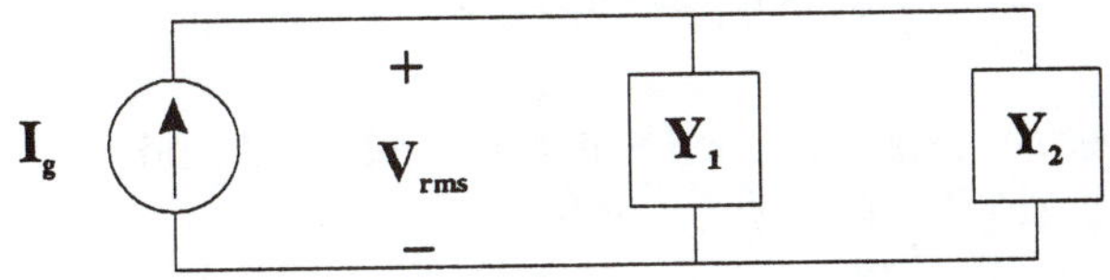

PROBLEM 10.20

10.21 The source voltage in the circuit shown is $v(t) = 50\sqrt{2}\sin(9t)$ V. If the load absorbs complex power $\mathbf{S} = 500 + j1000$, find (a) current i_1 and (b) $\mathbf{Z}_1$. (c) If $\mathbf{Z}_1$ is a single element, what is it?

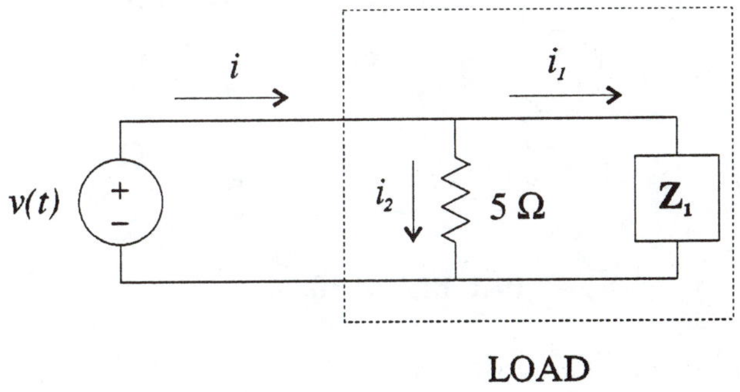

PROBLEM 10.21

10.22 (a) Find the average power P absorbed by the load in the circuit below.
(b) Find the reactive power Q absorbed by the load in the circuit below.
(c) What do these two values tell us in relation to the elements in the load?

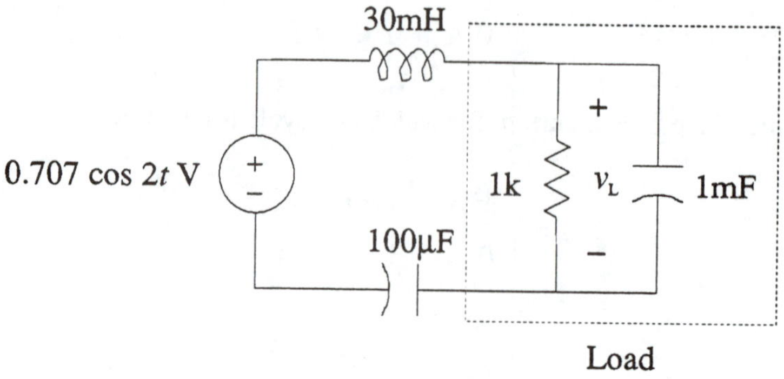

PROBLEM 10.22

10.23 Find the complex power delivered to a load which absorbs 40 W and (a) $\theta = \phi_v - \phi_i = 45°$, (b) $\theta = \phi_v - \phi_i = -45°$.

10.24 Find the complex power delivered to a load which absorbs 15 var and (a) $\theta = \phi_v - \phi_i = 45°$, (b) $\theta = \phi_v - \phi_i = -45°$.

10.25 Find the complex power delivered to a load which absorbs 2 VA and (a) $\theta = \phi_v - \phi_i = 45°$, (b) $\theta = \phi_v - \phi_i = -45°$.

10.26 Find the complex power absorbed by each component in the circuit shown.

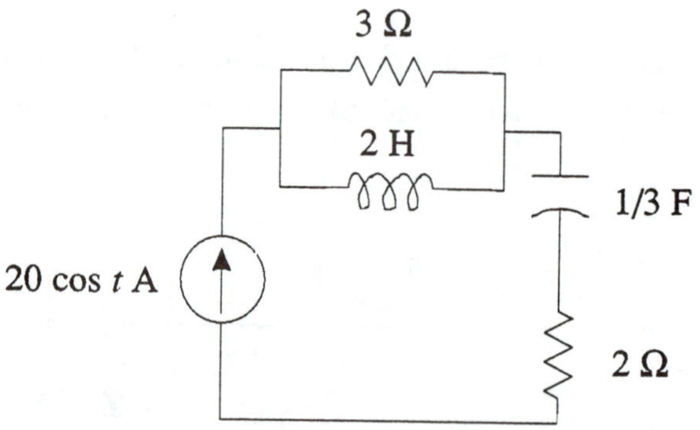

PROBLEM 10.26

10.27 Find the complex power absorbed by each component in the circuit shown.

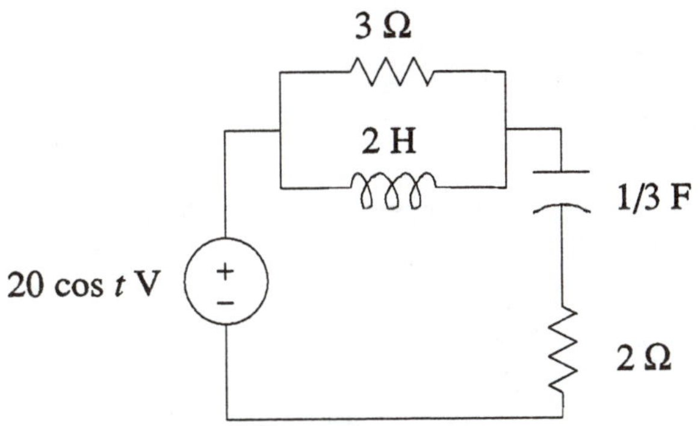

PROBLEM 10.27

10.28 Find the complex power absorbed by each component in the circuit shown.

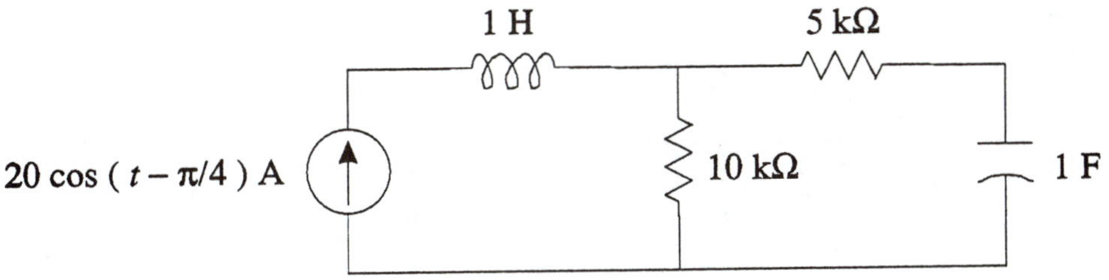

PROBLEM 10.28

10.4 Superposition and Power

10.29 Find the average power delivered to the resistor if $R = 1\text{k}\Omega$ and $v_{g1} = 5\cos(10t)$ V and $v_{g2} = 5\sin(10t)$ V.

10.30 Find the average power delivered to the resistor if $R = 1\text{k}\Omega$ and $v_{g1} = 5\cos(10t)$ V and $v_{g2} = 5\sin(20t)$ V.

10.31 Find the average power delivered to the resistor if $R = 1\text{k}\Omega$ and $v_{g1} = 5\cos(10t)$ V and $v_{g2} = 5$ V.

10.32 Find the complex power supplied to the load.

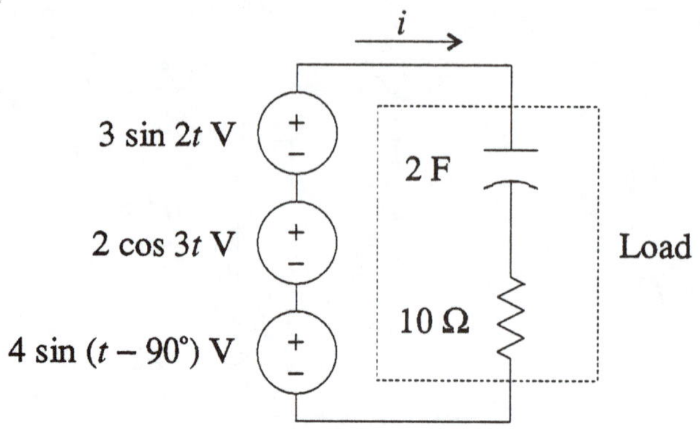

PROBLEM 10.32

10.33 Three passive loads $\mathbf{Z_1}$, $\mathbf{Z_2}$, and $\mathbf{Z_3}$ are absorbing complex power values of $\mathbf{S_1} = 2 + j3$, $\mathbf{S_2} = 3 - j1$, and $\mathbf{S_3} = 1 + j6$, respectively. $\mathbf{V_{rms}} = 10\angle 0$ V rms. Find the current $\mathbf{I_{rms}}$.

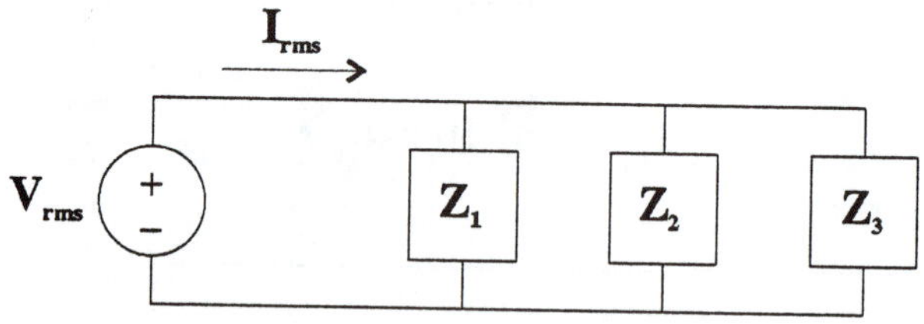

PROBLEM 10.33

10.34 If $\mathbf{I_{1,rms}} = 10\cos(t)$ A rms and $\mathbf{I_{2,rms}} = 10\cos(t)$ A rms, then find the complex power supplied to the load.

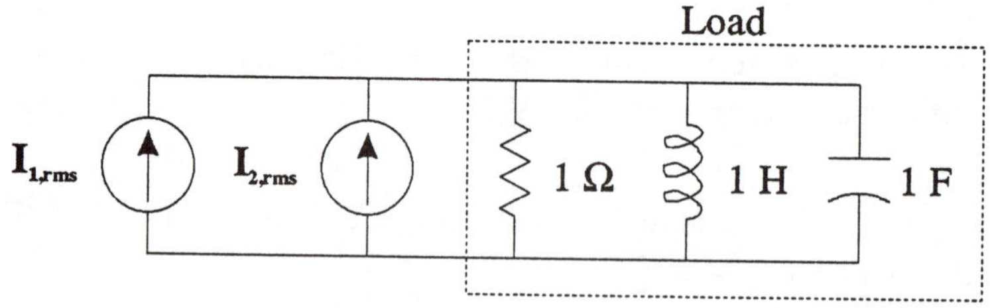

PROBLEM 10.34

10.5 Maximum Power Transfer

10.35 (a) Select $\mathbf{Z_L}$ to provide maximum (average) power transfer to the load.
(b) Find the complex power delivered to the load.

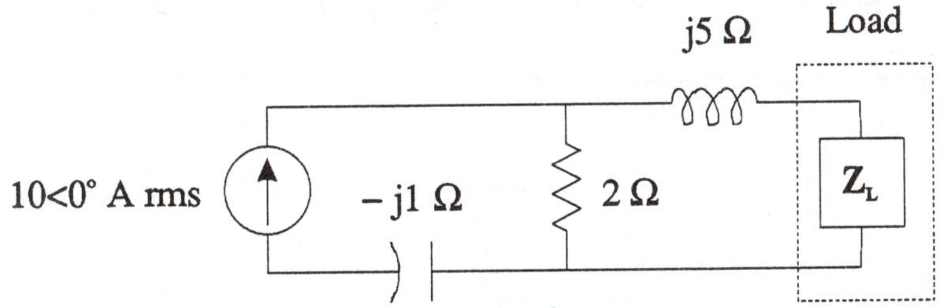

PROBLEM 10.35

10.36 A company is building an audio amplifier with an output impedance $\mathbf{Z_T} = 16 + j\,0.5\ \Omega$ at $\omega = 6000$ rad/s. (a) What should the impedance of the load be in order to achieve maximum power transfer at this frequency? (b) If the load is a speaker which may be modeled as a series RL subcircuit, what should the value of R and L be to have this load impedance?

10.37 Find $\mathbf{Z_L}$ to have maximum power transfer in the circuit below if $v(t) = 50\sin(10t)$ mV.

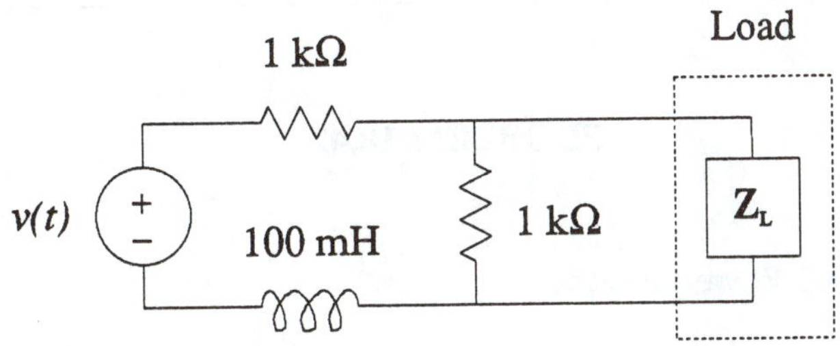

PROBLEM 10.37

10.6 Conservation of Power

10.38 Show that the sum of the complex power absorbed by each element in the circuit of Prob. 10.26 equals zero.

10.39 Show that the sum of the complex power absorbed by each element in the circuit of Prob. 10.27 equals zero.

10.40 Find the complex power supplied by each source in Prob. 10.34 using conservation of power.

10.41 Use conservation of power to find the power delivered to the resistor by first finding the real part of the complex power supplied by the source.

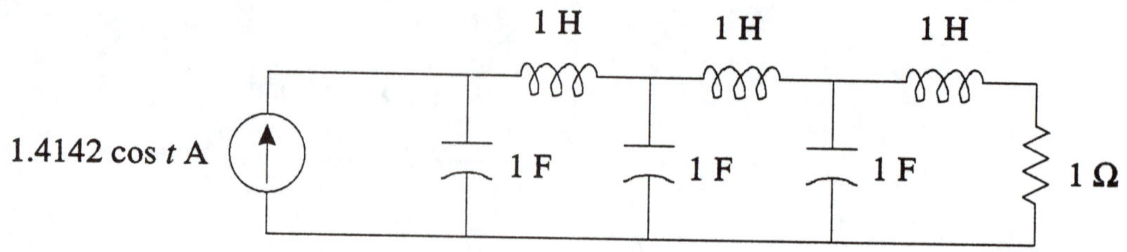

PROBLEM 10.41

10.42 Use conservation of power to find the power delivered to the resistor by first finding the real part of the complex power supplied by the source.

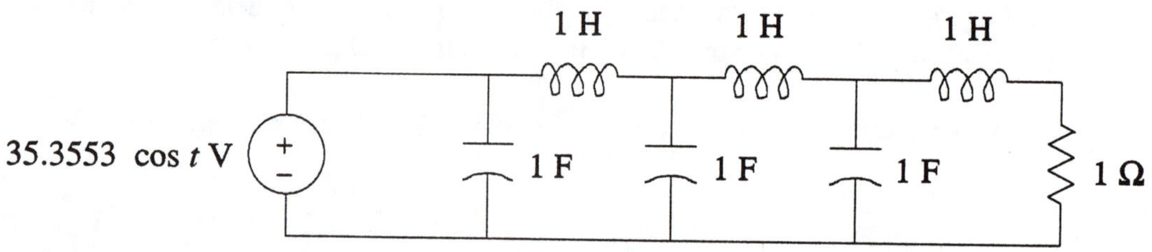

PROBLEM 10.42

10.7 Reactive Power and Power Factor

10.43 A load consists of a 200 Ω resistor in series with a 0.1 H inductor. Find the parallel capacitance necessary to adjust to a *pf* of 0.95 lagging if $\omega = 1000$ rad/s.

10.44 A load consists of a 200 Ω resistor in series with a 0.1 H inductor. Find the parallel capacitance necessary to adjust to a *pf* of 0.95 leading if $\omega = 1000$ rad/s.

10.45 A load consists of a 200 Ω resistor in series with a 0.1 H inductor. Find the parallel capacitance necessary to adjust to a unity *pf* if $\omega = 1000$ rad/s.

10.46 Find the power factor of the parallel connection of a 16 Ω resistor, a 2 H inductor, and a 1/32 F capacitor at a frequency of $\omega = 2$ rad/s. At what frequency does a unity *pf* occur?

10.47 Repeat Prob. 10.46 for the elements connected in series.

10.48 Find the power factor seen from the terminals for the source in the circuit below. What reactance connected in parallel with the source will change the power factor to unity?

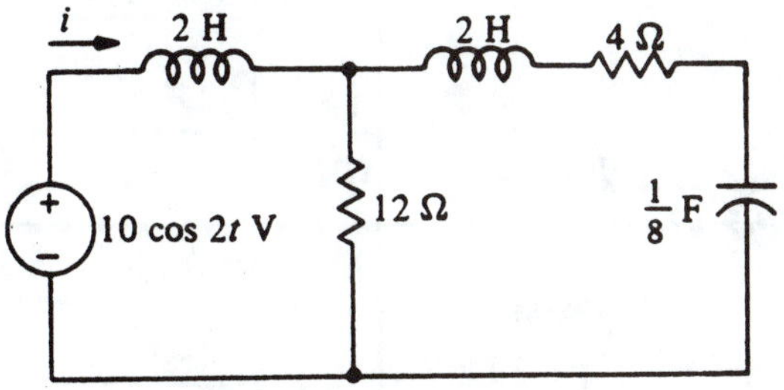

PROBLEM 10.48

10.8 SPICE and AC Steady State Power

10.49 Solve Prob. 10.22 (a) and (b) using SPICE.

10.50 Solve Prob. 10.26 using SPICE.

10.51 Solve Prob. 10.27 using SPICE.

10.52 Solve Prob. 10.28 using SPICE.

CHAPTER 10 SOLUTIONS

10.1 (a) $T = 0.75\,ms = 7.5\times10^{-4}\,s$

(b) $p(t) = v^2/R \Rightarrow T_p = T = 7.5\times10^{-4}\,s$

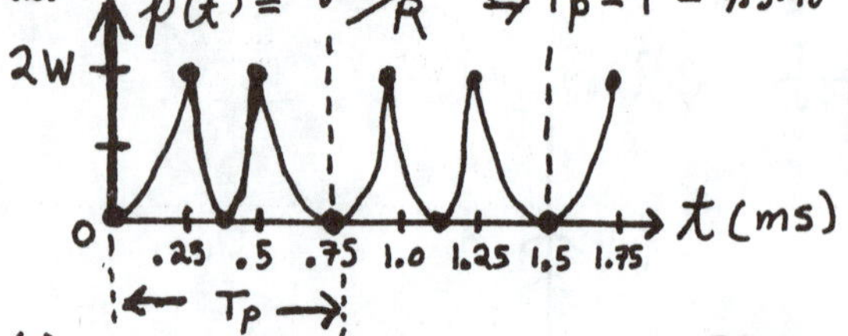

(c)
$$P = \frac{1}{T}\int_{t_1}^{t_1+T} p(t)\,dt = \frac{1}{7.5\times10^{-4}}\int_0^{.75ms} p(t)\,dt$$

$$= 1.3\bar{3}\times10^3\left[\int_0^{.25ms}(3.2\times10^7)t^2dt + \int_{.25ms}^{.375ms}(1.28\times10^8)(t-.375ms)^2dt\right.$$

$$\left. + \int_{.375ms}^{.5ms}(1.28\times10^8)(t-0.375ms)^2dt + \int_{0.5ms}^{0.75ms}(3.2\times10^7)(t-.75)^2dt\right]$$

$$= 1.3\bar{3}\times10^3\left[2\int_0^{.25ms}(3.2\times10^7)t^2dt + 2\int_{.25ms}^{.375ms}(1.28\times10^8)\cdot(t-.375)^2dt\right]$$

$$= 0.4\bar{4} + 0.2\bar{2} = \underline{0.667\,W}$$

10.2 (a) $T = 0.75s$ (b) $T_p = T = 0.75s$

$$p(t) = i^2R = \begin{cases} 1\times10^5\,W, & (0 \pm nT) \le t < (.25 \pm nT) \\ 0\,W, & (.25 \pm nT) \le t < (.75 \pm nT) \end{cases}$$

(c) $P = \frac{1}{0.75}\int_0^{0.75} p\,dt = \frac{4}{3}\int_0^{0.25}(1\times10^5)\,dt$

$= 3.3\bar{3}\times10^4\,W = \underline{33.\bar{3}\,kW}$

10.3 $P = \frac{\omega}{2\pi}\int_0^{2\pi/\omega} R I_m^2(1 + 2\cos\omega t + \cos^2\omega t)\,dt$

$\Rightarrow P = \frac{\omega}{2\pi} R I_m^2\left[\frac{2\pi}{\omega} + \frac{\pi}{\omega}\right] = \frac{3RI_m^2}{2}$

10.4 (a) $\underline{I} = 5\angle0^\circ A \Rightarrow \underline{V} = R\underline{I} = 25\angle0^\circ\,V$

$P_L = \frac{V_m I_m}{2}\cos(\phi_v - \phi_i) = \frac{(25)5}{2}(1) = \underline{62.5W}$

(b) ϕ does not affect average power

$\Rightarrow P_L = 62.5\,W$

10.5 (a) $\underline{I} = 5\angle0^\circ A,\ \underline{V} = j377(1)\underline{I} = 1885\angle90^\circ V$

$P_L = \frac{(1885)(5)}{2}(\cos 90^\circ) = 0\,W$

(b) (same answer) $P_L = 0\,W$

10.6 (a) $\underline{I} = 5\angle0^\circ A,\ \underline{V} = \frac{1}{j377}\underline{I} = .01326\angle-90^\circ$

$P_L = \frac{V_m I_m}{2}\cos(-90^\circ) = 0$

(b) $P_L = 0$

10.7

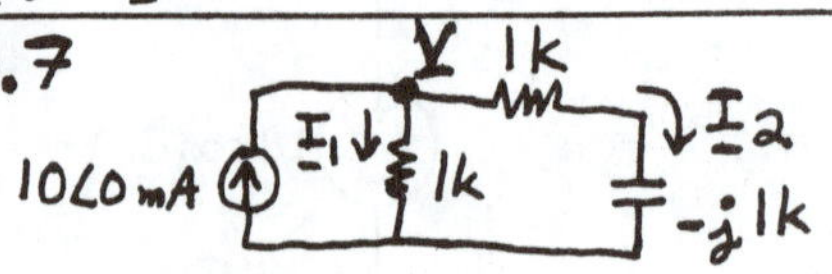

By current division:

$\underline{I}_1 = \frac{1-j}{1+1-j}(10mA) = 6 - j2 = \sqrt{40}\angle-18.43^\circ mA$

$\underline{I}_2 = 10 - \underline{I}_1 = 10 - 6 + j2 = \sqrt{20}\angle26.57^\circ mA$

$\underline{V} = \underline{I}_1 R_1 = \sqrt{40}\angle-18.43^\circ\,V$

$P_{R1} = \frac{I_{1m}^2(1k)}{2} = 20\,mW$

$P_{R2} = \frac{I_{2m}^2(1k)}{2} = 10\,mW$

$P_{capacitor} = \frac{I_{m2}^2}{2}\cdot Re(\underline{Z}) = 0\,W$

$P_{source} = -\frac{V_m(10mA)}{2}\cos(-18.43^\circ - 0^\circ)$

$= -30\,mW$

10.8 $\underline{V}$ = voltage across 2Ω resistor

KCL: $\underline{V}\left(+j\frac{1}{2} + \frac{1}{3+j4}\right) = \frac{10}{3+j4}$

$\Rightarrow \underline{V} = \frac{10}{0.5 + j3.5} = 2.828\angle-81.87^\circ\,V$

$P = \frac{V_m^2}{2(2\Omega)} = \underline{2W}$

10.9 v = voltage of right $\frac{1}{2}F$ capacitor

KCL: $j2(\underline{V} - 5\angle-45^\circ) + \frac{\underline{V}}{.5+j.5} + j\underline{V} = 2\underline{V}_1$

Where $\underline{V}_1 = \frac{(1)(5\angle-45^\circ)}{1-j1} = \frac{5}{\sqrt{2}}\,V.$

Then, $\underline{V} = 5\sqrt{2}\angle-36.9^\circ\,V$

$|\underline{I}_{\frac{1}{2}\Omega}| = \frac{|V|}{|0.5 + j0.5|} = 10A$

$\therefore P = \left(|\underline{I}_{\frac{1}{2}\Omega}|^2/2\right)\cdot\left(\frac{1}{2}\right) = \underline{25\,W}$

10.10

$\frac{5}{\sqrt{2}}\angle -45° \, V_{rms}$

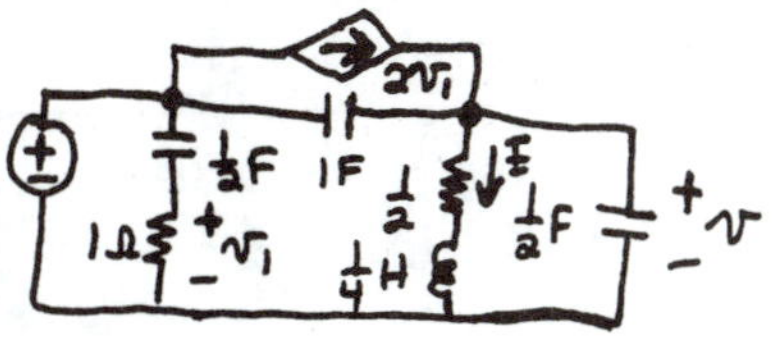

Using analysis from 10.9:

$$\underline{V}_{1rms} = \frac{(1)\left(\frac{5}{\sqrt{2}}\angle -45°\right)}{(1-j1)} = \frac{5}{2}\angle 0 \text{ V}$$

Using same KCL equation: $\underline{V} = 5\angle -36.9° \text{ V}$

$P = |\underline{I}_{rms}|^2 R = (5\sqrt{2})^2 \left(\frac{1}{2}\right) = \underline{25W}$ ✓

10.11 $\underline{V}_1 = -\underline{I}_1(10)$ then KVL gives

$\underline{I}(10 + j100) + 11\underline{V}_1 = 100 \Rightarrow \underline{I} = \frac{1}{\sqrt{2}}\angle -135° A$

$$\underline{I}_{rms} = \frac{(1/\sqrt{2})\angle -135°}{\sqrt{2}} = \frac{1}{2}\angle -135° A_{rms}$$

10.12 $\underline{I}^2_{rms} = \frac{1}{2}(5)^2 + (8)^2 = 76.5$

$\therefore \underline{I}_{rms} = \underline{8.746A}$

10.13 $\underline{I}^2_{rms} = \frac{1}{2}(6)^2 + \frac{1}{2}(2)^2 = 20$

$\therefore \underline{I}_{rms} = \underline{4.472}A$

10.14 $\underline{I} = 4 + 5\angle 15° = 8.924\angle 8.34° A$

$\underline{I}_{rms} = \frac{8.924}{\sqrt{2}} = \underline{6.31} \text{ A}$

10.15 $\underline{V}^2_{rms} = (12)^2 + \frac{1}{2}(5\sqrt{2})^2 = 169$

$\therefore \underline{V}_{rms} = \underline{13V}$

10.16 $\underline{V}^2_{rms} = \frac{1}{2}(9\sqrt{2})^2 + \frac{1}{2}(3\sqrt{2})^2 + \frac{1}{2}(2\sqrt{5})^2$

$= 100 \quad \therefore \underline{V}_{rms} = \underline{10}V$

10.17 $\underline{V}^2_{rms} = \frac{1}{2}(1)^2 + \frac{1}{2}(-6)^2 + \frac{1}{2}(3\sqrt{2})^2$

$= 27.5 \quad \therefore \underline{V}_{rms} = \underline{5.244}V$

10.18 $\underline{I}^2_{rms} = \frac{1}{3}\left[\int_0^1 (\sqrt{t})^2 dt + \int_1^3 0\, dt\right]$

$= \frac{1}{3}\left(\frac{1}{2}\right) = \frac{1}{6}$

$\therefore \underline{I}_{rms} = \frac{1}{\sqrt{6}} = 0.408A$

10.19 $\underline{I}^2_{rms} = \frac{1}{4}\left[\int_0^3 (t^2)^2 dt + \int_3^4 0\, dt\right]$

$= \frac{1}{4}\left(\frac{243}{5}\right) = \frac{243}{20} = 12.15$

$\therefore \underline{I}_{rms} = 3.486A$

10.20

$\underline{I}_{1rms} = \underline{I}_{grms}\left(\frac{\underline{Y}_1}{\underline{Y}_1 + \underline{Y}_2}\right) = 2.5\angle 51.87° A_{rms}$

$\underline{I}_{2rms} = \underline{I}_{grms}\left(\frac{\underline{Y}_2}{\underline{Y}_1 + \underline{Y}_2}\right) = 2.5\angle -51.87° A_{rms}$

$\underline{S}_{Z1} = \underline{V}_{rms}\underline{I}^*_{1rms} = \underline{Z}_1|\underline{I}_{1rms}|^2 = \frac{|\underline{I}_{1rms}|^2}{\underline{Y}_1}$

$\Rightarrow \underline{S}_{Z1} = 1.25\angle \underline{-36.87°}$

$\underline{S}_{Z2} = \underline{V}_{rms}\underline{I}^*_{2rms} = \underline{Z}_2|\underline{I}_{2rms}|^2 = \frac{|\underline{I}_{2rms}|^2}{\underline{Y}_2}$

$\Rightarrow \underline{S}_{Z2} = 1.25\angle \underline{36.87°}$

10.21 (a) $\underline{V}_{rms} = 50\angle -90° \text{ V}_{rms}$

$\Rightarrow \underline{I}_{2rms} = \frac{\underline{V}_{rms}}{5\Omega} = 10\angle -90° A_{rms}$

$\underline{S} = \underline{V}_{rms}\underline{I}^*_{rms} = \underline{V}_{rms}(\underline{I}^*_{1rms} + \underline{I}^*_{2rms})$

$\Rightarrow 500 + j1000 = 50\angle -90°(\underline{I}^*_{1rms} + 10\angle +90°)$

$= 500\angle 0° + 50\angle -90°(\underline{I}^*_{1rms})$

$\Rightarrow j1000 = -j50(\underline{I}^*_{1rms})$

$\Rightarrow \underline{I}^*_{1rms} = -20 A_{rms} = 20\angle \pm 180° = \underline{I}_{1rms}$

$\Rightarrow i_1(t) = 20\cos(9t - 180°) A_{rms}$

$= 20\sin(9t - 90°) A_{rms}$

(b) $\underline{Z}_1 = \frac{\underline{V}_{rms}}{\underline{I}_{1rms}} = \frac{50\angle -90°}{20\angle -180°} = 2.5\angle 90°$

$= j2.5\Omega$

(c) Inductor, $Z_1 = j\omega L \Rightarrow L = 0.278H$

10.22 (a) $\underline{Z}_L = \dfrac{(1000\,\Omega)(-j500\,\Omega)}{1000 - j500\,\Omega} = \begin{matrix}200\\ -j400\end{matrix}$

$\underline{V}_{Lrms} = 5\angle 0°\left(\dfrac{\underline{Z}_L}{\underline{Z}_L + j0.06 - j5000}\right)$

$= \dfrac{2236\angle -63.43°}{5403.6\angle -87.88°} = 0.414\angle 24.45°\ \text{Vrms}$

$\underline{S}_L = \underline{V}_{Lrms}\underline{I}^*_{Lrms} = \dfrac{|V_{Lrms}|^2}{\underline{Z}_L^*} = \dfrac{0.1714}{447.2\angle 63.4°}$

$= 3.83\times 10^{-4}\angle -63.43°$

$\Rightarrow P_L = \text{Re}\{\underline{S}_L\} = 1.713\times 10^{-4}\ \text{W}$

(b) $Q_L = \text{Im}\{\underline{S}_L\} = -3.426\times 10^{-4}\ \text{var}$

(c) The resistor absorbs average power P_L since this is the only resistive device in the load. The capacitor in the load stores an average Q_L var of power.

10.23 (a) $P = 40\,\text{W} \Rightarrow Q = P\tan\theta = 40\,\text{var}$

$\therefore\ \underline{S} = P + jQ = 40 + j40 = 40\sqrt{2}\angle 45°$

(b) $Q = 40\tan(-45°) = -40\ \text{var}$

$\therefore\ \underline{S} = 40 - j40 = 40\sqrt{2}\angle -45°$

10.24 (a) $Q = 15\,\text{var} \Rightarrow P = \dfrac{Q}{\tan\theta} = 15\,\text{W}$

$\therefore\ \underline{S} = P + jQ = 15 + j15 = 15\sqrt{2}\angle 45°$

(b) $P = \dfrac{15\text{var}}{\tan(-45°)} = -15\,\text{W}$

$\therefore\ \underline{S} = 15 - j15 = 15\sqrt{2}\angle -45°$

10.25

(a) $\underline{S} = 2\angle 45°\,\text{VA} = \sqrt{2} + j\sqrt{2}$

(b) $\underline{S} = 2\angle -45°\,\text{VA} = \sqrt{2} - j\sqrt{2}$

10.26

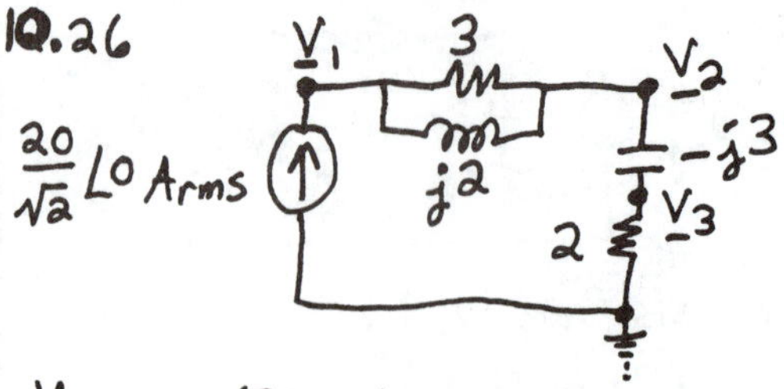

$\underline{V}_{3rms} = \left(\dfrac{20}{\sqrt{2}}\angle 0\right)(2) = \dfrac{40}{\sqrt{2}}\angle 0\ \text{V rms}$

$\underline{V}_{2rms} = \left(\dfrac{20}{\sqrt{2}}\angle 0\right)(2 - j3) = \dfrac{20\sqrt{13}}{\sqrt{2}}\angle -56.3°$ Vrms

$\underline{V}_{1rms} - \underline{V}_{2rms} = \left(\dfrac{20}{\sqrt{2}}\angle 0\right)\left(\dfrac{3(j2)}{3 + j2}\right)$

$= \dfrac{120}{\sqrt{26}}\angle 56.31°$

$\Rightarrow \underline{V}_{1rms} = 47.226\angle -28.937°$

$\underline{S}_{Source} = -\underline{V}_{1rms}\underline{I}^*_{rms} = 667.9\angle 151.06°$

$\underline{S}_{2\Omega} = \underline{V}_{3rms}\underline{I}^*_{rms} = 400\angle 0°$

$\underline{S}_{Cap} = (\underline{V}_{2rms} - \underline{V}_{3rms})\underline{I}^*_{rms} = 600\angle -90°$

$\underline{S}_{3\Omega} = \dfrac{|\underline{V}_{1rms} - \underline{V}_{2rms}|^2}{\underline{Z}^*} = 184.08\angle 0°$

$\underline{S}_{2H} = \dfrac{|\underline{V}_{1rms} - \underline{V}_{2rms}|^2}{\underline{Z}^*} = 276.13\angle 90°$

10.27

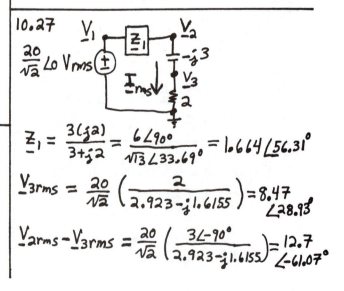

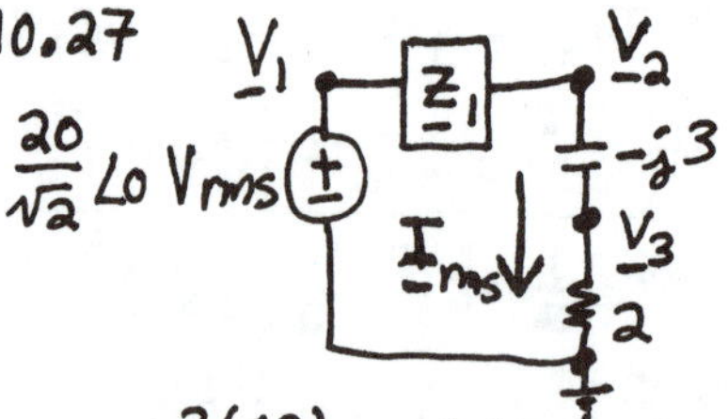

$\underline{Z}_1 = \dfrac{3(j2)}{3 + j2} = \dfrac{6\angle 90°}{\sqrt{13}\angle 33.69°} = 1.664\angle 56.31°$

$\underline{V}_{3rms} = \dfrac{20}{\sqrt{2}}\left(\dfrac{2}{2.923 - j1.6155}\right) = 8.47\angle 28.93°$

$\underline{V}_{2rms} - \underline{V}_{3rms} = \dfrac{20}{\sqrt{2}}\left(\dfrac{3\angle -90°}{2.923 - j1.6155}\right) = 12.7\angle -61.07°$

10.27 (cont.)

$$\underline{V}_{1rms} - \underline{V}_{2rms} = \frac{20}{\sqrt{2}}\left(\frac{1.664\angle 56.31^\circ}{2.923 - j1.6155}\right)$$

$$= 7.046\angle 85.24^\circ \text{ V rms}$$

$$\underline{I}_{rms} = \frac{\underline{V}_{rms}}{2.923 - j1.6155} = 4.234\angle 28.93^\circ$$

$$\underline{S}_{source} = -\underline{V}_{rms}\underline{I}^*_{rms} = 59.88\angle 151.07^\circ$$

$$\underline{S}_{2\Omega} = \underline{V}_{3rms}\underline{I}^*_{rms} = 35.86\angle 0^\circ$$

$$\underline{S}_{cap} = (\underline{V}_{2rms} - \underline{V}_{3rms})\underline{I}^*_{rms} = 53.77\angle -90^\circ$$

$$\underline{S}_{3\Omega} = \frac{|\underline{V}_{1rms} - \underline{V}_{2rms}|^2}{3} = 16.55\angle 0^\circ$$

$$\underline{S}_{2H} = \frac{|\underline{V}_{1rms} - \underline{V}_{2rms}|^2}{2\angle -90^\circ} = 24.82\angle 90^\circ$$

10.28

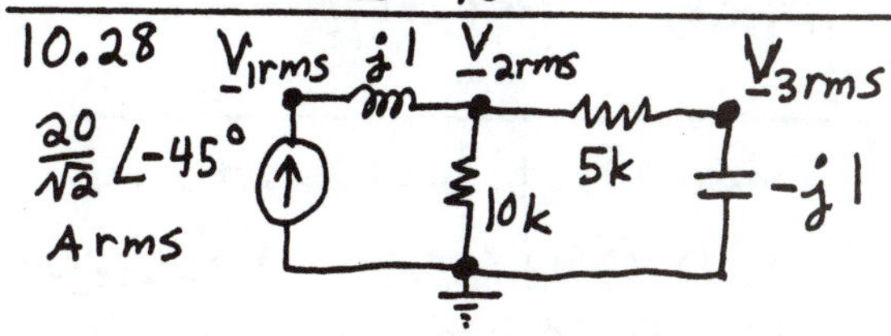

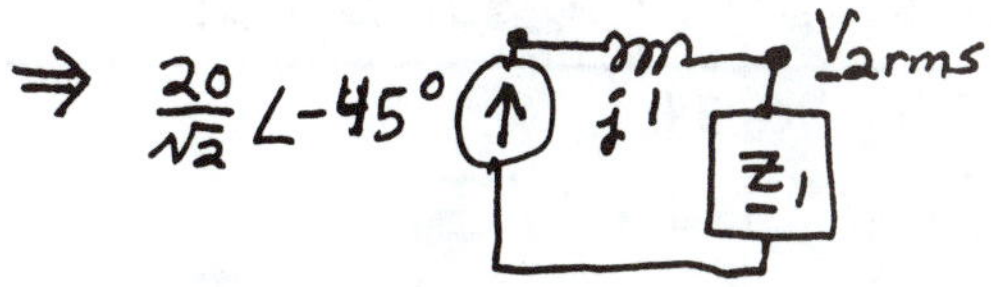

$$\underline{Z}_1 = \frac{10k(5k - j1)}{10k + 5k - j1} = 3333.\overline{3}\angle -.00764^\circ$$

$$\underline{V}_{2rms} = \left(\frac{20}{\sqrt{2}}\angle -45^\circ\right)(3333.\overline{3}\angle -.00764^\circ)$$

$$= 47140.45\angle -45.00764^\circ \text{ Vrms}$$

$$\underline{V}_{1rms} - \underline{V}_{2rms} = \left(\frac{20}{\sqrt{2}}\angle -45^\circ\right)(1\angle 90^\circ) = \frac{20}{\sqrt{2}}\angle 45^\circ$$

$$\underline{V}_{1rms} = 47140.4\angle -44.98^\circ \text{ Vrms}$$

$$\underline{V}_{3rms} = \underline{V}_{2rms}\left(\frac{-j1}{5000 - j1}\right) = 9.428\angle -135^\circ$$

$$\underline{S}_{source} = -\underline{I}^*_{rms}\underline{V}_{rms} = 666666\angle -179.99^\circ$$

$$\underline{S}_{1H} = (\underline{V}_{1rms} - \underline{V}_{2rms})\underline{I}^*_{rms} = 200\angle 90^\circ$$

$$\underline{S}_{10k} = \frac{|\underline{V}_{2rms}|^2}{10k} = 222222\angle 0^\circ$$

$$\underline{S}_{1F} = \frac{|\underline{V}_{3rms}|^2}{1\angle 90^\circ} = 88.887\angle -90^\circ$$

$$\underline{S}_{5k} = \frac{|\underline{V}_{2rms} - \underline{V}_{3rms}|^2}{5000} = 44444.4\angle 0^\circ$$

10.29 $\underline{I} = (\underline{V}_{g1} - \underline{V}_{g2})/R = \frac{5 + j5}{1k}$

$$= 5\sqrt{2}\angle 45^\circ \text{mA}$$

$$P = \frac{I_m^2(1k\Omega)}{2} = 25\text{mW}$$

10.30

$$P = \frac{|\underline{V}_{g1}|^2}{2R} + \frac{|\underline{V}_{g2}|^2}{2R} = \frac{5^2 + 5^2}{2(1k)} = 25\text{mW}$$

10.31

$$P = \frac{|\underline{V}_{g1}|^2}{2R} + \frac{v_{g2}^2}{R} = \frac{25}{2k} + \frac{25}{1k} = 37.5\text{mW}$$

10.32 Using superposition to find $\underline{I}$:

(i)

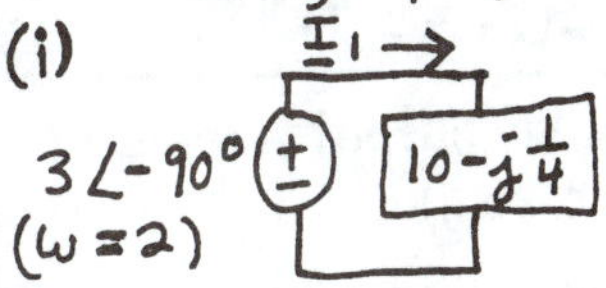

$$\underline{I}_1 = \frac{3\angle -90^\circ}{10.0\angle -1.43^\circ} = 0.30\angle -88.57^\circ$$

(ii)

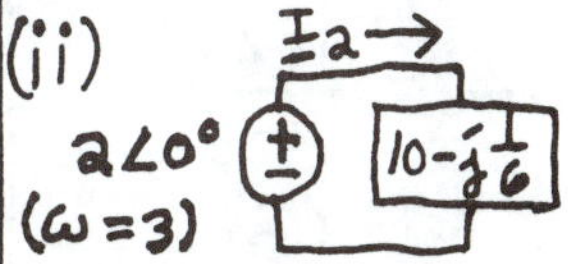

$$\underline{I}_2 = \frac{2\angle 0^\circ}{10.0\angle -0.955^\circ} = 0.20\angle 0.955^\circ$$

(iii)

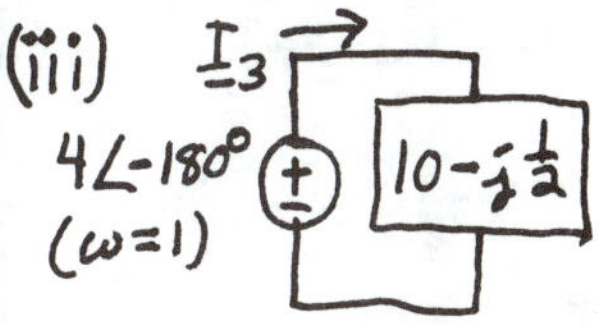

$$\underline{I}_3 = \frac{4\angle -180^\circ}{10.0\angle -2.86^\circ} = 0.40\angle -177.14^\circ$$

Now use superposition to find $\underline{S}$:

$$\underline{S} = \underline{S}_{L1} + \underline{S}_{L2} + \underline{S}_{L3}$$

$$\underline{S}_{L1} = \frac{1}{2}\underline{V}_1\underline{I}_1^* = \frac{(3\angle -90^\circ)(0.3\angle 88.57^\circ)}{2}$$

$$= 0.45\angle -1.43^\circ = 0.45 - j0.011$$

$$\underline{S}_{L2} = \frac{1}{2}\underline{V}_2\underline{I}_2^* = \frac{2(0.2\angle -0.955^\circ)}{2}$$

$$= 0.2\angle -0.955^\circ = 0.20 - j0.0033$$

$$\underline{S}_{L3} = \frac{1}{2}\underline{V}_3\underline{I}_3^* = \frac{(4\angle -180^\circ)(0.4\angle +177.14^\circ)}{2}$$

$$= 0.80 - j0.040$$

$$\therefore\ \underline{S} = 1.45 - j0.054$$

10.33 $\underline{S}_{LOAD} = \underline{S}_1 + \underline{S}_2 + \underline{S}_3$

$\Rightarrow \underline{S}_{LOAD} = (2+j3)+(3-j1)+(1+j6) = 6+j8$

$\underline{S}_{LOAD} = \underline{S}_{SUPPLIED} = \underline{V}_{rms}\underline{I}^*_{rms}$

$\Rightarrow \underline{I}^*_{rms} = \dfrac{6+j8}{10\angle 0^\circ} = 1\angle 53.13^\circ$

$\Rightarrow \underline{I}_{rms} = 1\angle -53.13^\circ$ A rms

---

10.34 Different frequencies - use superposition

(i)

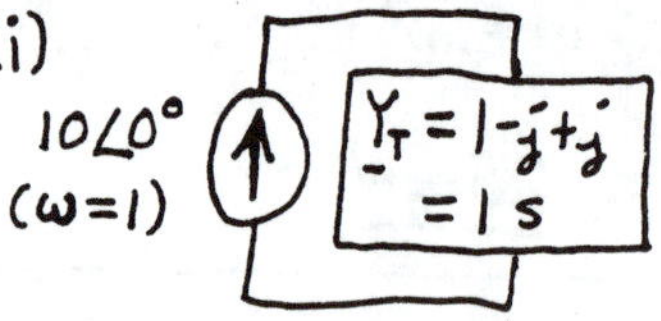

$\underline{S}_{L1} = \underline{Z}_T |\underline{I}_{rms}|^2 = 1(10)^2 = 100\angle 0^\circ$

(ii)

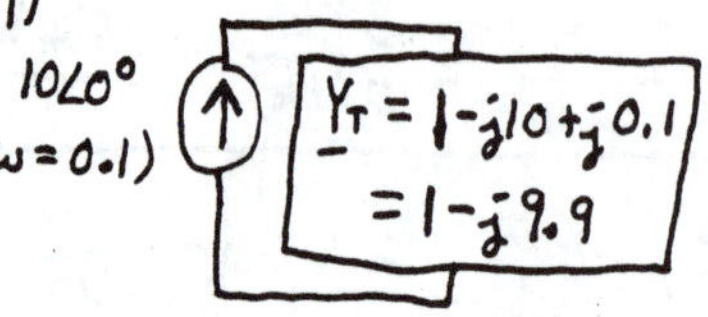

$\underline{S}_{L2} = \dfrac{|\underline{I}_{rms}|^2}{\underline{Y}_T} = \dfrac{10^2}{9.95\angle -84.23^\circ} = 10.05\angle 84.23^\circ$

$\therefore \underline{S} = \underline{S}_{L1} + \underline{S}_{L2} = 100 + (1.01 + j9.999) = 101.1 + j9.999$

10.35 (a) Find the equivalent output impedance using Thevenin equiv. circuit:

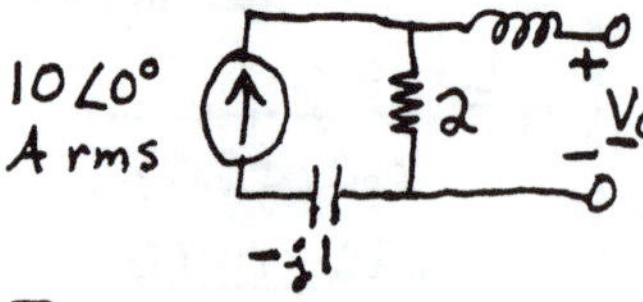

$\underline{V}_{ocrms} = \dfrac{10\angle 0^\circ}{2-j1} = 4.47\angle 26.57^\circ$

Turn off current source to find $\underline{Z}_T$

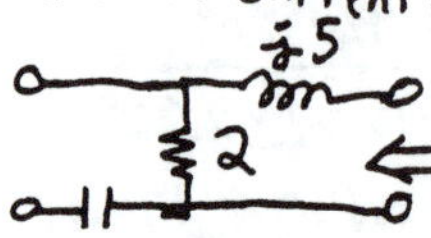

$\Leftarrow \underline{Z}_T = 2+j5\ \Omega$

$\Rightarrow$ Choose $\underline{Z}_L = \underline{Z}^*_T = 2-j5\ \Omega$

(b) Redraw using Thevenin equivalent

$\underline{V}_{Lrms} = \underline{V}_{Trms}\left(\dfrac{\underline{Z}^*_T}{\underline{Z}^*_T + \underline{Z}_T}\right)$

$\Rightarrow \underline{V}_{Lrms} = 6.02\angle -41.63^\circ$ Vrms

$\underline{S}_L = \underline{V}_{Lrms}\underline{I}^*_{rms} = \dfrac{|\underline{V}_{Lrms}|^2}{\underline{Z}_L^*} = \dfrac{(6.02)^2}{2+j5} = 6.73\angle -68.20^\circ$

10.36 (a) $\underline{Z}_L = \underline{Z}^*_T = 16 + j0.5\ \Omega$

(b) $R = 16\ \Omega$, $j(6000)L = j0.5$

$\Rightarrow L = \frac{1}{12000}\text{ H} = 8.\bar{3}\times 10^{-5}\text{ H} = 83.\bar{3}\ \mu\text{H}$

10.37 Find the (Thevenin) equivalent output impedance by turning off $\underline{V}$ source:

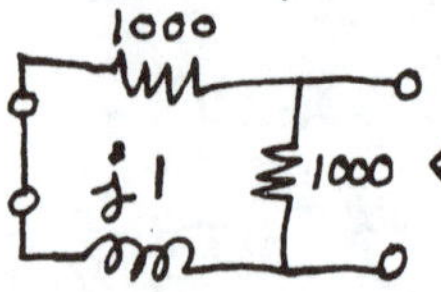

$\Leftarrow \underline{Z}_T = \dfrac{1000(1000+j1)}{2000+j1} = 500\angle 0.0287^\circ$

$\therefore$ Choose $\underline{Z}_L = \underline{Z}^*_T = 499.9999 - j0.25\ \Omega$

10.38 $\underline{S}_{source} + \underline{S}_{2\Omega} + \underline{S}_{cap} + \underline{S}_{3\Omega} + \underline{S}_{2H}$

$= (667.9\angle 151.06^\circ) + (400\angle 0^\circ) + (600\angle -90^\circ) + (184.08\angle 0^\circ) + (276.13\angle 90^\circ)$

≈ 0

10.39 $\underline{S}_{source} + \underline{S}_{2\Omega} + \underline{S}_{cap} + \underline{S}_{3\Omega} + \underline{S}_{2H}$

$= (59.88\angle 151.07^\circ) + (35.86\angle 0^\circ) + (53.77\angle -90^\circ) + (16.55\angle 0^\circ) + (24.82\angle 90^\circ) \approx 0$

10.40 From 10.34,

$\underline{S}_{L1} = 100\angle 0^\circ$ is due to $\underline{I}_1$

$\underline{S}_{L2} = 10.05\angle 84.23^\circ$ is due to $\underline{I}_2$

Since sources are at different frequencies, we can use superposition. Thus

$\underline{I}_1$ supplies $100\angle 0^\circ$

$\underline{I}_2$ supplies $10.05\angle 84.23^\circ$

10.41

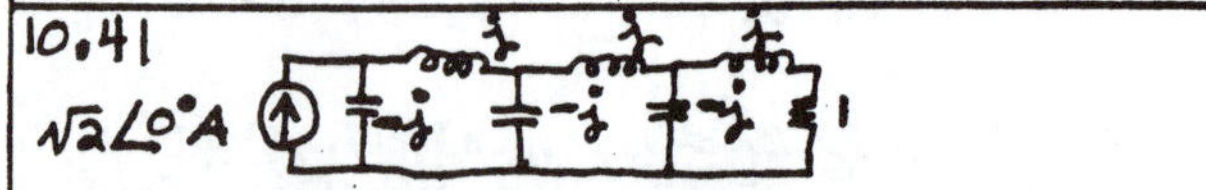

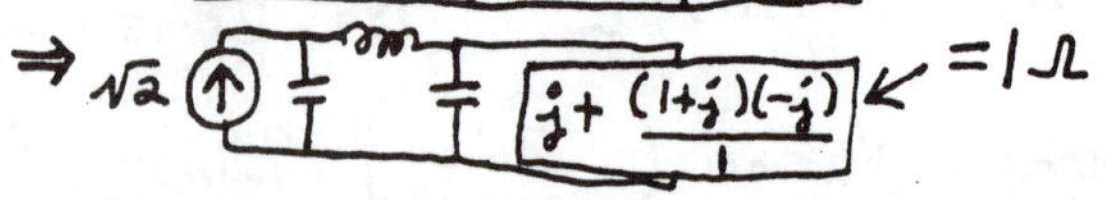

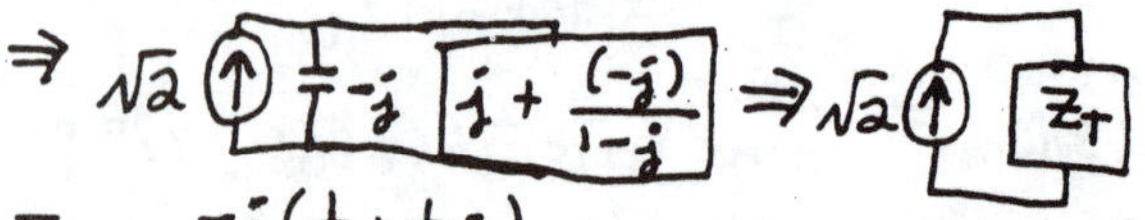

$\underline{Z}_T = \dfrac{-j\left(\frac{1}{2}+\frac{1}{2}j\right)}{\frac{1}{2}-\frac{1}{2}j} = 1,\quad \underline{S} = \underline{Z}\cdot|\underline{I}_{rms}|^2 = 1\cdot 1^2 = 1\angle 0^\circ$

$\Rightarrow P_R = Re\{\underline{S}\} = 1\text{W}$

10.42 After simplifying the circuit (see 10.41) it becomes:

$\underline{S} = \dfrac{|\underline{V}_{rms}|^2}{\underline{Z}^*} = 625\angle 0^\circ$

$P_R = Re\{\underline{S}\} = 625\ W$

10.43 $X_1 = \dfrac{R^2 + X^2}{R\tan(\cos^{-1}PF) - X}$

$= \dfrac{(200)^2 + (100)^2}{200\tan(\cos^{-1}0.95) - 100}$

$= -1459$

since negative $C = -1/\omega X_1$

$= \underline{0.685\ \mu F}$

10.44 $X_1 = \dfrac{R^2 + X^2}{-R\tan(\cos^{-1}PF) - X}$

$= -301.7$ since negative

$C = -1/\omega X_1 = \underline{3.315}\ \mu F$

10.45 $X_1 = \dfrac{(200)^2 + (100)^2}{-(100)} = -500$

$C = -1/\omega X_1 = \underline{2.0\ \mu F}$

10.46 $\underline{Y}_T = \frac{1}{16} + j\frac{\omega}{32} - j\frac{1}{2\omega}$, at $\omega = 2$ rad/s

$\underline{Z}_T = \frac{1}{\underline{Y}} = 1.6 + j4.8 = 5.06\angle 71.57^\circ$

$PF = \cos(71.57^\circ) = \underline{0.316}$ (lagging)

for $PF = 1$, $\frac{\omega}{32} = \frac{1}{2\omega}$ or $\omega^2 = 16$

$\therefore\ \omega = \underline{4\ \text{rad/s}}$

10.47 $\underline{Z}_T = 16 + j2\omega - j\frac{32}{\omega}$, at $\omega = 2$ rad/s

$\underline{Z}_T = 16 - j12 = 20\angle -36.87$

$PF = \cos(-36.87^\circ) = \underline{0.80}$ (leading)

for $PF = 1$, $2\omega = \frac{32}{\omega}$ or $\omega^2 = 16$

$\therefore\ \omega = \underline{4\ \text{rad/s}}$

10.48 (a) The impedance seen by the source is $\underline{Z}_T = j4 + \dfrac{(4 + j4 - j4)(12)}{12 + 4}$

$= 3 + j4 = 5\angle 53.13^\circ$

$\Rightarrow PF = \cos(53.13^\circ) = 0.60$ (lagging)

$\underline{S} = \frac{1}{2}|\underline{V}|^2 / \underline{Z}_T^* = 10\angle -53.13^\circ$

$\Rightarrow$ Source delivers 10 VA

10.48 (cont.)

(b) $\underline{Z}_T = 3 + j4$

For $PF = 1$, then $X_1 = \dfrac{3^2 + 4^2}{-4}$

$\therefore\ X_1 = -\frac{25}{4}\ \Omega$

(c) $X_1 = -\frac{25}{4}\Omega \Rightarrow$

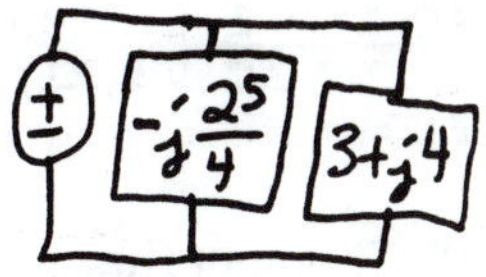

$\Rightarrow$

$\underline{S} = \frac{1}{2}|\underline{V}|^2 / 8.\overline{3} = 6\angle 0^\circ$

$\Rightarrow$ Source delivers 6 VA (or 6 W real average power)

10.49 SPICE input file

```
PROBLEM 10.49
V1 1 0 AC 7.071 0
L1 1 2 30M
R1 2 3 1K
C1 2 3 1M
C2 3 0 100U
RDUMMY 3 0 1G
.AC LIN 1 0.31831 0.31831
.PRINT AC VM(R1) VP(R1) IM(L1) IP(L1)
.END
```

SPICE output file

```
****      AC ANALYSIS                          TEMPERATURE =    27.000 DEG C
******************************************************************************

  FREQ          VM(R1)        VP(R1)        IM(L1)        IP(L1)


   3.183E-01    5.852E-01    2.444E+01    1.309E-03    8.788E+01
```

Thus, $\mathbf{S}_L = \mathbf{V}_{Lrms}\mathbf{I}^*_{Lrms} = \dfrac{(0.5852\angle 24.44°)(0.001309\angle -87.88°)}{2} = (3.83\times 10^{-4})\angle -63.44°$

which agrees with Prob. 10.22.

10.50 SPICE input file

```
PROBLEM 10.50
I1 0 1 AC 20 0
R1 1 2 3
L1 1 2 2
C1 2 3 0.3333
RDUMMY 2 3 1G
R2 3 0 2
.AC LIN 1 0.159155 0.159155
.PRINT AC VM(1) VP(1) IM(I1) IP(I1)
.PRINT AC VM(1,2) VP(1,2) IM(R1) IP(R1)
.PRINT AC VM(1,2) VP(1,2) IM(L1) IP(L1)
.PRINT AC VM(2,3) VP(2,3) IM(C1) IP(C1)
.PRINT AC VM(3) VP(3) IM(R2) IP(R2)
.END
```

SPICE output file

```
****     AC ANALYSIS                              TEMPERATURE =    27.000 DEG C
**********************************************************************************

  FREQ          VM(1)         VP(1)         IM(I1)        IP(I1)

   1.592E-01    6.680E+01   -2.893E+01    2.000E+01    0.000E+00
****     AC ANALYSIS                              TEMPERATURE =    27.000 DEG C
**********************************************************************************

  FREQ          VM(1,2)       VP(1,2)       IM(R1)        IP(R1)

   1.592E-01    3.328E+01    5.631E+01    1.109E+01    5.631E+01

****     AC ANALYSIS                              TEMPERATURE =    27.000 DEG C
**********************************************************************************

  FREQ          VM(1,2)       VP(1,2)       IM(L1)        IP(L1)

   1.592E-01    3.328E+01    5.631E+01    1.664E+01   -3.369E+01
****     AC ANALYSIS                              TEMPERATURE =    27.000 DEG C
**********************************************************************************

  FREQ          VM(2,3)       VP(2,3)       IM(C1)        IP(C1)

   1.592E-01    6.001E+01   -9.000E+01    2.000E+01    1.719E-07
****     AC ANALYSIS                              TEMPERATURE =    27.000 DEG C
**********************************************************************************

  FREQ          VM(3)         VP(3)         IM(R2)        IP(R2)

   1.592E-01    4.000E+01    3.229E-15    2.000E+01    3.229E-15
```

Thus, $$\mathbf{S}_{\text{Source}} = -\mathbf{V}_{\text{rms}}\mathbf{I}^*_{\text{rms}} = \frac{-(66.80\angle -28.93^\circ)(20.0\angle 0^\circ)}{2} = 668\angle 151.07^\circ$$

$$\mathbf{S}_{\text{R1}} = \mathbf{V}_{\text{rms}}\mathbf{I}^*_{\text{rms}} = \frac{(33.88\angle 56.31^\circ)(11.09\angle -56.31^\circ)}{2} = 187.86\angle 0^\circ$$

$$\mathbf{S}_{\text{L1}} = \mathbf{V}_{\text{rms}}\mathbf{I}^*_{\text{rms}} = \frac{(33.88\angle 56.31^\circ)(16.64\angle 33.69^\circ)}{2} = 281.88\angle 90^\circ$$

$$\mathbf{S}_{\text{R2}} = \mathbf{V}_{\text{rms}}\mathbf{I}^*_{\text{rms}} = \frac{(40.0\angle 0^\circ)(20.0\angle 0^\circ)}{2} = 400\angle 0^\circ$$

$$\mathbf{S}_{\text{C1}} = \mathbf{V}_{\text{rms}}\mathbf{I}^*_{\text{rms}} = \frac{(60.0\angle -90^\circ)(20.0\angle 0^\circ)}{2} = 600\angle -90^\circ$$

which agree with Prob. 10.26.

10.51 SPICE input file

```
PROBLEM 10.51
V1 1 0 AC 20 0
R1 1 2 3
L1 1 2 2
C1 2 3 0.3333
RDUMMY 2 3 1G
R2 3 0 2
.AC LIN 1 0.159155 0.159155
.PRINT AC VM(1) VP(1) IM(V1) IP(V1)
.PRINT AC VM(1,2) VP(1,2) IM(R1) IP(R1)
.PRINT AC VM(1,2) VP(1,2) IM(L1) IP(L1)
.PRINT AC VM(2,3) VP(2,3) IM(C1) IP(C1)
.PRINT AC VM(3) VP(3) IM(R2) IP(R2)
.END
```

SPICE output file

```
****      AC ANALYSIS                              TEMPERATURE =   27.000 DEG C
*******************************************************************************

  FREQ          VM(1)         VP(1)         IM(V1)        IP(V1)

   1.592E-01    2.000E+01     0.000E+00     5.988E+00    -1.511E+02

****      AC ANALYSIS                              TEMPERATURE =   27.000 DEG C
*******************************************************************************

  FREQ          VM(1,2)       VP(1,2)       IM(R1)        IP(R1)

   1.592E-01    9.965E+00     8.524E+01     3.322E+00     8.524E+01

****      AC ANALYSIS                              TEMPERATURE =   27.000 DEG C
*******************************************************************************

  FREQ          VM(1,2)       VP(1,2)       IM(L1)        IP(L1)

   1.592E-01    9.965E+00     8.524E+01     4.983E+00    -4.759E+00

****      AC ANALYSIS                              TEMPERATURE =   27.000 DEG C
*******************************************************************************

  FREQ          VM(2,3)       VP(2,3)       IM(C1)        IP(C1)

   1.592E-01    1.797E+01    -6.107E+01     5.988E+00     2.893E+01

****      AC ANALYSIS                              TEMPERATURE =   27.000 DEG C
*******************************************************************************

  FREQ          VM(3)         VP(3)         IM(R2)        IP(R2)

   1.592E-01    1.198E+01     2.893E+01     5.988E+00     2.893E+01
```

Thus, $\mathbf{S}_{\text{Source}} = -\mathbf{V}_{\text{rms}}\mathbf{I}^*_{\text{rms}} = \dfrac{(20.0\angle 0°)(5.988\angle 151.1°)}{2} = 59.88\angle 151.1°$

$$\mathbf{S}_{R1} = \mathbf{V}_{\text{rms}}\mathbf{I}^*_{\text{rms}} = \frac{(9.965\angle 85.24°)(3.322\angle -85.24°)}{2} = 16.55\angle 0°$$

$$\mathbf{S}_{L1} = \mathbf{V}_{\text{rms}}\mathbf{I}^*_{\text{rms}} = \frac{(9.965\angle 85.24°)(4.983\angle 4.759°)}{2} = 24.83\angle 90°$$

$$\mathbf{S}_{R2} = \mathbf{V}_{\text{rms}}\mathbf{I}^*_{\text{rms}} = \frac{(11.98\angle 28.93°)(5.988\angle -28.93°)}{2} = 35.87\angle 0°$$

$$\mathbf{S}_{C1} = \mathbf{V}_{\text{rms}}\mathbf{I}^*_{\text{rms}} = \frac{(17.97\angle -61.07°)(5.988\angle -28.93°)}{2} = 53.8\angle -90°$$

which agree with Prob. 10.27.

10.52 SPICE input file

```
PROBLEM 10.52
I1 0 1 AC 20 -45
L1 1 2 1
R1 2 0 10K
R2 2 3 5K
C1 3 0 1
.AC LIN 1 0.159155 0.159155
.PRINT AC VM(1) VP(1) IM(I1) IP(I1)
.PRINT AC VM(1,2) VP(1,2) IM(L1) IP(L1)
.PRINT AC VM(2) VP(2) IM(R1) IP(R1)
.PRINT AC VM(2,3) VP(2,3) IM(R2) IP(R2)
.PRINT AC VM(3) VP(3) IM(C1) IP(C1)
.END
```

SPICE output file

```
****      AC ANALYSIS                               TEMPERATURE =    27.000 DEG C
******************************************************************************

  FREQ          VM(1)        VP(1)        IM(I1)       IP(I1)

  1.592E-01   6.667E+04   -4.499E+01   2.000E+01   -4.500E+01

****      AC ANALYSIS                               TEMPERATURE =    27.000 DEG C
******************************************************************************

  FREQ          VM(1,2)      VP(1,2)      IM(L1)       IP(L1)

  1.592E-01   2.000E+01    4.500E+01   2.000E+01   -4.500E+01
```

```
****     AC ANALYSIS                              TEMPERATURE =   27.000 DEG C
******************************************************************************

 FREQ          VM(2)        VP(2)        IM(R1)       IP(R1)

  1.592E-01   6.667E+04   -4.501E+01   6.667E+00   -4.501E+01

****     AC ANALYSIS                              TEMPERATURE =   27.000 DEG C
******************************************************************************

 FREQ          VM(2,3)      VP(2,3)      IM(R2)       IP(R2)

  1.592E-01   6.667E+04   -4.500E+01   1.333E+01   -4.500E+01

****     AC ANALYSIS                              TEMPERATURE =   27.000 DEG C
******************************************************************************

 FREQ          VM(3)        VP(3)        IM(C1)       IP(C1)

  1.592E-01   1.333E+01   -1.350E+02   1.333E+01   -4.500E+01
```

Thus, $$\mathbf{S}_{\text{Source}} = -\mathbf{V}_{\text{rms}}\mathbf{I}^*_{\text{rms}} = \frac{-(6.667\times10^4\angle-44.99°)(20.0\angle45°)}{2} = 6.667\times10^5\angle179.99°$$

$$\mathbf{S}_{\mathbf{R1}} = \mathbf{V}_{\text{rms}}\mathbf{I}^*_{\text{rms}} = \frac{(6.667\times10^4\angle-45.01°)(6.667\angle45.01°)}{2} = 222244\angle0°$$

$$\mathbf{S}_{\mathbf{L1}} = \mathbf{V}_{\text{rms}}\mathbf{I}^*_{\text{rms}} = \frac{(20.0\angle45°)(20.0\angle45°)}{2} = 200\angle90°$$

$$\mathbf{S}_{\mathbf{R2}} = \mathbf{V}_{\text{rms}}\mathbf{I}^*_{\text{rms}} = \frac{(6.667\times10^4\angle-45°)(13.33\angle45°)}{2} = 44437\angle0°$$

$$\mathbf{S}_{\mathbf{C1}} = \mathbf{V}_{\text{rms}}\mathbf{I}^*_{\text{rms}} = \frac{(13.33\angle-135.0°)(13.33\angle45°)}{2} = 88.84\angle-90°$$

which agree with Prob. 10.28.

Chapter 11

Three-Phase Circuits

11.1 Single-Phase, Three-Wire Systems

11.1 If $\mathbf{V}_{an} = \mathbf{V}_{nb} = 120\angle 0°$ V rms, the impedance between terminals A-N is $\mathbf{Z}_{AN} = 8\angle 15°\ \Omega$, and the impedance between terminals N-B is $\mathbf{Z}_{NB} = 8\angle 0°\ \Omega$, find the neutral current $\mathbf{I}_{nN}$.

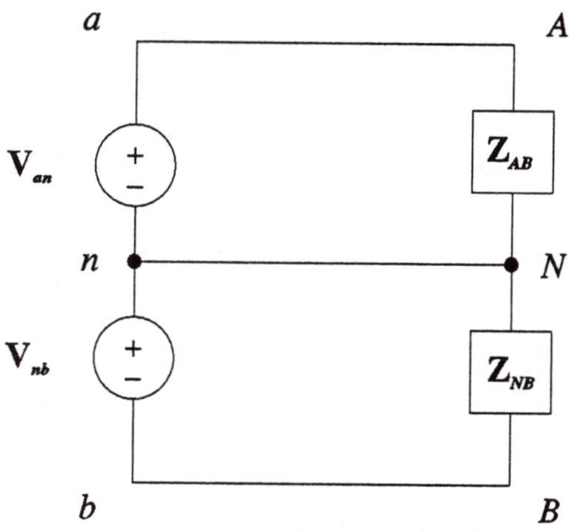

PROBLEM 11.1

11.2 Let $\mathbf{V}_1 = 15\angle 0°$ V rms, $\mathbf{Z}_1 = 2 + j1\ \Omega$, $\mathbf{Z}_2 = 5\ \Omega$, $\mathbf{Z}_3 = 2\ \Omega$, and $\mathbf{Z}_4 = 8\ \Omega$. Find the average power absorbed by the loads, lost in the lines, and delivered by the sources.

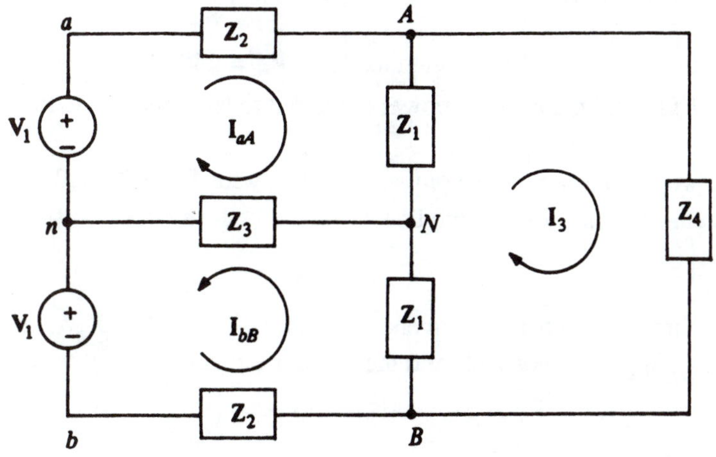

PROBLEM 11.2

11.2 Three-Phase Wye-Wye Systems

11.3 $\mathbf{V}_{ab} = 100\angle 30°$ V rms is a line voltage of a balanced Y-connected three-phase source. If the phase sequence is *abc*, find the phase voltages.

11.4 The source voltages are determined by Prob. 11.3 and the load in each phase is a series combination of a 10 Ω resistor and a 20 μF capacitor. The frequency for the sources is $\omega = 500$ rad/s. Find the line currents and the average power delivered to the load.

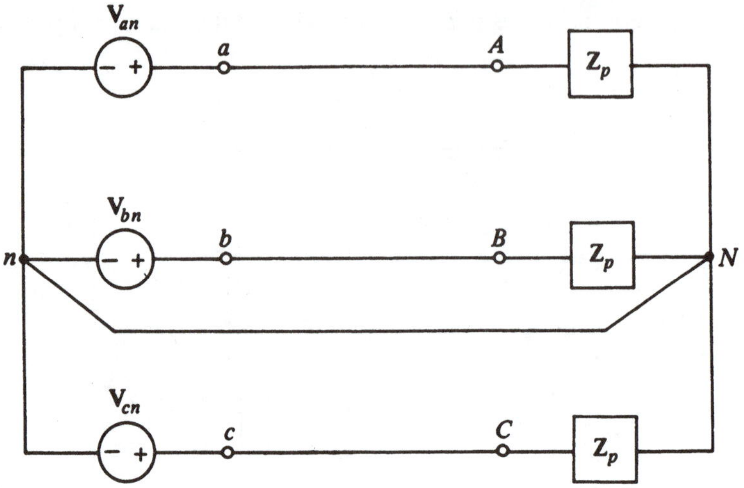

PROBLEM 11.4

11.5 In Prob. 11.4 the line currents form a balanced, positive sequence set with $\mathbf{I}_{aA} = 20\angle 0°$ A rms and $\mathbf{V}_{ab} = 60\angle 60°$ V rms. Find $\mathbf{Z}_p$ and the power delivered to the three-phase load.

11.6 A balanced Y-connected load is present on a 240 V rms (line-to-line) balanced three-phase system. If the phase impedance is $3\angle 60°$ Ω, find the total average power delivered to the load.

11.7 A balanced, positive-sequence Y-connected load has $\mathbf{V}_{ab} = 240\angle 0°$ V rms and $\mathbf{Z}_p = 8\sqrt{3}\angle -30°$ Ω. Find the average power delivered to the three-phase load.

11.8 A balanced Y-Y system has a positive sequence source with $\mathbf{V}_{an} = 200\angle 0°$ V rms and $\mathbf{Z}_p = 5\angle 60°$ Ω. Find the line voltage, the line current, and the average power delivered to the load.

11.9 A balanced Y-Y three-wire, positive sequence system has $\mathbf{V}_{an} = 80\angle 0°$ V rms and $\mathbf{Z}_p = 3 + j4$ Ω. If the lines have an impedance of 1 Ω, find the line current and the total power delivered to the load.

11.10 A balanced Y-connected, negative sequence source $\mathbf{V}_{an} = 100\angle 0°$ V rms is connected with perfect conductors to an unbalanced Y-connected load with $\mathbf{Z}_{AN} = 5$ Ω, $\mathbf{Z}_{BN} = j15$ Ω, and $\mathbf{Z}_{CN} = -j10$ Ω. Find the four line currents.

11.11 If $Z_1 = 3 + j3\ \Omega$, $Z_2 = 3 - j3\ \Omega$, and the line voltage is $V_L = 300$ V rms, find the current I_L in each line.

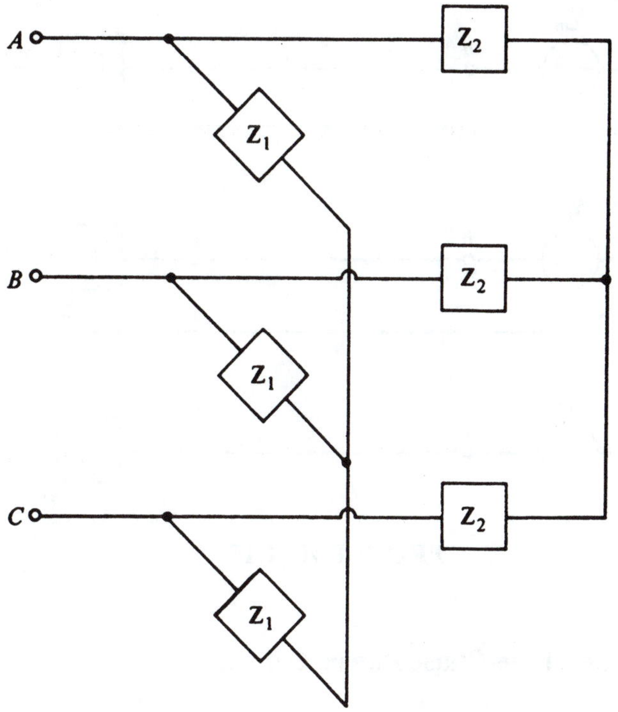

PROBLEM 11.11

11.12 A balanced three-phase Y-connected load draws 1 kW at a power factor of 0.707 leading. A balanced Y of capacitors is to be placed in parallel with the load so that the power factor for the combination is 1.0. If the frequency is 60 Hz and the line voltages are a balanced 200 V rms set, find the value of the required capacitors.

11.13 In Prob. 11.4, if the source is balanced with positive phase sequence, and $V_{an} = 200\angle 0°$ V rms, and if the source provides 9 kW at a $pf = 0.707$ lagging, find Z_p.

11.14 A balanced three-phase Y-connected load draws 6 kW at a power factor of 0.8 lagging. if the line voltages are a balanced 200 V rms set, find the line current I_L.

11.15 In the diagram, the line currents form a balanced, positive sequence set with $\mathbf{I}_{aA} = 20\angle{-30°}$ A rms and $\mathbf{V}_{ab} = 60\angle 30°$ V rms. Find $\mathbf{Z}_p$ and the power delivered to the three-phase load.

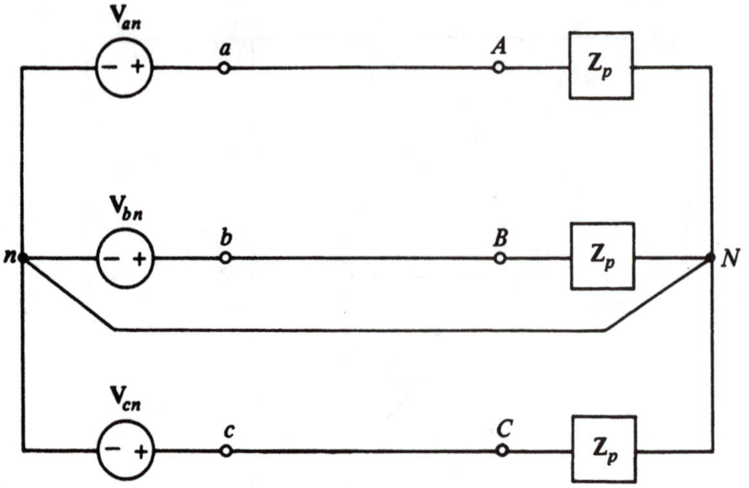

PROBLEM 11.15

11.3 Single-Phase Versus Three-Phase Power Delivery

11.16 In the figure, let V = 100 V rms, $\mathbf{Z}_L = 5\ \Omega$, and the three-phase load impedance be $2 + j10\ \Omega$. Determine (a) the single-phase load impedance $\mathbf{Z}_1$ for equal power from the single-phase and three-phase sources at equal power factors, and (b) the power delivered to the lines and loads.

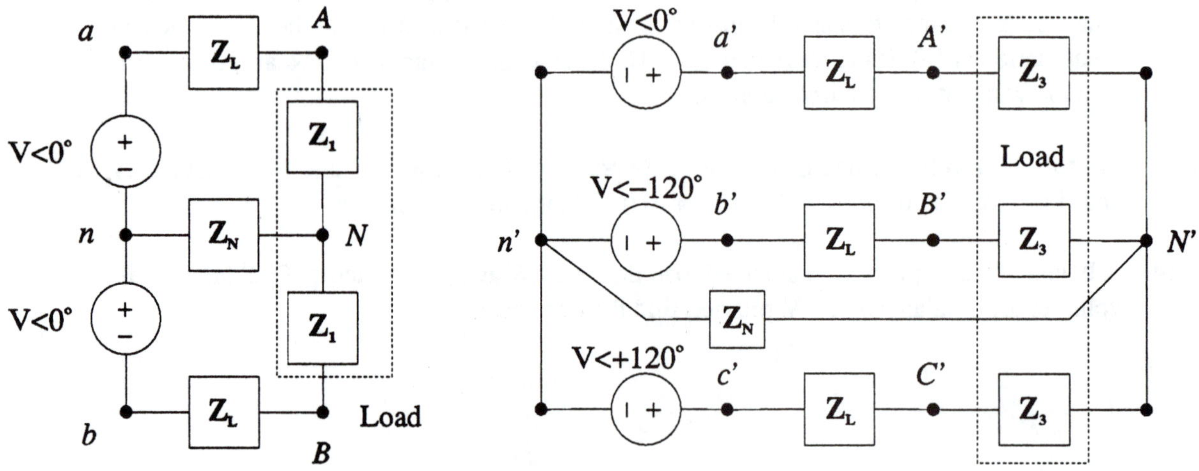

PROBLEM 11.16

11.17 (a) What is the *efficiency* of the single-phase system in Prob. 11.16, defined as power delivered to the load divided by total power produced by the source?
(b) What is the efficiency of the three-phase system in Prob. 11.16 using the same definition?
(c) Which system is more efficient?

11.18 Repeat Prob. 11.16 with $\mathbf{Z_L} = 0$ (perfectly conducting lines).

11.4 Delta Connection

11.19 Solve Prob. 11.4 if the source and load are unchanged except that the load is Δ-connected.

11.20 A balanced Δ-connected load has $\mathbf{Z_p} = 2 - j1\ \Omega$ and the line voltage is $\mathbf{V_L} = 100$ V rms at the load terminals. Find the total power delivered to the load.

11.21 A balanced Δ-connected load has a line voltage $\mathbf{V_L} = 100$ V rms at the load terminals and absorbs a total power of 4.8 kW. If the power factor of the load is 0.8 leading, find the phase impedance.

11.22 Repeat Prob. 11.13 if the load is a balanced Δ.

11.23 Repeat Prob. 11.6 if the load is Δ-connected and the phase impedance is $12\angle 60°\ \Omega$.

11.24 A balanced three-phase, positive-sequence source with $\mathbf{V_{ab}} = 200\angle 0°$ V rms is supplying a Δ-connected load, $\mathbf{Z_{AB}} = 10\angle 30°\ \Omega$, $\mathbf{Z_{BC}} = 10\angle 30°\ \Omega$, $\mathbf{Z_{CA}} = 20\angle 30°\ \Omega$. Find the line currents (assume lines are perfect conductors).

11.25 In the Y-Δ system shown, the source is positive sequence with $\mathbf{V_{an}} = 200\angle 0°$ V rms and the phase impedance is $\mathbf{Z_p} = 4 + j3\ \Omega$. Find the line voltage $\mathbf{V_L}$, the line current $\mathbf{I_L}$, and the power delivered to the load.

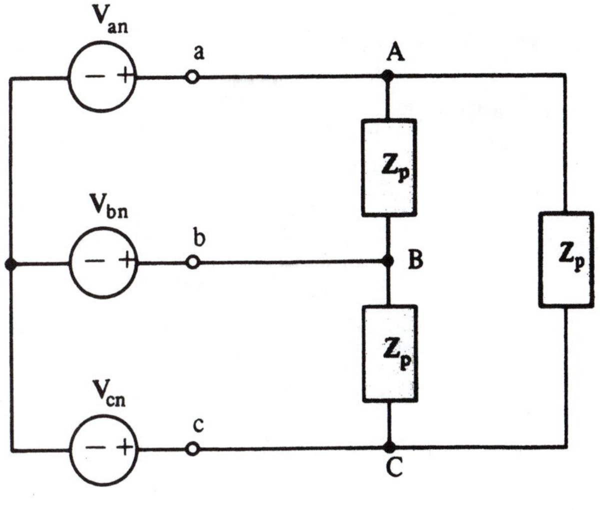

PROBLEM 11.25

11.26 A balanced three-phase, positive-sequence source with $\mathbf{V_{ab}} = 200\angle 0°$ V rms is supplying a Δ-connected load, $\mathbf{Z_{AB}} = 3 - j4\ \Omega$, $\mathbf{Z_{BC}} = 20\angle 60°\ \Omega$, $\mathbf{Z_{CA}} = 50\angle 30°\ \Omega$. Find the phasor line currents (assume lines are perfect conductors)

11.5 Wye-Delta Transformations

11.27 A balanced three-phase source with $\mathbf{V_L} = 100$ V rms is delivering power to a balanced Y-connected load with phase impedance $\mathbf{Z_1} = 5 + j12\ \Omega$ in parallel with a balanced Δ-connected load with a phase impedance $\mathbf{Z_2} = 15\ \Omega$. Find the power delivered by the source.

11.28 If the lines in Prob. 11.20 each have a resistance of 0.1 Ω, find the power lost in the lines.

11.29 Repeat Prob. 11.22 if each line contains an impedance of 1 Ω.

11.30 A balanced Y-Δ system with $\mathbf{V_{an}} = 100\angle 0°$ V rms, positive phase sequence, has $\mathbf{Z_p} = 6 - j12\ \Omega$ and an impedance of 1 Ω in the lines. Find the power delivered to the load.

11.31 A balanced three-phase positive-sequence source with $\mathbf{V_{ab}} = 100\angle 0°$ V rms is supplying a parallel combination of a Y-connected load and a Δ-connected load. If these Y and Δ loads are each balanced with phase impedances of $3 - j3\ \Omega$ and $9 + j9\ \Omega$, respectively, find the line current $\mathbf{I_L}$ and the power supplied by the source assuming perfectly conducting lines.

11.6 SPICE and Three-Phase Circuits

11.32 Solve Prob. 11.1 using SPICE.

11.33 Solve Prob. 11.7 using SPICE.

11.34 Solve Prob. 11.20 using SPICE.

11.35 Solve Prob. 11.24 using SPICE.

CHAPTER 11 SOLUTIONS

11.1 $\mathbf{I}_{AN} = \dfrac{120\angle 0^\circ}{8\angle 15^\circ} = 15\angle -15^\circ$ A rms

$\mathbf{I}_{NB} = \dfrac{120\angle 0^\circ}{8\angle 0^\circ} = 15\angle 0^\circ$ A rms.

$\mathbf{I}_{nN} = -(\mathbf{I}_{AN} + \mathbf{I}_{BN}) = -(15\angle -15^\circ - 15\angle 0^\circ)$

$= \underline{3.916\angle 82.5^\circ \text{ A rms.}}$

11.2 Since $\mathbf{I}_{aA} + \mathbf{I}_{bB} = 0$, the first mesh equation can be written, using $\mathbf{I}_{aA} = \mathbf{I}_1$,

$(\mathbf{Z}_1 + \mathbf{Z}_2)\mathbf{I}_1 - \mathbf{Z}_1\mathbf{I}_3 = \mathbf{V}_1$

The equation for the right mesh can be written

$-2\mathbf{Z}_1\mathbf{I}_1 + (2\mathbf{Z}_1 + \mathbf{Z}_4)\mathbf{I}_3 = 0$

Then $\mathbf{I}_1 = \dfrac{(2\mathbf{Z}_1 + \mathbf{Z}_4)\mathbf{V}_1}{(2\mathbf{Z}_1 + \mathbf{Z}_4)\mathbf{Z}_2 + \mathbf{Z}_1\mathbf{Z}_4}$

For $\mathbf{Z}_1 = 2 + j$, $\mathbf{Z}_2 = 5$, $\mathbf{Z}_3 = 2$ and $\mathbf{Z}_4 = 8$

$\mathbf{I}_1 = \mathbf{I}_{aA} = -\mathbf{I}_{bB} = \dfrac{(4 + 2j + 8)(15)}{(12 + j2)5 + 16 + j8}$

$= 2.336\angle -3.86^\circ$ A rms

$\mathbf{I}_3 = \dfrac{2\mathbf{Z}_1}{2\mathbf{Z}_1 + \mathbf{Z}_4}\mathbf{I}_1 = \dfrac{2(2+j)}{12 + j2}(2.34\angle -3.9^\circ)$

$= 0.429\angle 13.24$ A rms

$P_{Z4} = 8|\mathbf{I}_3|^2 = 1.475$ W

$P_{Z1} = \text{RE}(2+j)|\mathbf{I}_3 - \mathbf{I}_1|^2$

$\mathbf{I}_3 - \mathbf{I}_1 = (0.429\angle 13.24^\circ - 2.336\angle -3.86^\circ)$

$= 1.93\angle 172.4^\circ$ A rms

$P_{Z1} = 2(1.93)^2 = 7.451$ W,

$P_{aA} = 5|\mathbf{I}_{aA}|^2 = 27.30$ W,

$P_{bB} = 5|\mathbf{I}_{aA}|^2 = 27.30$ W

$P_{\text{Top source}} = |\mathbf{V}_1| \cdot |\mathbf{I}_{aA}| \cos[\text{ang}\,\mathbf{V}_1 - \text{ang}\,\mathbf{I}_{aA}]$

$= 15(2.34)\cos(3.86^\circ)$

$= 34.97$ W

11.2 (continued)

$P_{\text{BOTTOM source}} = |\mathbf{V}_1| \cdot |\mathbf{I}_{bB}| \cos[3.86^\circ]$

$= 15(2.34)\cos(3.86^\circ)$

$= 34.97$ W

$P_{LOAD} = 2P_{Z_1} + P_{Z_4} = \underline{16.38 \text{ W}}$

$P_{LOSS} = P_{aA} + P_{bB} = \underline{54.59 \text{ W}}$

$P_{delivered} = P_{LOAD} + P_{LOSS} = \underline{70.97 \text{ W}}$

11.3 $\mathbf{V}_{ab} = 100\angle 30^\circ$ V rms

$\mathbf{V}_{an} = \dfrac{100}{\sqrt{3}}\angle 30^\circ - 30^\circ = \underline{57.74\angle 0^\circ}$ V rms

$\mathbf{V}_{bn} = \underline{57.74\angle -120^\circ}$ V rms

$\mathbf{V}_{cn} = \underline{57.74\angle 120^\circ}$ V rms

11.4 $\mathbf{Z}_p = 10 - j\dfrac{1}{(500)(20\times 10^{-6})} = 100.5\angle -84.3^\circ\ \Omega$

$\mathbf{I}_{aA} = \dfrac{\mathbf{V}_{an}}{\mathbf{Z}_p} = \dfrac{57.74\angle 0^\circ}{100.5\angle -84.3^\circ} = \underline{0.574\angle 84.3^\circ \text{ A rms}}$

$\mathbf{I}_{bB} = 0.574\angle 84.3^\circ - 120^\circ = \underline{0.574\angle -35.71^\circ \text{ A rms}}$

$\mathbf{I}_{cC} = \underline{0.574\angle -155.7^\circ}$ A rms

$P = 3V_pI_p\cos\theta$

$= 3(57.74)(0.574)\cos(-84.3)^\circ$

$= \underline{9.90 \text{ W}}$

11.5 $\mathbf{V}_{aN} = \dfrac{60\angle 30^\circ}{\sqrt{3}}$ Vrms

$\mathbf{I}_{aA} = 20\angle 0^\circ$ A rms

$\mathbf{Z}_p = \mathbf{V}_{aN}/\mathbf{I}_{aA} = \dfrac{60}{20\sqrt{3}}\angle 30^\circ$

$= \underline{\sqrt{3}\angle 30^\circ\ \Omega}$

$P = 3V_pI_p\cos\theta = 3\left(\frac{60}{\sqrt{3}}\right)20\cos 30^\circ$

$= \underline{1{,}800 \text{ W}}$

11.6 $V_p = \dfrac{240}{\sqrt{3}} = 80\sqrt{3}$ Vrms

$I_p = \dfrac{V_p}{|\mathbf{Z}_p|} = \dfrac{80\sqrt{3}}{3} = \dfrac{80}{\sqrt{3}}$ Arms

$P = 3V_pI_p\cos\theta$

$= 3(80\sqrt{3})(80/\sqrt{3})\cos 60^\circ$

$= \underline{9.6 \text{ KW}}$

11.7 $V_p = \frac{V_L}{\sqrt{3}} = 80\sqrt{3}$ Vrms

$I_p = \frac{V_p}{|Z_p|} = \frac{80\sqrt{3}}{8\sqrt{3}} = 10$ A rms

$P = 3V_p I_p \cos\theta$

$= 3(80\sqrt{3})(10)\cos(-30°) = \underline{3.6\text{ KW}}$

11.8 $V_L = \sqrt{3}\,V_p = \underline{200\sqrt{3}\text{ V rms}}$

$I_p = I_L = \frac{V_p}{|Z_p|} = \frac{200}{5} = \underline{40\text{ A rms}}$

$P = 3V_pI_p\cos\theta = 3(200)(40)\cos 60°$

$= \underline{12\text{ KW}}$

11.9 $Z'_p = Z_p + 1 = 4 + j4 = 4\sqrt{2}\angle 45°\,\Omega$

$V_p = 80$ Vrms, $I_p = \frac{V_p}{|Z'_p|} = \frac{80}{4\sqrt{2}} = 10\sqrt{2}$ A rms

$P = 3(I_p^2)\,\text{Re}\,Z_p = 3(10\sqrt{2})^2(3)$

$= \underline{1.8\text{ KW}}$

11.10 $I_{AN} = \frac{V_{an}}{Z_{AN}} = \frac{100}{5} = \underline{20\text{ A rms}}$

$I_{BN} = \frac{V_{bn}}{Z_{BN}} = \frac{100\angle 120°}{j15} = \underline{6.67\angle 30°\text{ A rms}}$

$I_{CN} = \frac{V_{cn}}{Z_{CN}} = \frac{100\angle -120°}{-j10} = \underline{10\angle -30°\text{ A rms}}$

$I_{nN} = -(I_{aN} + I_{bN} + I_{cN})$

$= -(20 + 5.77 + j3.33 + 8.66 - j5)$

$= \underline{34.47\angle 177.2°\text{ A rms}}$

11.11 $Z_p = \frac{Z_1 Z_2}{Z_1 + Z_2} = \frac{(3+j3)(3-j3)}{3+j3+3-j3} = 3\Omega$

$V_p = \frac{V_L}{\sqrt{3}} = \frac{300}{\sqrt{3}} = 173.2$ V rms

$I_L = I_p = \frac{V_p}{|Z_p|} = \frac{173.2}{3} = \underline{57.74\text{ A rms}}$

11.12 $P_1 = \frac{1}{3}$ KW, $Q_1 = P_1\tan(\cos^{-1}0.707)$

$= P_1 = \frac{1}{3}$ KW

$Q_T = Q_1 + Q_2 = P_T\tan(\cos^{-1}1) = 0$

$Q_2 = -Q_1 = -\frac{1}{3}$ KVAR

$C = \frac{-Q_2}{2\pi f V_p^2} = \frac{\frac{1}{3}\times 10^3}{377(200/\sqrt{3})^2} = \underline{66.3\ \mu\text{F}}$

11.13 $P = 3V_pI_p\cos\theta = 3\frac{V_p^2}{|Z_p|}\cos\theta$

$|Z_p| = \frac{3V_p^2\cos\theta}{P} = \frac{3(200)^2(0.707)}{9000}$

$= 9.43\ \Omega$

$Z_p = 9.43\angle\cos^{-1}(0.707) = \underline{9.43\angle 45°\,\Omega}$

11.14 $I_p = I_L$, $V_p = V_L/\sqrt{3} = 200/\sqrt{3}$

$I_L = \frac{P}{3V_p\cos\theta} = \frac{6000}{3(\frac{200}{\sqrt{3}})(0.8)} = \underline{21.65\text{ A rms}}$

11.15 $V_{an} = 60/\sqrt{3}\angle 0°$ Vrms,

$I_{aA} = 20\angle -30°$ A rms,

$Z_p = V_{an}/I_{aA} = \left(\frac{60}{\sqrt{3}}/20\right)\angle{+30°} = \sqrt{3}\angle 30°\,\Omega$

$P = 3V_pI_p\cos\theta = 3(\frac{60}{\sqrt{3}})(20)\cos 30°$

$= \underline{1.8\text{ KW}}$.

11.16 In the three-phase circuit,

$I_{a'N'} = \frac{100\angle 0°}{7 + j10} = 8.19\angle -55°$ Arms

Power delivered to load $= 3|I_{a'N'}|^2\,\text{Re}\,Z_3$

$= 3(8.19)^2(2) = 402.5$ W

Power dissipated in lines $= P'_L$

$= 3|I_{a'N'}|^2\,\text{Re}\,Z_L = 3(8.19)^2(5)$

$= 1006.1$ W

In the single-phase circuit,

$P_L = \frac{3}{2}P'_L = 1509.2$ W

Power delivered to the load is equal

$\Rightarrow P_{TOT} = 1509.2 + 402.5$ W

$= 1911.7$ W

(a) (Single-phase circuit)

$2|I_{aN}|^2\,\text{Re}\,Z_1 = 402.5$ W

$|I_{aN}| = \frac{3}{2}|I_{a'N'}| = 12.285$ A rms

$\Rightarrow \text{Re}\,Z_1 = \frac{402.5}{301.8} = 1.33\,\Omega$

PF angle $\theta = -55°$

$\Rightarrow \text{Im}\,Z_1 = (\text{Re}\,Z_1)\tan(-55°) = -1.90$

$\underline{Z_1 = 1.33 - j1.90\ \Omega}$

(b) See above calculations.

11.17 (a) $\dfrac{402.5}{1911.7} = 0.21$

(b) $\dfrac{402.5}{1408.6} = 0.29$

11.18 (a) $\mathbf{I}_{a'N'} = \dfrac{100\angle 0^\circ}{2+j10} = 9.806\angle -78.69^\circ$ A rms

$\Rightarrow |\mathbf{I}_{aN}| = \frac{3}{2}|\mathbf{I}_{a'N'}| = 14.71$ A rms

Since $\mathbf{Z}_L = 0$, $\mathbf{Z}_1 = \mathbf{Z}_3$

(b) $P_{L'} = P_L = 0$ since $\mathbf{Z}_L = 0$

Power delivered to 3-phase load

$= 3|\mathbf{I}_{a'N'}|^2 \operatorname{Re}\mathbf{Z}_3 = 3(9.806)^2(2)$
$= 576.9$ W

The same power is delivered to the single phase load: 576.9 W

11.19 $\mathbf{V}_{ab} = 100\angle 30^\circ$ V rms

$\mathbf{Z}_p = 10 - j100\,\Omega = 100.5\angle -84.3^\circ\,\Omega$

$\mathbf{I}_{AB} = \dfrac{\mathbf{V}_{ab}}{\mathbf{Z}_p} = \dfrac{100\angle 30^\circ}{100.5\angle -84.3^\circ} = 0.995\angle 114.3^\circ$ A rms

$\mathbf{I}_{aA} = 0.995\sqrt{3}\angle 114.3^\circ - 30^\circ$
$= \underline{1.723\angle 84.3^\circ \text{ A rms}}$

$\mathbf{I}_{bB} = \underline{1.723\angle -35.7^\circ \text{ A rms}}$

$\mathbf{I}_{cC} = \underline{1.723\angle -155.7^\circ \text{ A rms}}$

11.20 $\mathbf{Z}_p = 2 - j\,\Omega = \sqrt{5}\angle -26.57^\circ\,\Omega$

$I_p = V_L/|\mathbf{Z}_p| = \dfrac{100}{\sqrt{5}}$ A rms

$P = 3(100)\left(\dfrac{100}{\sqrt{5}}\right)\cos(-26.57^\circ)$
$= \underline{12 \text{ kW}}$

11.21 $|\mathbf{Z}_p| = \dfrac{3V_p^2\cos\theta}{P} = \dfrac{3(100)^2(0.8)}{4800} = 5\,\Omega$

$\mathbf{Z}_p = 5\angle -\cos(0.8) = 5\angle -36.9^\circ\,\Omega$
$= \underline{4 - j3\,\Omega}$

11.22 $\mathbf{V}_{ab} = 200\sqrt{3}\angle 0^\circ$ V rms

$|\mathbf{Z}_p| = \dfrac{3V_p^2\cos\theta}{P} = \dfrac{3(200\sqrt{3})^2(0.707)}{9000}$
$= 28.28\,\Omega$

$\mathbf{Z}_p = \underline{28.28\angle 45^\circ\,\Omega}$

11.23 $V_p = 240$ V rms

$P = \dfrac{3V_p^2\cos\theta}{|\mathbf{Z}_p|} = \dfrac{3(240)^2\cos 60^\circ}{12}$
$= \underline{7.2 \text{ W}}$

11.24 $\mathbf{I}_{AB} = \dfrac{\mathbf{V}_{ab}}{\mathbf{Z}_{AB}} = \dfrac{200\angle 0^\circ}{10\angle 30^\circ} = 20\angle -30^\circ$ A rms

$\mathbf{I}_{BC} = \dfrac{\mathbf{V}_{bc}}{\mathbf{Z}_{BC}} = \dfrac{200\angle -120^\circ}{10\angle 30^\circ} = 20\angle -150^\circ$ A rms

$\mathbf{I}_{CA} = \dfrac{\mathbf{V}_{ca}}{\mathbf{Z}_{CA}} = \dfrac{200\angle 120^\circ}{20\angle 30^\circ} = 10\angle 90^\circ$ A rms

$\mathbf{I}_{aA} = \mathbf{I}_{AB} - \mathbf{I}_{CA} = (17.32 - j10) - (j10)$
$= 17.32 - j20 = \underline{26.46\angle -49.1^\circ \text{ A rms}}$

$\mathbf{I}_{bB} = \mathbf{I}_{BC} - \mathbf{I}_{AB} = (-17.32 - j10) - (17.32 - j10)$
$= \underline{-34.64 \text{ A rms}}$

$\mathbf{I}_{cC} = \mathbf{I}_{CA} - \mathbf{I}_{BC} = (j10) - (-17.32 - j10)$
$= 17.32 + j10 = \underline{26.46\angle 49.1^\circ \text{ A rms}}$

11.25 $V_L = \underline{200\sqrt{3} \text{ V rms}}$, $\mathbf{Z}_p = 5\angle 36.9^\circ\,\Omega$

$I_p = \dfrac{200\sqrt{3}}{5} = 40\sqrt{3}$ A rms

$I_L = \sqrt{3}\,I_p = \underline{120 \text{ A rms}}$

$P = 3(200\sqrt{3})(40\sqrt{3})\cos 36.9^\circ$
$= \underline{57.6 \text{ W}}$

11.26 $\mathbf{I}_{AB} = \dfrac{\mathbf{V}_{ab}}{\mathbf{Z}_{AB}} = \dfrac{200}{3-j4} = 20\angle 53.1^\circ$ A rms

$\mathbf{I}_{BC} = \dfrac{\mathbf{V}_{bc}}{\mathbf{Z}_{BC}} = \dfrac{200\angle -120^\circ}{20\angle 60^\circ} = -10$ A rms

$\mathbf{I}_{CA} = \dfrac{\mathbf{V}_{ca}}{\mathbf{Z}_{CA}} = \dfrac{200\angle 120^\circ}{50\angle 30^\circ} = j4$ A rms

$\mathbf{I}_{aA} = \mathbf{I}_{AB} - \mathbf{I}_{CA} = \underline{12 + j12 \text{ A rms}}$

$\mathbf{I}_{bB} = \mathbf{I}_{BC} - \mathbf{I}_{AB} = \underline{-22 - j16 \text{ A rms}}$

$\mathbf{I}_{cC} = \mathbf{I}_{CA} - \mathbf{I}_{BC} = \underline{10 + j4 \text{ A rms}}$

11.27 $\underline{Z}_{2Y} = \frac{1}{3}\underline{Z}_2 = \frac{1}{3}(15) = 5\,\Omega$

$$\underline{Z}_p = \frac{\underline{Z}_1 \underline{Z}_{2Y}}{\underline{Z}_1 + \underline{Z}_{2Y}} = \frac{(5+j12)(5)}{5+5+j12} = 4.16\angle 17.19^\circ\,\Omega$$

$$I_p = \frac{V_p}{|\underline{Z}_p|} = \frac{100/\sqrt{3}}{4.16} = 13.87 \text{ A rms}$$

$$P = \sqrt{3}\,V_L I_L \cos\theta = \sqrt{3}(100)(13.87)\cos 17.19^\circ$$

$$= \underline{2{,}296 \text{ W}}$$

11.28 $P_L = 3(|I|^2 R) = 3(I_p\sqrt{3})^2 R$

$$= 3\left(\frac{100}{\sqrt{3}}\right.$$

11.29 $\underline{Z}_{TY}$ = Total Impedance of Y load

$$= 9.43\angle 45^\circ = \frac{20}{3} + j\frac{20}{3}\,\Omega$$

from Prob. 13.13

$$\underline{Z}_{pY} = \underline{Z}_{TY} - 1 = \frac{17}{3} + j\frac{20}{3}\,\Omega$$

$$\underline{Z}_{p\Delta} = 3\underline{Z}_{pY} = 17 + j20\,\Omega$$

$$= \underline{26.25\angle 49.64^\circ}\,\Omega$$

11.30 $\underline{Z}_y = \frac{1}{3}\underline{Z}_p = \frac{6-j12}{3} = 2 - j4\,\Omega$

$$\underline{Z}'_y = \underline{Z}_y + 1 = 3 - j4 = 5\angle\underline{53}.13^\circ$$

$$I_p = I_L = \frac{V_p}{|\underline{Z}'_p|} = \frac{100}{5} = 20 \text{ A rms}$$

$$P_L = 3 I_p^2 \operatorname{Re}\underline{Z}_y = 3(20)^2 2 = \underline{2.4 \text{ kW}}$$

11.31 Δ-Y: $\underline{Z}_y = \frac{1}{3}\underline{Z}_d = \frac{1}{3}(9+j9) = 3+j3\,\Omega$

$$\underline{Z}_{eq} = \frac{(3-j3)(3+j3)}{3-j3+3+j3} = 3\,\Omega$$

$$I_L = \frac{V_p}{|\underline{Z}_{eq}|} = \frac{100/\sqrt{3}}{3} = 19.25 \text{ A rms}$$

$$P = 3 I_L^2 \operatorname{Re}\underline{Z}_{eq} = 3(19.25)^2(3) = \underline{\tfrac{10}{3} \text{ kW}}$$

11.32 SPICE input file

```
PROBLEM 11.32
VS1 1 0 AC 169.7056 0
VS2 0 2 AC 169.7056 0
RNB 4 2 8
* NOTE: CHOOSING F = 60 HZ ARBITRARILY (W = 377 RAD/S)
RAN 1 3 7.7274
LAN 3 4 5.49M
* DUMMY VOLTAGE SRC TO MEASURE CURRENT THROUGH (SPICE ONLY)
VDUMMY 0 4 0
.AC LIN 1 60 60
.PRINT AC IM(VDUMMY) IP(VDUMMY)
.END
```

SPICE output file

```
****     AC ANALYSIS                              TEMPERATURE =    27.000 DEG C
******************************************************************************

  FREQ          IM(VDUMMY)   IP(VDUMMY)

   6.000E+01    5.536E+00    8.251E+01
```

This agrees with Prob. 11.1 as follows:

$$\mathbf{I}_{aN} = \frac{\text{IM}}{\sqrt{2}}\angle\text{IP} = \frac{5.536}{\sqrt{2}}\angle 82.51° = 3.915\angle 82.51° \text{ A rms.}$$

11.33 SPICE input file

```
PROBLEM 11.33
* NOTE: SOURCES ARE NOT IN RMS VALUES
VAN 1 0 AC 138.56 0
VBN 2 0 AC 138.56 -120
VCN 3 0 AC 138.56 120
* NOTE: CHOOSING F = 60 HZ ARBITRARILY (W = 377 RAD/S)
* ALSO CHOOSING CAPACITORS IN PARALLEL AVOIDS FLOATING NODES
RAN 1 0 16.01
CAN 1 0 95.59U
RBN 2 0 16.01
CBN 2 0 95.59U
RCN 3 0 16.01
CCN 3 0 95.59U
.AC LIN 1 60 60
.PRINT AC IM(VAN) IP(VAN)
.END
```

SPICE output file

```
****     AC ANALYSIS                         TEMPERATURE =   27.000 DEG C
******************************************************************************

  FREQ          IM(VAN)       IP(VAN)

   6.000E+01    9.992E+00   -1.500E+02
```

This agrees with Prob. 11.7 as follows:

$$I_p = IM = 9.992 \approx 10 \text{ A rms}$$

$$\theta = 180° - IP = -30°$$

11.34 SPICE input file

```
PROBLEM 11.34
* NOTE: SOURCES (PHASE VOLTAGES) ARE NOT IN RMS VALUES
VAN 1 0 AC 100 0
VBN 2 0 AC 100 -120
VCN 3 0 AC 100 120
* NOTE: CHOOSING F = 60 HZ ARBITRARILY (W = 377 RAD/S)
* ALSO CHOOSING CAPS IN PARALLEL AVOIDS FLOATING NODES
RAB 1 2 2.5
CAB 1 2 0.5305M
RBC 2 3 2.5
CBC 2 3 0.5305M
RCA 3 1 2.5
CCA 3 1 0.5305M
.AC LIN 1 60 60
.PRINT AC IM(VAN) IP(VAN)
.END
```

SPICE output file

```
****     AC ANALYSIS                         TEMPERATURE =   27.000 DEG C
******************************************************************************

  FREQ          IM(VAN)       IP(VAN)

   6.000E+01    1.342E+02   -1.534E+02
```

This agress with Prob. 11.20 as follows:

$$I_p = \frac{IM}{3} = \frac{134.2}{3} = 44.73 \approx 44.72 \text{ A rms}$$

$$\theta = IP - 180° = -26.6°$$

11.35 SPICE input file

```
PROBLEM 11.35
* NOTE: SOURCES (PHASE VOLTAGES) ARE NOT IN RMS VALUES
VAN 1 0 AC 200 -30
VBN 2 0 AC 200 -150
VCN 3 0 AC 200 90
* NOTE: CHOOSING F = 60 HZ ARBITRARILY (W = 377 RAD/S)
RAB 1 4 8.66
LAB 4 2 0.013263
RBC 2 5 8.66
LBC 5 3 0.013263
RCA 3 6 17.32
LCA 6 1 0.026526
.AC LIN 1 60 60
.PRINT AC IM(RAB) IP(RAB) IM(RBC) IP(RBC) IM(RCA) IP(RCA)
.END
```

SPICE output file

```
****      AC ANALYSIS                         TEMPERATURE =    27.000 DEG C
******************************************************************************

FREQ        IM(RAB)     IP(RAB)      IM(RBC)     IP(RBC)      IM(RCA)     IP(RCA)

6.000E+01   3.464E+01   -3.000E+01   3.464E+01   -1.500E+02   1.732E+01   9.000E+01
```

This agrees with Prob. 11.24 as follows:

$$\mathbf{I}_{AB} = \frac{\text{IM(RAB)}}{\sqrt{3}}\angle\text{IP(RAB)} = 19.98\angle -30° \text{ A rms}$$

$$\mathbf{I}_{BC} = \frac{\text{IM(RBC)}}{\sqrt{3}}\angle\text{IP(RBC)} = 19.98\angle -150° \text{ A rms}$$

$$\mathbf{I}_{CA} = \frac{\text{IM(RCA)}}{\sqrt{3}}\angle\text{IP(RCA)} = 10.00\angle 90° \text{ A rms}$$

Chapter 12

The Laplace Transform

12.1 The s-Domain

12.1 Using the definition of the Laplace transform, find the transform **F**(s) of the following time functions:

(a) $f(t) = 3\cos(10t)\,u(t)$

(b) $f(t) = 4\sin(5t)\,u(t)$

(c) $f(t) = 4\sin(5t)$

(d) $f(t) = 2e^{-6t}\,u(t)$

(e) $f(t) = 10e^{-t}$

(f) $f(t) = \frac{1}{2}\,u(t)$

(g) $f(t) = \frac{1}{2}$

(h) $f(t) = \cosh 8t$

(i) $f(t) = \sinh 15t\,u(t)$

(j) $f(t) = (5-t)u(t) - (5-t)u(t-5)$

12.2 Find the region of convergence for which the Laplace transform exists for each of the functions in Prob. 12.1.

12.3 For the following time functions, find the Laplace transform if it exists. If it does not exist, explain why not.

(a) $f(t) = \dfrac{1}{t-4}$

(b) $f(t) = t^{3t}$

(c) $f(t) = \dfrac{1}{t}$

(d) $f(t) = t^2$

12.4 Find the Laplace transform of the following functions:

(a)

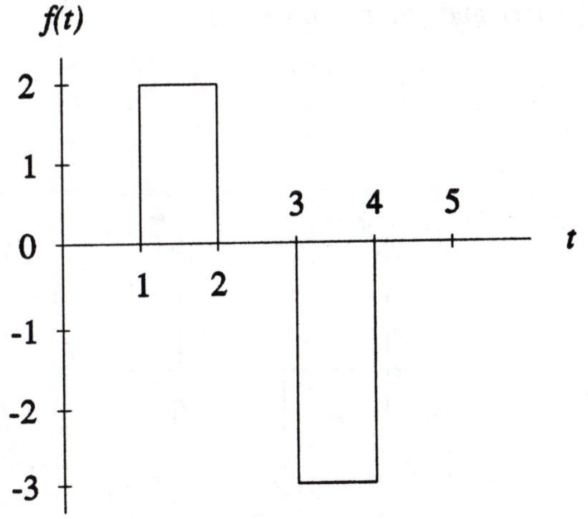

(b)

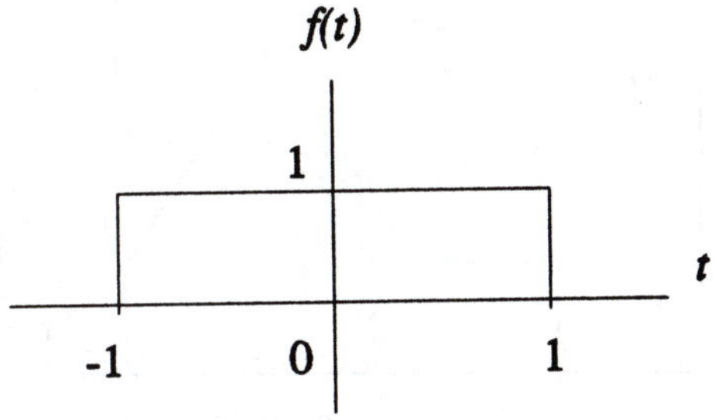

(c)

$$f(t) = \begin{cases} 3e^{-(t-1)}, & 1 \le t \le 1 + \ln(3/2) \\ 3e^{-(t-1)} - 1, & t \ge 1 + \ln(3/2) \end{cases}$$

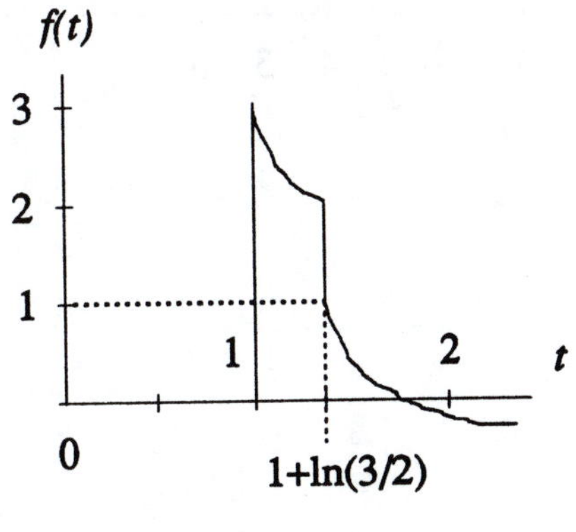

12.2 Singularity Functions

12.5 **Express the following functions as linear combinations of time-shifted impulses $\delta(t)$, unit steps $u(t)$, unit ramps $r(t)$, and unit parabolas $p(t)$:**

(a)

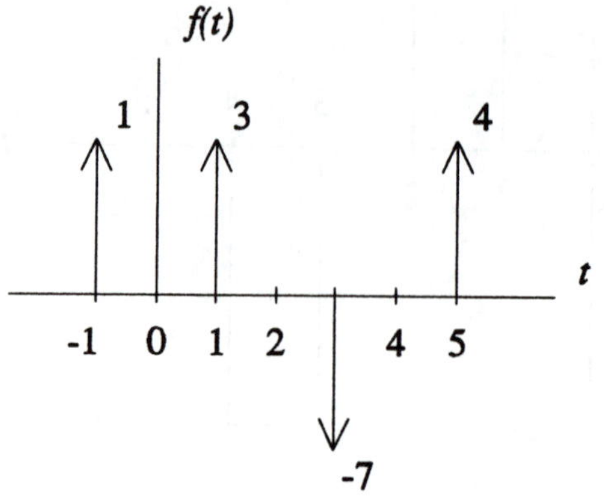

(b)

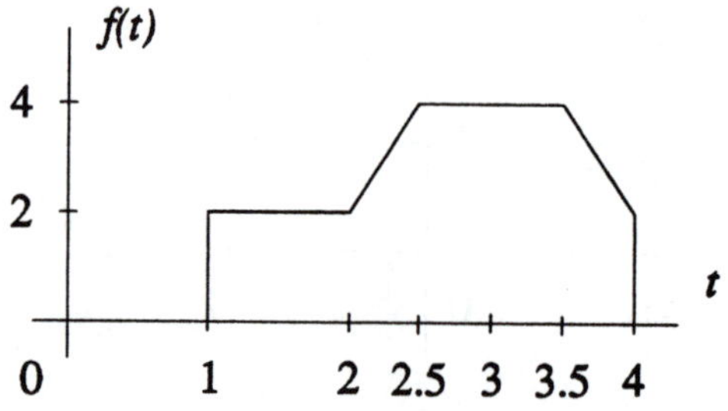

(c)

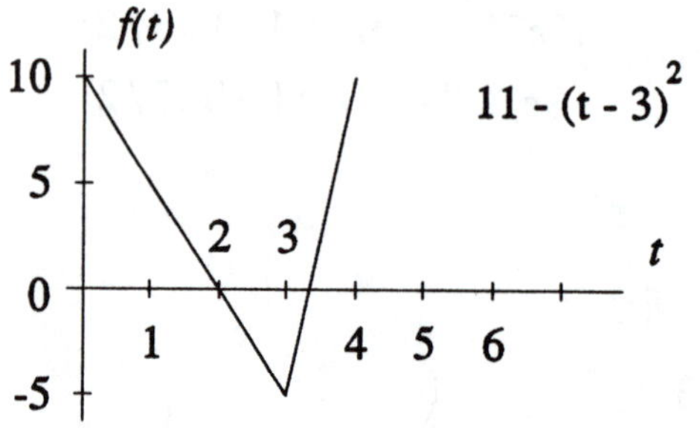

(d)

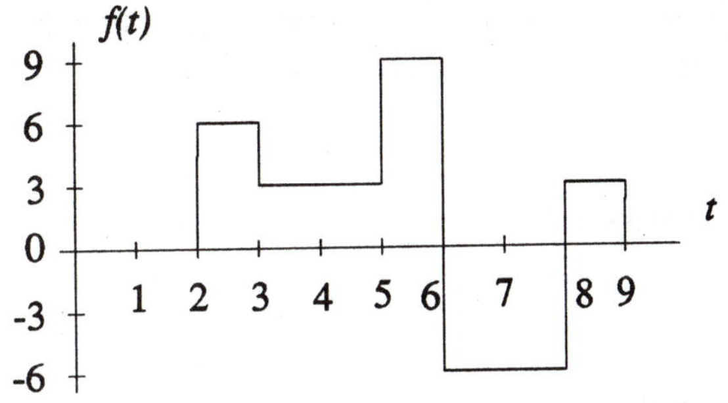

12.6 Find the Laplace transforms of the functions shown in Prob. 12.5.

12.7 Find the first and second derivatives of the functions shown in Prob. 12.5.

12.8 Sketch the following functions:

(a) $f(t) = 3p(t) - 3p(t-4)$
(b) $f(t) = r(t) + \frac{3}{2}r(t-3) + u(t-4) - 5r(t-5) + \frac{5}{2}r(t-6)$
(c) $f(t) = 3[u(t-3) + u(t-4) - 3u(t-5)]$
(d) $f(t) = u(t) + p(t-2) + 3\delta(t-4)$

12.9 Find the Laplace transforms of the functions in Prob. 12.8.

12.10 Find the first and second derivatives of the functions in Prob. 12.8.

12.11 Find $\int_{-\infty}^{t} f(t)\,dt$ for each of the functions in Prob. 12.8.

12.12 Find the Laplace transform of the first derivative of each of the functions in Prob. 12.8 (Hint: use the differentiation property).

12.13 Find the Laplace transform of the integral $\int_{0-}^{t} f(t)\,dt$ for each of the functions in Prob. 12.8

(Hint: use the integration property).

12.14 Evaluate the following expressions:

(a) $\int_{-\infty}^{\infty} \frac{5}{6} t^3 \delta(t-6)\, dt$

(b) $\int_{14}^{37} 13 \ln(t^{0.1t}) \delta(t-20)\, dt$

(c) $\int_{-\infty}^{\infty} \cos(3t) \frac{d\delta(t-6)}{dt}\, dt$

12.15 Evaluate $\int_{a}^{\infty} e^{-3t} \delta(t-2)\, dt$ for (a) $a = 3$ and (b) $a = 1$.

12.3 Other Transform Properties and Pairs

12.16 Find the Laplace transforms of the following:

(a) $f(t) = e^{-5t}\, u(t)$
(b) $f(t) = \sin 5(t-3)\, u(t-3)$
(c) $f(t) = p(3t) + tr(t) + (t-5)^2\, p(t-5)$
(d) $f(t) = 8te^{-5t}$
(e) $f(t) = t^3 e^{-3t}\, u(t)$

12.17 Find the inverse Laplace transforms:

(a) $\mathbf{F}(s) = \frac{s}{(s+2)^2 + 4}$

(b) $\mathbf{F}(s) = \frac{e^{-3s}}{s^2}$

(c) $\mathbf{F}(s) = \frac{4}{4s+8}$

(d) $\mathbf{F}(s) = -\frac{1}{(s+a)^3}$

12.18 Find $\mathbf{F}(s)$ if $f(t) = \sum_{n=0}^{\infty} \delta\,(t-n)$.

12.4 Partial Fraction Expansion

12.19 Find the inverse Laplace transform $f(t)$ of

$$\mathbf{F}(s) = \frac{5(s+5)}{s(s+3)(s+4)}$$

12.20 Find the inverse Laplace transform $f(t)$ of

$$\mathbf{F}(s) = \frac{11s^2 - 10s + 11}{(s^2+1)(s^2 - 2s + 5)}$$

12.21 Find the inverse Laplace transform $f(t)$ of

$$\mathbf{F}(s) = \frac{s^3 + 2s^2 + 4s + 5}{(s+1)^2(s+2)^2}$$

12.22 Find the inverse Laplace transform $f(t)$ of

$$\mathbf{F}(s) = \frac{2 - 2e^{-2s} - 4se^{-2s} + 2s^2 e^{-2s}}{s^3}$$

12.5 Solving Integrodifferential Equations

12.23 Use Laplace transforms to solve for $y(t)$ for $t > 0$

$$y'' + 2y' + y = 3te^{-t}$$
$$y(0) = 4, \quad y'(0) = 2$$

12.24 Use Laplace transforms to solve for $y(t)$ for $t > 0$

$$y'' - 4y' + 4y = 4\cos 2t$$
$$y(0) = 2, \quad y'(0) = 5$$

12.25 Use Laplace transforms to solve for $y(t)$ for $t > 0$

$$\int_0^t y(\alpha)\sin(t-\alpha)\,d\alpha = y(t) + \sin t - \cos t$$

12.26 Solve for $x(t)$ for $t > 0$

$$x'' + x = f(t)$$

$$f(t) = 1, \quad 0 \le t < \pi/2$$
$$= 0, \quad t > \pi/2$$

$$x(0) = 0, \quad x'(0) = 1$$

12.27 Use Laplace transforms to solve for $x(t)$ for $t > 0$

$$x' = t + \int_0^t x(t-\alpha)\cos\alpha \, d\alpha$$
$$x(0) = 4$$

12.28 Write a mesh equation for i_2 in the circuit and solve using Laplace transforms if $v_c(0-) = -2$ V and $i_2(0-) = 2$ A.

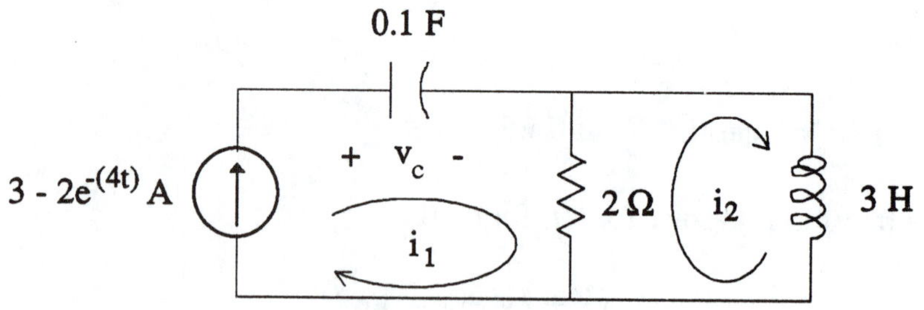

PROBLEM 12.28

12.29 Write and solve the nodal equation for $v(t)$ assuming all initial conditions are zero.

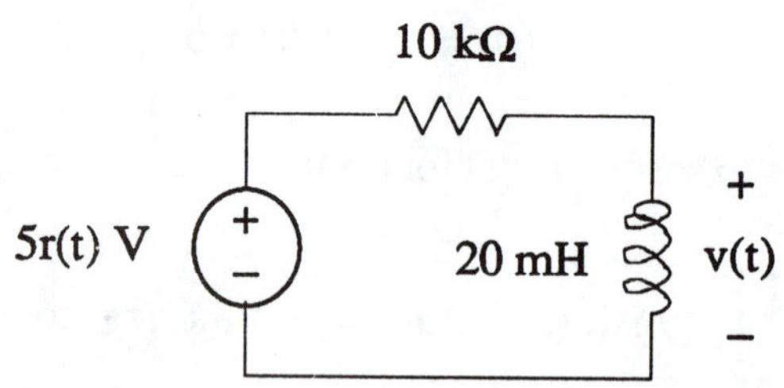

PROBLEM 12.29

CHAPTER 12 SOLUTIONS

12.1 (a) $\mathbf{F}(s)=\int_0^\infty 3(\cos 10t)e^{-st}dt$

$$=\frac{3e^{-st}}{s^2+100}[-s\cos 10t+10\sin 10t]_0^\infty$$

$$=0-\frac{-3s}{s^2+100}=\frac{3s}{s^2+100}$$

(b) $\mathbf{F}(s)=\int_0^\infty 4(\sin 5t)e^{-st}dt$

$$=\frac{4e^{-st}}{s^2+25}[-s\sin 5t-5\cos 5t]_0^\infty$$

$$=0-\frac{-20}{s^2+25}=\frac{20}{s^2+25}$$

(c) Same answer as part (b)

(d) $\mathbf{F}(s)=\int_0^\infty 2e^{-6t}e^{-st}dt$

$$=\left.\frac{2e^{-(s+6)t}}{-(s+6)}\right]_0^\infty=\frac{2}{s+6}$$

(e) $\mathbf{F}(s)=\int_0^\infty 10e^{-t}e^{-st}dt$

$$=\int_0^\infty 10e^{-(s+1)t}dt=\left.\frac{10e^{-(s+1)t}}{-(s+1)}\right]_0^\infty$$

$$=\frac{10}{s+1}$$

(f) $\mathbf{F}(s)=\int_0^\infty \tfrac{1}{2}u(t)e^{-st}dt=\int_0^\infty \tfrac{1}{2}e^{-st}dt$

$$=\left.\frac{\tfrac{1}{2}e^{-st}}{-s}\right]_0^\infty=\frac{1}{2s}$$

(g) Same answer as part (f)

(h) $\mathbf{F}(s)=\int_0^\infty \cosh(8t)e^{-st}dt$

$$=\int_0^\infty \tfrac{1}{2}\left(e^{8t}+e^{-8t}\right)e^{-st}dt$$

$$=\tfrac{1}{2}\int_0^\infty e^{-(s-8)t}dt+\tfrac{1}{2}\int_0^\infty e^{-(s+8)t}dt$$

$$=\frac{1}{2}\left[\frac{e^{-(s-8)t}}{-(s-8)}\right]_0^\infty+\frac{1}{2}\left[\frac{e^{-(s+8)}}{-(s+8)}\right]_0^\infty$$

$$=\frac{1}{2}\left(\frac{1}{s-8}+\frac{1}{s+8}\right)=\frac{s}{s^2-64}$$

(i) $\mathbf{F}(s)=\int_0^\infty \sinh(15t)e^{-st}dt$

$$=\int_0^\infty \tfrac{1}{2}\left(e^{15t}-e^{-15t}\right)e^{-st}dt$$

$$=\tfrac{1}{2}\int_0^\infty e^{-(s-15)t}dt-\tfrac{1}{2}\int_0^\infty e^{-(s+15)t}dt$$

$$=\frac{1}{2}\left[\frac{e^{-(s-15)t}}{-(s-15)}\right]_0^\infty-\frac{1}{2}\left[\frac{e^{-(s+15)}}{-(s+15)}\right]_0^\infty$$

$$=\frac{1}{2}\left(\frac{1}{s-15}-\frac{1}{s+15}\right)=\frac{15}{s^2-225}$$

(j) $\mathbf{F}(s)=\int_0^\infty (5-t)[u(t)-u(t-5)]\,e^{-st}dt$

$$=\int_0^5 (5-t)e^{-st}dt=\left[\frac{e^{-st}}{s^2}+(t-5)\frac{e^{-st}}{s}\right]_0^5$$

$$=\frac{e^{-5s}-1}{s^2}+\frac{5}{s}$$

12.2 (a) $\text{Re}\{s\} > 0$
(b) $\text{Re}\{s\} > 0$
(c) $\text{Re}\{s\} > 0$
(d) $\text{Re}\{s\} > -6$
(e) $\text{Re}\{s\} > -1$
(f) $\text{Re}\{s\} > 0$
(g) $\text{Re}\{s\} > 0$
(h) $\text{Re}\{s\} > 8$
(i) $\text{Re}\{s\} > 15$
(j) $\text{Re}\{s\} > 0$

12.3 (a) **F**(s) does not exist since *f(t)* has an infinite discontinuity at $t = 4$. Thus, the Laplace integral cannot be computed.

(b) **F**(s) does not exist because *f(t)* grows too fast.

(c) **F**(s) does not exist since *f(t)* has an infinite discontinuity at $t = 0$. Thus, the Laplace integral cannot be computed.

(d) **F**(s) does exist: t^2 has the same Laplace transform as $2\,p(t)$.

$$\mathbf{F}(s) = \frac{2}{s^3}$$

12.4 (a) $f(t) = 2[u(t-1) - u(t-2)] - 3[u(t-3) - u(t-4)]$

$$\mathbf{F}(s) = \int_0^\infty f(t)e^{-st}dt = \int_1^2 2e^{-st}dt - \int_3^4 3e^{-st}dt$$

$$= \left[\frac{2e^{-st}}{-s}\right]_1^2 - \left[\frac{3e^{-st}}{-s}\right]_3^4$$

$$= \frac{2e^{-s}}{s} - \frac{2e^{-2s}}{s} + \frac{3e^{-4s}}{s} - \frac{3e^{-3s}}{s}$$

(b) $f(t) = u(t+1) - u(t-1)$

$$\mathbf{F}(s) = \int_0^1 1e^{-st}dt = \left[\frac{e^{-st}}{-s}\right]_0^1 = \frac{1}{s} - \frac{e^{-s}}{s}$$

(c) $f(t) = 3e^{-(t-1)}u(t-1) - u(t - [1 + \ln\frac{3}{2}])$

$$\mathbf{F}(s) = \int_1^\infty 3e^{-(s+t-1)}dt - \int_{1.4055}^\infty 1e^{-st}dt$$

$$= \frac{3e^{-s}}{s+1} - \frac{e^{-1.4055s}}{s}$$

12.5 (a) $f(t) = \delta(t+1) + 3\delta(t-1) - 7\delta(t-3) + 4\delta(t-5)$

(b) $f(t) = 2u(t-1) + 4r(t-2) - 4r(t-2.5) - 4r(t-3.5) + 4r(t-4) - 2u(t-4)$

(c)
$$f(t) = 10u(t) - 5r(t) + 5r(t-3) + 15r(t-3) + \left[11 - (t-3)^2\right]u(t-4)$$
$$= 10u(t) - 5r(t) + 20r(t-3) + 2u(t-4) + \left[6t - t^2\right]u(t-4)$$

(d) $f(t) = 6u(t-2) - 3u(t-3) + 6u(t-5) - 15u(t-6) + 9u(t-8) - 3u(t-9)$

12.6 (a) $\mathbf{F}(s) = 0 + 3e^{-s} - 7e^{-3s} + 4e^{-5s}$

(b)

$$\mathbf{F}(s) = \frac{2e^{-s}}{s} + \frac{4e^{-2s}}{s^2} - \frac{4e^{-2.5s}}{s^2} - \frac{4e^{-3.5s}}{s^2} + \frac{4e^{-4s}}{s^2} - \frac{2e^{-4s}}{s}$$

(c)

$$\mathbf{F}(s) = \frac{10}{s} - \frac{5}{s^2} + \frac{20e^{-3s}}{s^2} + \frac{2e^{-4s}}{s} - \int_4^\infty t^2 e^{-st}\,dt + 6\int_4^\infty t e^{-st}\,dt$$

Using integration by parts,

$$= \frac{10}{s} - \frac{5}{s^2} + \frac{20e^{-3s}}{s^2} + \frac{2e^{-4s}}{s} - \left[\frac{t^2 e^{-st}}{-s} + \frac{2}{s}\left[\frac{te^{-st}}{-s} - \frac{e^{-st}}{s^2}\right]\right]_4^\infty + 6\left[\frac{te^{-st}}{-s} - \frac{e^{-st}}{s^2}\right]_4^\infty$$

$$= \frac{10}{s} - \frac{5}{s^2} + \frac{20e^{-3s}}{s^2} + \frac{2e^{-4s}}{s} - \frac{16e^{-4s}}{s} + \frac{8e^{-4s}}{s^2} - \frac{2e^{-4s}}{s^3} + \frac{24e^{-4s}}{s} + \frac{6e^{-4s}}{s^2}$$

$$= \frac{10 + 10e^{-4s}}{s} + \frac{20e^{-3s} + 14e^{-4s} - 5}{s^2} - \frac{2e^{-4s}}{s^3}$$

(d)

$$\mathbf{F}(s) = \frac{6e^{-2s} - 3e^{-3s} + 6e^{-5s} - 15e^{-6s} + 9e^{-8s} - 3e^{-9s}}{s}$$

12.7 (a)

$$\frac{df}{dt} = \delta^{(1)}(t+1) + 3\delta^{(1)}(t-1) - 7\delta^{(1)}(t-3) + 4\delta^{(1)}(t-5)$$

$$\frac{d^2 f}{dt^2} = \delta^{(2)}(t+1) + 3\delta^{(2)}(t-1) - 7\delta^{(2)}(t-3) + 4\delta^{(2)}(t-5)$$

(b)

$$\frac{df}{dt} = 2\delta(t-1) + 4u(t-2) - 4u(t-2.5) - 4u(t-3.5) + 4u(t-4) - 2\delta(t-4)$$

$$\frac{d^2 f}{dt^2} = 2\delta^{(1)}(t-1) + 4\delta(t-2) - 4\delta(t-2.5) - 4\delta(t-3.5) + 4\delta(t-4) - 2\delta^{(1)}(t-4)$$

(c)

$$\frac{df}{dt} = 10\delta(t) - 5u(t) + 20u(t-3) + 2\delta(t-4) - 2t\,u(t-4) - t^2\delta(t-4) + 6u(t-4) + 6t\,\delta(t-4)$$

$$= 10\delta(t) - 5u(t) + 20u(t-3) + 10\delta(t-4) + (6-2t)u(t-4)$$

$$\frac{d^2 f}{dt^2} = 10\delta^{(1)}(t) - 5\delta(t) + 20\delta(t-3)$$
$$+10\delta^{(1)}(t-4) + (6-2t)\delta(t-4)$$
$$-2u(t-4)$$

(d)

$$\frac{df}{dt} = 6\delta(t-2) - 3\delta(t-3) + 6\delta(t-5)$$
$$-15\delta(t-6) + 9\delta(t-8) - 3\delta(t-9)$$

$$\frac{d^2 f}{dt^2} = 6\delta^{(1)}(t-2) - 3\delta^{(1)}(t-3) + 6\delta^{(1)}(t-5)$$
$$-15\delta^{(1)}(t-6) + 9\delta^{(1)}(t-8) - 3\delta^{(1)}(t-9)$$

12.9 (a) $\mathbf{F}(s) = \frac{3}{s^3} - \frac{3e^{-4s}}{s^3}$

(b)

$$\mathbf{F}(s) = \frac{1}{s^2} + \frac{3}{2}\frac{e^{-3s}}{s^2} + \frac{e^{-4s}}{s}$$
$$-\frac{5e^{-5s}}{s^2} + \frac{5}{2}\frac{e^{-6s}}{s^2}$$

(c) $\mathbf{F}(s) = 3\mathbf{L}[u(t-3) + u(t-4) - 3u(t-5)]$

$$= \frac{3e^{-3s} + 3e^{-4s} - 9e^{-5s}}{s}$$

(d) $\mathbf{F}(s) = \frac{1}{s} + \frac{e^{-2s}}{s^3} + 3e^{-4s}$

12.10 (a) $\frac{df}{dt} = 3r(t) - 3r(t-4)$

$$\frac{d^2 f}{dt^2} = 3u(t) - 3u(t-4)$$

(b)

$$\frac{df}{dt} = u(t) + \tfrac{3}{2}u(t-3) + \delta(t-4) - 5u(t-5) + \tfrac{5}{2}u(t-6)$$

$$\frac{d^2 f}{dt^2} = \delta(t) + \tfrac{3}{2}\delta(t-3) + \delta^{(1)}(t-4) - 5\delta(t-5) + \tfrac{5}{2}\delta(t-6)$$

(c) $\frac{df}{dt} = 3[\delta(t-3) + \delta(t-4) - 3\delta(t-5)]$

$$\frac{d^2 f}{dt^2} = 3[\delta^{(1)}(t-3) + \delta^{(1)}(t-4) - 3\delta^{(1)}(t-5)]$$

(d) $\frac{df}{dt} = \delta(t) + r(t-2) - 3\delta^{(1)}(t-4)$

$$\frac{d^2 f}{dt^2} = \delta^{(1)}(t) + u(t-2) - 3\delta^{(2)}(t-4)$$

12.11 Let $c(t)$ be defined as the unit cubic function, $c(t) \equiv \frac{1}{6}t^3 u(t)$.

(a) $3c(t) - 3c(t-4)$

(b)

$p(t) + \frac{3}{2}p(t-3) + r(t-4) - 5p(t-5) + \frac{5}{2}p(t-6)$

(c) $3[r(t-3) + r(t-4) - 3r(t-5)]$

(d) $r(t) + c(t-2) + 3u(t-6)$

12.12 Use the differentiation property:

$$\mathbf{L}\left\{\frac{df}{dt}\right\} = s\mathbf{F}(s) - f(0-)$$

(a)

$$s\mathbf{F}(s) - 0 = s\left[\frac{3}{s^3} - \frac{3e^{-4s}}{s^3}\right]$$

$$= \frac{3 - 3e^{-4s}}{s^2}$$

(b)

$$s\mathbf{F}(s) - 0 = s\left[\frac{1}{s^2} + \frac{3}{2}\frac{e^{-3s}}{s^2} + \frac{e^{-4s}}{s} - \frac{5e^{-5s}}{s^2} + \frac{5}{2}\frac{e^{-6s}}{s^2}\right]$$

$$= \frac{1}{s} + \frac{3}{2}\frac{e^{-3s}}{s} + e^{-4s} - \frac{5e^{-5s}}{s} + \frac{5}{2}\frac{e^{-6s}}{s}$$

(c) $s\mathbf{F}(s) - 0 = 3\left[e^{-3s} + e^{-4s} - 3e^{-5s}\right]$

(d) $s\mathbf{F}(s) - 0 = 1 + \dfrac{e^{-2s}}{s^2} + 3s\,e^{-4s}$

12.13 Use the integration property:

$$\mathbf{L}\left\{\int_{0-}^{t} f(\alpha)d\alpha\right\} = \frac{\mathbf{F}(s)}{s}$$

(a) $\dfrac{3 - 3e^{-4s}}{s^4}$

(b) $\dfrac{1}{s^3} + \dfrac{3}{2}\dfrac{e^{-3s}}{s^3} + \dfrac{e^{-4s}}{s^2} - \dfrac{5e^{-5s}}{s^3} + \dfrac{5}{2}\dfrac{e^{-6s}}{s^3}$

(c) $3\left[\dfrac{e^{-3s}}{s^2} + \dfrac{e^{-4s}}{s^2} - \dfrac{3e^{-5s}}{s^2}\right]$

(d) $\dfrac{1}{s^2} + \dfrac{e^{-2s}}{s^4} + \dfrac{3e^{-4s}}{s}$

12.14 Use the sifting property of $\delta(t)$:

(a) $\left[\frac{5}{6}t^3\right]_{t=6} = 180$

(b) $\left[13\ln\left(t^{[t/10]}\right)\right]_{t=20} = 13\ln\left(20^2\right) = 77.89$

(c) $\left[-\dfrac{d}{dt}\cos 3t\right]_{t=-1} = \left[3\sin 3t\right]_{t=-1} = -0.4234$

12.15 (a) 0

(b) $e^{-3(2)} = e^{-6}$

12.16 (a) $\dfrac{1}{s+5}$

(b) $\dfrac{5e^{-3s}}{s^2+25}$

(c)

$$\frac{1}{3}\frac{27}{s^3} - \frac{d}{ds}\left[\frac{1}{s^2}\right] + e^{-5s}\left[\frac{d^2}{ds^2}\frac{1}{s^3}\right]$$

$$= \frac{11}{s^3} + \frac{12e^{-5s}}{s^5}$$

(d) $-8\dfrac{d}{ds}\left[\dfrac{1}{s+5}\right] = \dfrac{8}{(s+5)^2}$

(e) $(-1)^3\dfrac{d^3}{ds^3}\left[\dfrac{1}{s+3}\right] = \dfrac{6}{(s+3)^4}$

12.17

$$f(t) = \mathbf{L}^{-1}\left\{\frac{s}{(s+2)^2+4}\right\}$$

(a) $$= \mathbf{L}^{-1}\left\{\frac{s+2}{(s+2)^2+2^2} - \frac{2}{(s+2)^2+2^2}\right\}$$

$$= e^{-2t}[\cos 2t - \sin 2t]\,u(t)$$

$$f(t) = \mathbf{L}^{-1}\left\{\frac{e^{-3s}}{s^2}\right\}$$

(b) $$\mathbf{L}^{-1}\left\{\frac{1}{s^2}\right\} = tu(t)$$

$$\therefore\ f(t) = (t-3)u(t-3)$$

$$\mathbf{L}^{-1}\{\mathbf{F}(as)\} = \tfrac{1}{a} f(\tfrac{t}{a})$$

(c) $$\therefore\ \mathbf{L}^{-1}\left\{\frac{4}{4s+8}\right\} = \frac{1}{4}(4e^{-8t/4})\,u(t)$$

$$= e^{-2t}u(t)$$

(d) Using the frequency shift property,

$$f(t) = -e^{-at}p(t)$$

12.18

$$\mathbf{F}(s) = \int_0^\infty e^{-st}\left[\sum_{n=0}^{\infty}\delta(t-n)\right]dt$$

$$= \sum_{n=0}^{\infty}\int_0^\infty e^{-st}\,\delta(t-n)\,dt$$

$$= \sum_{n=0}^{\infty} e^{-sn}$$

12.19

$$F(s) = \frac{5(s+5)}{s(s+3)(s+4)} = \frac{A}{s} + \frac{B}{s+3} + \frac{C}{s+4}$$

$$A = s\,F(s)\big|_{s=0} = \frac{5(0+5)}{(0+3)(0+4)} = \frac{25}{12}$$

$$B = (s+3)\,F(s)\big|_{s=-3} = \frac{5(5-3)}{(-3)(4-3)} = -\frac{10}{3}$$

$$C = (s+4)F(s)\big|_{s=-4} = \frac{5(5-4)}{(-4)(3-4)} = \frac{5}{4}$$

Thus, $$F(s) = \frac{25}{12}\left(\frac{1}{s}\right) - \frac{10}{3}\left(\frac{1}{s+3}\right) + \frac{5}{4}\left(\frac{1}{s+4}\right)$$

$$\Rightarrow f(t) = \frac{25}{12}u(t) - \frac{10}{3}e^{-3t}u(t) + \frac{5}{4}e^{-4t}u(t)$$

12.20

$$F(s) = \frac{11s^2 - 10s + 11}{(s^2+1)(s^2-2s+5)}$$

$$= \frac{As+B}{s^2+1} + \frac{Cs+D}{s^2-2s+5}$$

$$11s^2 - 10s + 11 = (As+B)(s^2-2s+5) + (Cs+D)(s^2+1)$$

s^3: $0 = A + C$

s^2: $11 = -2A + B + D$

s : $-10 = 5A - 2B + C$

s^0: $11 = 5B + D$

which yields

$A = -2,\ B = 1,\ C = 2,\ D = 6$

$$F(s) = \frac{-2s+1}{s^2+1} + \frac{2s+6}{(s-1)^2+4}$$

$$= \frac{-2s}{s^2+1} + \frac{1}{s^2+1} + \frac{2(s-1)}{(s-1)^2+4} + \frac{8}{(s-1)^2+4}$$

$$f(t) = -2\cos t + \sin t + 2e^t(\cos 2t + 2\sin 2t)$$

12.21

$$\frac{s^3+2s^2+4s+5}{(s+1)^2(s+2)^2}=\frac{A}{(s+1)^2}+\frac{B}{s+1}+\frac{C}{(s+2)^2}+\frac{D}{s+2}$$

$$A=\left.\frac{s^3+2s^2+4s+5}{(s+2)^2}\right|_{s=-1}=2$$

$$C=\left.\frac{s^3+2s^2+4s+5}{(s+1)^2}\right|_{s=-2}=-3$$

$$s^3+2s^2+4s+5=2(s+2)^2+B(s+1)(s^2+4s+4)$$
$$-3(s+1)^2+D(s+2)(s^2+2s+1)$$

s^3: $1=B+D$

s^0: $5=8+4B-3+2D$

which yields $B=-1, D=2$

$$F(s)=\frac{2}{(s+1)^2}-\frac{1}{s+1}-\frac{3}{(s+2)^2}+\frac{2}{s+2}$$

$$f(t)=\underline{(2t-1)e^{-t}+(-3t+2)e^{-2t}}$$

12.22

$$F(s)=\frac{2}{s^3}-\frac{2e^{-2s}}{s^3}-\frac{4e^{-2s}}{s^2}+\frac{2e^{-2s}}{s}$$

$$f(t)=t^2u(t)-(t-2)^2u(t-2)$$
$$-4(t-2)u(t-2)+2u(t-2)$$

or $f(t)=t^2, \quad 0<t<2$

$\underline{=6, \quad t>2}$

12.23 $s^2Y(s)-4s-2+2[sY(s)-4]+Y(s)$

$$=\frac{3}{(s+1)^2};$$

$$Y(s)=\frac{4s+10+\frac{3}{(s+1)^2}}{s^2+2s+1}$$

$$=\frac{4s+10}{(s+1)^2}+\frac{3}{(s+1)^4}$$

$$=\frac{4(s+1)+6}{(s+1)^2}+\frac{3}{(s+1)^4}$$

$$=\frac{4}{s+1}+\frac{6}{(s+1)^2}+\frac{3}{(s+1)^4}$$

$$y(t)=\underline{\left(4e^{-t}+6te^{-t}+\tfrac{1}{2}t^3e^{-t}\right)u(t)}$$

12.24

$$s^2Y(s)-2s-5-4[sY(s)-2]+4Y(s)$$

$$=\frac{4s}{s^2+4}$$

$$Y(s)=\frac{2s-3+\frac{4s}{s^2+4}}{s^2-4s+4}$$

$$=\frac{(2s-3)(s^2+4)+4s}{(s-2)^2(s^2+4)}=\frac{A}{(s-2)^2}+\frac{B}{s-2}+\frac{Cs+D}{s^2+4}$$

$$A=\left.\frac{(2s-3)(s^2+4)+4s}{s^2+4}\right|_{s=2}=2$$

$$(2s-3)(s^2+4)+4s=2(s^2+4)+B(s-2)(s^2+4)$$
$$+(Cs+D)(s^2-4s+4)$$

s^3: $2=B+C$

s^2: $-3=2-2B-4C+D$

s^0: $-12=8-8B+4D$

which yields

$B=2, \ C=0, \ D=-1$

$$Y(s)=\frac{2}{(s-2)^2}+\frac{2}{s-2}-\frac{1}{s^2+4}$$

$$y(t)=\underline{\left[2(t+1)e^{2t}-\tfrac{1}{2}\sin 2t\right]u(t)}$$

12.25

$$\mathcal{L}\left[\int_0^t y(\tau)\sin(t-\tau)d\tau\right]$$

$$=Y(s)+\frac{1-s}{s^2+1}$$

$$Y(s)\frac{1}{s^2+1}=Y(s)+\frac{1-s}{s^2+1}$$

$$Y(s)[1-(s^2+1)]=1-s$$

$$Y(s)=\frac{s-1}{s^2}=\frac{1}{s}-\frac{1}{s^2}$$

$$y(t)=\underline{(1-t)u(t)}$$

12.26 $x'' + x = u(t) - u(t - \pi/2)$

$$s^2 X(s) - 1 + X(s) = \frac{1}{s} - \frac{1}{s} e^{-\pi s/2}$$

$$X(s) = \frac{\frac{1}{s} + 1}{s^2+1} - \frac{\frac{1}{s} e^{-s\pi/2}}{s^2+1}$$

$$= \frac{s+1}{s(s^2+1)} - \frac{1}{s(s^2+1)} e^{-s\pi/2}$$

$$\frac{1}{s(s^2+1)} = \frac{A}{s} + \frac{Bs+C}{s^2+1}$$

$$A = \left.\frac{1}{s^2+1}\right|_{s=0} = 1$$

$$1 = s^2 + 1 + (Bs+c)s$$

s^2: $0 = 1 + B$, $B = -1$

s: $0 = c$

$$X(s) = \frac{1}{s^2+1} + \frac{1}{s(s^2+1)} - \frac{1}{s(s^2+1)} e^{-s\pi/2}$$

$$= \frac{1}{s^2+1} + \frac{1}{s} - \frac{s}{s^2+1} - \left[\frac{1}{s} - \frac{s}{s^2+1}\right] e^{-s\pi/2}$$

$$x(t) = (\sin t + 1 - \cos t) u(t) - \left[1 - \cos\left(t - \frac{\pi}{2}\right)\right] u\left(t - \frac{\pi}{2}\right)$$

$$x(t) = (\sin t + 1 - \cos t) u(t) - (1 - \sin t) u\left(t - \frac{\pi}{2}\right)$$

12.27

$$sX(s) - 4 = \frac{1}{s^2} + \frac{s}{s^2+1} X(s)$$

$$X(s)\left[s - \frac{s}{s^2+1}\right] = \frac{1}{s^2} + 4$$

$$X(s) = \frac{\left(\frac{1}{s^2} + 4\right)(s^2+1)}{s(s^2+1) - s}$$

$$= \frac{(4s^2+1)(s^2+1)}{s^5}$$

$$= \frac{4}{s} + \frac{5}{s^3} + \frac{1}{s^5}$$

$$x(t) = \left(4 + \frac{5}{2} t^2 + \frac{1}{24} t^4\right) u(t)$$

12.28 $i_1 = 3 - 2e^{-4t} \Rightarrow I_1(s) = \frac{3}{s} - \frac{2}{s+4}$

$$3\frac{di_2}{dt} + 2i_2 = 2i_1 \Rightarrow 3[sI_2 - i_2(0-)] + 2I_2 = 2I_1$$

$$\Rightarrow 3sI_2(s) + 2I_2(s) = 6 + \frac{6}{s} - \frac{4}{s+4}$$

$$\Rightarrow I_2(s) = \frac{6}{3s+2} + \frac{6}{s(3s+2)} - \frac{4}{(3s+2)(s+4)}$$

Partial fraction expansion:

$$\frac{6}{s(3s+2)} = \frac{A}{s} + \frac{B}{3s+2}$$

$$A = \frac{6}{2} = 3 \qquad B = 6/(-2/3) = -9$$

$$\frac{-4}{(3s+2)(s+4)} = \frac{C}{3s+2} + \frac{D}{s+4}$$

$$C = \frac{-4}{\left(-\frac{2}{3} + 4\right)} = -\frac{6}{5} \qquad D = \frac{-4}{3(-4)+2} = \frac{2}{5}$$

$$\therefore I_2(s) = \frac{6}{3s+2} + \frac{3}{s} - \frac{9}{3s+2} - \frac{6}{5}\left(\frac{1}{3s+2}\right) + \frac{2}{5}\left(\frac{1}{s+4}\right)$$

$$= \frac{3}{s} + \frac{2}{5}\left(\frac{1}{s+4}\right) - \frac{21}{5}\left(\frac{1}{3s+2}\right)$$

$$i_2(t) = 3u(t) + \frac{2}{5} e^{-4t} u(t) - \frac{21}{5}\left(\frac{1}{3} e^{-\frac{1}{3}(2t)}\right) u(t)$$

12.29

$$\frac{5r(t) - v(t)}{10000} = \frac{1}{20\times10^{-3}} \int_0^t v(\tau)\,d\tau$$

$$\text{Transform} \Rightarrow \frac{1}{10000}\left[\frac{5}{s^2} - V(s)\right] = 50\frac{V(s)}{s}$$

$$\Rightarrow V(s) = \frac{5}{s(s+500000)}$$

$$= \frac{A}{s} + \frac{B}{s+500000}$$

$$A = \frac{5}{500000} = 1\times10^{-5}$$

$$B = \frac{5}{-500000} = -1\times10^{-5}$$

$$\therefore v(t) = (1\times10^{-5}) u(t) - (1\times10^{-5}) e^{-500000t} u(t)$$

Chapter 13

Circuit Analysis in the s-Domain

13.1 Element and Kirchhoff's Laws

13.1 Find the equivalent Laplace impedance Z(s) looking into terminals *ab*:

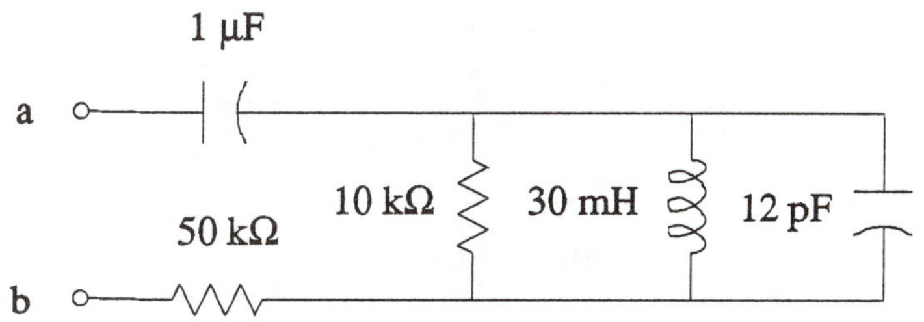

PROBLEM 13.1

13.2 Find the equivalent Laplace admittance Y(s) looking into terminals *ab*:

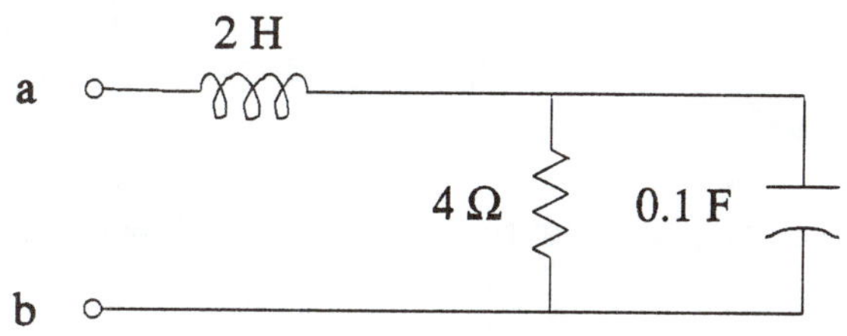

PROBLEM 13.2

13.3 Find *v(t)* using Laplace transforms assuming all initial conditions at $t = 0-$ are zero.

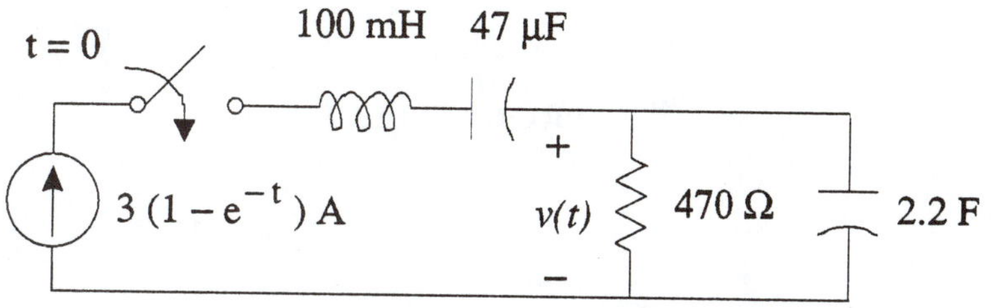

PROBLEM 13.3

13.4 Find $i(t)$ using Laplace transforms assuming all initial conditions at $t = 0-$ are zero.

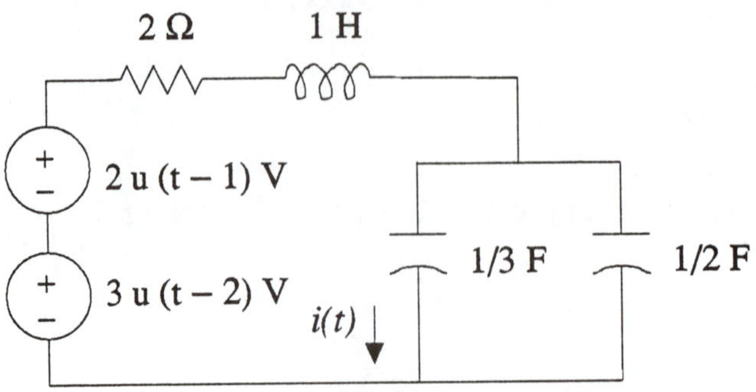

PROBLEM 13.4

13.5 Find the Thevenin and Norton equivalents in the s-domain at the terminals of the parallel resistor and capacitor in Prob. 13.3 looking into the left part of the circuit shown (across the inductor, capacitor, and source in series) for $t > 0$.

13.6 Find the Thevenin and Norton equivalents in the s-domain at the terminals of the 1/2 F capacitor (looking into the remaining part of the circuit).

13.2 The s-Domain Circuit

13.7 Draw the s-domain circuit and use it to find i for $t > 0$ if $i(0) = 4$ A and $v(0) = 6$ V.

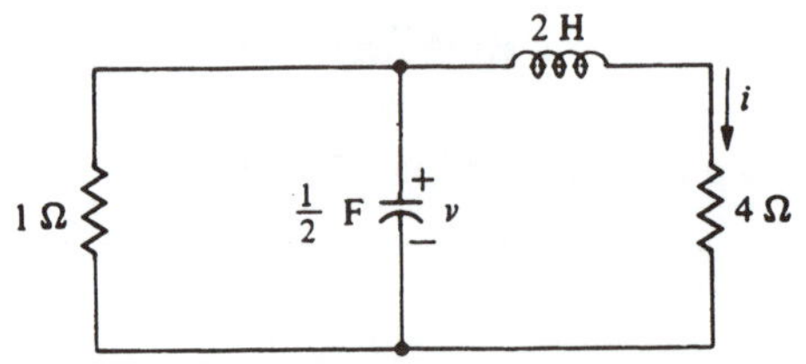

PROBLEM 13.7

13.8 Draw the s-domain circuit and use it to find i for $t > 0$ if $L = 3$ H and the circuit is in steady-state at $t = 0-$.

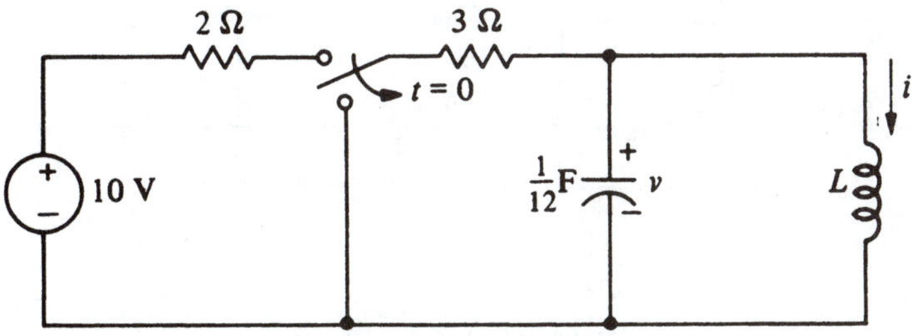

PROBLEM 13.8

13.9 Draw the s-domain circuit and use it to find i for $t > 0$ if $L = 3$ H and the circuit is in steady-state at $t = 0-$.

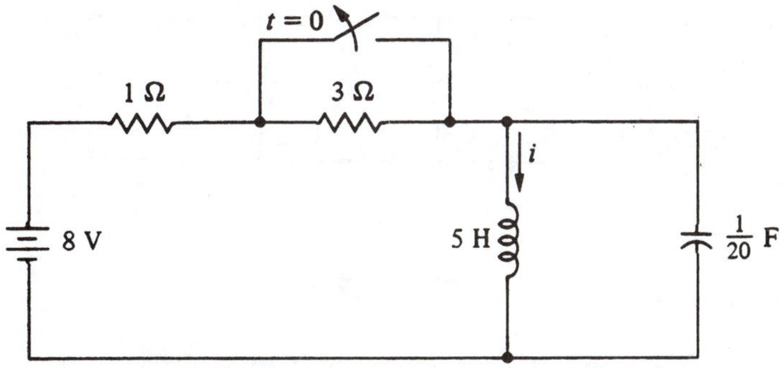

PROBLEM 13.9

13.10 Draw the s-domain circuit and use it to find v for $t > 0$ if $i_g = 2\,u(t)$ A.

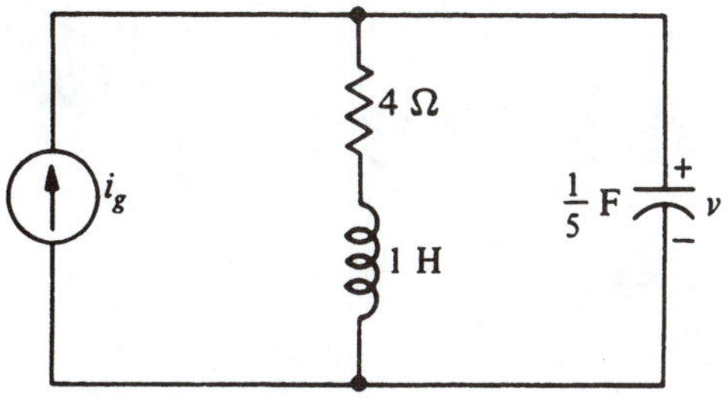

PROBLEM 13.10

13.11 Draw the s-domain circuit and use it to find $i(t)$ for $t > 0$ if $v(0) = 4$ V and $i(0) = 10$ A.

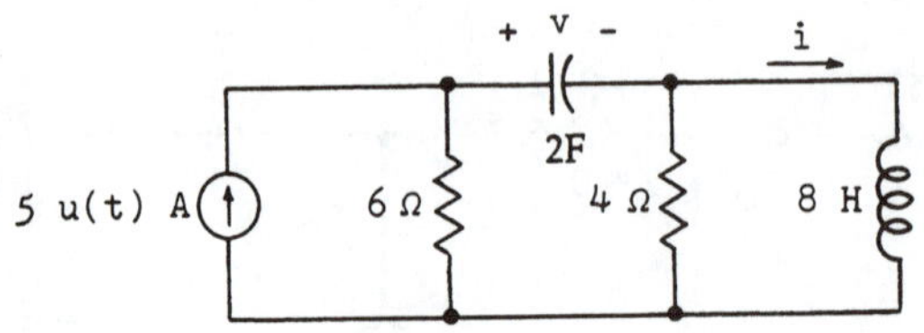

PROBLEM 13.11

13.12 Draw the s-domain circuit and use it to find $i(t)$ for $t > 0$ if $v(0) = 4$ V and $i(0) = 2$ A.

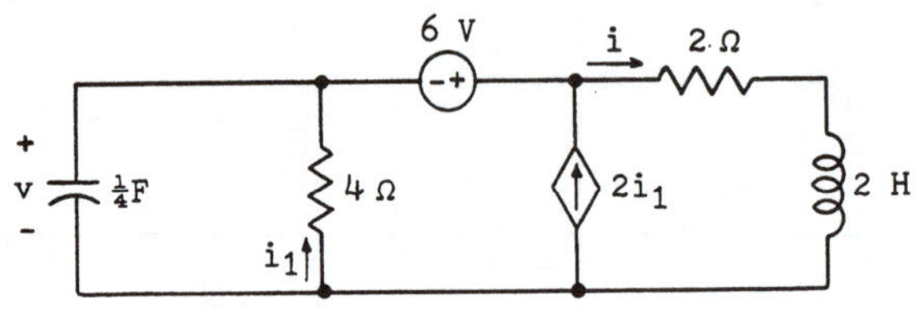

PROBLEM 13.12

13.13 Draw the s-domain circuit and use it to find v if $v_1(0) = 4$ V and $v_2(0) = 2$ V.

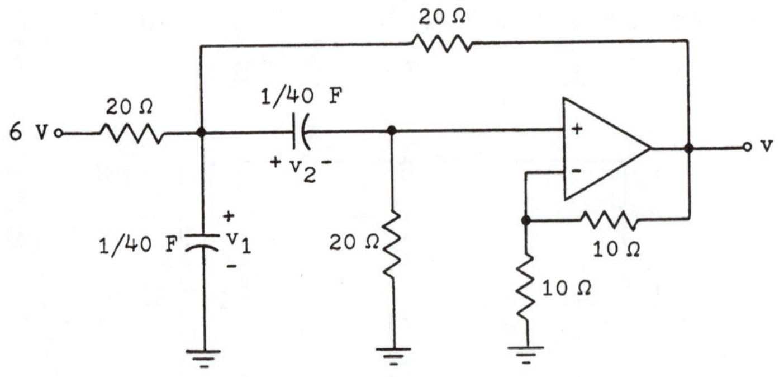

PROBLEM 13.13

13.3 Transfer Functions

13.14 Find **H**(s) if $v_o(t)$ is the output and $v_i(t)$ is the input, and $R = 20\ \Omega$, $L = 2$ H, $C = 1/80$ F.

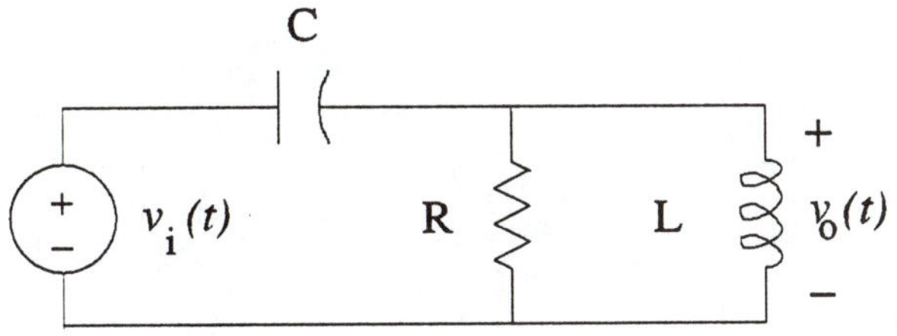

PROBLEM 13.14

13.15 For $R = 7\ \Omega$, $L = 8$ H, $C = 1/56$ F, find **H**(s) in Prob. 13.14.

13.16 Find $\mathbf{H_1}(s) = \mathbf{V_o}(s) / \mathbf{I_1}(s)$ and $\mathbf{H_2}(s) = \mathbf{V_o}(s) / \mathbf{V_1}(s)$.

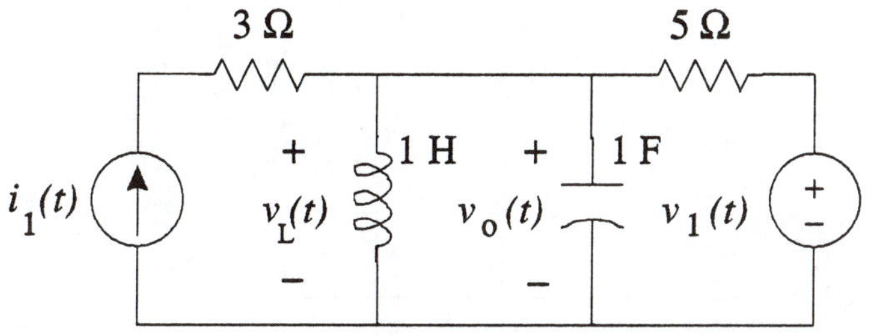

PROBLEM 13.16

13.17 (a) Find the transfer function between output $\mathbf{V}_3(s)$ and input $\mathbf{V}_1(s)$
(b) Find the transfer function between output $\mathbf{V}_3(s)$ and input $\mathbf{V}_2(s)$
(c) Find the transfer function between the output $\mathbf{V}_3(s)$ and input $\mathbf{I}_1(s)$

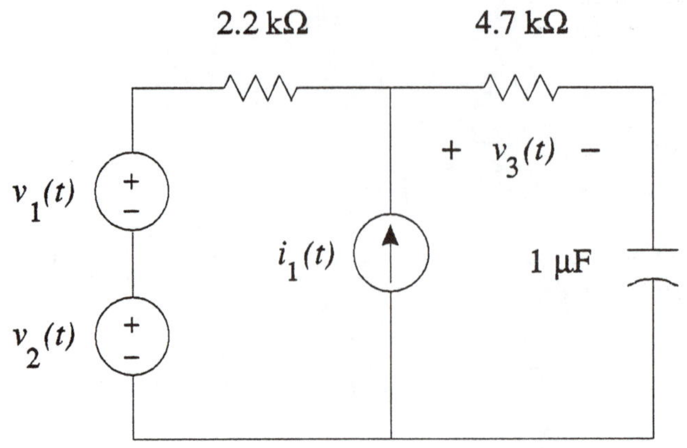

PROBLEM 13.17

13.18 Find (a) $\mathbf{H}_1(s) = \mathbf{V}_o(s) / \mathbf{V}_1(s)$ and (b) $\mathbf{H}_2(s) = \mathbf{V}_o(s) / \mathbf{V}_2(s)$.

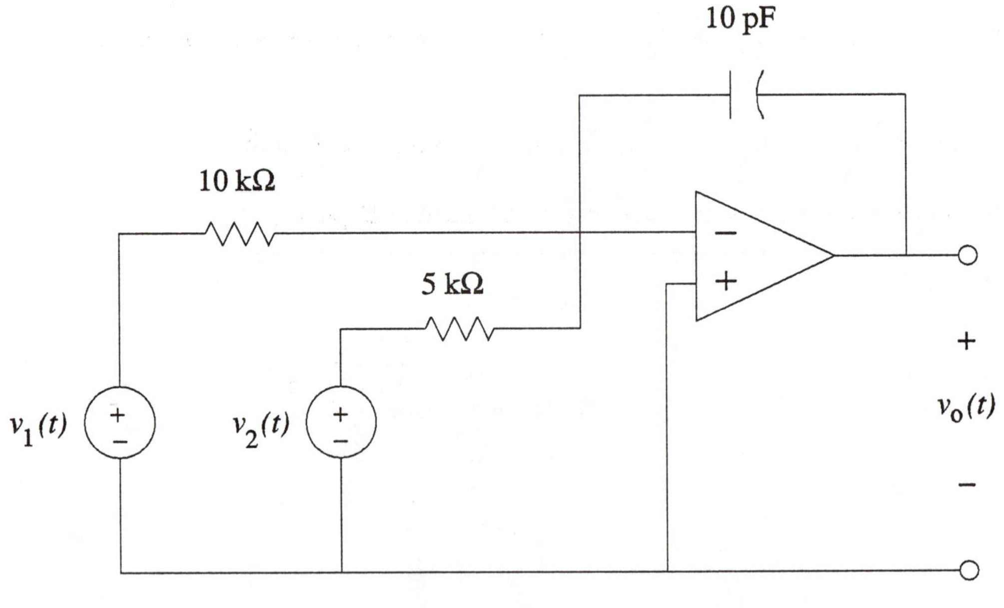

PROBLEM 13.18

13.4 Poles and Stability

13.19 For each of the following transfer functions (which may describe a circuit, control system, mechanical device, etc.), find the poles of the system and whether or not it is stable.

(a) $\mathbf{H}(s) = \dfrac{s}{s^3 + 6s^2 + 11s + 6}$

(b) $\mathbf{H}(s) = \dfrac{s}{s^3 + 4s^2 + s - 6}$

(c) $\mathbf{H}(s) = \dfrac{s^3}{\left(s^2 + 1\right)^2 (s + 4)}$

(d) $\mathbf{H}(s) = \dfrac{s}{s^3 + 2s^2 + s + 2}$

(e) $\mathbf{H}(s) = \dfrac{s^4}{(s + 1)(s + 3)}$

(f) $\mathbf{H}(s) = \dfrac{s^4}{s^4 + 4s^2 + 3}$

13.20 If the poles of a transfer function of a certain circuit are $s_{1,2} = \dfrac{\mu - 5 \pm \sqrt{(5 - \mu)^2 - 1}}{2}$, find a range of values for the circuit parameter μ that will make the circuit stable.

13.21 Determine whether or not the circuit in Prob. 13.14 is stable. Explain.

13.22 Determine whether or not the following circuit is stable. Explain.

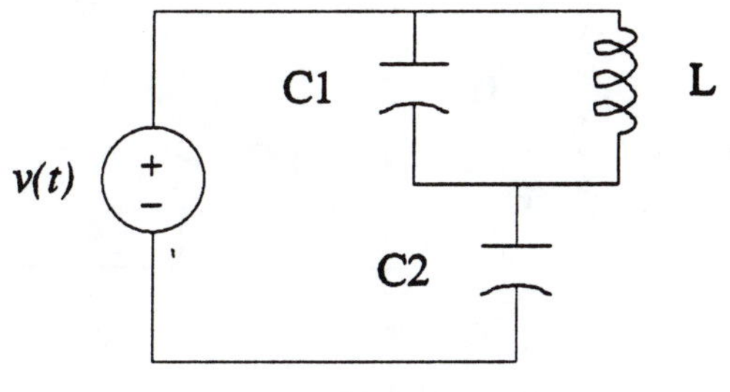

PROBLEM 13.22

13.23 Determine whether or not the following circuit is stable. Explain.

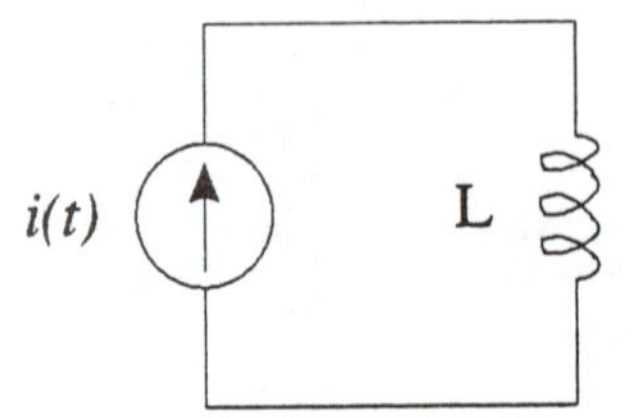

PROBLEM 13.23

13.24 Find a range of values for g that will make the following circuit stable.

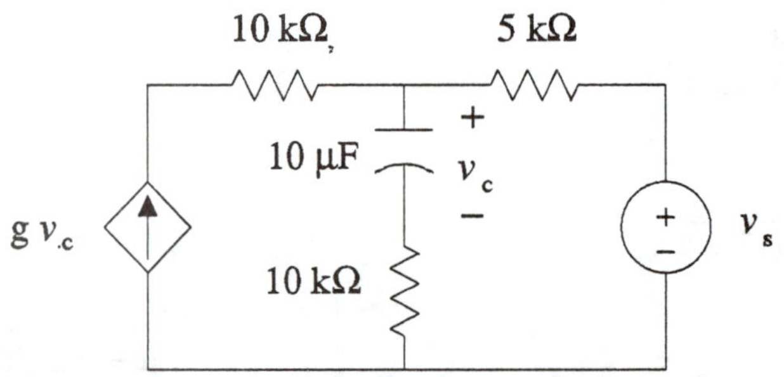

PROBLEM 13.24

13.25 Find a range of values for r that will make the following circuit stable.

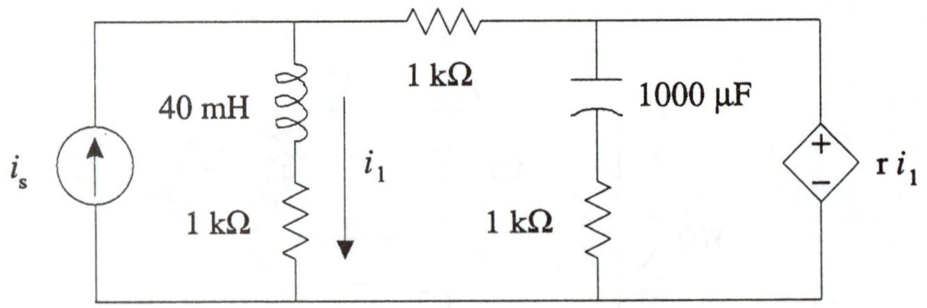

PROBLEM 13.25

13.5 Initial and Final Value Theorems

13.26 Find $f(0+)$ and $f(\infty)$ using the initial and final value theorems if $\mathbf{F}(s) = \dfrac{12}{(s+1)(s^2+4s+3)}$.

13.27 Find the initial value (at $t = 0+$) and steady-state value (at $t = \infty$) of $v(t)$ in Prob. 13.3 using the initial and final value theorems.

13.28 Find the initial value (at $t = 0+$) and steady-state value (at $t = \infty$) of $i(t)$ in Prob. 13.4 using the initial and final value theorems.

13.29 A circuit has a transfer function $\mathbf{H}(s) = \dfrac{\mathbf{V}_o(s)}{\mathbf{V}_i(s)} = \dfrac{s}{(s+1)(s+3)}$.

(a) If an input $v_i(t) = u(t)$ V is applied to the circuit, find the initial and final values of the output using Laplace transforms and the initial and final value theorems.

(b) If an input $v_i(t) = \left(e^{-2t} + 2\sin 5t\right) u(t)$ V is applied to the circuit, find the initial and final values of the output using Laplace transforms and the initial and final value theorems.

13.6 Impulse Response and Convolution

13.30 A circuit has a transfer function $\mathbf{H}(s) = \dfrac{5s-3}{s^2+2s}$.

Find (a) its impulse response $h(t)$ and (b) its step response $s(t)$.

13.31 The step response of a causal, linear time-invariant circuit is $s(t) = 3r(t) - 4p(t) + \delta(t) + 1$. Find the impulse response of the circuit, $h(t)$.

13.32 Using the transfer function $\mathbf{H}(s)$ for the circuit in Prob. 13.14, find the output $v_o(t)$ if the input is:

(a) $v_i(t) = \delta(t)$ V
(b) $v_i(t) = u(t)$ V
(c) $v_i(t) = \delta(t-3)$ V
(d) $v_i(t) = u(t-3)$ V
(e) $v_i(t) = 5e^{-4t}\, u(t)$ V
(f) $v_i(t) = \cos t\, u(t)$ V

13.33 Find the convolution $f(t)*g(t)$ (a) graphically and (b) by multiplication in the s-domain.

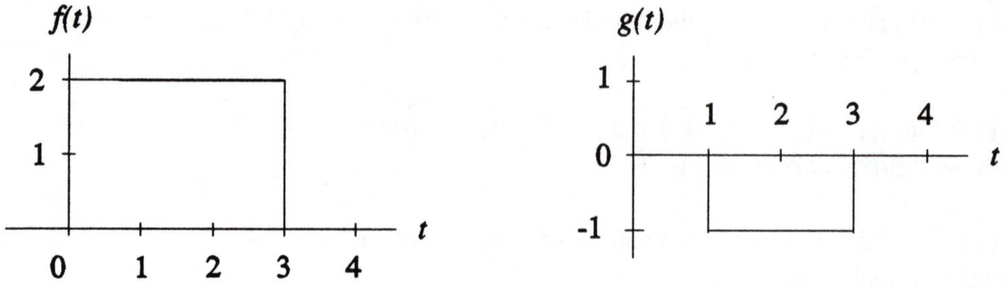

PROBLEM 13.33

13.34 Find the convolution $f(t)*g(t)$ (a) graphically and (b) by multiplication in the s-domain.

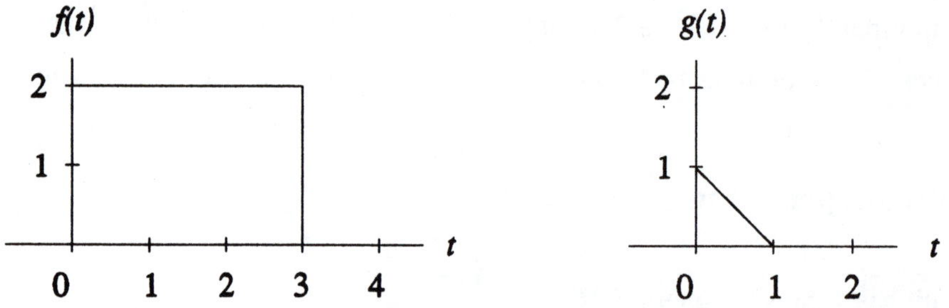

PROBLEM 13.34

13.35 Using Laplace transforms, find the convolution of $v(t)$ and $h(t)$ if $v(t)=e^{-3t}u(t)$ V and $h(t)=\delta(t-1)+2e^{-(t-4)}u(t-4)$.

CHAPTER 13 SOLUTIONS

13.1 $1\times10^6/s$

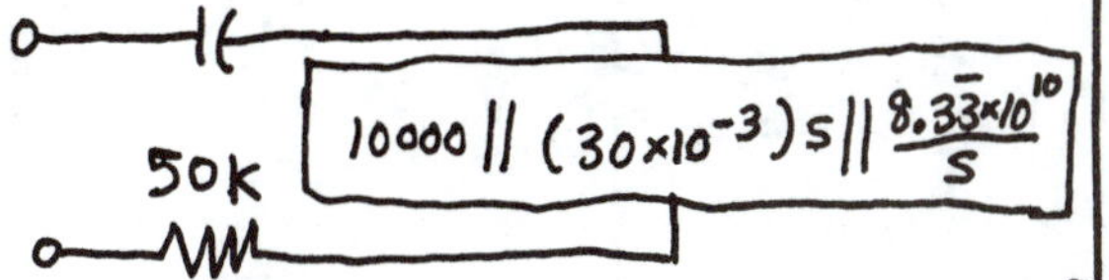

$$Z(s) = \frac{(1\times10^{+6})}{s} + (50\times10^3) + \left(\frac{300}{10K+.03s} \Big\| \frac{8.3\times10^{8}}{s}\right)$$

$$= 5\times10^4 + 1\times10^6\left(\frac{1}{s}\right) + \frac{(2.5\times10^4)}{2.5s + 8.3\bar{3}\times10^5}$$

13.2 $Y_L = \frac{1}{2s}$

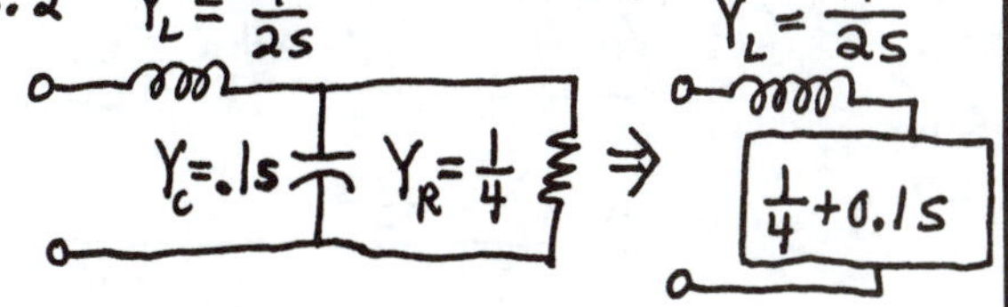

$$Y(s) = \frac{\frac{1}{2s}\left(\frac{1}{4}+0.1s\right)}{0.1s+\frac{1}{4}+\frac{1}{2s}} = \frac{10s+25}{20s^2+50s+100}$$

13.3 s-Domain circuit:

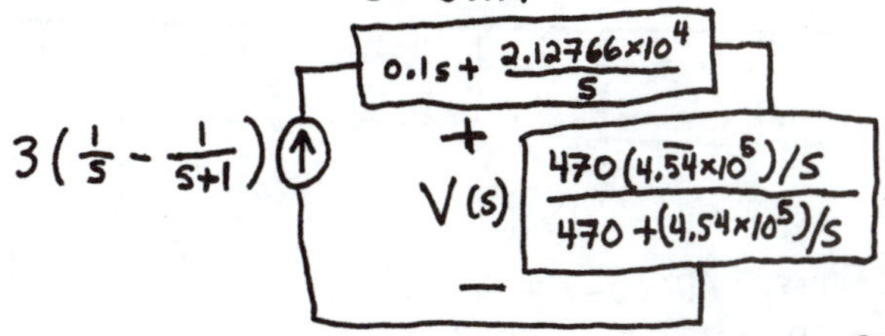

$$V(s) = I(s)Z(s) = 3\left(\frac{1}{s}-\frac{1}{s+1}\right)\left[\frac{2.136\bar{3}\times10^8}{470s+4.54\times10^5}\right]$$

$$= 1.3\bar{6}\times10^6\left[\frac{1}{s(s+967.118)} - \frac{1}{(s+1)(s+967.118)}\right]$$

PFE:

$$\frac{1}{s(s+967.118)} = \frac{A}{s} + \frac{B}{s+967.118}$$

$$A = \frac{1}{967.118} = 0.001034,\quad B = \frac{1}{-967.118} = -A$$

$$\frac{-1}{(s+1)(s+967.118)} = \frac{C}{s+1} + \frac{D}{s+967.118}$$

$$C = \frac{-1}{966.118} = -.00103507$$

$$D = \frac{-1}{-966.118} = -C = .00103507$$

$$\therefore V(s) = \frac{1410}{s} - \frac{1410}{s+967} - \frac{1411}{s+1} + \frac{1411}{s+967}$$

$$= \frac{1410}{s} - \frac{1411}{s+1} + \frac{1}{s+967}$$

$$\Rightarrow \boxed{v(t) = 1410 - 1411e^{-t} + e^{-967t}}$$

Note: at $t=0$, $v(t)=0$

at $t\to\infty$, $v(t) = 1410 = 3A(470)$

13.4 s-Domain circuit:

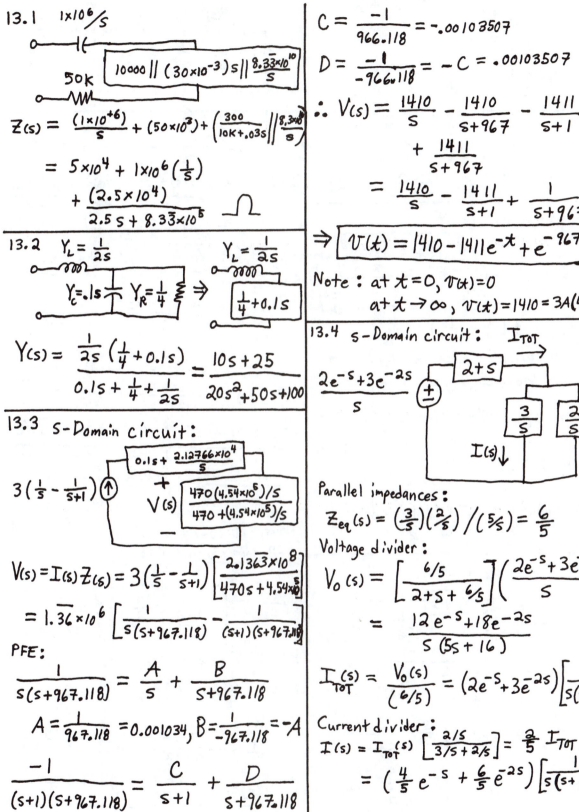

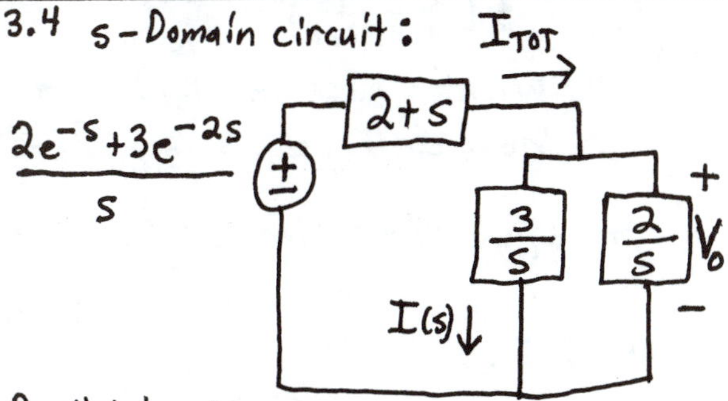

Parallel impedances:

$$Z_{eq}(s) = \left(\frac{3}{s}\right)\left(\frac{2}{s}\right)\Big/\left(\frac{5}{s}\right) = \frac{6}{5}$$

Voltage divider:

$$V_o(s) = \left[\frac{6/5}{2+s+6/5}\right]\left(\frac{2e^{-s}+3e^{-2s}}{s}\right)$$

$$= \frac{12e^{-s}+18e^{-2s}}{s(5s+16)}$$

$$I_{TOT}(s) = \frac{V_o(s)}{(6/5)} = (2e^{-s}+3e^{-2s})\left[\frac{1}{s(s+\frac{16}{5})}\right]$$

Current divider:

$$I(s) = I_{TOT}(s)\left[\frac{2/s}{3/s+2/s}\right] = \frac{2}{5}I_{TOT}$$

$$= \left(\frac{4}{5}e^{-s} + \frac{6}{5}e^{-2s}\right)\left[\frac{1}{s(s+\frac{16}{5})}\right]$$

13.4 cont'd:

$$\frac{1}{s(s+\frac{16}{5})} = \frac{A}{s} + \frac{B}{s+16/5}$$

$$A = 5/16, \quad B = \frac{1}{(-16/5)} = -5/16$$

$$\therefore I(s) = \frac{e^{-s}}{4s} - \frac{e^{-s}}{4(s+16/5)} + \frac{3e^{-2s}}{8s} - \frac{3e^{-2s}}{8(s+16/5)}$$

$$i(t) = \left[\frac{1}{4} - \frac{1}{4}e^{-\frac{16}{5}(t-1)}\right]u(t-1) + \left[\frac{3}{8} - \frac{3}{8}e^{-\frac{16}{5}(t-2)}\right]u(t-2)$$

13.5 s-Domain circuit

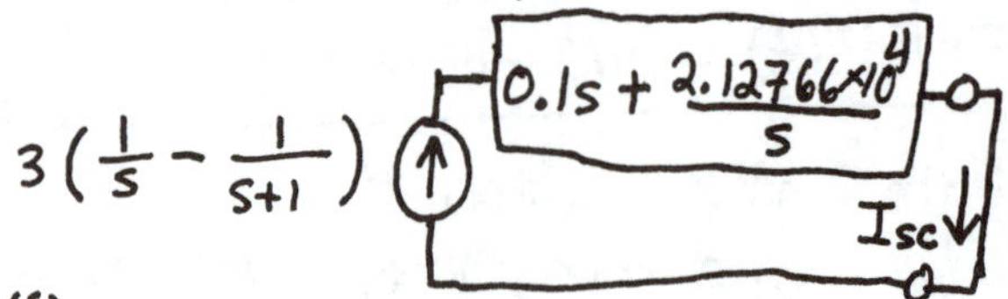

(i) $I_{sc} = 3(\frac{1}{s} - \frac{1}{s+1}) = I_N$

(ii) kill current source and look into terminals $\Rightarrow Z_N = \infty$ (OPEN CKT)

Norton equivalent: ideal current src

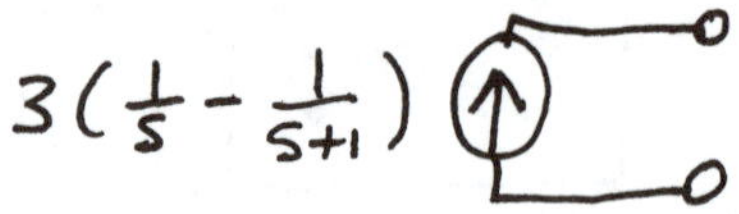

Thevenin equiv. is undefined since $Z_{TH} = Z_N = \infty$

13.6 s-Domain circuit:

$$I_{sc} = \frac{2e^{-s}+3e^{-2s}}{s(2+s)} = I_N$$

$$Z_N = Z_{TH} = [2+s] \,||\, \frac{3}{s} = \frac{3s+6}{s^2+2s+3}$$

$$V_{TH} = I_{sc} Z_N = \frac{6e^{-s}+9e^{-2s}}{s^3+2s^2+3s}$$

13.7 s-Domain circuit:

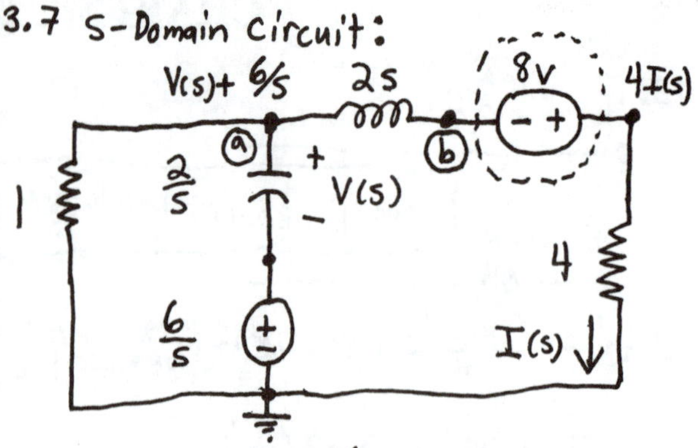

KCL at ⓐ:
$$\frac{V(s)+\frac{6}{s}}{1} + \frac{V(s)}{2/s} + \frac{V(s)+\frac{6}{s}-4I(s)+8}{2s} = V(s)\left[1+\frac{s}{2}+\frac{1}{2s}\right] + \frac{10-2I(s)}{s} + \frac{3}{s^2} = 0$$

KCL at ⓑ:
$$\frac{V(s)+\frac{6}{s}-4I(s)+8}{2s} = I(s)$$

Solve...

$$\Rightarrow I(s) = \frac{22s^2+8s^3}{(2s+4)(2s^2+s^3+s)-4s} = \frac{4s+11}{s^2+4s+5} = \frac{4s+11}{(s+2)^2+1}$$

$$\Rightarrow I(s) = \frac{4(s+2)}{(s+2)^2+1} + \frac{3}{(s+2)^2+1}$$

$$\therefore i(t) = e^{-2t}[4\cos t + 3\sin t]\ A$$

13.8 First find steady-state values at $t=0-$

Initial circuit:

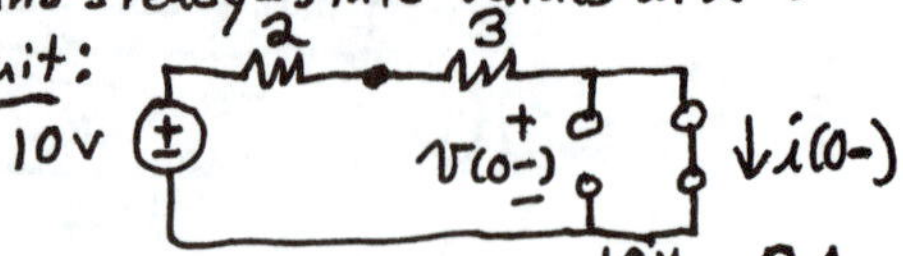

Thus, $v(0-) = 0\,V$, $i(0-) = \frac{10V}{2+3} = 2A$

S-Domain circuit ($t>0$):

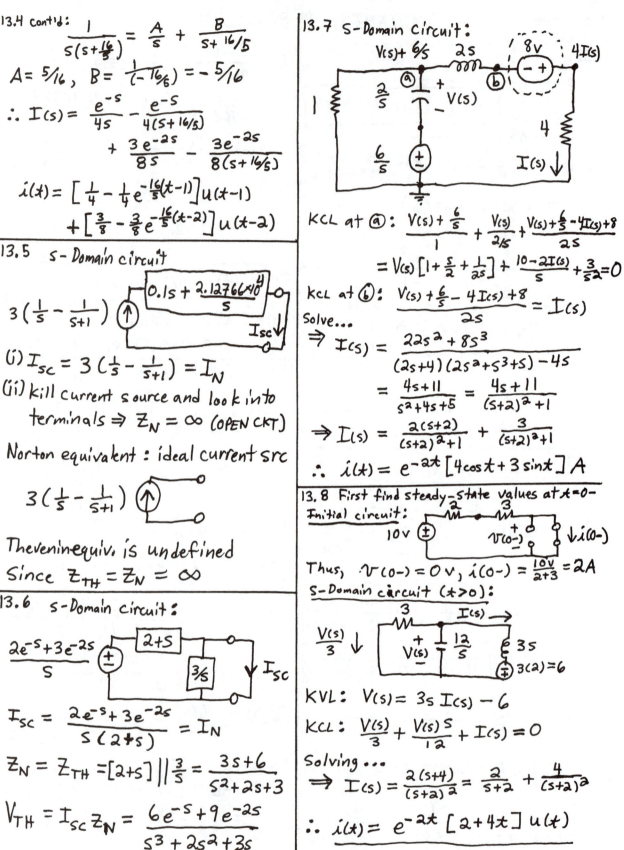

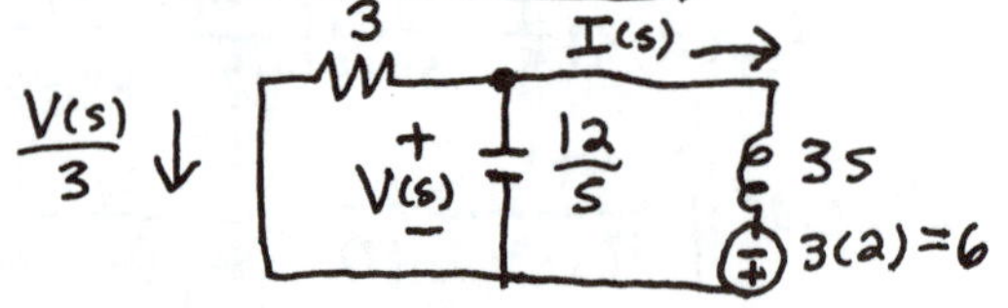

KVL: $V(s) = 3s\,I(s) - 6$

KCL: $\frac{V(s)}{3} + \frac{V(s)\,s}{12} + I(s) = 0$

Solving...

$$\Rightarrow I(s) = \frac{2(s+4)}{(s+2)^2} = \frac{2}{s+2} + \frac{4}{(s+2)^2}$$

$$\therefore i(t) = e^{-2t}[2+4t]\,u(t)$$

13.9 First find steady-state values at $t=0^-$

$t=0^-$:

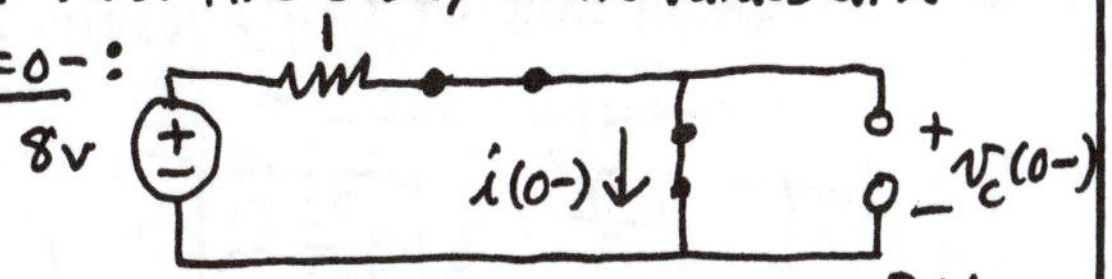

Thus, $i(0^-) = 8A$, $v_c(0^-) = 0\,V$

$t>0$: s-Domain

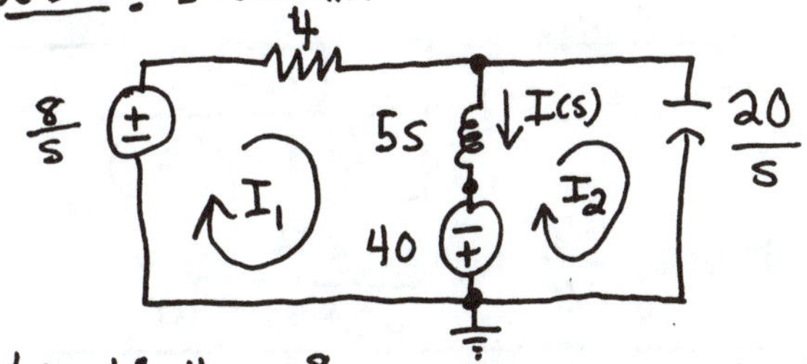

Loop 1: $40 + \frac{8}{s} = I_1(4+5s) - I_2(5s)$

Loop 2: $I_2\left(\frac{20}{s} + 5s\right) - I_1(5s) = -40$

$$\begin{bmatrix} (4+5s) & -5s \\ -5s & \left(\frac{20}{s}+5s\right) \end{bmatrix} \begin{bmatrix} I_1 \\ I_2 \end{bmatrix} = \begin{bmatrix} 40 + 8/s \\ -40 \end{bmatrix}$$

$$\Delta(s) = (4+5s)\left(\frac{20}{s}+5s\right) - 25s^2$$

$$I_1 = (40 + 8/s)(20/s + 5s - 200s)/\Delta(s)$$

$$I_2 = \frac{(4+5s)([redacted] - 40) + 5s(40 + 8/s)}{\Delta(s)}$$

$$\Rightarrow I(s) = I_1 - I_2 = \frac{8s^2+40s+8}{s(s+1)(s+4)}$$

$$= \frac{A}{s} + \frac{B}{s+1} + \frac{C}{s+4}$$

$A = \frac{8}{4} = 2$, $B = \frac{8-40+8}{(-1)3} = 8$, $C = -2$

$$\therefore\ i(t) = 2 + 8e^{-t} - 2e^{-4t}\,A$$

13.10 s-Domain circuit:

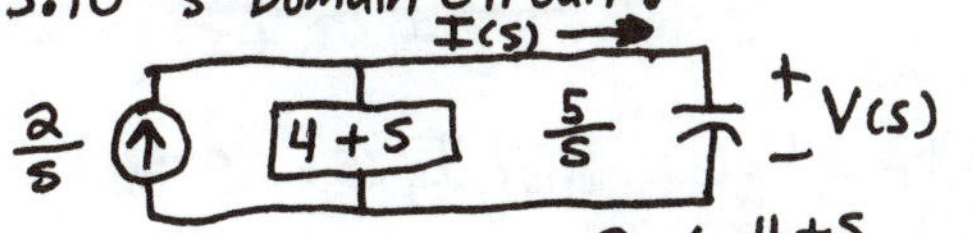

Current div: $I(s) = \frac{2}{s}\left(\frac{4+s}{4+s+5/s}\right)$

$$= \frac{8+2s}{s^2+4s+5}$$

$$\Rightarrow V(s) = \frac{5}{s} I(s) = \frac{10s+40}{s(s^2+4s+5)} = \frac{A}{s} + \frac{Bs+C}{s^2+4s+5}$$

$A = \frac{40 [redacted]}{5} = 8$

Substituting,

$$\frac{8(s^2+4s+5)}{s} + Bs + C = \frac{10s+40}{s}$$

$$\Rightarrow 8s^2 + 22s = -(Bs^2 + Cs)$$

$$\Rightarrow B = -8,\ C = -22$$

$$\Rightarrow V(s) = \frac{8}{s} - \frac{8s+22}{s^2+4s+5}$$

$$= \frac{8}{s} - \frac{8(s+2)}{(s+2)^2+1} - \frac{6}{(s+2)^2+1}$$

$$\therefore\ v(t) = 8 - e^{-2t}[8\cos t + 6\sin t]\,V$$

13.11

The transformed circuit is

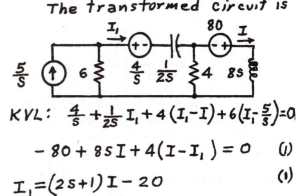

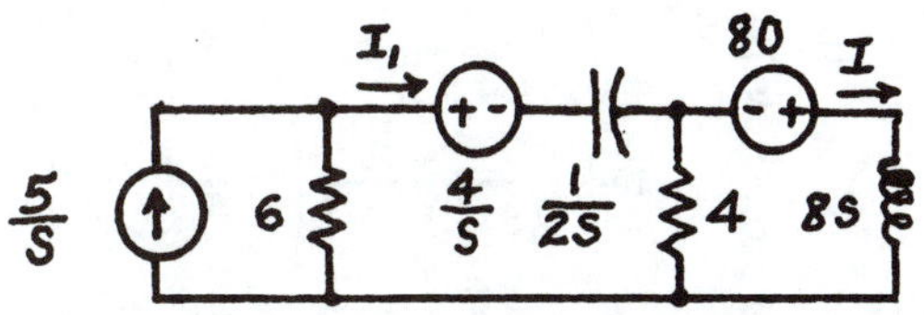

KVL: $\frac{4}{s} + \frac{1}{2s} I_1 + 4(I_1 - I) + 6\left(I_1 - \frac{5}{s}\right) = 0$

$$-80 + 8sI + 4(I - I_1) = 0 \quad (1)$$

$$I_1 = (2s+1)I - 20 \quad (1)$$

$$\frac{4}{s} + \left(\frac{1}{2s} + 4 + 6\right)\left[(2s+1)I - 20\right] - 4I - \frac{30}{s} = 0$$

$$I = \frac{\frac{30}{s} + 20\left(\frac{1}{2s} + 10\right) - \frac{4}{s}}{\left(\frac{1}{2s}+10\right)(2s+1) - 4}$$

$$= \frac{400s + 72}{40s^2 + 14s + 1} = \frac{10s + \frac{9}{5}}{\left(s+\frac{1}{10}\right)\left(s+\frac{1}{4}\right)}$$

$$= \frac{\frac{16}{3}}{s + \frac{1}{10}} + \frac{\frac{14}{3}}{s + \frac{1}{4}}$$

$$i(t) = \frac{16}{3} e^{-t/10} + \frac{14}{3} e^{-t/4}\ A$$

13.12 Transformed circuit:

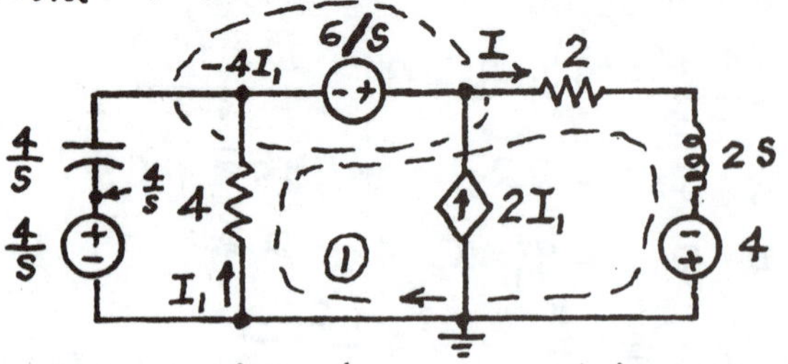

KCL equation (supernode):

$$\frac{-4I_1-\frac{4}{S}}{\frac{4}{S}}-I_1-2I_1+I=0 \qquad (1)$$

KVL equation (Loop 1):

$$-\frac{6}{S}+(2S+2)I-4+4I_1=0 \qquad (2)$$

From (1) and (2),

$$I=\frac{2S^2+11S+9}{S(S^2+4S+5)}=\frac{9/5}{S}+\frac{\frac{1}{5}S+\frac{19}{5}}{S^2+4S+5}$$

$$=\frac{9/5}{S}+\frac{1}{5}\left[\frac{S+2}{(S+2)^2+1}\right]+\frac{17}{5}\left[\frac{1}{(S+2)^2+1}\right]$$

$$i(t)=\frac{9}{5}+\frac{1}{5}e^{-2t}(\cos t+17\sin t)\ \text{A}$$

13.13

Transformed circuit:

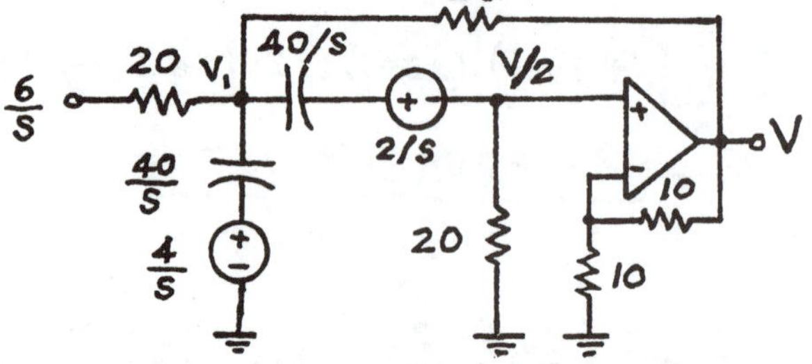

$$\text{KCL: } \frac{V_1-\frac{6}{S}}{20}+\frac{V_1-\frac{4}{S}}{\frac{40}{S}}+\frac{V_1-\frac{V}{2}-\frac{2}{S}}{\frac{40}{S}}$$

$$+\frac{V_1-V}{20}=0 \qquad (1)$$

$$\frac{\frac{V}{2}+\frac{2}{S}-V_1}{\frac{40}{S}}+\frac{V/2}{20}=0 \qquad (2)$$

from which

$$V(S)=\frac{4(S+2)}{S^2+4S+8}=\frac{4(S+2)}{(S+2)^2+4}$$

$$v(t)=4e^{-2t}\cos 2t\ \ \text{V}$$

13.14 s-Domain circuit (simplified):

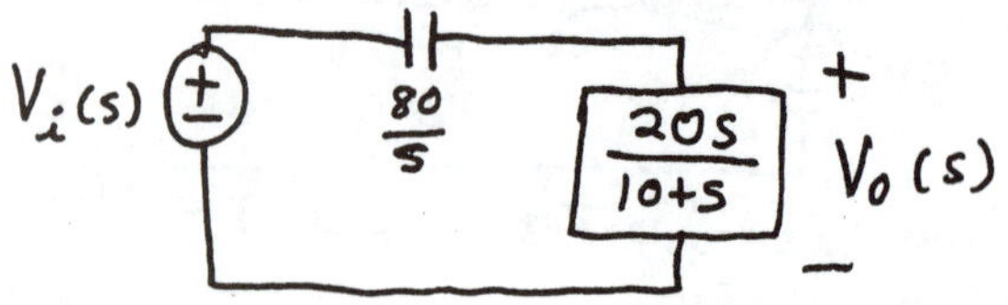

Voltage divider:

$$V_0(s)=V_i(s)\left[\frac{\left(\frac{20s}{10+s}\right)}{\left(\frac{80}{s}+\frac{20s}{10+s}\right)}\right]$$

$$H(s)\equiv\frac{V_0}{V_i}=\frac{s^2}{s^2+4s+40}$$

13.15 Tranformed circuit:

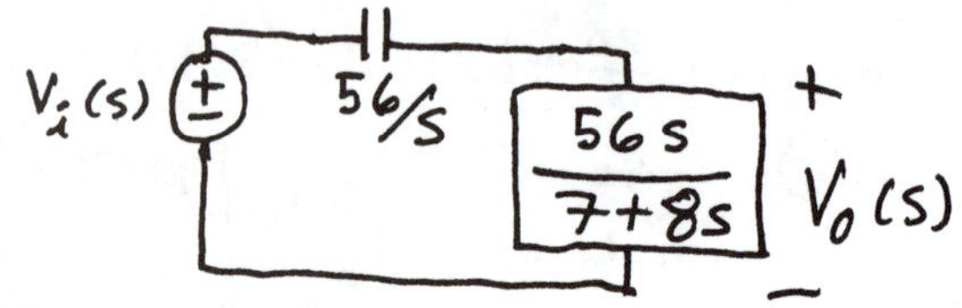

Voltage divider:

$$V_0(s)=V_i(s)\left[\frac{56s/(7+8s)}{56/s+56s/(7+8s)}\right]$$

$$H(s)\equiv\frac{V_0}{V_i}=\frac{s^2}{56(s^2+8s+7)}$$

13.16 (i) To find $H_1(s)$, kill $V_1(s)$ (SHORT)

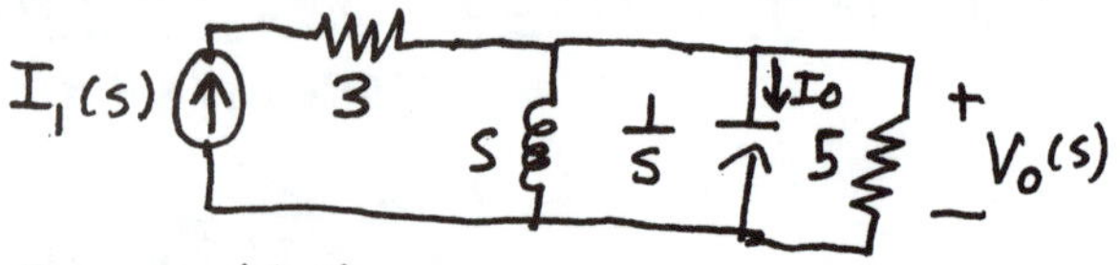

Current divider:

$$I_0(s)=I_1(s)\ \frac{5s/(5+s)}{5s/(5+s)+\frac{1}{s}}=\frac{I_1 5s^2}{5s^2+5+s}$$

$$\Rightarrow H_1(s)=\frac{V_0}{I_1}=\frac{5s}{5s^2+5+s}$$

(ii) To find $H_2(s)$, kill $I_1(s)$ (OPEN)

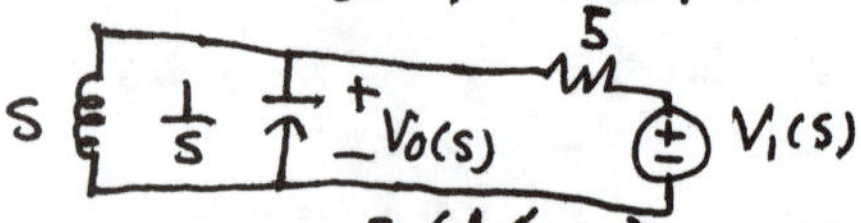

$$V_0(s)=V_1(s)\left[\frac{(1/s+1/s)}{5+1/(s+1/s)}\right]=V_1\left[\frac{s}{5s^2+s+5}\right]$$

$$\Rightarrow H_2(s)=V_0/V_1=s/5s^2+s+5$$

13.17 (a) Kill $V_2(s)$ and $I_1(s)$:

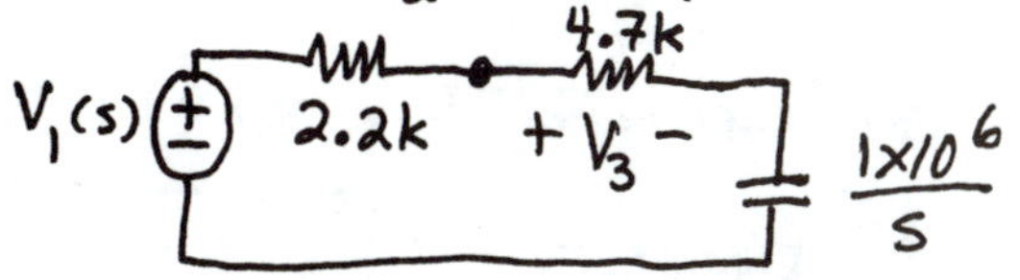

$$H_1(s) \equiv \frac{V_3}{V_1} = \frac{4700}{2200+4700+(1\times10^6)/s}$$

$$= \frac{47s}{69s+10000}$$

(b) Kill $V_1(s)$ and $I_1(s)$:

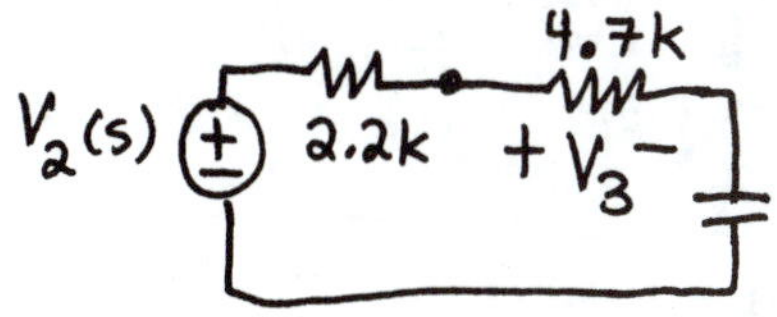

$$\therefore\ H_2(s) \equiv \frac{V_3}{V_2} = H_1(s) = \frac{47s}{69s+10000}$$

(c) Kill $V_1(s)$ and $V_2(s)$:

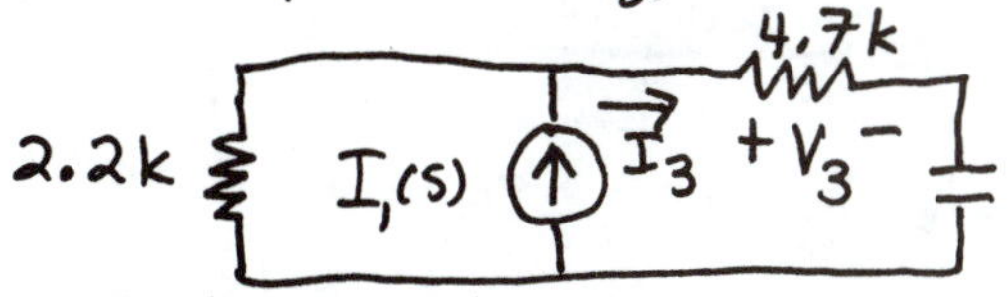

Current divider:

$$I_3 = I_1\left[\frac{2.2k}{2.2k+4.7k+1\times10^6/s}\right]$$

$$= I_1\left[\frac{22s}{69s+10000}\right]$$

$$V_3 = 4700 I_3 = I_1\left[\frac{103400s}{69s+10000}\right]$$

$$H_3 \equiv V_3/I_1 = \frac{103400\,s}{69s+10000}$$

13.18 (i) Kill V_2:

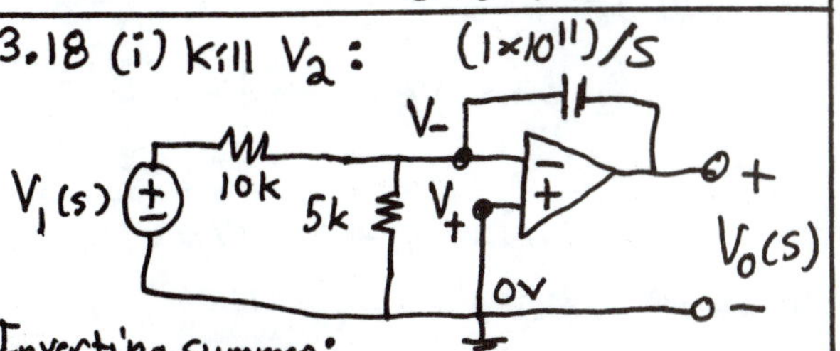

Inverting summer:

$$V_0(s) = -\frac{1\times10^{11}}{s(10k)}V_1 - \frac{1\times10^{11}}{s(5k)}\,0\text{v}$$

$$\Rightarrow\ H_1(s) \equiv \frac{V_0}{V_1} = -\frac{1\times10^7}{s}$$

(ii) Kill V_1:

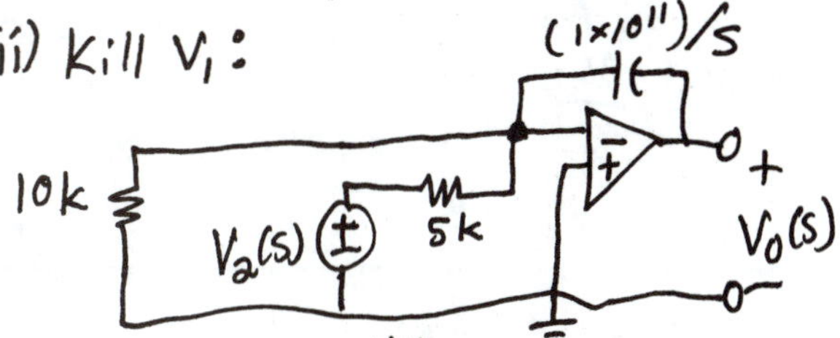

$$V_0(s) = -\frac{(1\times10^{11})V_2}{s(5k)} = \frac{-2\times10^7}{s}V_2$$

$$\Rightarrow\ H_2(s) \equiv \frac{V_0}{V_2} = -\frac{2\times10^7}{s}$$

(The virtual short principle could also be used with KCL at V_-)

13.19 (a) Denominator $= (s+1)(s+2)(s+3)$

(i) Poles all in LHP: $p = -1, -2, -3$

(ii) Stable system

(b) Denominator $= (s-1)(s+2)(s+3)$

(i) Poles: $p = +1, -2, -3$

(ii) Not stable since $p=1$ is in RHP (unstable)

(c) (i) Double poles on $j\omega$ axis: $p = \pm 1j, -4$

(ii) Not stable (conditionally stable)

(d) $H(s) = \dfrac{s}{(s+2)(s^2+1)}$ (i) Poles at $p = -2, \pm 1j$

(ii) Not stable since poles on $j\omega$ axis (conditionally stable)

(e) (i) Poles at $p = -1, -3$, and double poles at ∞

(ii) Not stable since poles at ∞ in RHP

(f) $H(s) = \dfrac{s^4}{(s^2+1)(s^2+3)}$

(i) Poles at $p = \pm 1j, \pm j\sqrt{3}$

(ii) Not stable (conditionally stable) Since poles on $j\omega$ axis

13.20 If $\mu < 5$, the poles are in the LHP and the network is stable.

13.21 In Prob. 13.14, it was found that

$$H(s) = \frac{V_o}{V_i} = \frac{s^2}{s^2+4s+40}$$

Poles at $s = \frac{-4 \pm \sqrt{16-160}}{2} = -2 \pm j\sqrt{3}$

in LHP → circuit is stable.

(Note that it depends on values of L + C)

13.22 s-Domain circuit:

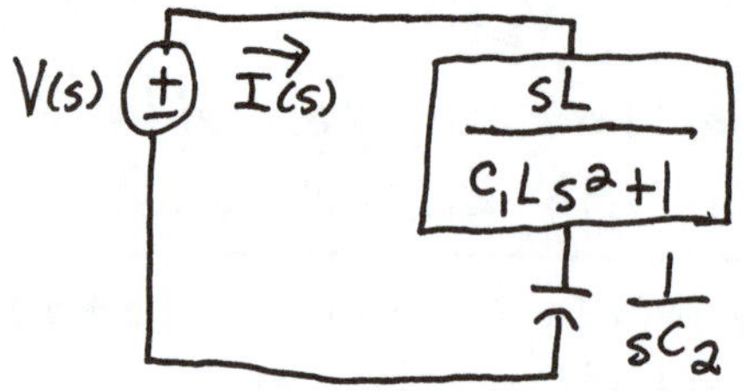

$$\frac{V(s)}{I(s)} = \frac{1}{sC_2} + \frac{sL}{C_1Ls^2+1} = \frac{C_1Ls^2+Ls+1}{sC_2(C_1Ls^2+1)}$$

Pole at $s = 0, \pm j\sqrt{\frac{1}{LC_1}}$ ⇒ Not stable

13.23 $Z = Ls$, If $H(s) = \frac{V_L}{I(s)} = Ls$

⇒ Pole at $s = \infty$ in RHP ⇒ Not stable

13.24 s-Domain circuit:

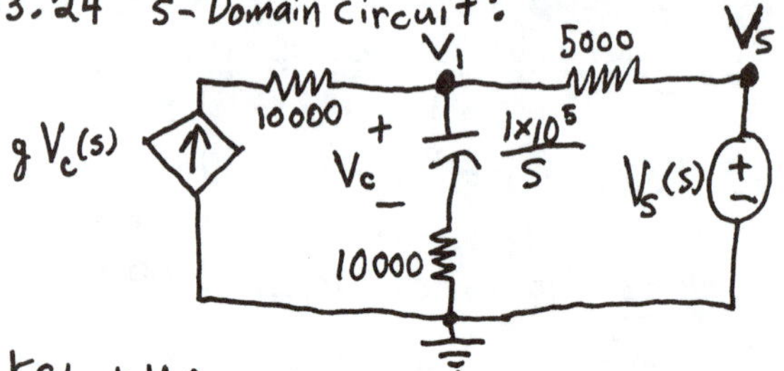

KCL at V_1:

$$g V_c(s) + \frac{V_s - V_1}{5000} = \frac{V_1}{10000 + 1\times10^5/s}$$

Voltage divider:

$$V_c = V_1 \left(\frac{1\times10^5}{s}\right) / \left[\frac{1\times10^5}{s} + 10000\right]$$

$$= V_1 \left[\frac{1}{1+0.1s}\right]$$

$$\Rightarrow \left[\left(\frac{g}{1+0.1s}\right) - \left(\frac{1}{5000}\right) - \left(\frac{1}{10000 + 1\times10^5/s}\right)\right] V_1$$

$$= V_s\left(-\frac{1}{5000}\right)$$

simplifying....

$$\frac{V_1}{V_s} = \frac{2s^2 + 40s + 200}{3s^2 + (50 - g\,100000)s + (200 - g\,1\times10^6)}$$

Poles at

$$p = \frac{g(1\times10^5) - 50 \pm \sqrt{(50 - g\times10^5)^2 + 12\times10^6 g - 2400}}{6}$$

Stable when $Re\{p\} < 0$

⇒ $\boxed{g < 5\times10^{-4}}$

13.25 s-Domain circuit:

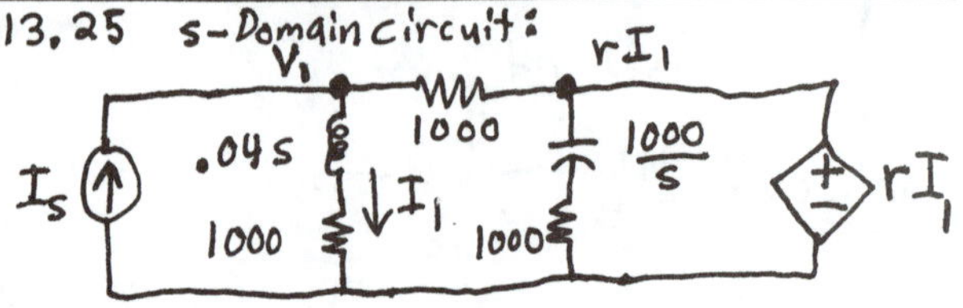

$V_1 = I_1(.04s + 1000)$

KCL at V_1: $I_s = \frac{V_1}{1000 + .04s} + \frac{V_1 - r\left[\frac{V_1}{1000+.04s}\right]}{1000}$

simplifying...

$$\frac{V_1}{I_s} = \frac{1000(s + 25000)}{s + [50000 - 25000r]}$$

Single pole at $s = 25000r - 50000$

Circuit stable when $s < 0$ ⇒ $\boxed{r < 2}$

13.26

(a) $f(0+) = \lim_{s\to\infty} sF(s) = \lim_{s\to\infty} \frac{12s}{(s+1)(s^2+4s+3)} = 0$

$f(\infty) = \lim_{s\to0} sF(s) = \frac{0}{3} = 0$

(b) $f(0+) = \lim_{s\to\infty} sF(s) = \lim_{s\to\infty} \frac{3s^2+2s}{s^2+5s+6} = 3$

$f(\infty) = \lim_{s\to0} sF(s) = \frac{0}{6} = 0$

(c) $f(0+) = \lim_{s\to\infty} sF(s) = \frac{s^2}{s^3+6s^2+11s+6} = 0$

$f(\infty) = \lim_{s\to0} sF(s) = \frac{0}{6} = 0$

13.27 $V(s) = \frac{1410}{s} - \frac{1411}{s+1} + \frac{1}{s+967.12}$

$$v(t=0+) = \lim_{s\to\infty} sV(s) = \lim_{s\to\infty}\left[1410 - \frac{1411s}{s+1} + \frac{s}{s+967}\right]$$

$$= 1410 - 1411 + 1 = 0\text{ v}$$

$$v(t=\infty) = \lim_{s\to0} sV(s) = \lim_{s\to0}\left[1410 - \frac{1411s}{s+1} + s/(s+967)\right]$$

$$= 1410 - 0 + 0 = 1410\text{ V}$$

13.28

$$i(t=0+) = \lim_{s\to\infty} sI(s) = \lim_{s\to\infty}\left(\frac{4}{5}e^{-s}+\frac{6}{5}e^{-2}\right)\frac{1}{s+16/5}$$
$$= 0\,A$$
$$i(t=\infty) = \lim_{s\to 0} sI(s) = \lim_{s\to 0}\left[\frac{4}{5}+\frac{6}{5}\right]\left[\frac{5}{16}\right]$$
$$= \frac{5}{8}A$$

13.29 (a) $V_i(s) = \frac{1}{s} \Rightarrow V_o(s) = \frac{1}{(s+1)(s+3)}$

$$v_o(t=0+) = \lim_{s\to\infty} sV_o(s) = \lim_{s\to\infty}\frac{s}{(s+1)(s+3)} = 0$$
$$v_o(t=\infty) = \lim_{s\to 0}\frac{s}{(s+1)(s+3)} = 0$$

(b) $V_i(s) = \frac{1}{s+2} + \frac{10}{s^2+25}$

$$\Rightarrow V_o(s) = \frac{s}{(s+1)(s+2)(s+3)} + \frac{10s}{(s^2+25)(s+1)(s+3)}$$
$$v_o(t=0+) = \lim_{s\to\infty}\left[\frac{10s^2}{(s^2+25)(s+1)(s+3)} + \frac{s^2}{(s+1)(s+2)(s+3)}\right]$$
$$= 0$$
$$v_o(t=\infty) = \lim_{s\to 0}[\quad] = 0$$

13.30

(a) $h(t) = \mathcal{L}^{-1}\left\{\frac{5s}{s^2+25} - \frac{3}{s^2+25}\right\}$
$$= \left[5\cos 5t - \frac{3}{5}\sin 5t\right]u(t)$$

(b) $S(s) = \frac{H(s)}{s} = \frac{5}{s^2+25} - \frac{3}{s(s^2+25)}$

$$\Rightarrow s(t) = [\sin 5t]u(t) + \mathcal{L}^{-1}\left\{-\frac{3}{s(s^2+25)}\right\}$$
$$\frac{-3}{s(s^2+25)} = \frac{A}{s} + \frac{Bs+C}{s^2+25} \quad \Rightarrow A = -\frac{3}{25}$$
$$Bs+C - \frac{3(s^2+25)}{25s} = -\frac{3}{s}$$
$$\Rightarrow Bs^2 + Cs = \frac{3}{25}s^2$$
$$\Rightarrow C = 0,\ B = \frac{3}{25}$$
$$\therefore s(t) = \left[\sin 5t - \frac{3}{25} + \frac{3}{25}\cos 5t\right]u(t)$$

13.31

$$h(t) = \frac{d}{dt}s(t) = 3u(t) - 4r(t) + \delta^{(1)}(t)$$

13.32

$$H(s) = \frac{s^2}{s^2+4s+40}$$

(a) $V_i(s) = 1 \Rightarrow V_o(s) = V_i(s)H(s) = H(s)$

Long division:

$$\frac{s^2}{s^2+4s+40} = 1 - \frac{4s+40}{s^2+4s+40}$$

$\Rightarrow$ Poles at $s = -2 \pm j\sqrt{3}$ for fraction

$$\frac{-4s-40}{(s+2-j\sqrt{3})(s+2+j\sqrt{3})} = \frac{A}{(s+2-j\sqrt{3})} + \frac{A^*}{(s+2+j\sqrt{3})}$$
$$A = \left.\frac{-4s-40}{s+2+j\sqrt{3}}\right|_{s=-2+j\sqrt{3}} = -2 + \frac{16}{\sqrt{3}}j$$
$$= 9.45\angle 102^\circ$$
$$\Rightarrow v(t) = \delta(t) + 2(9.45)e^{-2t}\cos(\sqrt{3}t + 102^\circ)\,V$$

$$\boxed{v(t) = \delta(t) + 18.9e^{-2t}\cos(\sqrt{3}t + 102^\circ)}$$

(b) $V_o(s) = \frac{H(s)}{s} = S(s)$

$$= \frac{s}{(s+2-j\sqrt{3})(s+2+j\sqrt{3})} = \frac{A}{s+2-j\sqrt{3}} + \frac{A^*}{s+2+j\sqrt{3}}$$
$$A = \left.\frac{s}{s+2+j\sqrt{3}}\right|_{s=-2+j\sqrt{3}} = \frac{1}{2} + \frac{j}{\sqrt{3}}$$
$$= 0.764\angle 49.1^\circ$$
$$\Rightarrow v(t) = 2(0.764)e^{-2t}\cos(\sqrt{3}t + 49.1^\circ)$$

$$\boxed{v(t) = 1.528e^{-2t}\cos(\sqrt{3}t + 49.1^\circ)\,V}$$

(c) $v_i(t) = \delta(t-3)$

$$\Rightarrow v_o(t) = v_{o,\text{Part (a)}}(t-3)$$
$$= \delta(t-3) + 18.9e^{-2(t-3)}\cos(\sqrt{3}(t-3) + 102^\circ)$$

(d) $v_o(t) = v_{o,\text{Part (b)}}(t-3) = 1.528e^{-2(t-3)}\cos(\sqrt{3}(t-3) + 49^\circ)$

13.32 cont'd

(e) $v_i(t) = 5e^{-4t}u(t) \Rightarrow V_i(s) = \frac{5}{s+4}$

$$V_o(s) = V_i(s)H(s) = \frac{5s^2}{(s+4)(s+2-j\sqrt{3})(s+2+j\sqrt{3})}$$

$$= \frac{A}{s+4} + \frac{B}{s+2-j\sqrt{3}} + \frac{B^*}{s+2+j\sqrt{3}}$$

$$A = \frac{5(16)}{(-4+2+j\sqrt{3})(-4+2-j\sqrt{3})} = \frac{80}{7}$$

$$B = \left.\frac{5s^2}{(s+4)(s+2+j\sqrt{3})}\right|_{s=-2+j\sqrt{3}}$$

$$= -3.2143 + j\,2.0620$$
$$= 3.8188 \angle 147.3^\circ$$

$$v(t) = \frac{80}{7}e^{-4t} + 2(3.8188)e^{-2t}\cos(\sqrt{3}t + 147.3^\circ)$$

$$= \frac{80}{7}e^{-4t} + 7.6376e^{-2t}\cos(\sqrt{3}t + 147.3^\circ)$$

(f) $V_i(s) = \frac{s}{s^2+1}$

$$V_o(s) = V_i(s)H(s) = \frac{s^3}{(s+j)(s-j)(s^2+4s+40)}$$

$$= \frac{A}{s-j} + \frac{A^*}{s+j} + \frac{B}{s+2-j\sqrt{3}} + \frac{B^*}{s+2+j\sqrt{3}}$$

$$A = \left.\frac{s^3}{(s+j)(s^2+4s+40)}\right|_{s=j} = 0.0693\angle 146^\circ$$

$$B = \left.\frac{s^3}{(s^2+1)(s+2+j\sqrt{3})}\right|_{s=-2+j\sqrt{3}}$$

$$= 0.298\angle 41.2^\circ$$

$$\boxed{v_o(t) = 0.596e^{-2t}\cos(\sqrt{3}t + 41.2^\circ) + 0.1386\cos(t + 146^\circ)}$$

13.33 (a) Flip $g(t)$ and slide across $f(t)$

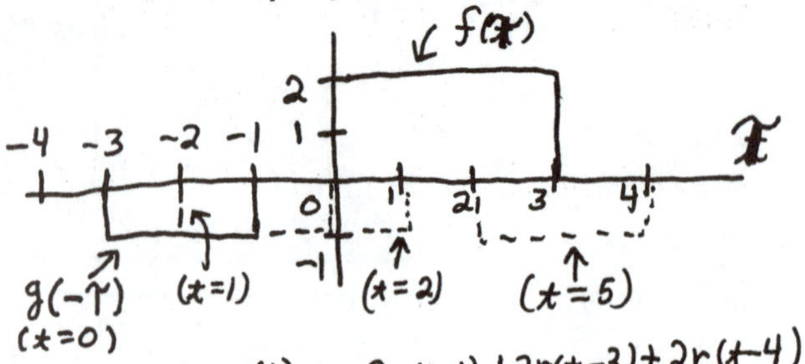

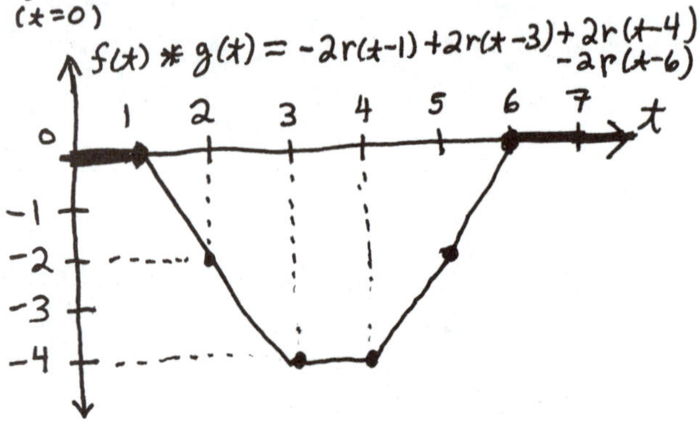

(b) $f(t) = 2[u(t) - u(t-3)]$

$g(t) = u(t-3) - u(t-1)$

$$\Rightarrow F(s) = \frac{2}{s} - \frac{2e^{-3s}}{s},\quad G(s) = \frac{e^{-3s}}{s} - \frac{e^{-s}}{s}$$

$$\mathcal{L}\{f*g\} = F(s)G(s) = \frac{2e^{-3s}}{s^2} - \frac{2e^{-s}}{s^2} - \frac{2e^{-6s}}{s^2} + \frac{2e^{-4s}}{s^2}$$

$$\Rightarrow f*g = 2r(t-3) - 2r(t-1) - 2r(t-6) + 2r(t-4)$$

13.34 (a) Flip $f(t)$ and slide:

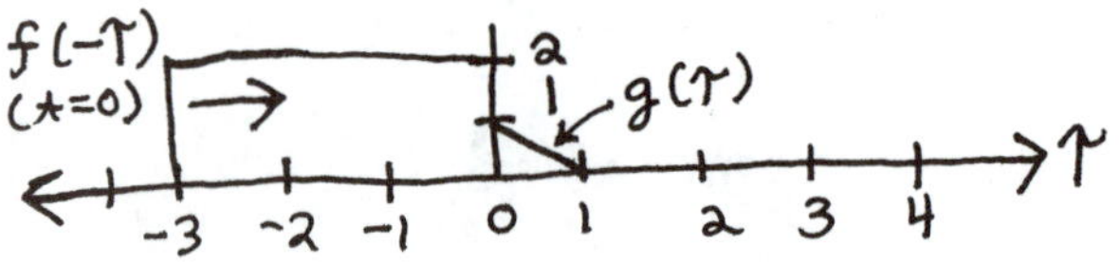

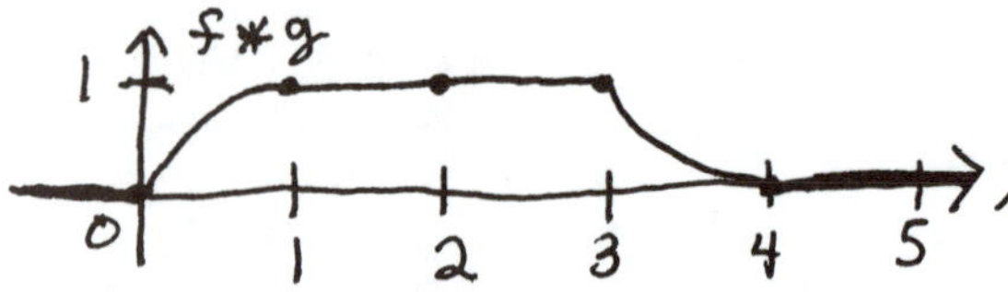

(b) $f(t) = 2[u(t) - u(t-3)] \Rightarrow F(s) = \frac{2}{s} - \frac{2e^{-3s}}{s}$

$g(t) = u(t) - r(t) + r(t-1)$

$$\Rightarrow G(s) = \frac{1}{s} - \frac{1}{s^2} + \frac{e^{-s}}{s^2}$$

13.34 (b) cont'd:

$$F(s)G(s) = \frac{2}{s^2} - \frac{2e^{-s}}{s^2} - \frac{2}{s^3} + \frac{2e^{-s}}{s^3} + \frac{2e^{-3s}}{s^3} - \frac{2e^{-4s}}{s^3}$$

$$\Rightarrow f(t) * g(t) = 2[r(t) - p(t)] + 2p(t-1) + 2[p(t-3) - r(t-3)] - 2p(t-4)$$

$$= 2[t\,u(t) - \tfrac{1}{2}t^2 u(t)] + (t-1)^2 u(t-1) + (t-3)^2 u(t-3) - 2(t-3)u(t-3) - (t-4)^2 u(t-4)$$

13.35

$$V(s) = \frac{1}{s+3}, \quad H(s) = e^{-s} + \frac{2e^{-4s}}{s+1}$$

$$V(s)H(s) = \frac{e^{-s}}{s+3} + \frac{2e^{-4s}}{(s+1)(s+3)}$$

$$\frac{2}{(s+1)(s+3)} = \frac{A}{s+1} + \frac{B}{s+3}$$

$$\Rightarrow A = \frac{2}{2} = 1, \quad B = \frac{2}{1-3} = -1$$

$$\therefore v(t) * h(t)$$

$$= \mathcal{L}^{-1}\left\{\frac{e^{-s}}{s+3} + e^{-4s}\left[\frac{1}{s+1} - \frac{1}{s+3}\right]\right\}$$

$$= e^{-3(t-1)} + e^{-(t-4)} - e^{-3(t-4)}$$

Chapter 14

Frequency Response

14.1 Frequency Response Function

14.1 Consider a circuit with a transfer function

$$\mathbf{H}(s) = \frac{3s^2}{s^3 + 2s^2 + 5s + 1}.$$

(a) Find the output phasor if the input is $\mathbf{V}_i = 10\angle 20°\,\mathrm{V}$ at $\omega = 9$ rad/s/

(b) What is the gain of the circuit at this frequency? What is the phase shift?

(c) Find $v_o(t)$ if $v_i(t) = 50\cos(9t - 40°)$ V.

14.2 A circuit has a transfer function $\mathbf{H}(s) = \dfrac{\mathbf{V}(s)}{\mathbf{I}(s)} = \dfrac{s}{s+1}$.

Find the forced response $v(t)$ if $i(t) = 2\cos t$ A.

14.3 (a) Find the frequency response function for the circuit below with input $i(t)$ and output $v_o(t)$.

(b) Find the forced response $v_o(t)$ to $i(t) = 8\cos(2t + 60°)$ A.

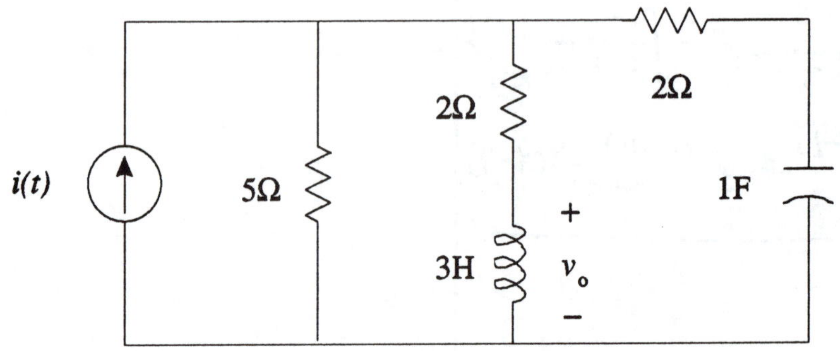

PROBLEM 14.3

14.4 Plot the gain $g(\omega) = |\mathbf{H}(j\omega)|$ and phase shift $\phi(\omega) = \angle\mathbf{H}(j\omega)$ versus ω for the circuit shown.

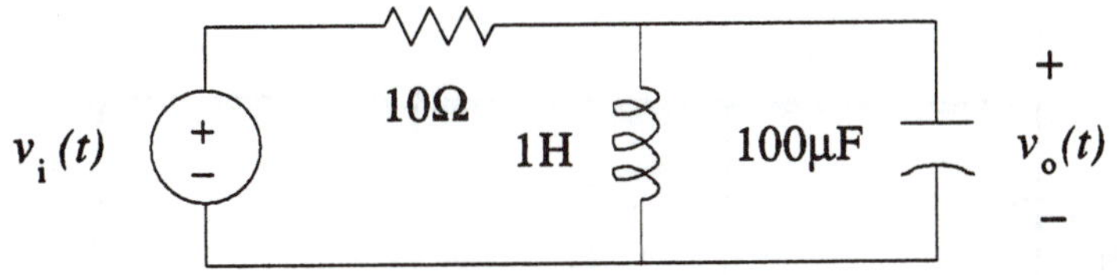

PROBLEM 14.4

14.5 Find the gain and phase shift for $\mathbf{H}(j\omega)$ in Prob. 14.4 at (a) $\omega = 10$ rad/s and (b) $f = 159.155$ Hz.

14.6 (a) Find all frequencies $\omega > 0$ which the amplitude of the output of the circuit in Prob. 14.4 is half the amplitude of the input. (b) Find all frequencies $\omega > 0$ which the phase of the output of the circuit in Prob. 14.4 is leading the input by 60°.

14.7 Find the gain $g(\omega)$ and phase shift $\phi(\omega)$ for the following frequency response functions. Make rough sketches.

(a) $\mathbf{H}(s) = \dfrac{2s+1}{2s-1}$

(b) $\mathbf{H}(s) = \dfrac{s}{2s-1}$

(c) $\mathbf{H}(s) = \dfrac{s}{2s+1}$

(d) $\mathbf{H}(s) = \dfrac{s+1}{(s+3)(s+5)}$

14.8 Let $R = 1\text{k}\Omega$, $L = 1\text{mH}$, and $C = 10\mu\text{F}$. Find the maximum amplitude of $|\mathbf{H}(j\omega)|$ and the frequency at which this occurs.

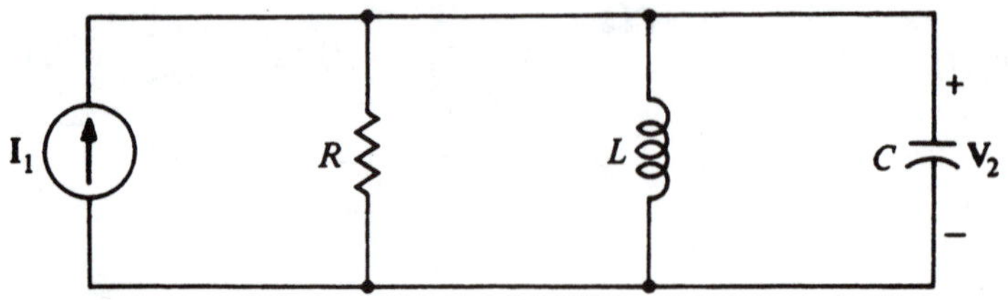

PROBLEM 14.8

14.9 For the circuit shown, $R = 20\Omega$, $L = 0.1\text{H}$, and $C = 0.001\text{F}$. If the input and output are $\mathbf{V}_1$ and $\mathbf{V}_2$, respectively, find the transfer function for the circuit and show that the peak amplitude and zero phase occur at $\omega = 100$ rad/s.

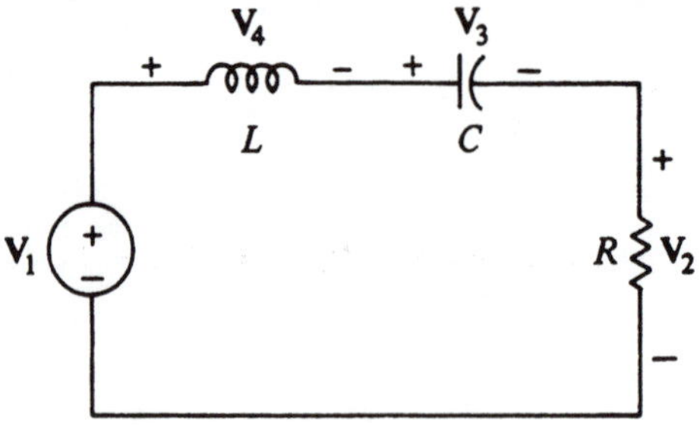

PROBLEM 14.9

14.10 For the circuit shown, find the network transfer function $\mathbf{H}(s) = \mathbf{V}_2(s) / \mathbf{V}_1(s)$. Show that the peak amplitude and zero phase occur at $\omega = 0$.

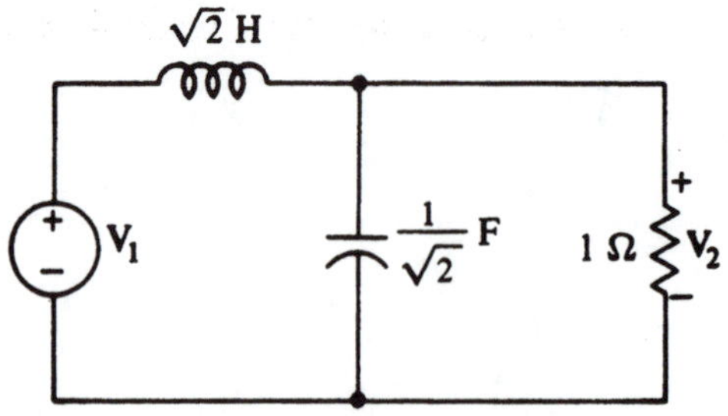

PROBLEM 14.10

14.2 The Decibel Scale

14.11 $5 \approx 14$ dB. Convert the following to decibels without use of a calculator.

(a) 2.5
(b) 20
(c) 25
(d) 2.5×10^4
(e) 2.5×10^{-7}
(f) 0.5
(g) 50
(h) $\sqrt{5}$
(i) $\sqrt[3]{5}$
(j) $\sqrt{10}$

14.12 Convert the following to decibels without using a calculator.

(a) $2^{9/2}$
(b) $\sqrt[4]{10}$
(c) (2)(10)
(d) $500^{4/3}$
(e) $(0.25)^{30}$
(f) $2/5$
(g) 4×10^{20}

14.13 $0.4 \approx -8$ dB. Convert the following decibel values to natural numbers without use of a calculator.

(a) 8 dB
(b) −16 dB
(c) 12 dB
(d) −4 dB
(e) −12 dB
(f) −5 dB
(g) −11 dB

14.14 Convert the following from decibels to natural numbers without using a calculator.

(a) 20 dB
(b) −20 dB
(c) 6 dB
(d) −6 dB
(e) 3 dB
(f) −3 dB
(g) 400 dB
(h) −17 dB

14.15 Convert the following numbers to decibels using a calculator.

(a) 3
(b) 0.3
(c) 8.7×10^{-13}
(d) 2.65×10^{5}
(e) $\sqrt{13}$

14.16 Convert the following decibel values to natural numbers using a calculator.

(a) −8.7 dB
(b) 2.4 dB
(c) −1 dB
(d) 25.105 dB
(e) 87 dB

14.3 Bode Gain (Amplitude) Plots

14.17 Sketch the uncorrected and corrected Bode gain plot for the following transfer functions.

(a) $\mathbf{H}(s) = -5(s+10)^2$

(b) $\mathbf{H}(s) = 5(s+10)^{-3}$

(c) $\mathbf{H}(s) = 17s^2$

(d) $\mathbf{H}(s) = \dfrac{20}{s}$

14.18 Sketch the uncorrected and corrected Bode gain plot for

$$\mathbf{H}(s) = \frac{2\,(s+3)^2}{(s+10)^2\,(s+100)^2}.$$

14.19 Sketch the uncorrected and corrected Bode gain plot for

$$\mathbf{H}(s) = \frac{3s^2 + 15s + 18}{s^2 + s}.$$

14.20 Sketch the uncorrected and corrected Bode gain plot for

$$\mathbf{H}(s) = (s^2 + s + 2500)^3$$

14.21 Sketch the uncorrected and corrected Bode gain plot for

$$\mathbf{H}(s) = \frac{10}{s^2 + 11s + 4}$$

14.22 Sketch the uncorrected and corrected Bode gain plot for $\mathbf{H}(s)$ of the circuit in Prob. 14.4.

14.4 Resonance

14.23 Design a series *RLC* resonant circuit with resonant frequency $f_r = 930$ kHz and $B = 95000$ rad/s. Fix $C = 0.01\mu\text{F}$.

14.24 Design a parallel *RLC* resonant circuit with resonant freqiency $f_r = 10$ kHz and $Q = 50$. Fix $C = 0.01\mu\text{F}$.

14.25 (a) Design a series *RLC* resonant circuit whose impedance (magnitude) goes through a minimum at $\omega_r = 377$ rad/s and that has a bandwidth $B = 1$ rad/s. Fix $L = 100\text{mH}$.
(b) Design a parallel *RLC* resonant circuit whose impedance (magnitude) goes through a maximum at $\omega_r = 377$ rad/s and that has a bandwidth $B = 1$ rad/s. Fix $L = 100\text{mH}$.
(c) What is the impedance of each circuit at the resonant requency?

14.5 Frequency Response of Op Amps

14.26 The inverting amp shown has a gain-bandwidth product of 500kHz. Find the gain-bandwidth product of the open-loop op amp.

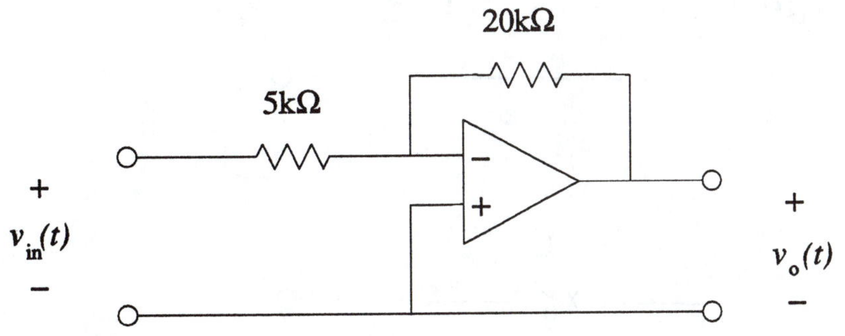

PROBLEM 14.26

14.27 An op amp with a 1.5 MHz gain-bandwidth product is used as an inverting amp. Find the bandwidth of this circuit (which equals the 3 dB cutoff frequency ω_c) if the gain R_F / R_A is set to: (a) 2, (b) 50, (c) 500.

14.6 Filters

14.28 For each of the following filters, identify the type of filter and the half-power (cutoff) frequencies.

(a)

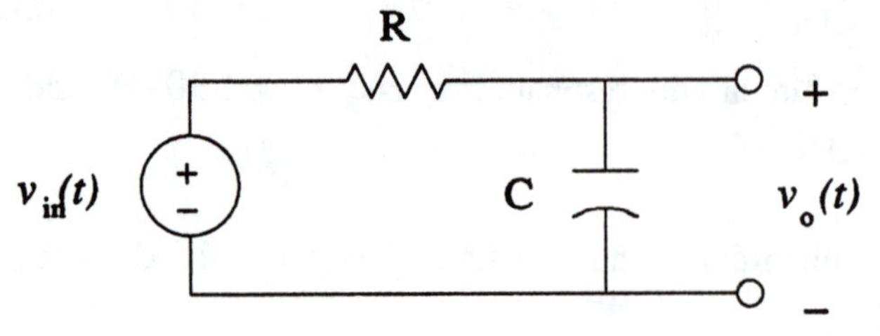

(b)

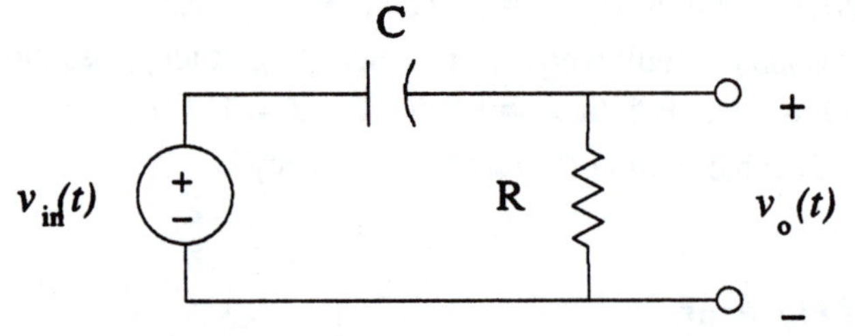

(c)

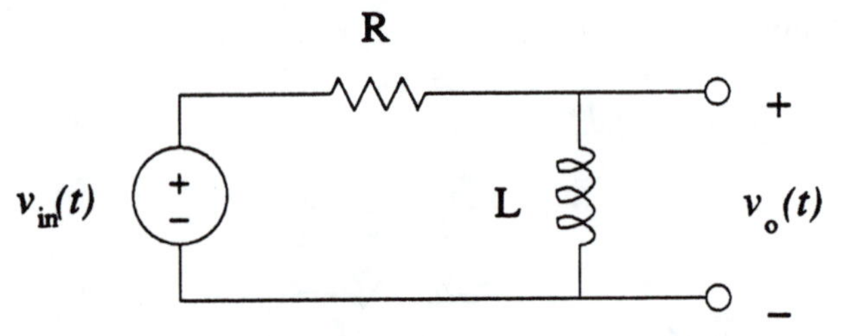

(d)

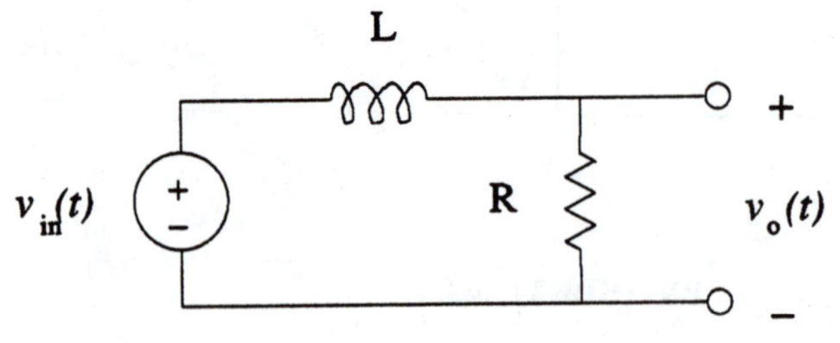

(e)

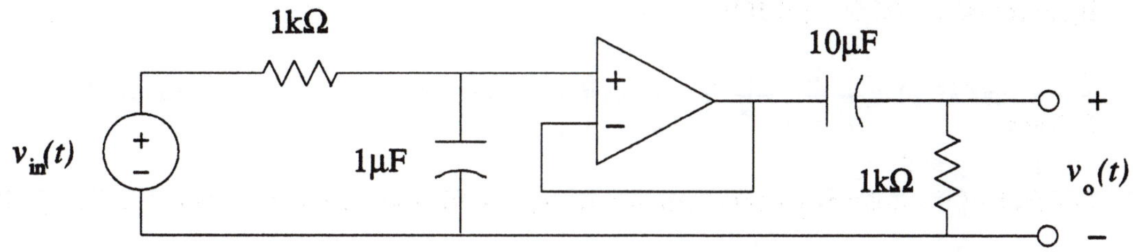

(f)

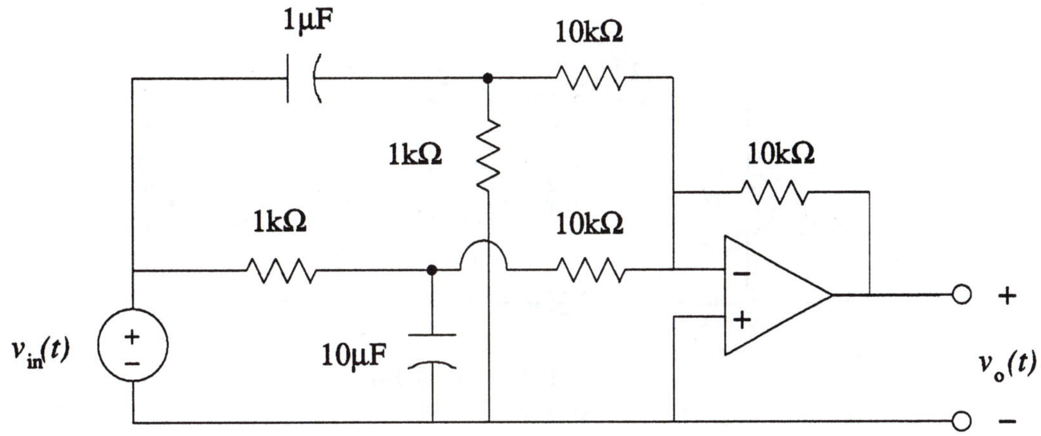

14.29 (a) Design a fourth-order low-pass filter out of resistors, capacitors, and voltage followers that has the transfer function

$$\mathbf{H}(s) = \frac{1}{(s+2)(s+10)(s+300)^2}.$$

(b) Design a fourth-order high-pass filter out of resistors, capacitors, and voltage followers that has the transfer function

$$\mathbf{H}(s) = (s+2)(s+10)(s+300)^2.$$

14.30 Show that $\mathbf{H}(s) = \dfrac{6s}{s^2 + s + 4}$ is the transfer function of a bandpass filter, and find ω_0, ω_{c1}, and ω_{c2}.

14.31 Show that $\mathbf{H}(s) = \dfrac{6s^2}{s^2 + s + 4}$ is the transfer function of a high-pass filter, and find $|\mathbf{H}(j\omega)|_{max}$ and ω_c.

14.32 Show that $\mathbf{H}(s) = \dfrac{6(s^2 + 4)}{s^2 + s + 4}$ is the transfer function of a bandstop filter, and find $|\mathbf{H}(j\omega)|_{max}$, ω_0, ω_{c1}, and ω_{c2}.

14.33 Show that the given circuit is a low-pass filter with $\omega_c = 1$ rad/s by finding $\mathbf{H}(s) = \mathbf{V}_2(s)/\mathbf{V}_1(s)$ and the amplitude response.

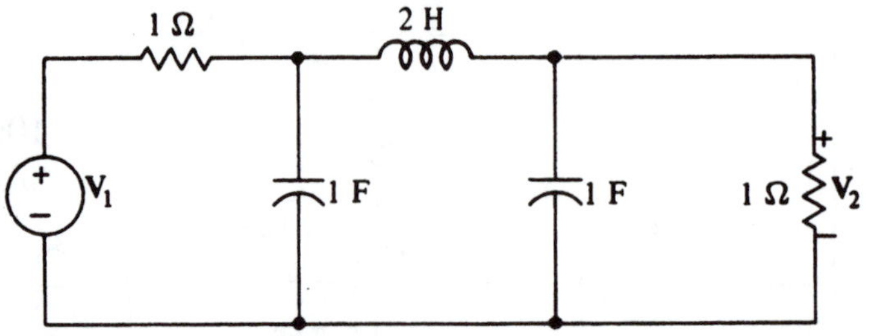

PROBLEM 14.33

14.34 Show that the given circuit is a bandpass filter by finding $\mathbf{H}(s) = \mathbf{V}_2(s)/\mathbf{V}_1(s)$. Find the maximum gain, the bandwidth, and the center frequency.

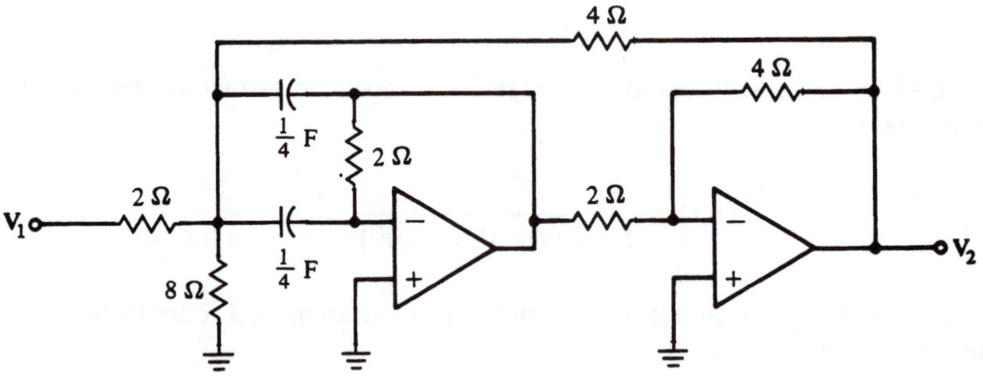

PROBLEM 14.34

14.35 Determine R_1 and R_2 so that the circuit is a first-order lowpass filter with $\omega_c = 10^5$ rad/s and $\mathbf{H}(0) = -2$. Use a 1nF capacitor.

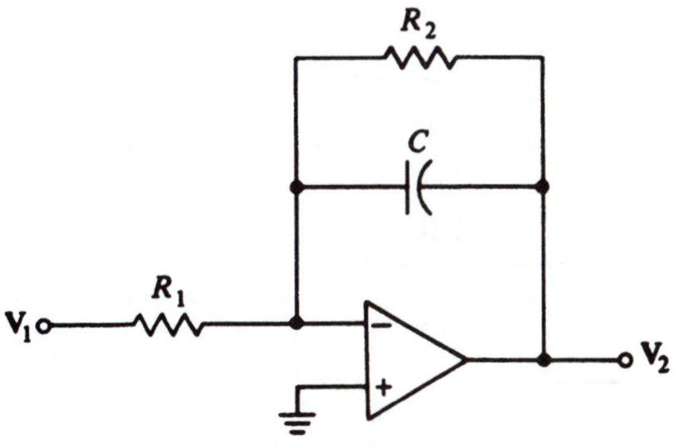

PROBLEM 14.35

14.7 Scaling

14.36 Suppose a circuit is impedance scaled by 1000. Five transfer functions in this circuit are:

$$\mathbf{H}_1(s) = \frac{\mathbf{I}_1(s)}{\mathbf{V}_1(s)},\ \mathbf{H}_2(s) = \frac{\mathbf{V}_2(s)}{\mathbf{I}_2(s)},\ \mathbf{H}_3(s) = \frac{\mathbf{V}_3(s)}{\mathbf{V}_4(s)},\ \mathbf{H}_4(s) = \frac{\mathbf{I}_5(s)}{\mathbf{I}_4(s)},\ \text{and}\ \mathbf{H}_5(s) = \frac{\mathbf{I}_3(s)}{\mathbf{V}_5(s)}.$$

Which will change due to impedance scaling and by what factor?

14.37 A series *RLC* circuit is frequency scaled by 400 and then impedance scaled by 25. What are the new *RLC* values?

14.38 A handbook lists a certain fitler transfer function with $\omega_c = 1$ as

$$\mathbf{H}(s) \equiv \frac{\mathbf{V}_2(s)}{\mathbf{V}_1(s)} = \frac{2s^2 + 3s + 14}{s^5 + 0.7s^4 + 17s^3 + 0.53s^2 + s + 1}.$$

(a) Find $\mathbf{H}(s)$ after impedance scaling by 1000.

(b) Find $\mathbf{H}(s)$ after frequency scaling by 10.

(c) Find $\mathbf{H}(s)$ after frequency scaling by 1/2.

14.39 (a) The circuit shown is a third-order lowpass Butterworth filter with $\omega_c = 1$ rad/s and a gain of 1. Scale the circuit so that the capacitors are 1μF each and $\omega_c = 2 \times 10^4$ rad/s.

(b) Impedance scale the original circuit by 5000.

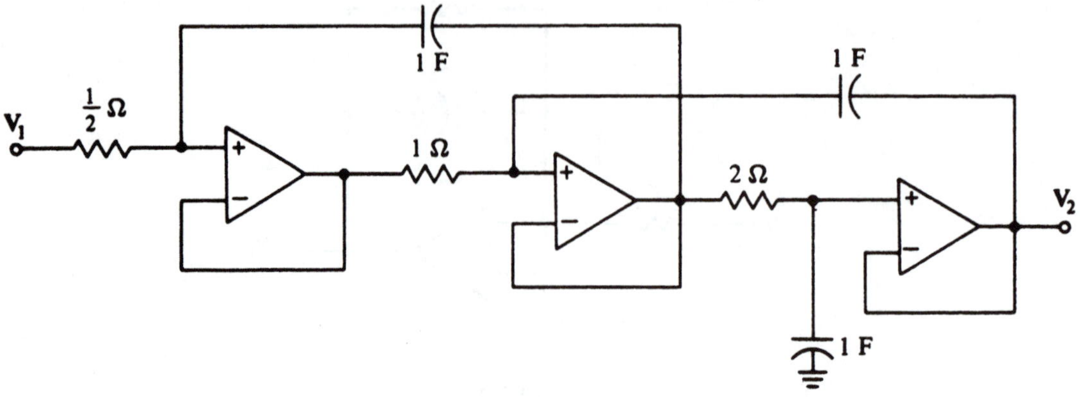

PROBLEM 14.39

14.8 SPICE and Frequency Response

14.40 Use SPICE to determine the frequency response (both gain and phase shift) of the circuit in Prob. 14.3. Use a starting frequency of 0.1 Hz and an ending frequency of 10Hz, with a frequency spacing of 0.1 decades.

14.41 Use SPICE to determine the frequency response (both gain and phase shift) of the circuit in Prob. 14.4. Use a starting frequency of 0.1 Hz and an ending frequency of 10kHz, with a frequency spacing of 0.2 decades.

14.42 Use SPICE to aid in designing a resonant series *RLC* circuit with a maximum peak at $\omega_n = 1884.96$ rad/s at which $20\log\left|\frac{\mathbf{I}(j\omega)}{\mathbf{V}_{in}(j\omega)}\right| = -6$ dB. The circuit should also have a bandwidth $B = 100$ rad/s. Assume that the input voltage is $1\angle 0°$ V for the simulation.

CHAPTER 14 SOLUTIONS

14.1 (a)

$$H(j\omega) = \frac{3(j\omega)^2}{(j\omega)^3 + 2(j\omega)^2 + 5j\omega + 1}$$

$$= \frac{-3\omega^2}{1 - 2\omega^2 + j[5\omega - \omega^3]}$$

$$= \frac{243}{161 + j684} = 0.346\angle -76.75^\circ$$

$$\therefore \underline{V}_o(j\omega) = H(j\omega)\,\underline{V}_i(j\omega)$$

$$= \underline{3.46\angle -56.75^\circ}\ \text{V}$$

(b) $g(\omega = 9) = |H(j9)| = 0.346$

$$\phi(\omega = 9) = \angle H(j9) = -76.75^\circ$$

(c) $v_o(t) = (50)(.346)\cos(9t - 76.75^\circ - 40^\circ)$

$$= \underline{17.3\cos(9t - 116.75^\circ)\ \text{V}}$$

14.2 $\omega = 1 \Rightarrow H(j1) = \dfrac{j}{j+1}$

$$\Rightarrow \underline{V}(j\omega) = H(j\omega)\,\underline{I}(j\omega)$$

$$= \frac{j}{j+1}(2\angle 0^\circ)$$

$$= \sqrt{2}\angle 45^\circ\ \text{V}$$

$$\therefore \underline{v(t) = \sqrt{2}\cos(t + 45^\circ)\ \text{V}}$$

14.3 (a) Freq. domain (phasor) circuit:

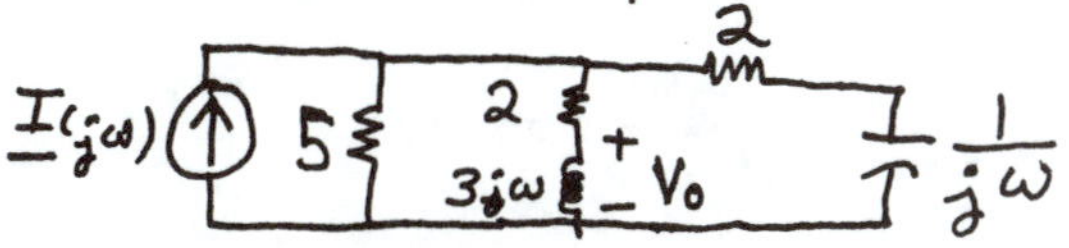

Source transformation: Norton → Thevenin

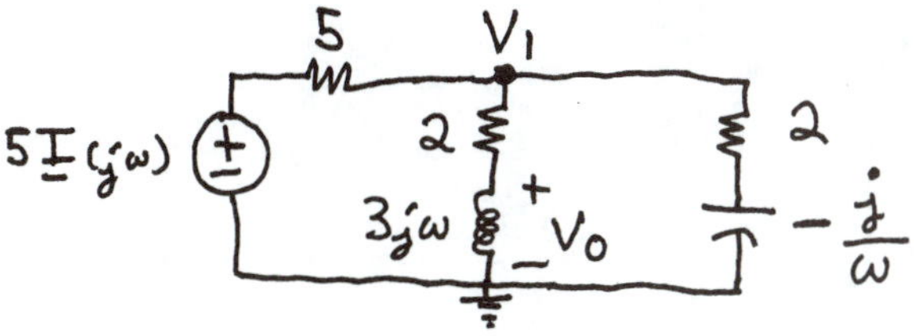

Voltage divider:

$$\underline{V}_1 = (5\underline{I})\frac{(2+3j\omega)\,||\,(2 - j/\omega)}{5 + (2+3j\omega)\,||\,(2 - \frac{j}{\omega})}$$

$$= \ldots\ldots = \frac{5\underline{I}\,[7 + j(6\omega - 2/\omega)]}{27 + j(21\omega - 7/\omega)}$$

Voltage divider:

$$\underline{V}_o = \underline{V}_1 \frac{(3j\omega)}{(2 + 3j\omega)} = \ldots.$$

$$= \frac{\underline{I}\,[30 - 90\omega^2 + j105\omega]}{75 - 63\omega^2 + j(123\omega - 14/\omega)}$$

$$\therefore H(j\omega) \equiv \frac{\underline{V}_o(j\omega)}{\underline{I}(j\omega)} = \frac{(30 - 90\omega^2) + j105\omega}{(75 - 63\omega^2) + j(123\omega - 14/\omega)}$$

(b) $\underline{I} = 8\angle 60^\circ$, $\omega = 2$

$$H(j2) = \frac{330 + 210j}{-177 + 239j} = 1.315\angle -94.05^\circ$$

$$\underline{V}_o = H(j2)\,\underline{I}(j2) = 10.52\angle -34.05^\circ$$

$$\therefore \underline{v_o(t) = 10.52\cos(2t - 34.05^\circ)}$$

14.4 Frequency domain circuit:

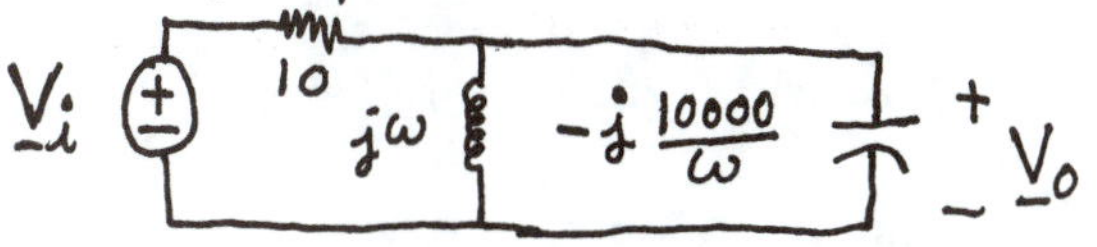

Parallel LC: $(j\omega)\,||\,(-j\frac{10000}{\omega}) = \dfrac{10000}{j(\omega - \frac{10000}{\omega})}$

Voltage divider:

$$\underline{V}_o = \underline{V}_i \frac{10000/[j(\omega - 10000/\omega)]}{10 + \left[\frac{10000}{j(\omega - 10000/\omega)}\right]}$$

Simplifying,

$$H(j\omega) \equiv \frac{\underline{V}_o}{\underline{V}_i} = \frac{1}{1 + j(\frac{\omega}{1000} - \frac{10}{\omega})}$$

14.4 (cont'd)

At natural freq. $\omega_0 = \frac{1}{\sqrt{LC}} = 100 \frac{rad}{s}$

$\Rightarrow H(j100) = \frac{1}{1+j0.09} \Rightarrow |H(j100)| = 1.004$

For $\omega \to 0$, $H(j\omega) \simeq \frac{1}{1-j\frac{10}{\omega}} \simeq \frac{1}{-j\frac{10}{\omega}}$

$\simeq j\left(\frac{\omega}{10}\right)$

For $\omega \to \infty$, $H(j\omega) \simeq \frac{1}{j\omega} \simeq -\frac{j}{\omega}$

Plotting (rough sketch):

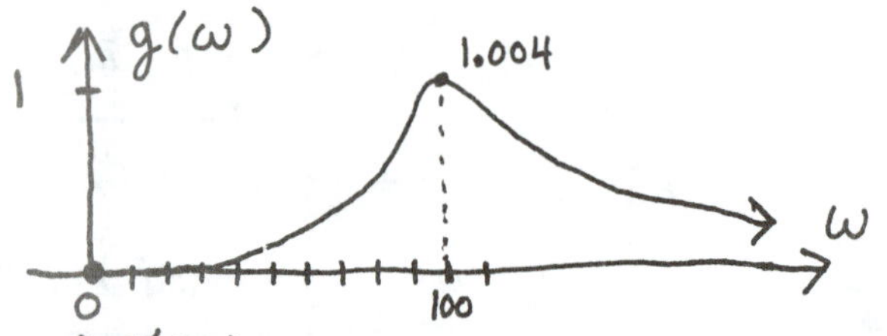

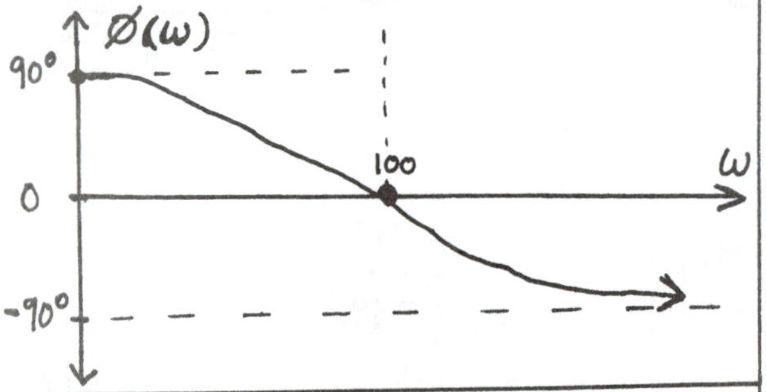

14.5 (a) $\omega = 10 \Rightarrow H(j10) = \frac{1}{1-j0.99}$

$\Rightarrow g(10) = \left|\frac{1}{1-j0.99}\right| = 0.71065$

$\Rightarrow \phi(10) = \angle 1 - \angle 1-j0.99 = 44.71^\circ$

(b) $\omega = 2\pi f = 1000 \Rightarrow H(j1000) = \frac{1}{1+j0.99}$

$\Rightarrow g(1000) = 0.71065$

$\Rightarrow \phi(1000) = -44.71^\circ$

14.6 (a)

$|H(j\omega)| = \frac{1}{2} = \frac{1}{\left|1+j\left(\frac{\omega}{1000} - \frac{10}{\omega}\right)\right|}$

$\Rightarrow 2 = \sqrt{1+\left(\frac{\omega}{1000} - \frac{10}{\omega}\right)^2}$

$\Rightarrow 4 = 1 + \frac{\omega^2}{1\times10^6} + \frac{100}{\omega^2} - 0.02$

$\Rightarrow \frac{\omega^2}{1\times10^6} + \frac{100}{\omega^2} = 3.02$

$\Rightarrow \underline{\omega = 5.7544 \frac{rad}{s}, \omega = 1737.8 \frac{rad}{s}}$

(b) $\angle H(j\omega) = +60^\circ$

$\Rightarrow \angle 1 - \angle\left[1+j\left(\frac{\omega}{1000} - \frac{10}{\omega}\right)\right] = 60^\circ$

$\Rightarrow -\angle\left[1+j\left(\frac{\omega}{1000} - \frac{10}{\omega}\right)\right] = 60^\circ$

$\Rightarrow \underline{\omega = 5.7544 \frac{rad}{s}}$

14.7 (a) $H(j\omega) = \frac{2j\omega+1}{2j\omega-1}$

$\Rightarrow g(\omega) = |H(j\omega)| = \frac{\sqrt{1+4\omega^2}}{\sqrt{1+4\omega^2}} = 1$

$\Rightarrow \phi(\omega) = \angle H(j\omega) = \angle(1+j2\omega) - \angle(-1+j2\omega)$

$= \tan^{-1}\frac{2\omega}{1} - \left[\tan^{-1}\frac{2\omega}{-1} + 180^\circ\right]$

$= \tan^{-1}2\omega - \tan^{-1}(-2\omega) - 180^\circ$

$= 2\tan^{-1}2\omega - 180^\circ$

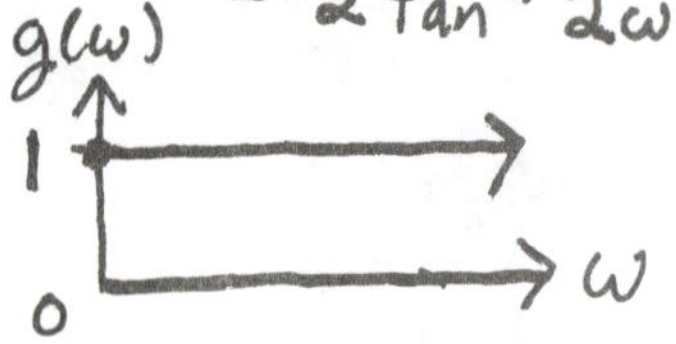

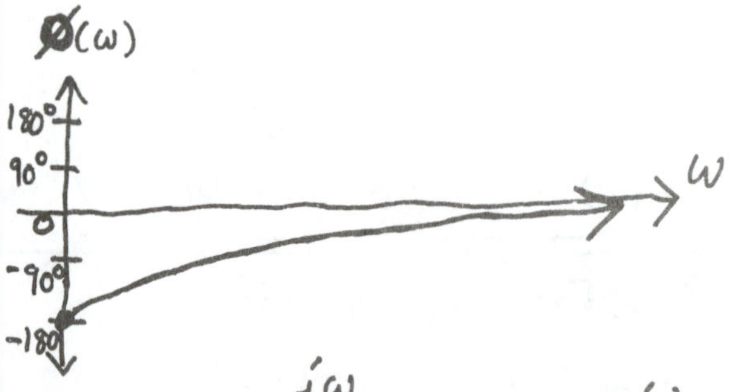

(b) $H(j\omega) = \frac{j\omega}{2j\omega-1} \Rightarrow g(\omega) = \frac{\omega}{\sqrt{1+4\omega^2}}$

As $\omega \to 0$, $g(\omega) \simeq \frac{\omega}{1}$

As $\omega \to \infty$, $g(\omega) \simeq \frac{\omega}{\sqrt{4\omega^2}} = \frac{1}{2}$

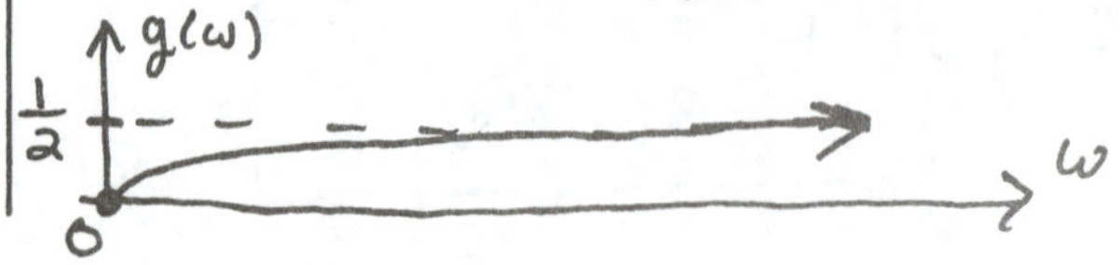

14.7(b) cont'd:

$$\phi(\omega) = \angle j\omega - \angle(-1 + j2\omega)$$
$$= 90^\circ - \tan^{-1}(-2\omega) + 180^\circ$$
$$= \tan^{-1}(2\omega) - 90^\circ$$

(c) $H(j\omega) = \frac{j\omega}{1+j2\omega} \Rightarrow g(\omega) = \frac{\omega}{\sqrt{1+4\omega^2}}$

$\Rightarrow g(\omega)$ same as part (b)

$$\phi(\omega) = \angle j\omega - \angle(1+j2\omega)$$
$$= 90^\circ - \tan^{-1} 2\omega$$

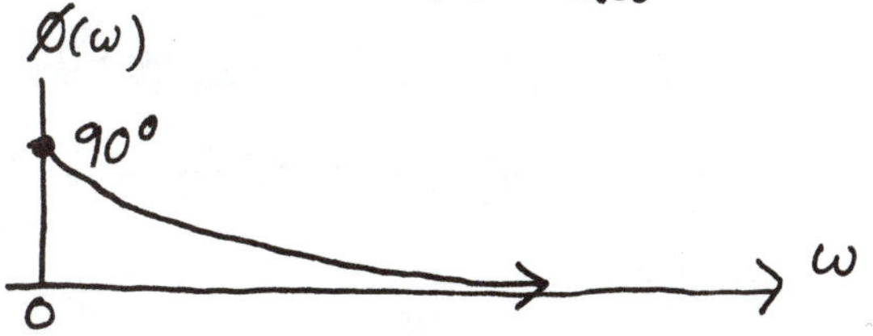

(d) $H(j\omega) = \frac{1+j\omega}{(3+j\omega)(5+j\omega)} = \frac{1+j\omega}{(15-\omega^2)+j8\omega}$

$$g(\omega) = |H(j\omega)| = \sqrt{\frac{\omega^2+1}{\omega^4+34\omega^2+225}}$$

$\Rightarrow g(0) = \sqrt{\frac{1}{225}} = \frac{1}{15}$

$g(\omega \to \infty) \simeq \sqrt{\omega^2/\omega^4} = \frac{1}{\omega}$

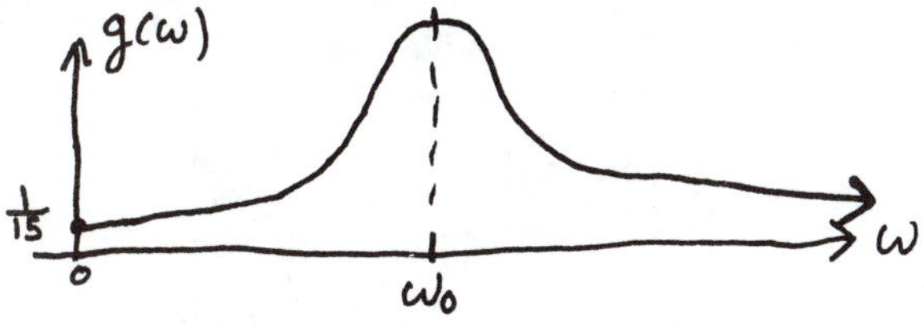

$$\phi(\omega) = \begin{cases} \tan^{-1}\omega - \tan^{-1}\frac{8\omega}{15-\omega^2} \\ \tan^{-1}\omega - \tan^{-1}\frac{8\omega}{15-\omega^2} - 180^\circ \end{cases}$$

Depending on the quadrant....

14.8 $\omega_0 = 1/\sqrt{LC} = 1/\sqrt{(10^{-3})(10^{-5})}$

$\Rightarrow \omega_0 = \underline{10000 \text{ rad/s}}$

$|H(j\omega)|_{max} = |H(j\omega_0)| = R = \underline{1000}$.

14.9

$$H(s) \equiv \frac{V_2}{V_1} = \frac{R}{R + sL + 1/sC}$$
$$= \frac{200s}{s^2 + 200s + 10^4}$$

$$|H(j\omega)| = \frac{R}{\sqrt{R^2 + (\omega L - \frac{1}{\omega C})^2}}$$

$|H(j\omega)|_{max}$ occurs when $\omega L - \frac{1}{\omega C} = 0$

$\Rightarrow$ at $\omega_0 = \frac{1}{\sqrt{LC}} = 100$ rad/s

$|H(j\omega_0)| = \underline{1} \Rightarrow \phi(\omega_0) = \underline{0^\circ}$

14.10 KCL yields:

$$\underline{V}_2\left(1 + \frac{s}{\sqrt{2}} + \frac{1}{s\sqrt{2}}\right) = \underline{V}_1\left(\frac{1}{s\sqrt{2}}\right)$$

$$H(s) \equiv \frac{\underline{V}_2}{\underline{V}_1} = \frac{1}{s^2 + s\sqrt{2} + 1}$$

$$|H(j\omega)| = \frac{1}{\sqrt{(1-\omega^2)^2 + 2\omega^2}} = \frac{1}{\sqrt{1+\omega^4}}$$

$|H(j\omega)|_{max}$ occurs when $\omega = 0$

$\Rightarrow \underline{|H(0)| = 1 \text{ and } \phi(0) = 0^\circ}$

14.11

(a) $2.5(dB) = \frac{5}{2}(dB) = 5(dB) - 2(dB)$
$= 14\,dB - 6\,dB = \underline{8dB}$

(b) $20(dB) = 5\cdot2\cdot2\,(dB) = 14dB + 6dB + 6dB = \underline{26\,dB}$

(c) $25(dB) = 5^2(dB) = 2[14\,dB] = \underline{28dB}$

(d) $2.5\times10^4 = 8dB + 4(20dB) = \underline{88dB}$

(e) $2.5\times10^{-7} = 8dB - 7(20dB) = \underline{-132dB}$

(f) $0.5 = 5\times10^{-1} = 14dB - 20dB = \underline{-6dB}$

(g) $50 = 5\times10 = 14dB + 20dB = \underline{34dB}$

(h) $\sqrt{5} = 5^{1/2} = \frac{1}{2}[14dB] = \underline{7dB}$

(i) $\sqrt[3]{5} = 5^{1/3} = \frac{1}{3}[14dB] = \underline{\frac{14}{3}dB}$

(j) $\sqrt{10} = 10^{1/2} = \frac{1}{2}[20dB] = \underline{10dB}$

14.12 (a) $2^{9/2} = \frac{9}{2}(6dB) = \underline{27dB}$

(b) $10^{1/4} = \frac{1}{4}(20dB) = \underline{5dB}$

(c) $6dB + 20dB = \underline{26dB}$

(d) $(5\times100)^{4/3} = \frac{4}{3}[14dB + 40dB] = \underline{72dB}$

(e) $(0.25)^{30} = ((0.5)^2)^{30} \simeq 60(-6dB)$
$= \underline{-360dB}$

(f) $2/5 = 6dB - 14dB = \underline{-8dB}$

(g) $4\times10^{20} = 2\times2\times10^{20} = 6dB + 6dB + 20(20dB)$
$= \underline{412dB}$

14.13 (a) $8dB = \frac{1}{0.4} = \frac{10}{4} = 2.5$

(b) $-16dB = -8dB - 8dB = (0.4)(0.4) = 0.16$

(c) $12dB = 20dB - 8dB = \frac{10}{2.5} = 4$

(d) $-48dB = -4(12dB) = 4^{-4} = \frac{1}{256}$

(e) $-12dB = \frac{1}{4}$ from part (c)

(f) $-5dB = -8dB + 3dB = 0.4\sqrt{2}$

(g) $-11dB = -8dB - 3dB = 0.4/\sqrt{2}$
$= 0.2\sqrt{2}$

14.14

(a) 10

(b) 0.1

(c) 2

(d) $1/2$

(e) $\sqrt{2}$

(f) $1/\sqrt{2} = \sqrt{2}/2$

(g) $400dB = 4(100dB) = (10^5)^4 = 10^{20}$

(h) $-17dB = -20dB + 3dB = 0.1\sqrt{2}$

14.15 (a) 9.54 (b) −10.46 (c) −241.21 (d) 108.46 (e) 11.14

14.16 (a) 0.367 (b) 1.318 (c) 0.891 (d) 18 (e) 22387.2

14.17 (a) $H(s) = H_1(s)\,H_2(s)$

$H_1(s) = -5,\ H_2(s) = (s+10)^2$

$\Rightarrow 20\log|H_1(j\omega)| = 20\log 5 = 14\,dB$

$20\log|H_2(j\omega)| = 20\log(\omega^2 + 100)$

As $\omega\to0$, $20\log|H_2(j\omega)| \simeq 20\log100 = 40dB$

As $\omega\to\infty$, $20\log|H_2(j\omega)| \simeq 40\log\omega$

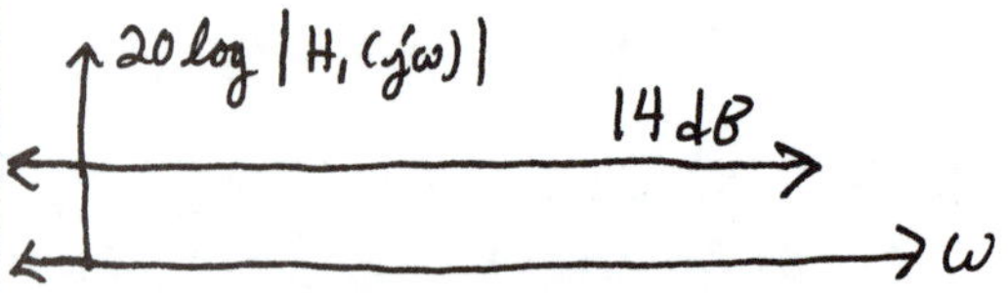

+

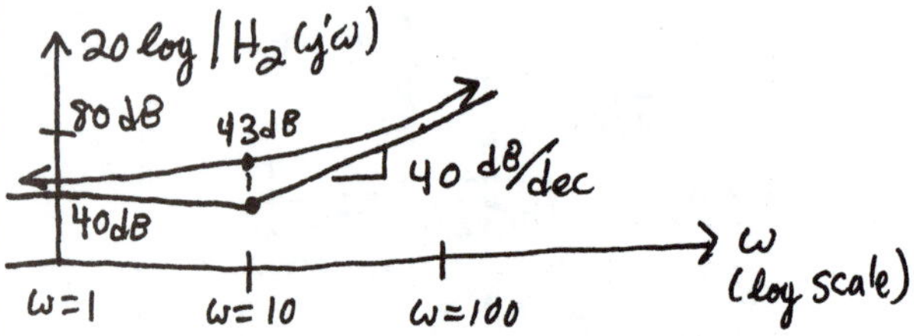

=

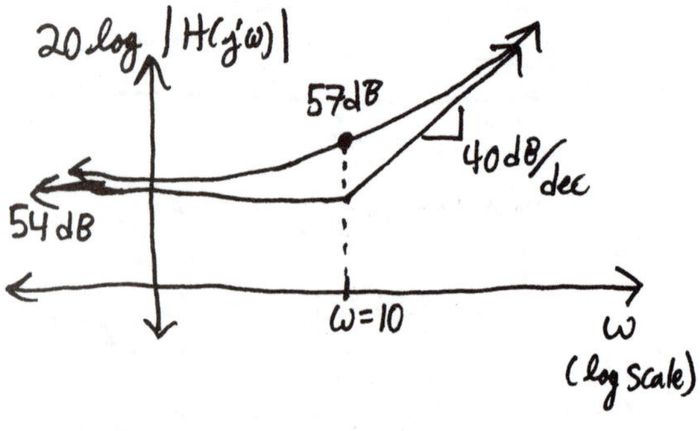

14.17 (b) $H(s) = 5\,(s+10)^{-3} = H_1(s)\,H_2(s)$

$H_1(s) = 5\,,\quad H_2(s) = (s+10)^{-3}$

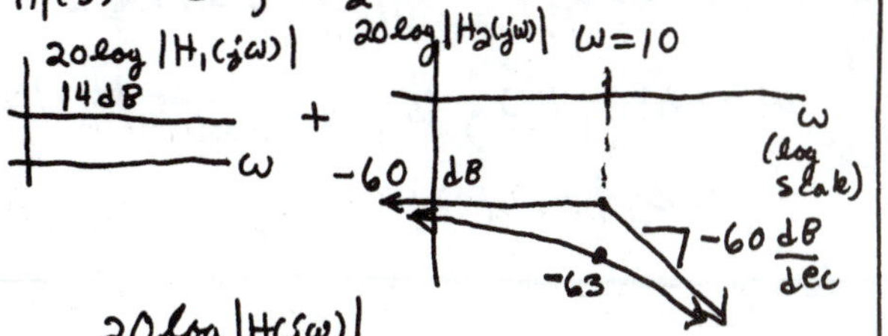

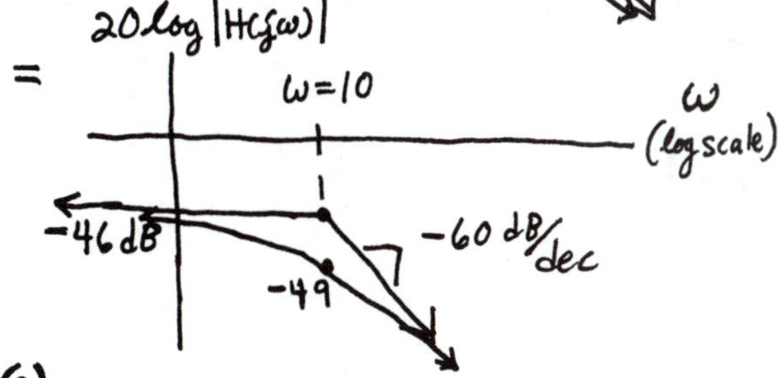

(c)

$|H(j\omega)| = 17\,\omega^2$

$20\log 17 = 24.6\text{ dB}\,,\quad 20\log\omega^2 = 40\log\omega$

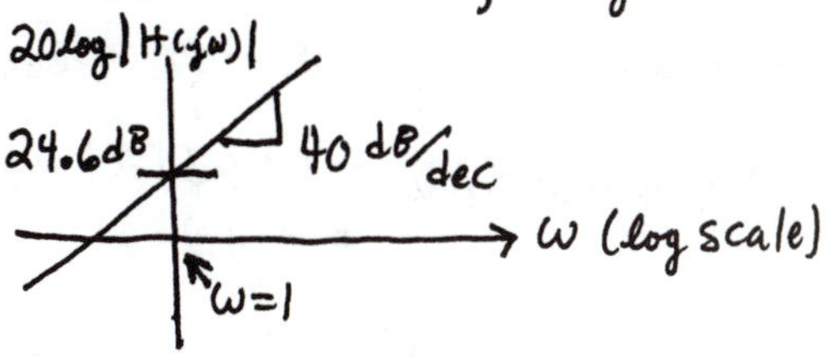

(d) $20\log|H(j\omega)| = 20\log 20 - 20\log\omega$
$= 26\text{dB} - 20\log\omega$

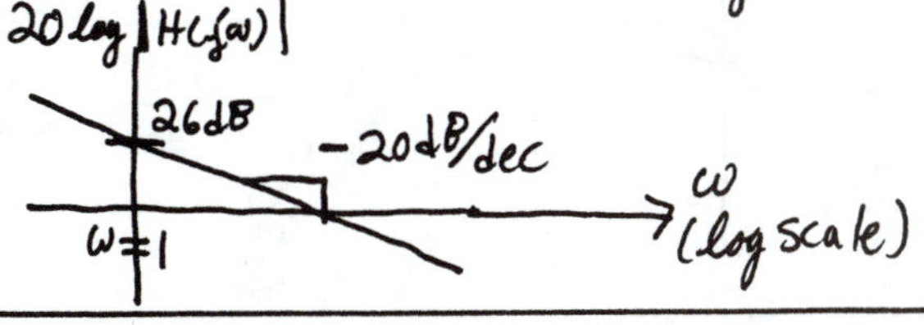

14.18 $H(s) = 2\cdot(s+3)^2\cdot(s+10)^{-2}\cdot(s+100)^{-3}$

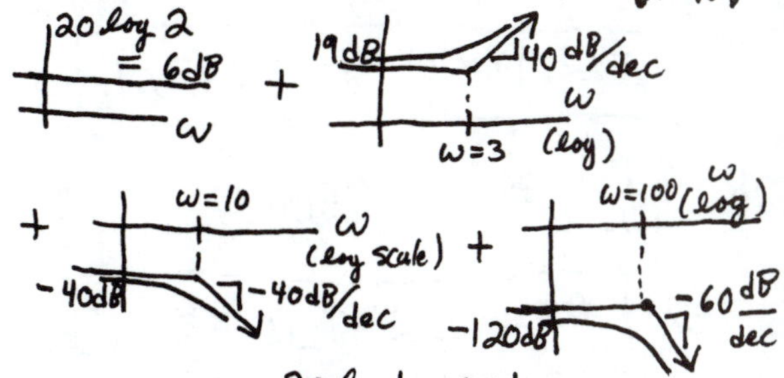

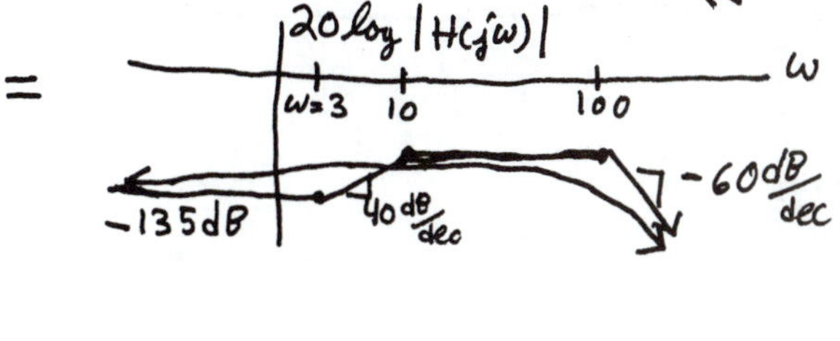

14.19 $H(s) = \dfrac{3(s+2)(s+3)}{s(s+1)}$

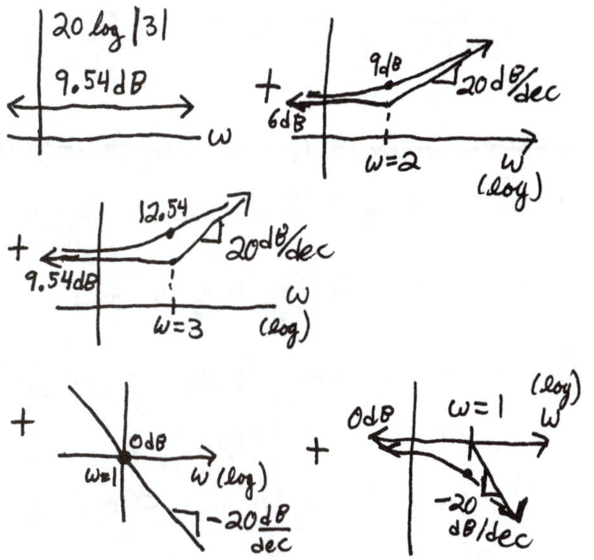

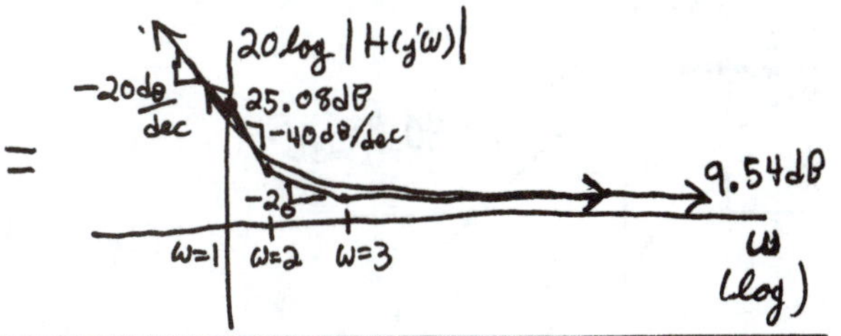

14.20 $\omega_n = \sqrt{2500} = 50\,,\ \zeta = \dfrac{1}{2(50)} = 0.01$

$< 0.5 \Rightarrow$ OUTSIDE KNEE

As $\omega \to 0$,

$20\log|H(j\omega)| = 20\log(2500)^3 = 203.88\text{dB}$

As $\omega \to \infty$,

$20\log|H(j\omega)| = 20\log(\omega^2)^3 = 120\log\omega$

At $\omega = \omega_n = 50$, $20\log|H(j\omega)| = 101.9\text{dB}$

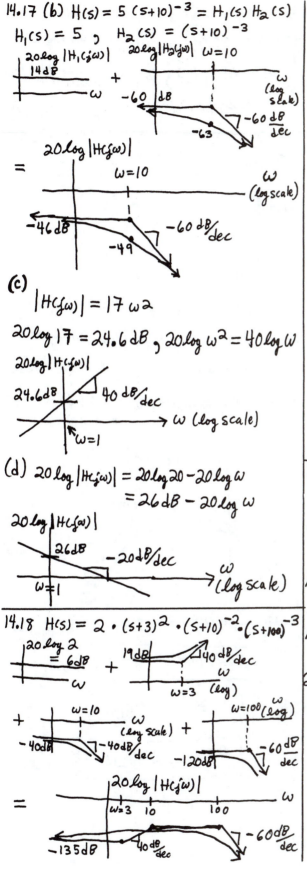

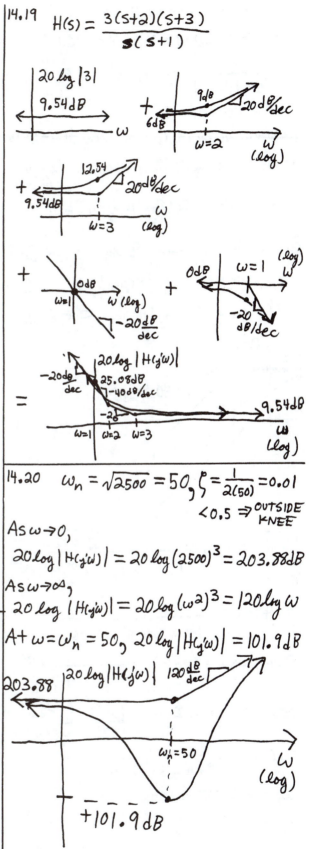

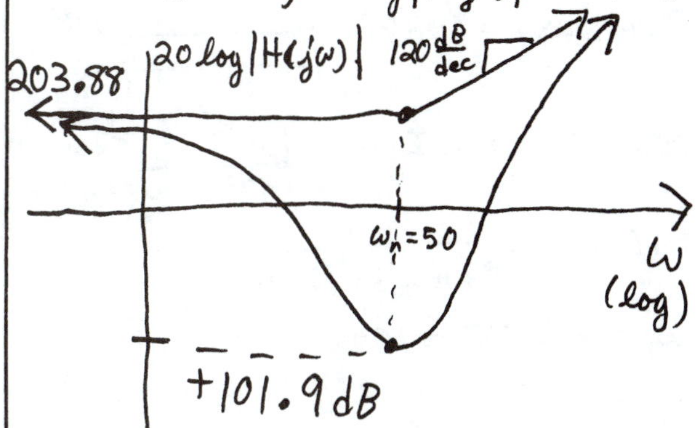

14.21 $H(s) = 10\cdot(s^2+11s+4)^{-1}$

$H_1 = 10 \Rightarrow 20\log|H_1| = 20\log 10 = 20\,dB$

$H_2 = \frac{1}{s^2+11s+4} \Rightarrow \omega_n = \sqrt{4} = 2$

$\zeta = \frac{11}{2(2)} = 2.75$

As $\omega \to 0$, $20\log|H_2(j\omega)| = 20\log\frac{1}{4} = -12\,dB$

As $\omega \to \infty$, $20\log|H_2(j\omega)| = 20\log\omega^{-2} = -40\,(\log\omega)$

At $\omega = \omega_n = 2$, $20\log|H_2(j2)| = -26.85\,dB$

For total $H(j\omega)$, $\omega \to 0$, $20\log|H(j\omega)| \to 8\,dB$

As $\omega \to \infty$, $20\log|H(j\omega)| \to -40\log\omega$

At ω_n, $20\log|H(j\omega_n)| = -6.85\,dB$

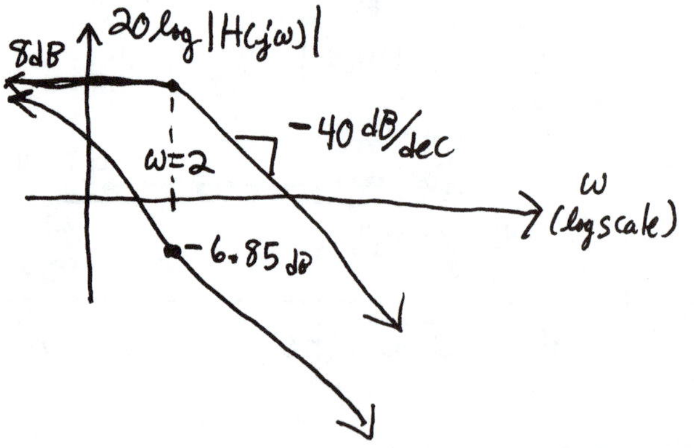

14.22 $H(j\omega) = \frac{1}{1 + \frac{j\omega}{1000} + \frac{10}{j\omega}} = \frac{1000j\omega}{(1000j\omega + (j\omega)^2 + 10000)}$

$\Rightarrow H(s) = \frac{1000s}{s^2 + 1000s + 10000}$

$20\log(1000) = 60\,dB$

$20\log|s| = 20\log\omega$

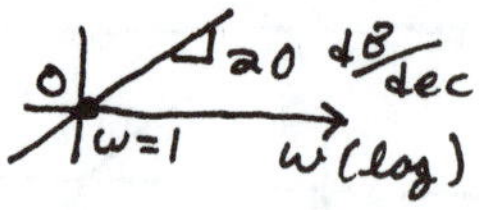

$H_3(s) = \frac{1}{s^2 + 1000s + 10000}$

$\omega_n = \sqrt{10000} = 100$

$\zeta = \frac{1000}{2(100)} = 5$

As $\omega \to 0$, $20\log|H_3(j\omega)| \to -60\,dB$

As $\omega \to \infty$, $20\log|H_3(j\omega)| \to -40\log\omega$

At $\omega = \omega_n = 100$, $20\log\left|\frac{1}{1000(100)}\right| = -100\,dB$

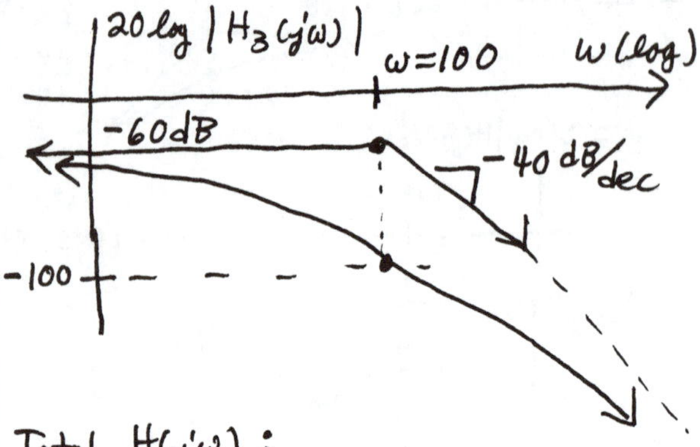

Total $H(j\omega)$:

At $\omega \approx 1$, $20\log|H(j\omega)| \approx 60 + 0 - 60 = 0\,dB$

As $\omega \to 0$, $20\log|H(j\omega)| \to 20\log\omega$

As $\omega \to \infty$, $20\log|H(j\omega)| \to -20\,dB/dec$

At $\omega = 100$, $20\log|H(j\omega)| = 60 + 2(20) - 100 = 0\,dB$

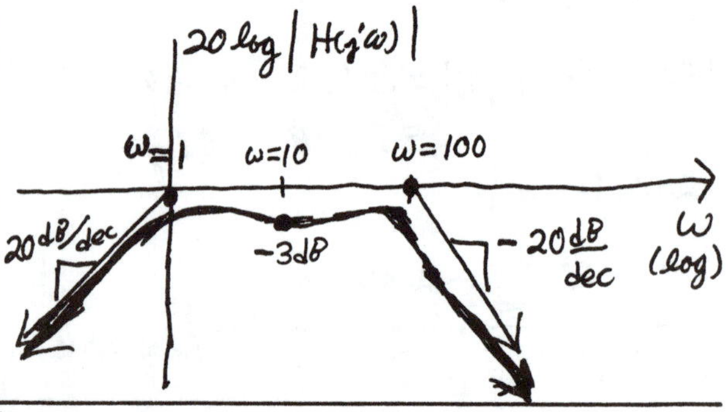

14.23 $\omega_r = 2\pi f_r = 5.8434\times10^6\,\frac{rad}{s} = \frac{1}{\sqrt{LC}}$

$\Rightarrow$ **$L = 2.93\times10^{-6}\,H = 2.93\,\mu H$**

$B = \frac{R}{L} \Rightarrow R = 95000L$

$\Rightarrow$ **$R = 0.278\,\Omega = 278\,m\Omega$**

14.24

$\omega_r = 2\pi f_r = 6.283\times10^4 \frac{rad}{s} = \frac{1}{\sqrt{LC}}$

$\Rightarrow \boxed{L = 0.0253H = 25.3H}$

$Q = R\sqrt{\frac{C}{L}} = 50 \Rightarrow R = 50\sqrt{\frac{L}{C}}$

$\Rightarrow \boxed{R = 79.577k\Omega}$

14.25

(a) $\omega_r = \frac{1}{\sqrt{LC}} = 377 \Rightarrow \underline{C = 70.36\mu F}$

$B = \frac{R}{L} = 1 \Rightarrow \underline{R = 100m\Omega}$

(b) $\omega_r = \frac{1}{\sqrt{LC}} = 377 \Rightarrow \underline{C = 70.36\mu F}$

$B = \frac{1}{RC} = 1 \Rightarrow \underline{R = \frac{1}{C} = 14.21k\Omega}$

(c) Series: $Z_{min} = R = \underline{100m\Omega}$

Parallel: $Z_{max} = R \,||\, [j\omega_r L \,||\, -\frac{j}{\omega_r C}]$

$= R \,||\, \infty = R$

$= \underline{14.21k\Omega}$

14.26

$2\pi(500kHz) = 3.14159\times10^6 \frac{rad}{s}$

$= (A_o\omega_o)\left[\frac{R_F/R_A}{1 + R_F/R_A}\right]$

$= \frac{4}{5}(A_o\omega_o)$

$\Rightarrow \boxed{GBW = A_o\omega_o = 3.927\times10^6\ rad/s = 625\ kHz}$

14.27

(a) $GBW = (A_o\omega_o)\frac{R_F/R_A}{1+R_F/R_A} = 2\pi\times(1.5\times10^6)\frac{2}{1+2}$

$= \underline{2\pi\times10^6 \frac{rad}{s}}$

$2\pi\times10^6 = \frac{R_F}{R_A}\omega_c = 2\,\omega_c$

$\Rightarrow \boxed{\omega_c = \pi\times10^6\ rad/s \quad f_c = 500\ kHz}$

(b) $GBW = 2\pi(1.5\times10^6)\frac{50}{1+50} = 9.24\times10^6 \frac{rad}{s}$

$= 50\omega_c$

$\Rightarrow \boxed{\omega_c = 1.848\times10^5 \frac{rad}{s} \Rightarrow f_c = 29.41kHz}$

(c) $GBW = 2\pi(1.5\times10^6)\frac{500}{501} = 9.405966\times10^6$

$= 500\,\omega_c$

$\Rightarrow \boxed{\omega_c = 1.8812\times10^4 \frac{rad}{s} \Rightarrow f_c = 2.994kHz}$

14.28

(a) $H(s) = \frac{V_{out}}{V_{in}} = \frac{\frac{1}{sC}}{R+\frac{1}{sC}} = \frac{1}{1+RCs}$

Bode gain plot:

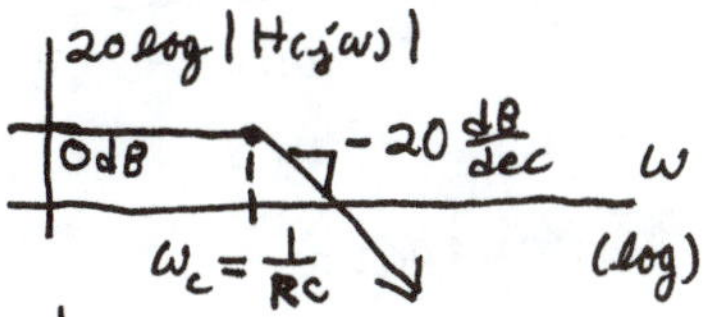

$\Rightarrow$ Lowpass, $\underline{\omega_c = \frac{1}{RC}}$

(b) $H(s) = \frac{R}{R+\frac{1}{sC}} = \frac{RCs}{RCs+1}$

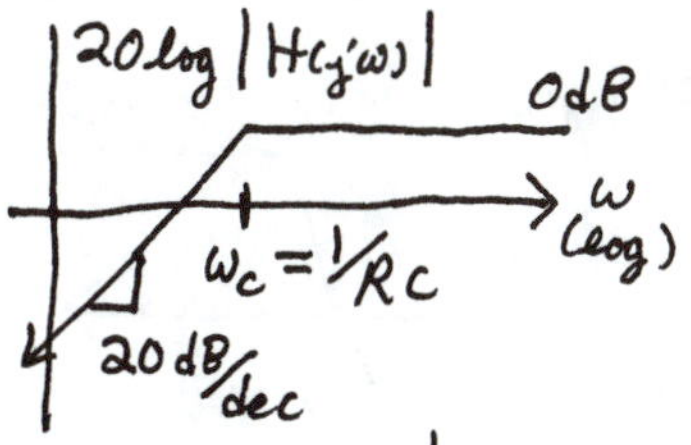

$\Rightarrow$ Highpass, $\omega_c = \frac{1}{RC}$

(c) $H(s) = \frac{Ls}{R+Ls} = \frac{(\frac{L}{R})s}{1+(\frac{L}{R})s}$

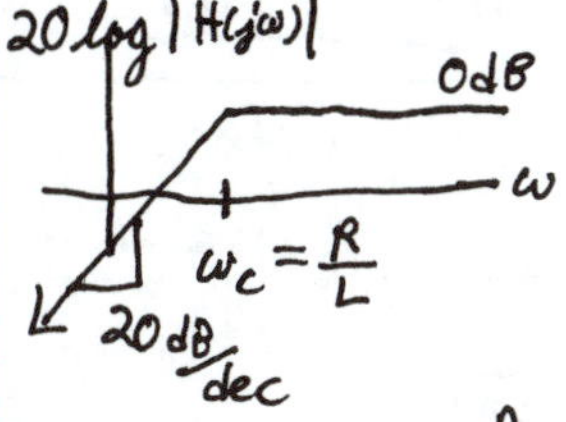

$\Rightarrow$ Highpass, $\omega_c = \frac{R}{L}$

(d) $H(s) = \frac{R}{R+Ls} = \frac{1}{1+(\frac{L}{R})s}$

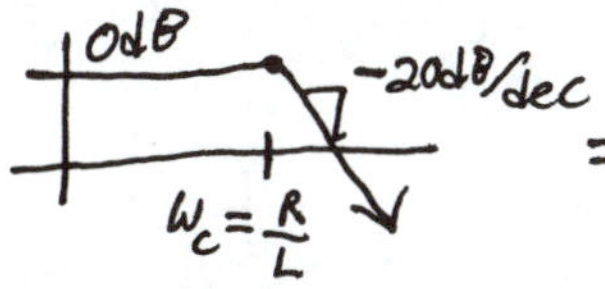

$\Rightarrow$ Lowpass, $\omega_c = \frac{R}{L}$

14.28 (cont'd)

(e) Left half : $H_1(s) = LPF$

with $\omega_{c1} = \frac{1}{R_1 C_1} = 1000 \frac{rad}{s}$

Right half: $H_2(s) = HPF$ with

$\omega_{c2} = \frac{1}{R_2 C_2} = 100 \frac{rad}{s}$

$H_{TOTAL}(s) = H_1(s) H_2(s)$

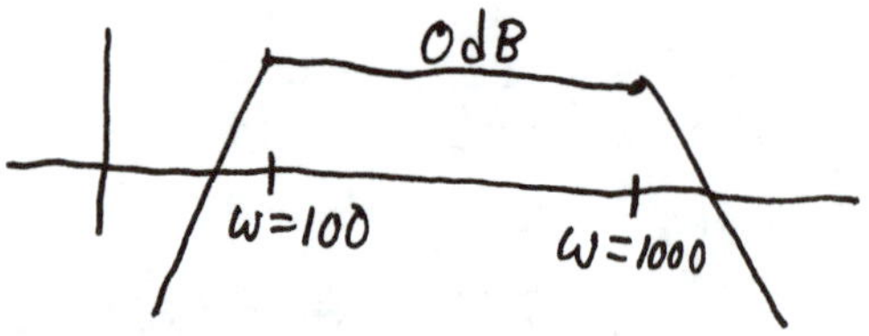

⇒ Bandpass

(f) Upper RC : HPF with $\omega_{c1} = \frac{1}{R_1 C_1} = 1000 \frac{rad}{s}$

Lower RC: LPF with $\omega_{c2} = \frac{1}{R_2 C_2} = 100 \frac{rad}{s}$

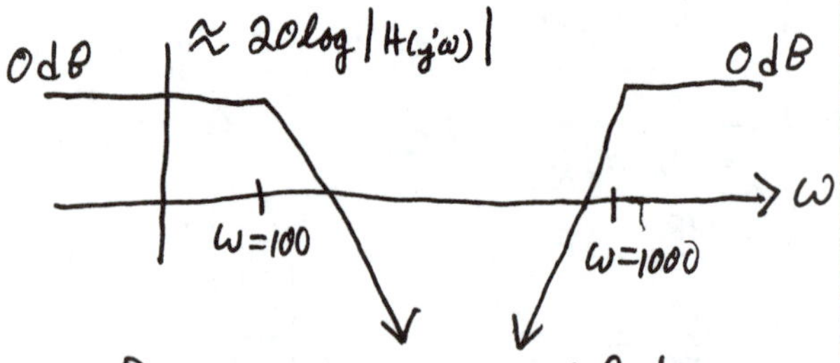

⇒ Bandstop or "notch" filter

14.29 (a) Cascade four filters with transfer fncs $H_1(s) = \frac{1}{s+2}$, $H_2(s) = \frac{1}{s+10}$

$H_3(s) = H_4(s) = \frac{1}{s+300}$

Each with a pole at $s = -\frac{1}{RC}$

Lets choose all $C_1 = C_2 = C_3 = C_4 = 0.01F$ (many solutions)

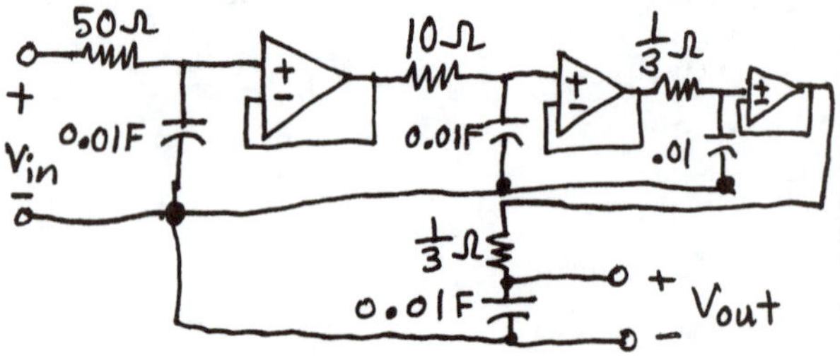

(b) Use same circuit, but swith R and C (or R and L for that case):

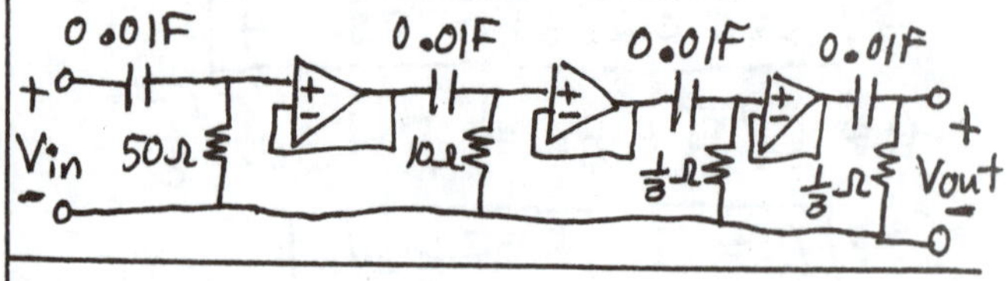

14.30 $|H(j\omega)|^2 = \frac{36\omega^2}{(4-\omega^2)^2+\omega^2} = \frac{36}{1+\left(\frac{4-\omega^2}{\omega}\right)^2}$

$|H(j0)| = |H(j\infty)| = 0$; ⇒ band pass

$|H(j\omega)|_{max} = |H(j2)| = 6$ ∴ $\omega_0 = 2 \frac{rad}{s}$

To find ω_{c1} and ω_{c2} set

$\frac{36}{2} = \frac{36}{1+\left(\frac{4-\omega^2}{\omega}\right)^2}$ ∴ $\frac{4-\omega^2}{\omega} = \pm 1$:

$\omega^2 \pm \omega - 4 = 0 \Rightarrow \omega_{c1} = 1.56$ rad/s;

$\omega_{c2} = 2.56$ rad/s; $B = \omega_{c2} - \omega_{c1} = 1$ rad/s

14.31 $|H(j\omega)|^2 = \frac{36\omega^4}{(4-\omega^2)+\omega^2} = \frac{36}{1+\left(\frac{16-7\omega^2}{\omega^4}\right)}$

$|H(j0)| = 0$; $|H(j\infty)| = 6$; ⇒ high pass

$|H(j\omega)|_{max} = |H(j\infty)| = 6$

$|H(j\omega_c)| = \left(\frac{6}{\sqrt{2}}\right)^2 = \frac{36}{1+\frac{16-7\omega^2}{\omega^4}}$

∴ $1 = \frac{16-7\omega^2}{\omega^4}$: $\omega^4 + 7\omega^2 - 16 = 0$

$\omega_c = 1.347$ rad/s .

14.32 $|H(j\omega)|^2 = \frac{36(4-\omega^2)^2}{(4-\omega^2)^2+\omega^2} = \frac{36}{1+\left(\frac{\omega}{4-\omega^2}\right)^2}$

$|H(j2)| = 0 \Rightarrow \omega_0 = 2$ rad/s

$|H(j0)| = |H(j\infty)| = 6 \Rightarrow$ band reject

ω_{c1} and ω_{c2} satisfy

$\frac{36}{2} = \frac{36}{1+\left(\frac{\omega}{4-\omega^2}\right)^2}$: $\pm 1 = \frac{\omega}{4-\omega^2}$

∴ $\omega^2 \pm \omega - 4 = 0$: $\omega_{c1} = 1.56$ rad/s

$\omega_{c2} = 2.56$ rad/s

14.33 V_3 = left side node voltage

KCL gives

$V_2(1+s+\frac{1}{2s}) - V_3(\frac{1}{2s}) = 0$ or

$V_3 = V_2(2s^2+2s+1)$

$V_3(1+s+\frac{1}{2s}) - V_2(\frac{1}{2s}) = V_1$ or

$V_2\left[\frac{(2s^2+2s+1)(2s^2+2s+1)}{2s} - \frac{1}{2s}\right] = V_1$

$H(s) = \frac{V_2}{V_1} = \frac{2s}{(2s^2+2s+1)^2-1}$

$= \frac{1}{2s^3+4s^2+4s^2+2}$

$|H(j\omega)|^2 = \frac{1}{(2-4\omega^2)^2+(4\omega-2\omega^3)^2}$

$= \frac{1/4}{1+\omega^6}$

$|H(j0)| = \frac{1}{2}$; $H(j\infty)=0$; $\Rightarrow$ low pass

$|H(j\omega)|_{max} = H(j0) = \frac{1}{2}$

$|H(j\omega_c)| = \frac{1/4}{2} = \frac{1/4}{1+\omega^6}$:

$\therefore \omega^6 = 1 \Rightarrow \omega_c = \underline{1 \text{ rad/s}}$.

14.34 V_3 = output voltage of first opamp

V_4 = voltage across 8-Ω resistor

At inverting input of second op amp KCL gives

$\frac{V_2}{4} + \frac{V_3}{2} = 0 \Rightarrow V_3 = -\frac{1}{2}V_2$

At inverting input of first opamp KCL gives

$\frac{V_3}{2} + \frac{s}{4}V_4 = 0 \Rightarrow V_4 = -\frac{2}{s}V_3 = \frac{1}{s}V_2$

KCL at V_4:

$V_4(\frac{1}{2}+\frac{1}{8}+\frac{s}{4}+\frac{s}{4}+\frac{1}{4}) - \frac{s}{4}V_3 - \frac{1}{4}V_2 = \frac{1}{2}V_1$

$\left[\frac{4s+7}{8s}(\frac{1}{s}) - \frac{s}{4}(-\frac{1}{2}) - \frac{1}{4}\right]V_2 = \frac{1}{2}V_1$

$\therefore H(s) = \frac{V_2}{V_1} = \frac{4s}{s^2+2s+7}$

$K = 4$, $B = \underline{2}$, $\omega_0^2 = \underline{7}$,

$(a=2)$, $G = \frac{K}{a} = \frac{4}{2} = \underline{2}$

14.35 KCL at inverting input yields

$\frac{V_1}{R_1} + V_2\left(\frac{1}{R_2}+sC\right) = 0$

$H(s) = \frac{V_2}{V_1} = \frac{-1/R_1}{\frac{1}{R_2}+sC} = \frac{-R_2/R_1}{1+sR_2C}$

$H(0) = -\frac{R_2}{R_1} = -2$ since:

$\frac{4}{2} = \frac{4}{1+(\omega_c R_2 C)^2} \Rightarrow \omega_c R_2 C = 1$

$R_2 = \frac{1}{\omega_c C} = \frac{1}{(10^5)(10^{-9})} = \underline{10\,k\Omega}$

$R_1 = R_2/2 = \underline{5\,k\Omega}$

14.36 $H_1(s)$ will be multiplied by $\frac{1}{1000}$

$H_2(s)$ will be multiplied by 1000

$H_5(s)$ will be multiplied by $\frac{1}{1000}$

Others unchanged.

14.37 $R_s = 25R$, $L_s = \left(\frac{L}{400}\right)25 = \frac{L}{16}$

$C_s = \left(\frac{C}{400}\right)\frac{1}{25} = \frac{C}{10\,000}$

14.38 (a) H(s) is unchanged after imp. scaling

(b) $H(s) = \frac{2000s^2+30000s+14\times10^5}{s^5+7s^4+1700s^3+530s^2+10000s+10^5}$

(c) $H(s) = \frac{0.25s^2+0.1875s+0.4375}{s^5+0.35s^4+4.25s^3+.06625s^2+\frac{s}{16}+\frac{1}{32}}$

14.39 (a) $K_f = \frac{\omega_c}{\Omega_c} = 2\times10^4$

$K_i = \frac{C'}{CK_f} = \frac{1}{(10^{-6})(2\times10^4)} = 50$

if $R_1' = \frac{1}{2}\Omega$, $R_2' = 1\Omega$, and $R_3' = 2\Omega$

then $R_1 = K_i R_1' = \underline{25\,\Omega}$

$R_2 = K_i R_2' = \underline{50\,\Omega}$

$R_3 = K_i R_3' = \underline{100\,\Omega}$

(b) Impedance scale original circuit by 5000.

$\Rightarrow$ Each $R_{new} = 5000\,R_{old}$

Each $C_{new} = \frac{C_{old}}{5000}$

14.40 SPICE input file:

```
PROBLEM 14.40
I1 0 1 AC 1 0
R1 1 0 5
R2 1 2 2
L1 2 0 3
R3 1 3 2
C1 3 0 1
.AC DEC 30 0.1 10
.PLOT AC VDB(2) VP(2)
.END
```

SPICE Output:

```
******************************************************************************

 LEGEND:

*: VM(2)
+: VP(2)

  FREQ          VM(2)


(*)----------     1.0000E-01   1.0000E+00   1.0000E+01   1.0000E+02   1.0000E+03
(+)----------     0.0000E+00   2.0000E+01   4.0000E+01   6.0000E+01   8.0000E+01
                       - - - - - - - - - - - - - - - - - - - - - - - - - - -
  1.000E-01  8.896E-01 .            *            .    +        .            .
  1.166E-01  9.633E-01 .            *            . +           .            .
  1.359E-01  1.034E+00 .            .*           +             .            .
  1.585E-01  1.101E+00 .            .*        +  .             .            .
  1.848E-01  1.162E+00 .            .*      +    .             .            .
  2.154E-01  1.215E+00 .            . *   +      .             .            .
  2.512E-01  1.260E+00 .            . * +        .             .            .
  2.929E-01  1.298E+00 .            . X          .             .            .
  3.415E-01  1.328E+00 .            + *          .             .            .
  3.981E-01  1.352E+00 .          + . *          .             .            .
  4.642E-01  1.371E+00 .         +  . *          .             .            .
  5.412E-01  1.385E+00 .       +    . *          .             .            .
  6.310E-01  1.396E+00 .      +     . *          .             .            .
  7.356E-01  1.405E+00 .     +      . *          .             .            .
  8.577E-01  1.411E+00 .    +       . *          .             .            .
  1.000E+00  1.415E+00 .    +       . *          .             .            .
  1.166E+00  1.419E+00 .   +        . *          .             .            .
  1.359E+00  1.421E+00 .  +         . *          .             .            .
  1.585E+00  1.423E+00 .  +         . *          .             .            .
  1.848E+00  1.425E+00 .  +         . *          .             .            .
  2.154E+00  1.426E+00 . +          .  *         .             .            .
  2.512E+00  1.426E+00 . +          .  *         .             .            .
  2.929E+00  1.427E+00 . +          .  *         .             .            .
  3.415E+00  1.427E+00 .+           .  *         .             .            .
  3.981E+00  1.428E+00 .+           .  *         .             .            .
  4.642E+00  1.428E+00 .+           .  *         .             .            .
  5.412E+00  1.428E+00 .+           .  *         .             .            .
  6.310E+00  1.428E+00 .+           .  *         .             .            .
  7.356E+00  1.428E+00 .+           .  *         .             .            .
  8.577E+00  1.428E+00 .+           .  *         .             .            .
  1.000E+01  1.428E+00 +            .  *         .             .            .
                       - - - - - - - - - - - - - - - - - - - - - - - - - - -
```

14.41 SPICE input file:

```
PROBLEM 14.41
V1 1 0 AC 1 0
R1 1 2 10
L1 2 0 1
C1 2 0 100U
.AC DEC 30 0.1 10K
.PLOT AC VM(2) VP(2)
.END
```

SPICE Output:

```
******************************************************************************

 LEGEND:

*: VM(2)
+: VP(2)

  FREQ          VM(2)

(*)----------     1.0000E-02   1.0000E-01   1.0000E+00   1.0000E+01  1.0000E+02
(+)----------    -1.0000E+02  -5.0000E+01   0.0000E+00   5.0000E+01  1.0000E+02
                      - - - - - - - - - - - - - - - - - - - - - - - - - - - - - -
1.000E-01  6.271E-02 .          * .            .            .        +   .
1.585E-01  9.910E-02 .            *            .            .        +   .
2.512E-01  1.559E-01 .            .  *         .            .       +    .
3.981E-01  2.428E-01 .            .     *      .            .      +     .
6.310E-01  3.690E-01 .            .       *    .            .    +       .
1.000E+00  5.335E-01 .            .         *  .            . +          .
1.585E+00  7.091E-01 .            .           *.           +.            .
2.512E+00  8.507E-01 .            .            *       +    .            .
3.981E+00  9.364E-01 .            .            *    +       .            .
6.310E+00  9.781E-01 .            .            *  +         .            .
1.000E+01  9.954E-01 .            .            *+           .            .
1.585E+01  1.000E+00 .            .            X            .            .
2.512E+01  9.956E-01 .            .           +*            .            .
3.981E+01  9.786E-01 .            .         +  *            .            .
6.310E+01  9.375E-01 .            .       +    *            .            .
1.000E+02  8.528E-01 .            .    +       *            .            .
1.585E+02  7.122E-01 .            .+          *.            .            .
2.512E+02  5.368E-01 .          + .         *  .            .            .
3.981E+02  3.717E-01 .       +    .       *    .            .            .
6.310E+02  2.447E-01 .     +      .     *      .            .            .
1.000E+03  1.572E-01 .    +       .  *         .            .            .
1.585E+03  9.993E-02 .   +        *            .            .            .
2.512E+03  6.324E-02 .   +      * .            .            .            .
3.981E+03  3.995E-02 .  +    *    .            .            .            .
6.310E+03  2.522E-02 .  +  *      .            .            .            .
1.000E+04  1.591E-02 .  X         .            .            .            .
                      - - - - - - - - - - - - - - - - - - - - - - - - - - - - - -
```

14.42 $\omega_n = \dfrac{1}{\sqrt{LC}} = 1884.96$, $B = \dfrac{R}{L} = 100$.

For the output current to be 1/2 the magnitude of the input voltage (6 dB down), choose $R = 2\,\Omega$. Then $L = 20\,\text{mH}$ and $C = 14.072\,\mu\text{F}$. Verify with SPICE.

SPICE input file:

```
PROBLEM 14.42
V 1 0 AC 1 0
R 1 2 2
L 2 3 20M
C 3 0 14.072U
RDUMMY 3 0 100T
.AC LIN 21 100 500
.PLOT AC IDB(R)
.END
```

SPICE Output:

```
******************************************************************************

 FREQ        IDB(R)


(*)----------  -6.0000E+01 -4.0000E+01 -2.0000E+01  0.0000E+00  2.0000E+01
                      -----------------------------------------
 1.000E+02 -4.005E+01 .         *         .         .         .
 1.200E+02 -3.797E+01 .         .*        .         .         .
 1.400E+02 -3.602E+01 .         . *       .         .         .
 1.600E+02 -3.409E+01 .         .  *      .         .         .
 1.800E+02 -3.210E+01 .         .   *     .         .         .
 2.000E+02 -2.996E+01 .         .    *    .         .         .
 2.200E+02 -2.755E+01 .         .     *   .         .         .
 2.400E+02 -2.465E+01 .         .       * .         .         .
 2.600E+02 -2.084E+01 .         .        *.         .         .
 2.800E+02 -1.493E+01 .         .         .  *      .         .
 3.000E+02 -6.021E+00 .         .         .      *  .         .
 3.200E+02 -1.443E+01 .         .         .  *      .         .
 3.400E+02 -1.971E+01 .         .         *         .         .
 3.600E+02 -2.290E+01 .         .        *.         .         .
 3.800E+02 -2.515E+01 .         .      *  .         .         .
 4.000E+02 -2.688E+01 .         .      *  .         .         .
 4.200E+02 -2.828E+01 .         .     *   .         .         .
 4.400E+02 -2.944E+01 .         .    *    .         .         .
 4.600E+02 -3.044E+01 .         .    *    .         .         .
 4.800E+02 -3.132E+01 .         .   *     .         .         .
 5.000E+02 -3.210E+01 .         .   *     .         .         .
                      -----------------------------------------
```

Chapter 15

Mutual Inductance and Transformers

15.1 Mutual Inductance

15.1 For the pair of coupled coils shown, the following experiment is performed. First, a sinusoidal test current, $i_1 = \sin t$ A, is injected into Coil 1 while holding Coil 2 open circuited, and the following voltages are measured: $v_1 = 5\cos t$ V and $v_2 = 20\cos(t + 180°)$ V. Next, a sinusoidal test current, $i_2 = \sin t$ A, is injected into Coil 2 while holding Coil 1 open circuited, and the following voltages is measured: $v_2 = 100\cos t$ V. Find L_1, L_2, and M for this pair of coils.

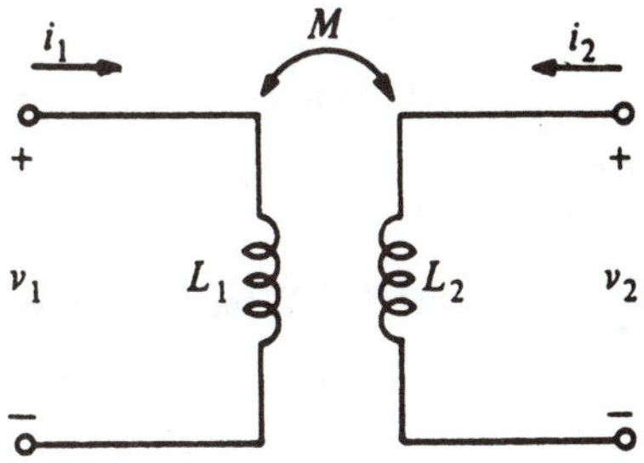

PROBLEM 15.1

15.2 Find the coefficient of coupling for the coils in Prob. 15.1. Are these coils tightly coupled?

15.3 At a certain time t_o, the coils in Prob. 15.1 have instantaneous coil currents $i_1(t_o) = 27$ A and $i_2(t_o) = 4.5$ A. What is the energy stored in the pair of coupled coils at t_o?

15.4 In a pair of coupled coils, find the net flux linked in Coil 1 if it is a 20-turn coil and $\phi_{21} = 5$ mWb, $\phi_{12} = 2$ mWb, and $\phi_{L1} = 10$ mWb (refer to figure in Prob. 15.1).

15.5 Find ϕ_{22} in Prob. 15.4 if the coefficient of coupling is $k = 0.85$.

15.6 In a pair of coupled coils (refer to figure in Prob. 15.1), the currents $i_1 = 2t$ A and $i_2 = 5\cos t$ A are applied simultaneously. A voltmeter measures $v_1 = (1 + 1.875\sin t)$ V and $v_2 = \left(-\frac{5}{3}\sin t - 0.75\right)$ V. Find L_1, L_2, M, and the energy stored in the coils at time $t = 1$ s.

15.2 Circuits with Mutual Inductance

15.7 In the circuit shown, $L_1 = L_2 = 0.1\,\text{H}$ and $M = 10\,\text{mH}$. Find v_1 and v_2 if $i_1 = 10\,\text{mA}$ and $i_2 = 0$.

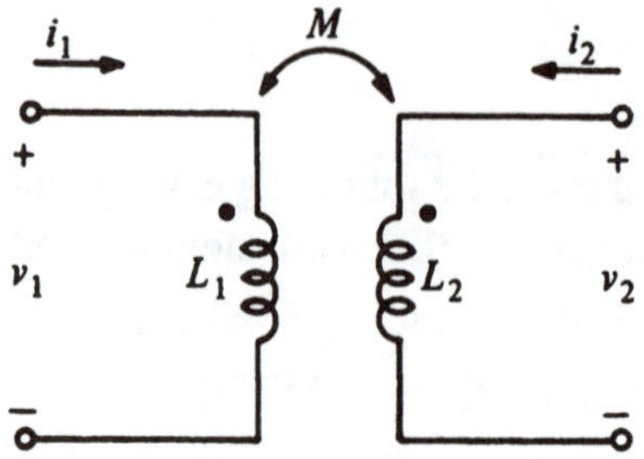

PROBLEM 15.7

15.8 In the circuit of Prob. 15.7, $L_1 = L_2 = 0.1\,\text{H}$ and $M = 10\,\text{mH}$. Find v_1 and v_2 if $i_1 = 0$ and $i_2 = 10\sin(100t)\,\text{mA}$.

15.9 Find $\mathbf{I}_2$ for a sinusoidal voltage $\mathbf{V}_1 = 26\angle 0°\,\text{V}$ having $\omega = 2$ rad/s.

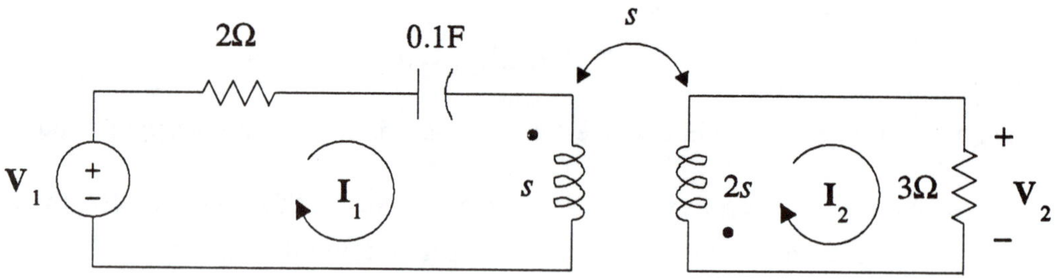

PROBLEM 15.9

15.10 If $N_2 = 500$ turns and $\phi_{21} = 30\ \mu\text{Wb}$ when $i_1 = 4$ A. Determine v_2 if $i_1 = 5\cos(10t)$ A.

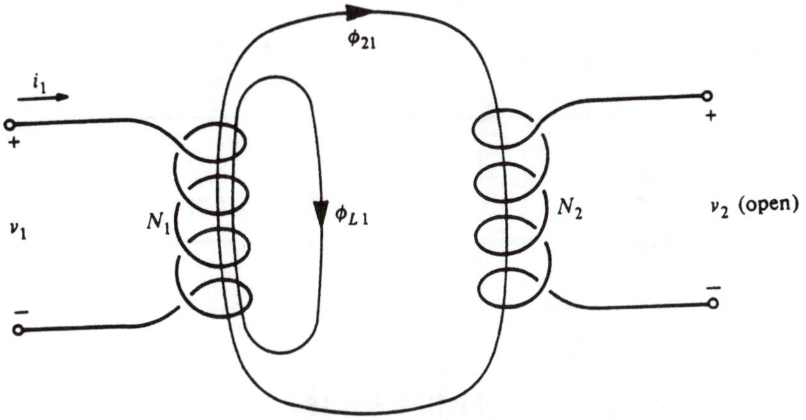

PROBLEM 15.10

15.11 If the inductance measured between terminals a and d is 4H when terminals b and c are connected, and the inductance measured between terminals a and c is 10H when terminals b and d are connected, find the mutual inductance M between the two coils and the position of the dots.

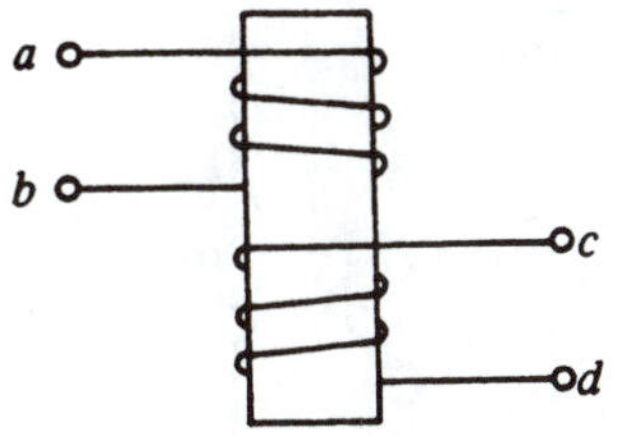

PROBLEM 15.11

15.12 Find v_1 and v_2 if $L_1 = 6$ H, $L_2 = 4$ H, $M = 2$ H, $i_1 = \sin(t)$ A, and $i_2 = -2\cos(2t)$ A.

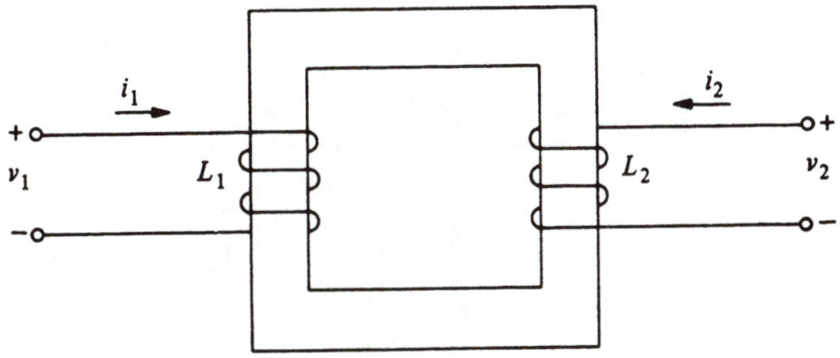

PROBLEM 15.12

15.13 Find v for $t > 0$ across the open circuit if $i = 4u(t)$ A.

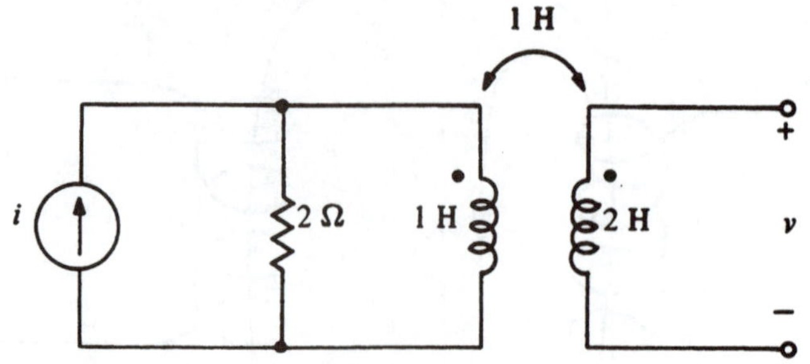

PROBLEM 15.13

15.14 Repeat Prob. 15.13 if $i = 2e^{-t}\, u(t)$ A.

15.15 Find the phasor currents $\mathbf{I}_1$ and $\mathbf{I}_2$.

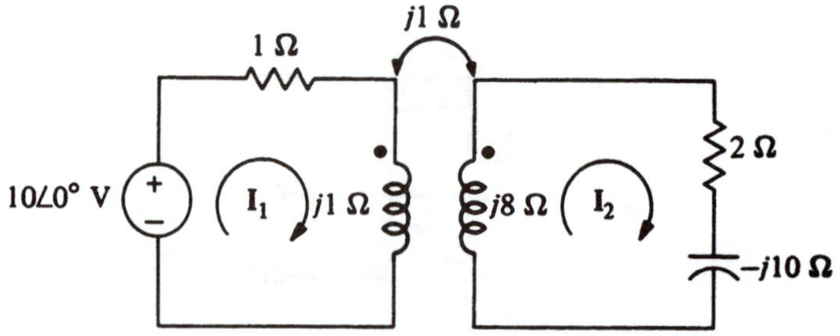

PROBLEM 15.15

15.16 Find v_1 and v_2 if $L_1 = 2$H, $L_2 = 5$H, $M = 3$H, and the currents i_1 and i_2 are changing at the rates −10 A/s and −2 A/s, respectively.

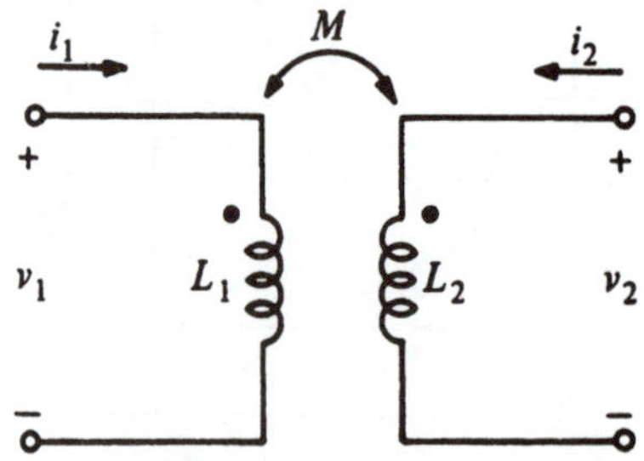

PROBLEM 15.16

15.17 Find v for $t > 0$ if $v_g = 4u(t)$ V.

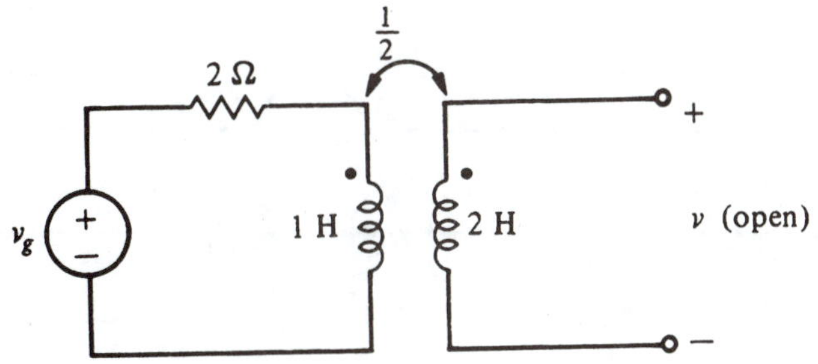

PROBLEM 15.17

15.18 Determine the energy stored in Prob. 15.7 at $t = 0$.

15.19 Determine the energy stored in Prob. 15.8 at $t = 0$.

15.20 Find the energy stored in the transformer of Prob. 15.15 at $t = 0$ if the frequency is $\omega = 10$ rad/s.

15.21 (a) Find the energy stored in the transformer at a time when $i_1 = 3$ A and $i_2 = 2$ A if $L_1 = 1$ H, $L_2 = 8$ H, and $M = 3$ H. (b) Repeat part (a) if one of the dots is moved to another terminal.

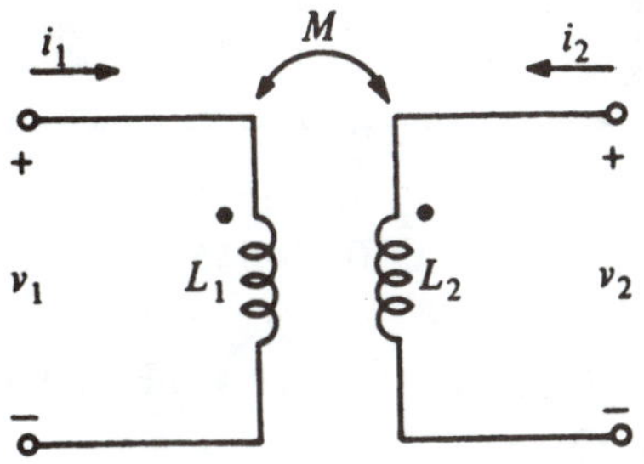

PROBLEM 15.21

15.22 Determine i_1 for $t > 0$ in the network, given $M = 1/2$ H. Assume the circuit is in steady state at $t = 0^-$.

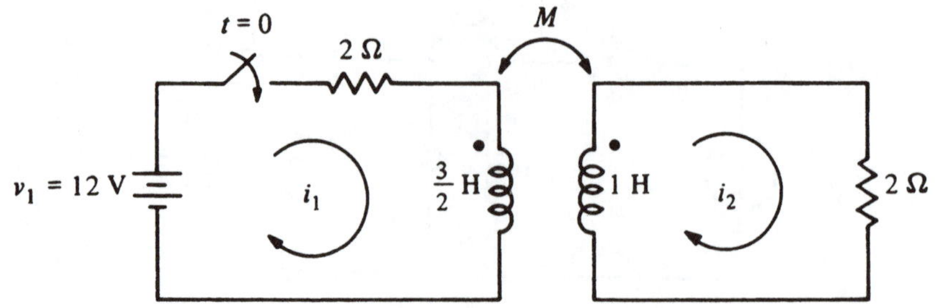

PROBLEM 15.22

15.23 Find the forced response v_2 if $v_1 = 3e^{-t}\cos(3t)$ V.

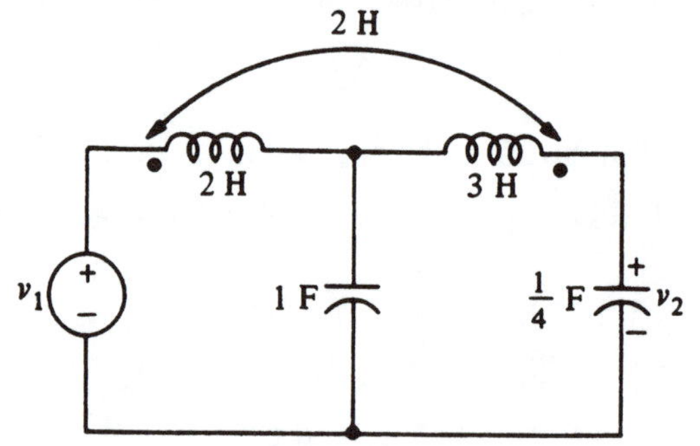

PROBLEM 15.23

15.24 Find the steady-state currents i_1 and i_2 if (a) $M = 1$ H and $R = 2\ \Omega$, and (b) $M = 2$ H and $R = 15/8\ \Omega$.

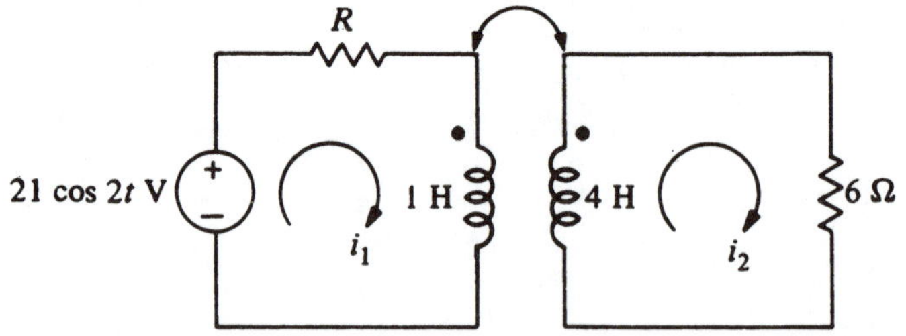

PROBLEM 15.24

15.3 Mutual Inductance and Transformers

15.25 In the circuit shown, $\mathbf{V}_g = 20\angle 0°$ V, $\mathbf{Z}_g = 20\ \Omega$, $L_1 = 1$H, $L_2 = 0.5$H, $M = 0.5$H, and $\omega = 10$ rad/s. If $\mathbf{Z}_2 = -(j10/\omega)\ \Omega$, find (a) $\mathbf{Z}_{in}$, (b) $\mathbf{I}_1$, (c) $\mathbf{I}_2$, (d) $\mathbf{V}_1$, and (e) $\mathbf{V}_2$.

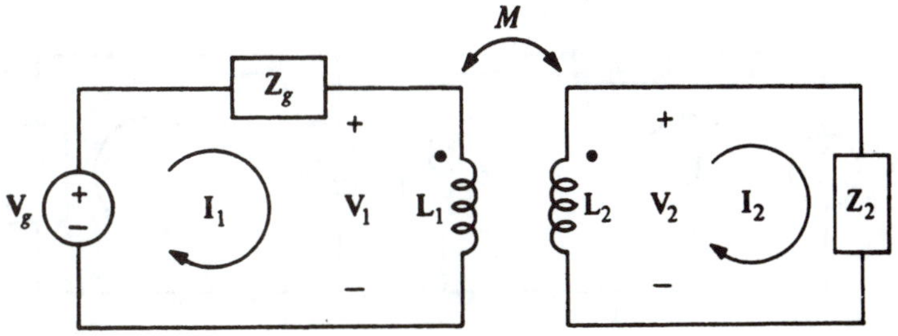

PROBLEM 15.25

15.26 Repeat Prob. 15.25 if the polarity dot is on the lower terminal of the secondary.

15.27 If $\mathbf{Z}_2 = (10 - j10/\omega)\ \Omega$ in Prob. 15.25, find the frequency for which the reflected impedance is real.

15.28 Find the steady-state current i_1 in Prob. 15.24(a) using reflected impedance.

15.29 Find the steady-state current i_2 in Prob. 15.24(a) by replacing everything in the corresponding phasor circuit except the $6\ \Omega$ resistor by its Thevenin equivalent circuit.

15.30 Find the steady-state current i_1 using reflected impedance.

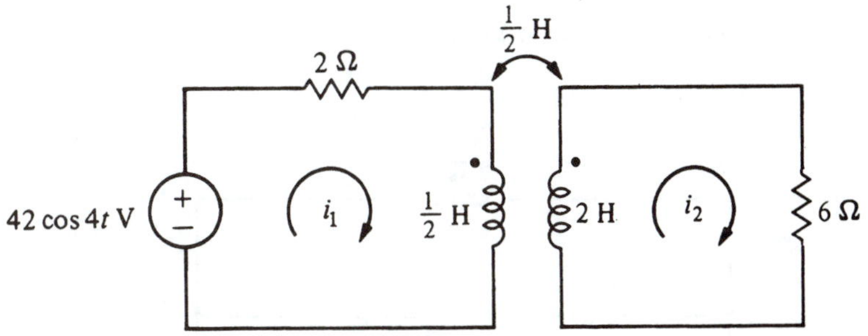

PROBLEM 15.30

15.31 Find the steady-state current i_2 in Prob. 15.30 by replacing everything in the corresponding phasor circuit except the $6\ \Omega$ resistor by its Thevenin equivalent circuit.

15.4 Ideal Transformer

15.32 $\mathbf{V}_g = 100\angle 0°$ V, $\mathbf{Z}_g = 100\Omega$, and $\mathbf{Z}_2 = 90\text{k}\Omega$. Find n such that $\mathbf{Z}_1 = \mathbf{Z}_g$, and then find the power delivered to $\mathbf{Z}_2$.

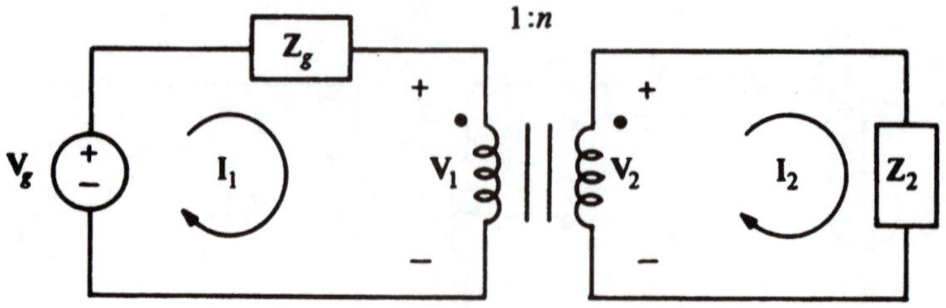

PROBLEM 15.32

15.33 Find the steady-state current i.

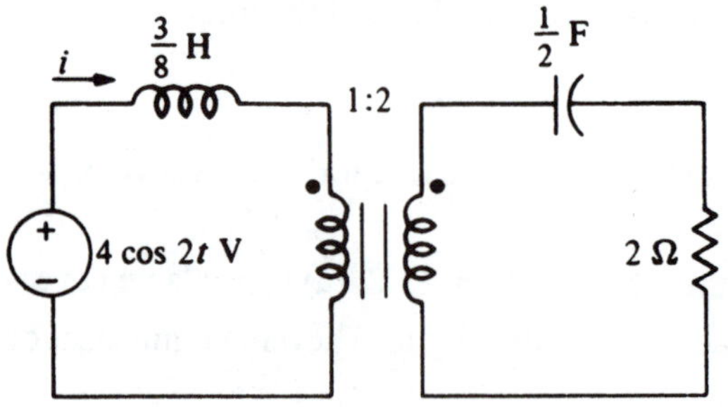

PROBLEM 15.33

15.34 Find the average power delivered to the 8 Ω resistor.

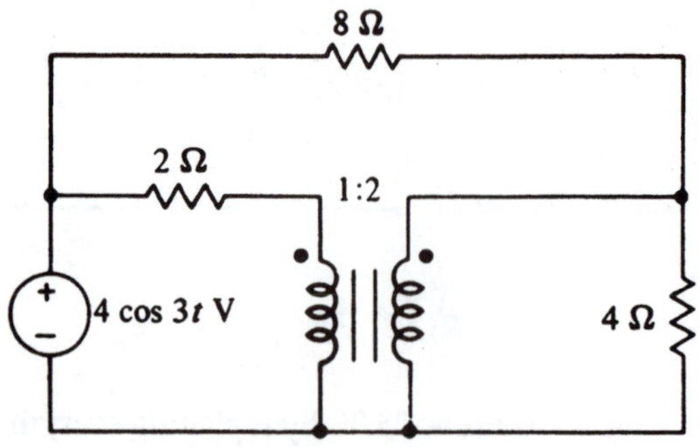

PROBLEM 15.34

15.35 Find $\mathbf{I}_1$, $\mathbf{I}_2$, $\mathbf{V}_1$, and $\mathbf{V}_2$ using the method of reflected impedance.

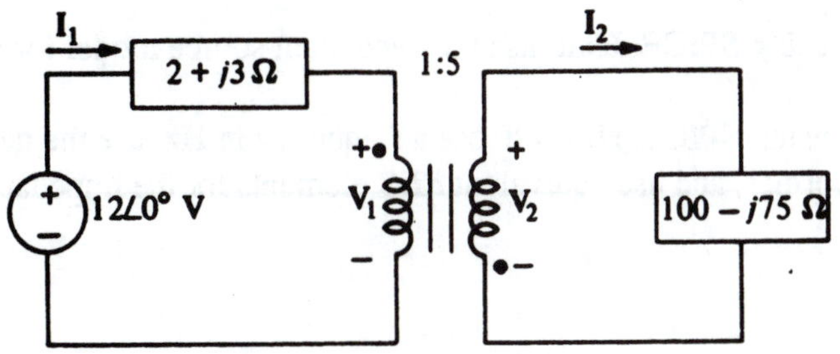

PROBLEM 15.35

15.36 Repeat Prob. 15.35 using the method illustrated in the figure below.

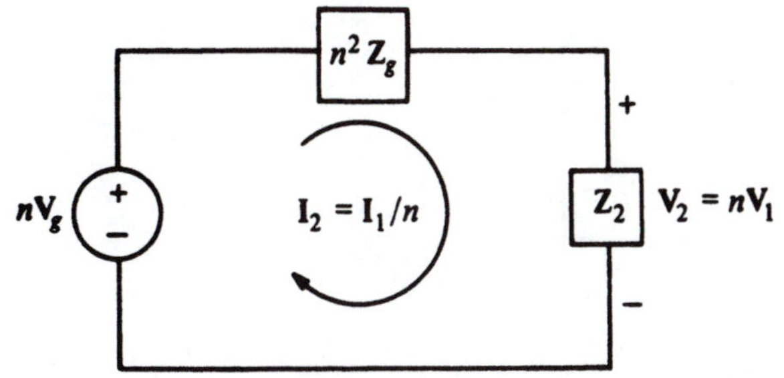

PROBLEM 15.36

15.37 Find the turns ratio n so that the maximum power is delivered to the 10 kΩ resistor.

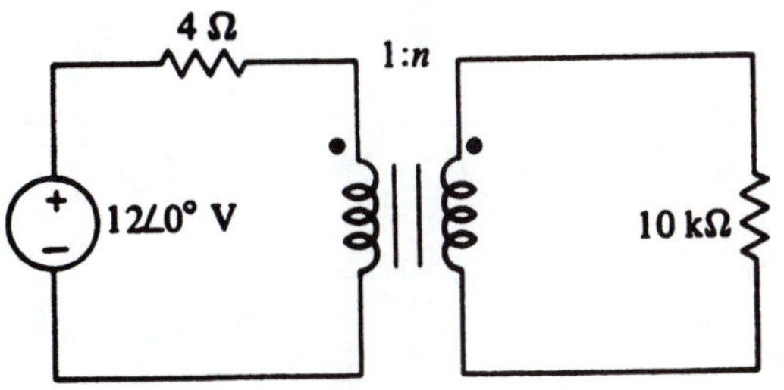

PROBLEM 15.37

15.5 SPICE and Coupled Coils

15.38 Solve Prob. 15.9 using SPICE.

15.39 Solve Prob. 15.33 using SPICE. Hint: use the dependent source model for the ideal transformer.

15.40 Solve Prob. 15.35 using SPICE. Hint: choose a frequency in Hz, use the dependent source model for the ideal transformer, and use equivalent *RLC* elements for the impedances given in the original problem.

CHAPTER 15 SOLUTIONS

15.1 First case: $i_1 = \sin t$, $i_2 = 0$
$v_1 = 5\cos t$, $v_2 = -20\cos t$

$$v_1 = L_1 \frac{d}{dt}(\sin t) \pm M\frac{d}{dt}(0)$$

$$\Rightarrow 5\cos t = L_1 \cos t \Rightarrow \underline{L_1 = 5\,H}$$

$$v_2 = \pm M\frac{d}{dt}(\sin t) + L_2\frac{d}{dt}(0)$$

$$\Rightarrow -20\cos t = \ominus M\cos t \Rightarrow \underline{M = 20\,H}$$

Second case: $i_1 = 0$, $i_2 = \sin t$,
$v_2 = 100\cos t$

$$v_2 = -M\frac{d}{dt}(0) + L_2\frac{d}{dt}(\sin t)$$

$$\Rightarrow 100\cos t = L_2\cos t \Rightarrow \underline{L_2 = 100\,H}$$

15.2 $k = M/\sqrt{L_1 L_2} = \frac{20}{\sqrt{500}} = \underline{0.89}$

$\Rightarrow$ Yes, tightly coupled (ideal is $k = 1$).

15.3

$$W = \tfrac{1}{2}L_1 i_1^2 + \tfrac{1}{2}L_2 i_2^2 - M i_1 i_2$$

$$= \tfrac{1}{2}(5)(27)^2 + \tfrac{1}{2}(100)(4.5)^2 - 20(27)(4.5)$$

$$= 1822.5 + 1012.5 - 2430$$

$$= \underline{405\,J}$$

15.4 $\phi_{21} = 5\,mWb$, $\phi_{12} = 2\,mWb$, $\phi_{L1} = 10\,mWb$

$$\lambda_1 = N_1\phi_1 = N_1[\phi_{11} + \phi_{12}]$$

$$= N_1[(\phi_{21} + \phi_{L1}) + \phi_{12}]$$

$$= 20[17\times10^{-3}\,Wb]$$

$$\lambda_1 = \underline{0.34\,Wb = 340\,mWb}$$

15.5

$$k = \sqrt{\frac{\phi_{21}}{\phi_{11}}}\sqrt{\frac{\phi_{12}}{\phi_{22}}} = 0.85$$

$$\Rightarrow \sqrt{\frac{(5\times10^{-3})(2\times10^{-3})}{(15\times10^{-3})\,\phi_{22}}} = 0.85$$

$$\Rightarrow \phi_{22} = \frac{(5\times10^{-3})(2\times10^{-3})}{(15\times10^{-3})(0.85)^2}$$

$$= \underline{9.227\times10^{-4}\,Wb}$$

15.6 $v_1 = L_1\frac{di_1}{dt} \pm M\frac{di_2}{dt}$

$$\Rightarrow 1 + 1.875\sin t = 2L_1 \pm M(-5\sin t)$$

Matching terms on both sides:

$1 = 2L_1$ and $1.875 = -(-5)M$

$$\Rightarrow \boxed{L_1 = \tfrac{1}{2}\,H} \text{ and } \boxed{M = 0.375 = \tfrac{3}{8}\,H}$$

$$v_2 = -M\frac{di_1}{dt} + L_2\frac{di_2}{dt}$$

$$\Rightarrow -\tfrac{5}{3}\sin t - 0.75 = -\tfrac{3}{8}(2) + L_2(-5\sin t)$$

$$\Rightarrow -5L_2 = -\tfrac{5}{3} \Rightarrow \boxed{L_2 = \tfrac{1}{3}\,H}$$

$$W(t=1) = \tfrac{1}{2}L_1[i_1(1)]^2 + \tfrac{1}{2}L_2[i_2(1)]^2 - M[i_1(1)\,i_2(1)]$$

$$= (\tfrac{1}{2})(\tfrac{1}{2})(2)^2 + (\tfrac{1}{2})(\tfrac{1}{3})(2.7015)^2 - (\tfrac{3}{8})(2)(2.7015)$$

$$= 1 + 1.21636 - 2.02613$$

$$= \underline{0.190\,J} = \underline{190\,mJ}$$

15.7 $\frac{di_1}{dt} = \frac{di_2}{dt} = 0 \therefore v_1 = v_2 = 0$

15.8 $v_1 = L_1 \frac{di_1}{dt} + M\frac{di_2}{dt}$
$= (0.1)(0) + (10^{-2})(10^3 \cos 100t)\text{mA}$
$= 10 \cos 100t$ mV

$v_2 = M\frac{di_1}{dt} + L_2\frac{di_2}{dt}$
$= 0 + (0.1)(10^3 \cos 100t \text{ mA})$
$= 100 \cos 100t$ mV.

15.9 KVL yields
$(12 + s + 10/s)\mathbf{I}_1 + s\mathbf{I}_2 = \mathbf{V}_1$
$s\mathbf{I}_1 + (3+2s)\mathbf{I}_2 = 0$

$$\mathbf{I}_2 = \frac{\begin{vmatrix} 12-j3 & 26 \\ j2 & 0 \end{vmatrix}}{\begin{vmatrix} 12-j3 & j2 \\ j2 & 3+j4 \end{vmatrix}} = \frac{-j52}{52+j39}$$
$= 0.8 \angle -126.9^\circ$ A

15.10 $M_{21} = \frac{N_2 \phi_{21}}{i_1} = \frac{(500)(30\times10^{-6})}{4}$
$= 3.75$ mH
$v_2 = M_{21}\frac{di_1}{dt} = (3.75\text{m})\frac{d}{dt}(5\cos 10t)$
$= -0.1875 \sin 10t$ V

15.11 $v_{ad} = (L_{ab} + L_{cd} - 2M)\frac{di}{dt} = L_{ad}\frac{di}{dt}$
with b & d connected
$v_{ac} = (L_{ab} + L_{cd} + 2M)\frac{di}{dt} = L_{ac}\frac{di}{dt}$
$\therefore L_{ab} + L_{cd} - 2M = L_{ad} = 4$ H
$L_{ab} + L_{cd} + 2M = L_{ac} = 10$H
subtracting equations
$4M = 10 - 4 \Rightarrow M = 3/2$ H

15.12 $v_1 = L_1\frac{di_1}{dt} - M\frac{di_2}{dt}$
$= 6\cos t - 2(4\sin 2t)$
$= 6\cos t - 8\sin 2t$ V.
$v_2 = -M\frac{di_1}{dt} + L_2\frac{di_2}{dt} = (-2)\cos t + 4(4\sin 2t)$
$= -2\cos t + 16\sin 2t$ V.

15.13 v_1 = voltage across the primary coil
$v_1 = \frac{di_1}{dt}$; $i = \frac{v_1}{2} + i_1 \Rightarrow \frac{di_1}{dt} + 2i_1 = 2i$
$v = \frac{di_1}{dt}$; $\frac{di_1}{dt} + 2i_1 = 8u(t) = 8$ for $t>0$
$i_1 = A_1 e^{-2t} + 4$; $i_1(0^+) = i(0^-) = 0$
$i_1(0) = A_1 + 4 \Rightarrow A_1 = -4$
$v = \frac{d}{dt}(-4e^{-2t} + 4) = 8e^{-2t}$ V for $t>0$

15.14 $\frac{di_1}{dt} + 2i = 4e^{-t}u(t) = 4e^{-t}$ for $t>0$
$i_{1f} = Ae^{-t} \Rightarrow -A + 2A = 4 \Rightarrow A = 4$
$i_1 = A_1e^{-2t} + 4e^{-t}$; $i(0) = i_1(0^-) = 0$
$i_1 = A_1 + 4 \Rightarrow A_1 = -4$
$v = \frac{d}{dt}(4e^{-t} - 4e^{-2t}) = 8e^{-2t} - 4e^{-t}$ V for $t>0$.

15.15 KVL yields
$(1+j)\mathbf{I}_1 - j\mathbf{I}_2 = 10$
$-j\mathbf{I}_1 + (2 + j8 - j10)\mathbf{I}_2 = 0$

$$\mathbf{I}_1 = \frac{\begin{vmatrix} 10 & -j \\ 0 & 2-j2 \end{vmatrix}}{\begin{vmatrix} 1+j & -j \\ -j & 2-j2 \end{vmatrix}} = \frac{20-j20}{5} = 4-j4 \text{ A};$$

$$\mathbf{I}_2 = \frac{\begin{vmatrix} 1+j & 10 \\ -j & 0 \end{vmatrix}}{5} = j2 \text{ A}.$$

15.16 $v_1 = 2(-10) + 3(-2) = -26$V
$v_2 = 3(-10) + 5(-2) = -40$V

15.17 v_1 = voltage across primary coil

$v_1 = \frac{di_1}{dt} \Rightarrow v_g = 2i_1 + \frac{di_1}{dt}$; $v = \frac{1}{2}\frac{di_1}{dt}$

$\frac{di_1}{dt} + 2i_1 = 4u(t) \Rightarrow \frac{di_1}{dt} + 2i_1 = 4$ for $t > 0$.

$i_1 = A_1 e^{-2t} + 2$; $i(0^+) = i(0^-) = 0$

$i_1(0^-) = A_1 + 2 \Rightarrow A_1 = -2$

$v = \frac{1}{2}\frac{d}{dt}(2 - 2e^{-2t}) = \underline{2e^{-2t}\text{ V}}$ for $t > 0$

15.18 $W = \frac{1}{2}L_1 i_1^2 + M i_1 i_2 + \frac{1}{2}L_2 i_2^2$

at $t = 0$, $i_1 = 10\text{mA}$, $i_2 = 0$

$W(0) = \frac{1}{2}(0.1)(10^{-2})^2 = \underline{5\mu\text{J}}$

15.19 at $t = 0$ $i_1 = 0$, $i_2 = 0$,

$W(0) = \underline{0\text{ J}}$.

15.20

$\mathbf{I}_1 = 4 - j4 = 4\sqrt{2}\angle -45°$

$i_1 = 4\sqrt{2}(10t - 45°)$, $i_1(0) = 4$

$\mathbf{I}_2 = j2$, $i_2 = 2\cos(10t + 90°)$,

$i_2(0) = 0$.

$W(0) = \frac{1}{2}L_1 i_1^2(0) = \frac{1}{2}(\frac{1}{10})(4)^2$

$= \underline{0.8\text{ J}}$ $[\omega L = 10L = 1]$

15.21 (a) $W = \frac{1}{2}L_1 i_1^2 + M i_1 i_2 + \frac{1}{2}L_2 i_2^2$

$= \frac{1}{2}(1)(3)^2 + (3)(3)(2) + \frac{1}{2}(8)(2)^2$

$= \underline{38.5\text{ J}}$

(b) $W = \frac{1}{2}L_1 i_1^2 - M i_1 i_2 + \frac{1}{2}L_2 i_2^2$

$= \underline{2.5\text{ J}}$

15.22 $V_1(s) = (\frac{3}{2}s + 2)\mathbf{I}_1 + \frac{s}{2}\mathbf{I}_2$

$0 = -\frac{s}{2}\mathbf{I}_1 + (s+2)\mathbf{I}_2$ or

$\mathbf{I}_2 = \frac{s}{2(s+2)}\mathbf{I}_1$, substitution

$H(s) = \frac{\mathbf{I}_1}{V_1} = \frac{4(s+2)}{5s^2 + 20s + 16}$

$= \frac{4/5(s+2)}{(s+1.11)(s+2.89)}$

$i_{1f} = H(0)V_1 = \frac{8}{16}(12) = 6$

$i_1 = 6 + A_1 e^{-1.11t} + A_2 e^{-2.89t}$

$i_1(0^+) = i_1(0^-) = 0 = 6 + A_1 + A_2$

From KVL

$\frac{3}{2}\frac{di_1(0^+)}{dt} - \frac{1}{2}\frac{di_2(0^+)}{dt} + 2i_1(0^+) = 12$

$-\frac{1}{2}\frac{di_1(0^+)}{dt} + \frac{di_2(0^+)}{dt} + 2i_2(0^+) = 0$

since $i_1(0^+) = i_2(0^+) = 0$ we obtain

$\frac{di_1(0^+)}{dt} = \frac{2di_2(0^+)}{dt} = \frac{4}{5}(12) = 9.6$

$\frac{di_1(0^+)}{dt} = 9.6 = -1.11A_1 - 2.89A_2$

$\therefore A_1 = -4.34$ & $A_2 = -1.66$

$\underline{i_1 = 6 - 4.34e^{-1.11t} - 1.66e^{-2.89t}\text{ A}}$

15.23 Let i_1 and i_2 be clockwise mesh currents then

$V_1 = (2s + 1/s)I_1 - (2s + \frac{1}{s})I_2$

$0 = -(2s + 1/s)I_1 + (3s + \frac{1}{s} + \frac{4}{s})I_2$

$V_2 = \frac{4}{s}I_2$

$\therefore I_2 = \frac{sV_1}{s^2+4}$, $V_2 = \frac{4V_1}{s^2+4}$,

$s = -1 + j3$, $V_1 = 3$,

$V_2 = \frac{4(3)}{(-1+j3)^2 + 4} = 1.66\angle 123.7°$

$\underline{v_{2f} = 1.66e^{-t}\cos(2t + 123.7°)\text{ V}}$

15.24

$$21 = (R + j2)I_1 - j2MI_2$$
$$0 = -j2MI_1 + (6+j8)I_2$$

(a)
$$I_1 = \frac{\begin{vmatrix} 21 & -j2(1) \\ 0 & 6+j8 \end{vmatrix}}{\begin{vmatrix} 2+j2 & -j2 \\ -j2 & 6+j8 \end{vmatrix}} = \frac{21(6+j8)}{j28}$$

$$= 7.5\angle -36.87^\circ$$

$$i_1 = \underline{7.5\cos(2t - 36.87^\circ)} \text{ A}$$

$$I_2 = \frac{\begin{vmatrix} 2+j2 & 21 \\ -j2 & 0 \end{vmatrix}}{j28} = \frac{j42}{j28} = 1.5$$

$$i_2 = \underline{1.5\cos 2t \text{ A}}$$

(b)
$$I_1 = \frac{\begin{vmatrix} 21 & -j4 \\ 0 & 6+j8 \end{vmatrix}}{\begin{vmatrix} 15/8 + j2 & -j4 \\ -j4 & 6+j8 \end{vmatrix}} = \frac{21(6+j8)}{11.25 + j27}$$

$$= 7.18\angle -14.25^\circ$$

$$i_1 = \underline{7.18\cos(2t - 14.25^\circ) \text{ A}}$$

$$I_2 = \frac{\begin{vmatrix} 15/8 + j2 & 21 \\ -j4 & 0 \end{vmatrix}}{11.25 + j27} = \frac{j84}{11.25 + j27}$$

$$= 2.87\angle 22.62^\circ$$

$$i_2 = \underline{2.87\cos(2t + 22.62^\circ) \text{ A}}$$

15.25

$$Z_1 = j\omega L_1 + \frac{\omega^2 M^2}{Z_2 + j\omega L_2}$$
$$= j\left(10 - \frac{25}{4}\right) = j3.75\ \Omega$$

(a) $Z_{in} = Z_g + Z_1 = \underline{20 + j3.75\ \Omega}$

(b) $I_1 = V_g/Z_{in} = \dfrac{20}{20 + j3.75} = 0.97 - j0.18$

$= \underline{0.983\angle -10.62^\circ \text{ A}}$

(c) $I_2 = \dfrac{j\omega M}{Z_2 + j\omega L_2} I_1 = \dfrac{j5(0.97 - j0.18)}{-j + j5}$

$= 1.21 - j0.23 = \underline{1.23\angle -10.6^\circ \text{A}}$

(d) $V_1 = Z_1 I_1 = (j3.75)(0.97 - j0.18)$

$= \underline{3.686\angle 79.38^\circ \text{ V}}$

(e) $V_2 = Z_2 I_2 = (-j)(1.23\angle -10.62^\circ)$

$= \underline{0.983\angle -100.62^\circ \text{V}}$

15.26

Z_1 is the same as in Prob. 15.25

(a) $Z_{in} = \underline{20 + j3.75\ \Omega}$

(b) $I_1 = \underline{0.983\angle -10.62^\circ \text{ A}}$

(c) $I_2 = \dfrac{(-j\omega M)}{Z_2 + j\omega L_2} = \underline{1.23\angle 169.4^\circ \text{A}}$

(d) $V_1 = \underline{3.686\angle 79.38^\circ} \text{ V}$

(e) $V_2 = Z_2 I_2 = \underline{0.983\angle 79.38^\circ \text{V}}$

15.27

$$Z_R = \frac{\omega^2 M^2}{Z_2 + j\omega L_2},$$

$$Z_2 + j\omega L_2 = 10 - j\,10/\omega + j\frac{\omega}{2}$$

Z_R is real when $Z_2 + j\omega L_2$ is real.

$$\therefore \frac{\omega}{2} - \frac{10}{\omega} = 0 \Rightarrow \omega^2 = 20$$

$$\omega = \underline{\sqrt{20} \text{ rad/s}}$$

15.28

$$Z_1 = j(2)(1) + \frac{(2)^2(1)^2}{6 + j(2)(a)} = 0.24 + j1.68$$

$$Z_{in} = R + Z_1 = 2.24 + j1.68\ \Omega$$

$$I_1 = V_g/Z_{in} = \frac{21}{2.24 + j1.68} = 7.5\angle -36.9^\circ$$

$$i_1 = \underline{7.5\cos(2t - 36.87^\circ)}\ \text{A}$$

15.29

open circuit secondary:

$$21 = (2 + j2)\mathbf{I}_1 \Rightarrow \mathbf{I}_1 = \frac{21}{2 + j2};$$

$$\mathbf{V}_{oc} = j2\mathbf{I}_1 = \frac{21}{2}(1 + j)\ \text{V}$$

short circuit secondary:

$$\left.\begin{aligned} 21 &= (2 + j2)\mathbf{I}_1 - j2\mathbf{I}_{sc} \\ 0 &= -j2\mathbf{I}_1 + j8\mathbf{I}_{sc} \end{aligned}\right\} \mathbf{I}_{sc} = \frac{21}{8 + j6}\ \text{A}$$

$$\mathbf{Z}_{th} = \frac{\mathbf{V}_{oc}}{\mathbf{I}_{sc}} = 1 + j7\ \Omega$$

From the Thevenin circuit.

$$\mathbf{I}_2 = \frac{\frac{21}{2}(1 + j)}{6 + 1 + j7} = 1.5\angle 0^\circ\ \text{A}$$

$$i_2 = \underline{1.5\cos 2t\ \text{A}}.$$

15.30

$$\mathbf{Z}_R = \frac{\omega^2 M^2}{\mathbf{Z}_2 + j\omega L_2} = \frac{(4)^2(\frac{1}{2})^2}{6 + j(4)(2)} = \frac{2}{3 + j4}$$

$$\mathbf{Z}_1 = j(4)(\tfrac{1}{2}) + \frac{2}{3 + j4} = \frac{6(-1 + j)}{3 + j4}$$

$$\mathbf{I}_1 = \frac{\mathbf{V}_g}{\mathbf{Z}_1 + \mathbf{Z}_g} = \frac{42(3 + j4)}{j14} = 15\angle -36.9^\circ\ \text{A}$$

$$i_1 = \underline{15\cos(4t - 36.9^\circ)}\ \text{A}$$

15.31

open circuit secondary:

$$42 = (2 + j2)\mathbf{I}_1 \Rightarrow \mathbf{I}_1 = \frac{42}{2 + j2}$$

$$\mathbf{V}_{oc} = j2\mathbf{I}_1 = 21(1 + j)\ \text{V}$$

short circuit secondary:

$$\left.\begin{aligned} 42 &= (2 + j2)\mathbf{I}_1 - j2\mathbf{I}_{sc} \\ 0 &= -j2\mathbf{I}_1 + j8\mathbf{I}_{sc} \end{aligned}\right\} \mathbf{I}_{sc} = \frac{21}{4 + j3}\ \text{A}$$

$$\mathbf{Z}_{th} = \frac{\mathbf{V}_{oc}}{\mathbf{I}_{sc}} = 1 + j7\ \Omega$$

From the Thevenin circuit

$$\mathbf{I}_2 = \frac{21(1 + j)}{6 + 1 + j7} = 3\angle 0^\circ\ \text{A}$$

$$i_2 = \underline{3\cos 4t\ \text{A}}.$$

15.32

$$Z_1 = \frac{Z_2}{n^2} = \frac{(90\times10^3)}{n^2};$$

$$Z_g = Z_1 \Rightarrow 100 = \frac{90\times10^3}{n^2} \Rightarrow n = \underline{30}$$

$$I_1 = \frac{V_g}{Z_g + Z_1} = \frac{V_g}{2Z_g} = \frac{100}{2(100)} = 0.5\ \text{A}$$

$$I_2 = -\frac{I_1}{n} = -\frac{0.5}{30} = -\frac{1}{60}\ \text{A}$$

$$P_{LOAD} = |I_2|^2(90\times10^3) = \underline{25\text{W}}.$$

15.33

$$Z_1 = \frac{Z_2}{n^2} = \frac{2 - j}{(2)^2} = \frac{1}{2} - j\frac{1}{4}\ \Omega$$

$$I_1 = \frac{V_g}{Z_g + Z_1} = \frac{4}{j\,6/8 + \frac{1}{2} - j\frac{1}{4}} = 4 - j4 = 4\sqrt{2}\angle -45^\circ$$

$$i_1 = \underline{4\sqrt{2}\cos(2t - 45^\circ)}\ \text{A}$$

15.34

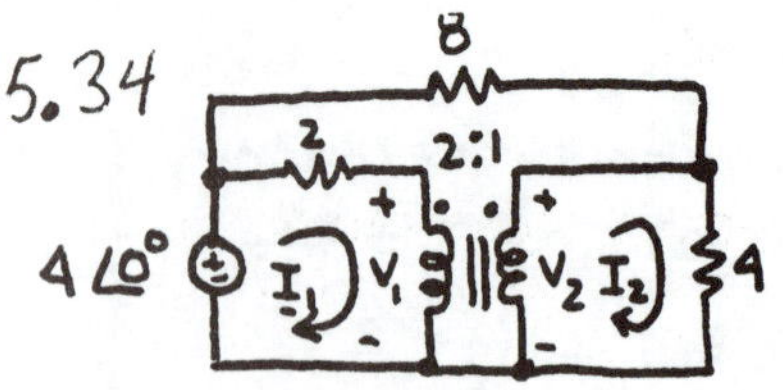

KVL: $4 = 2I_1 + V_1 = 2(2I_2) + \frac{V_2}{2}$ or

$$V_2 + 8I_2 = 8$$

KCL: $\frac{V_2 - 4}{8} - I_2 + \frac{V_2}{4} = 0$ or

$$3V_2 - 8I_2 = 4$$

Adding, $4V_2 = 12 \Rightarrow V_2 = 3\text{V}$

$$P_{8\Omega} = \frac{1}{2}\left|\frac{V_2 - 4}{8}\right|^2 (8) = \underline{\tfrac{1}{16}\text{ W}}$$

15.35 $\mathbf{Z}_1 = \frac{\mathbf{Z}_2}{n^2} = \frac{100 - j75}{(5)^2} = 4 - j3\ \Omega$

$$\mathbf{I}_1 = \frac{\mathbf{V}_g}{\mathbf{Z}_g + \mathbf{Z}_1} = \frac{12}{2 + j3 + 4 - j3} = \underline{2\angle 0°\text{A}}$$

$$\mathbf{I}_2 = -\frac{\mathbf{I}_1}{n} = \underline{-\tfrac{2}{5}\angle 0°}\ \text{A}$$

$$\mathbf{V}_1 = \mathbf{Z}_1\mathbf{I}_1 = (4 - j3)\,2 = \underline{10\angle -36.87°\text{V}}$$

$$\mathbf{V}_2 = -n\mathbf{V}_1 = \underline{50\angle 143.13°}\ \text{V}$$

15.36 $\mathbf{V}_2 = \frac{-\mathbf{Z}_2}{\mathbf{Z}_2 + n^2\mathbf{Z}_g}(n\mathbf{V}_g)$, voltage division

$$= \frac{-(100 - j75)(5)(12)}{100 - j75 + 25(2 + j3)}$$

$$= \underline{50\angle 143.1°\ \text{V}}$$

$$\mathbf{V}_1 = -\frac{\mathbf{V}_2}{n} = \underline{10\angle -36.87°\ \text{V}}$$

$$\mathbf{I}_2 = \frac{-n\mathbf{V}_g}{n^2\mathbf{Z}_g + \mathbf{Z}_2} = \frac{5(12)}{25(2 + j3) + 100 - j75}$$

$$= \underline{\tfrac{2}{5}\angle 180°\ \text{A}}$$

$$\mathbf{I}_1 = -\mathbf{I}_2 n = \underline{2\angle 0°\text{A}}$$

15.37 $\mathbf{Z}_g = \frac{\mathbf{Z}_2}{n^2} \Rightarrow 4 = \frac{10000}{n^2}$

$$n^2 = \frac{10^4}{4} \Rightarrow n = \underline{50}\ .$$

15.38 SPICE Code:

```
PROBLEM 15.38
V1 1 0 AC 26 0
R1 1 2 12
C1 2 3 0.1
* NOTE ORDER OF NODES FOR DOTS
LPRI 3 0 1
LSEC 0 4 2
K LPRI LSEC 0.70710678
VDUMMY 4 5 0
R2 5 0 3
.AC LIN 1 0.31831 0.31831
.PRINT AC IM(VDUMMY) IP(VDUMMY)
.END
```

SPICE Output:

```
*************************************

  FREQ          IM(VDUMMY)   IP(VDUMMY)

  3.183E-01     8.000E-01   -1.269E+02
```

15.39 SPICE Code:

```
PROBLEM 15.39
* DEFINE VOLTAGE SOURCE BACKWARDS
* FOR CURRENT MEASUREMENT OUTPUT
V1 0 1 AC 4 180
L1 1 2 0.375
* DUMMY VSRC FOR CURRENT MEASUREMENT
VDUMMY 3 4 0
* USE DEPENDENT SOURCE XFORMER MODEL
* FOR PRIMARY USE CURRENT
* THROUGH DUMMY VSRC
F1 2 0 VDUMMY 2
* FOR SECONDARY USE VOLTAGE
* ACROSS CCCS (2 0)
E1 3 0 2 0 2
C1 4 5 0.5
R1 5 0 2
.AC LIN 1 0.31831 0.31831
.PRINT AC IM(V1) IP(V1)
.END
```

SPICE Output:

```
*************************************

  FREQ          IM(V1)        IP(V1)

  3.183E-01     5.657E+00    -4.500E+01
```

15.40 SPICE Code:

```
PROBLEM 15.40
* DEFINE VOLTAGE SOURCE BACKWARDS
* FOR CURRENT MEASUREMENT OUTPUT
V1 0 1 AC 12 180
R1 1 2 2
* CHOOSE OMEGA=1 FOR CONVENIENCE
L1 2 3 3
* USE DEPENDENT SOURCE XFORMER MODEL
* FOR PRI USE CURRENT THRU DUMMY VSRC
F1 3 0 VD 5
* FOR SECONDARY USE VOLTAGE
* ACROSS CCCS
E1 0 4 3 0 5
* DUMMY VSRC FOR CURRENT MEASUREMENT
* NOTE: DOTS ARE OPPOSITE THAT OF
* MODEL -> REVERSE NODES FOR VD
VD 5 4 0
R2 5 6 100
C1 6 0 0.01333333333
* FREQ = OMEGA/(2*PI)
.AC LIN 1 0.159155 0.159155
.PRINT AC VM(3) VP(3) VM(4) VP(4)
.PRINT AC IM(V1) IP(V1) IM(VD) IP(VD)
.END
```

SPICE Output:

```
*************************************

  FREQ          VM(3)         VP(3)

  1.592E-01     1.000E+01    -3.687E+01

                VM(4)         VP(4)

                5.000E+01     1.431E+02

*************************************

  FREQ          IM(V1)        IP(V1)
  1.592E-01     2.000E+00    -2.048E-05

                IM(VD)        IP(VD)

                4.000E-01    -2.048E-05
```

Chapter 16

Two-Port Circuits

16.1 Two-Port Circuits

16.1 Find the impedance matrix (z-parameter matrix) and two-port description for the circuit shown.

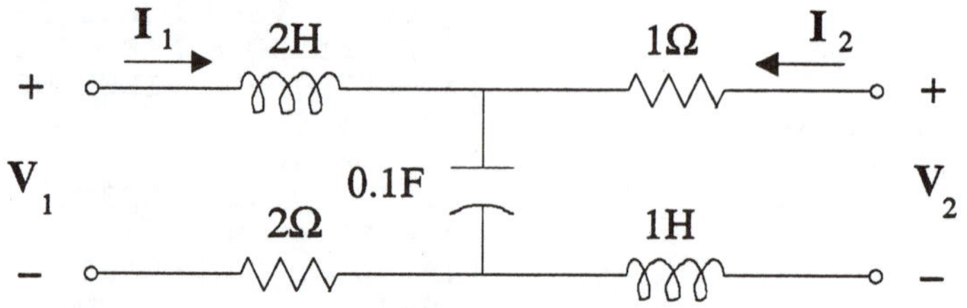

PROBLEM 16.1

16.2 Find the impedance matrix (z-parameter matrix) and two-port description for the circuit shown.

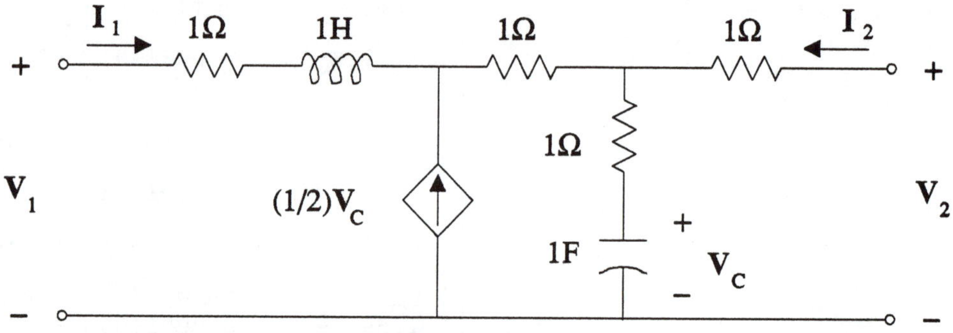

PROBLEM 16.2

16.3 For the two-port circuit shown, the z-parameter matrix is:

$$\mathbf{z} = \begin{bmatrix} s+1 & 2 \\ 2 & 3-2s \end{bmatrix}.$$

If $v_1 = 3\cos(2t)$ V and $v_2 = 2\sin(2t)$ V, find $i_1(t)$ and $i_2(t)$.

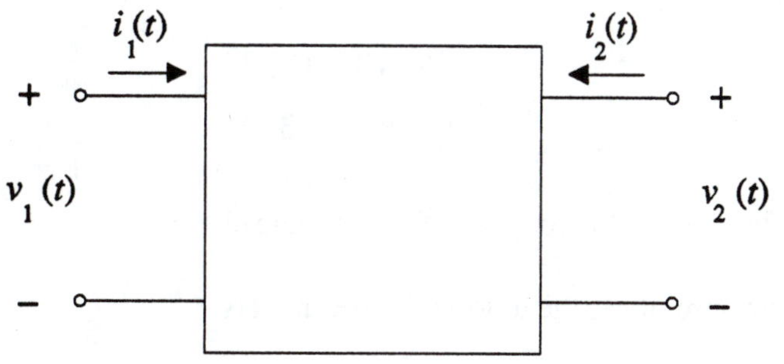

PROBLEM 16.3

16.4 (a) Find $\mathbf{Z}_1$, $\mathbf{Z}_2$, and $\mathbf{Z}_3$ in the 3-terminal circuit below that realizes the z-parameter matrix description of Prob. 16.3. (b) Can you design a circuit with this topology and z-parameter description using only passive *RLC* components?

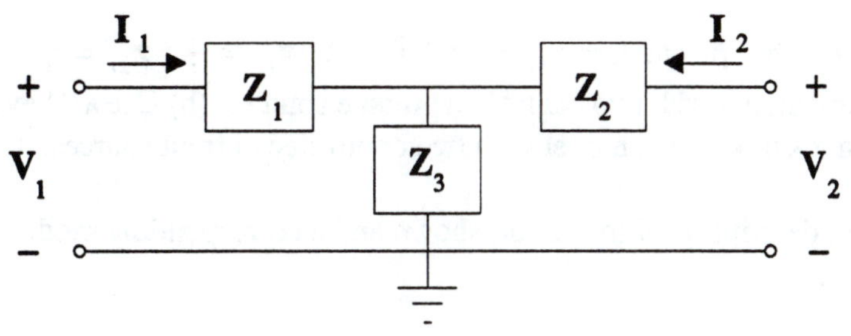

PROBLEM 16.4

16.2 Two-Port Parameters

16.5 Find the admittance matrix for the circuit in Prob. 16.1 directly by replacing each component with its associated admittance (or impedance) value, and using the technique described in the text.

16.6 For a two-port circuit with admittance matrix:

$$\mathbf{y} = \begin{bmatrix} 2 & 2s \\ 1+3s & 1/s \end{bmatrix},$$

find $i_1(t)$ and $i_2(t)$ if $v_1(t) = 6\cos(3t + 20°)$ V and $v_2(t) = 6\cos(2t + 20°)$ V.

16.7 A certain two-port circuit has the following *z*-parameters:

$$\mathbf{z} = \begin{bmatrix} 2s+1 & 1/s \\ 1/s & 3 \end{bmatrix}.$$

Find the other five two-port descriptions for this same circuit.

16.8 A certain two-port circuit has the following *t*-parameters:

$$\mathbf{t} = \begin{bmatrix} 2s+1 & 1/s \\ 1/s & 3 \end{bmatrix}.$$

Find the other five two-port descriptions for this same circuit.

16.3 Two-Port Models

16.9 A two-port circuit has parameters $\mathbf{z}_{11} = 2s + \frac{10}{s} + 2$, $\mathbf{z}_{12} = \frac{10}{s}$, $\mathbf{z}_{21} = \frac{10}{s}$, and $\mathbf{z}_{22} = 1 + \frac{10}{s} + s$. (a) Find a circuit model with two controlled voltage sources. (b) Use a Thevenin-Norton transformation to convert to a model with two controlled current sources.

16.10 Find a two-port description of the circuit shown and a corresponding model with two controlled voltage sources.

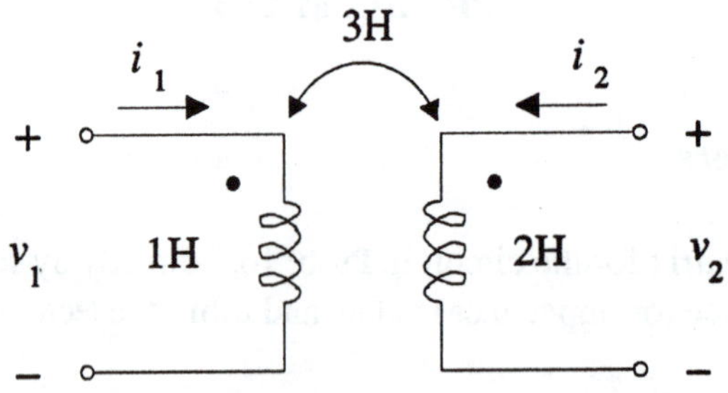

PROBLEM 16.10

16.11 Find a three-terminal two-port circuit model for the transformer circuit in Prob. 16.10.

16.12 A two-port circuit has *y*-parameters $\mathbf{y}_{11} = -\frac{s}{2}$, $\mathbf{y}_{12} = 3s+1$, $\mathbf{y}_{21} = 4$, and $\mathbf{y}_{22} = s^2+1$. Find a three-terminal two-port model which includes two dependent current sources that realizes this circuit description.

16.4 Interconnection of Two-Port Circuits

16.13 Two circuits A and B are cascaded as described in the text.

The *z*-parameter description of circuit A is:

$$\begin{bmatrix} \mathbf{V}_1 \\ \mathbf{V}_2 \end{bmatrix} = \begin{bmatrix} s & 3 \\ 3 & 1/s \end{bmatrix} \begin{bmatrix} \mathbf{I}_1 \\ \mathbf{I}_2 \end{bmatrix},$$

and the *y*-parameter description of circuit B is:

$$\begin{bmatrix} \mathbf{I}_1 \\ \mathbf{I}_2 \end{bmatrix} = \begin{bmatrix} 2s & 4 \\ -3 & 2 \end{bmatrix} \begin{bmatrix} \mathbf{V}_1 \\ \mathbf{V}_2 \end{bmatrix}.$$

Find the overall transmission matrix for the cascaded pair of two-port circuits.

16.14 The two circuits A and B described in Prob. 16.13 are connected in parallel (as described in the text). Find the overall admittance (*y*-parameter) description of the parallel combination.

16.15 The two circuits A and B described in Prob. 16.13 are connected in series (as described in the text). Find the overall impedance matrix of the series combination.

16.16 Simplify the circuit shown into interconnected two-port circuits. Find an input-output description for the overall circuit.

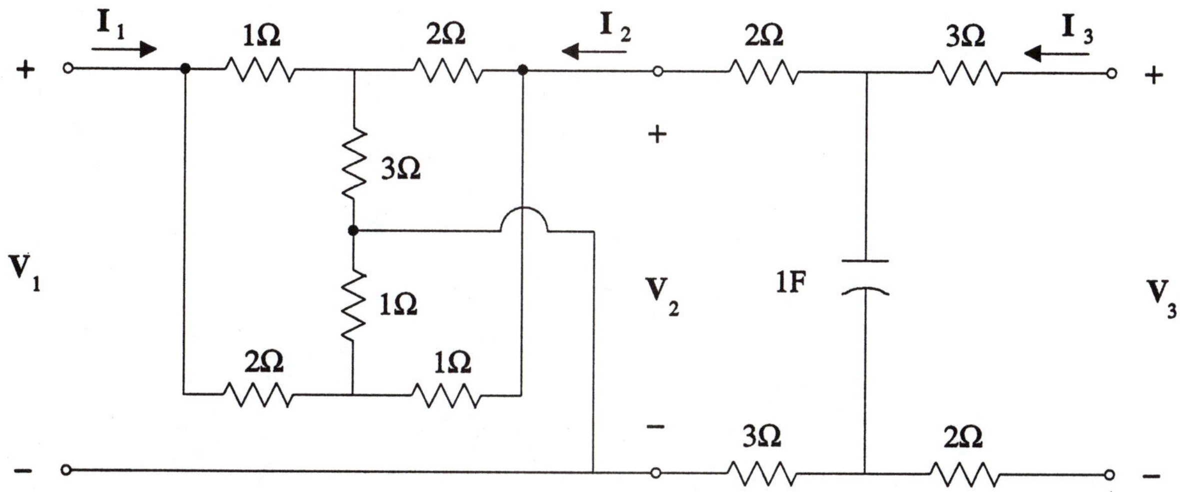

PROBLEM 16.16

16.5 SPICE and Two-Port Circuits

16.17 Using SPICE, verify Prob. 16.1 at $\omega = 1\,\text{rad/s}$ and $\omega = 10\,\text{rad/s}$ by using $\mathbf{I}_1 = \mathbf{I}_2 = 1\angle 0°$ A.

16.18 Using SPICE, verify Prob. 16.2 at $\omega = 1\,\text{rad/s}$ and $\omega = 10\,\text{rad/s}$ by using $\mathbf{I}_1 = \mathbf{I}_2 = 1\angle 0°$ A.

16.19 Using SPICE, verify Prob. 16.5 at $\omega = 1\,\text{rad/s}$ and $\omega = 10\,\text{rad/s}$ by using $\mathbf{V}_1 = \mathbf{V}_2 = 1\angle 0°$ V.

16.20 Using SPICE, find the open-circuit output $v_3(t)$ and the current $i_1(t)$ in Prob. 16.16 due to an input $v_1(t) = \cos(2t)$ V. Verify by hand.

CHAPTER 16 SOLUTIONS

16.1 $Z_{11} = \frac{V_1}{I_1}\Big|_{I_2=0}$ $\quad Z_{21} = \frac{V_2}{I_1}\Big|_{I_2=0}$

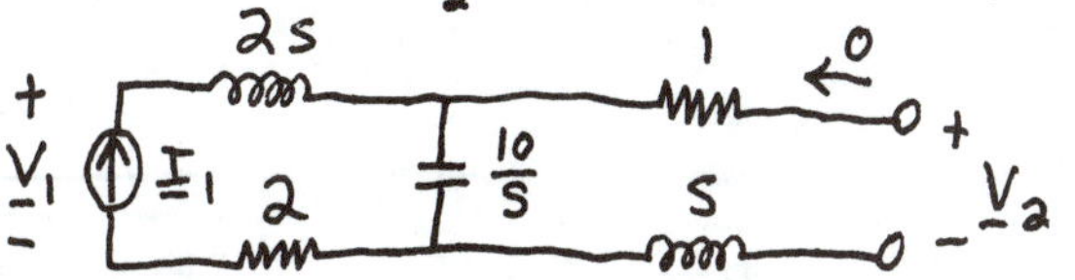

KVL: $V_1 = I_1\left(2s + \frac{10}{s} + 2\right) \Rightarrow Z_{11} = 2s + \frac{10}{s} + 2$

$V_2 = I_1\left(\frac{10}{s}\right) \Rightarrow Z_{21} = \frac{10}{s}$

$Z_{12} = \frac{V_1}{I_2}\Big|_{I_1=0}$ $\quad Z_{22} = \frac{V_2}{I_2}\Big|_{I_1=0}$

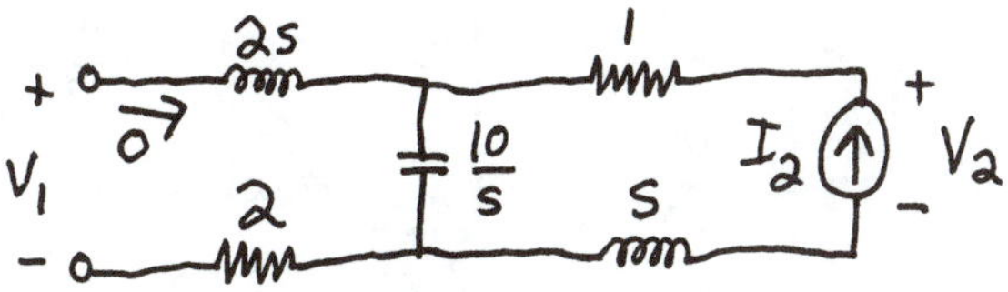

KVL: $V_2 = I_2\left(1 + \frac{10}{s} + s\right) \Rightarrow Z_{22} = 1 + \frac{10}{s} + s$

$V_1 = I_2\left(\frac{10}{s}\right) \Rightarrow Z_{12} = \frac{10}{s}$

$$\therefore \begin{bmatrix} V_1 \\ V_2 \end{bmatrix} = \begin{bmatrix} (2s + \frac{10}{s} + 2) & 10/s \\ 10/s & (1 + \frac{10}{s} + s) \end{bmatrix} \begin{bmatrix} I_1 \\ I_2 \end{bmatrix}$$

16.2 $Z_{11} = \frac{V_1}{I_1}\Big|_{I_2=0}$ $\quad Z_{21} = \frac{V_2}{I_1}\Big|_{I_2=0}$

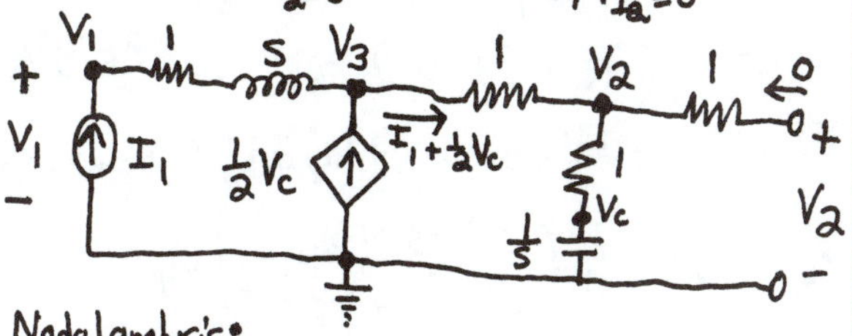

Nodal analysis:

$V_3 = V_1 - I_1(1+s) = [I_1 + \frac{1}{2}V_c][1 + 1 + \frac{1}{s}]$

$V_c = \frac{1}{s}(I_1 + \frac{1}{2}V_c) \Rightarrow V_c = I_1\left[\frac{2s}{2s^2 - s}\right]$

$\Rightarrow V_1 = I_1(1+s) + I_1(2 + \frac{1}{s}) + I_1\left(\frac{s}{2s^2-s}\right)(2 + \frac{1}{s})$

$= I_1\left(1 + s + 2 + \frac{1}{s} + \frac{2s+1}{2s^2-s}\right)$

$\Rightarrow Z_{11} = 3 + s + \frac{1}{s} + \frac{2s+1}{2s^2-s}$

$V_2 = [I_1 + \frac{1}{2}V_c][1 + \frac{1}{s}] = I_1\left[1 + \frac{s}{2s^2-s}\right][1 + \frac{1}{s}]$

$\Rightarrow Z_{21} = 1 + \frac{1}{s} + \frac{s+1}{2s^2-s}$

$Z_{12} = \frac{V_1}{I_2}\Big|_{I_1=0}$ $\quad Z_{22} = \frac{V_2}{I_2}\Big|_{I_1=0}$

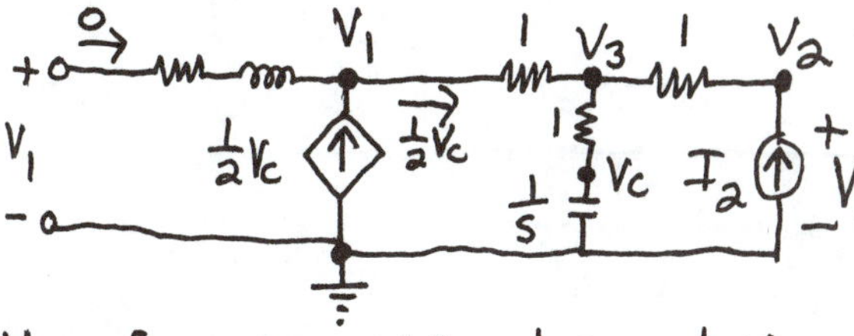

$V_c = [I_2 + \frac{1}{2}V_c]\left(\frac{1}{s}\right) = \frac{1}{s}I_2 + \frac{1}{2s}V_c$

$\Rightarrow V_c = I_2\left[\frac{1}{s(1 - \frac{1}{2s})}\right] = I_2\left[\frac{2}{2s-1}\right]$

$V_2 = I_2(1) + (I_2 + \frac{1}{2}V_c)(1 + \frac{1}{s})$

$= I_2\left[(1 + 1 + \frac{1}{s}) + \left(\frac{1}{2s-1}\right)(1 + \frac{1}{s})\right]$

$\Rightarrow Z_{22} = 2 + \frac{1}{s} + \frac{s+1}{2s^2-s}$

$V_1 = \frac{1}{2}V_c(1) + [I_2 + \frac{1}{2}V_c](1 + \frac{1}{s})$

$= I_2\left[\frac{1}{2s-1} + 1 + \frac{1}{s} + \frac{1}{2s-1} + \frac{1}{2s^2-s}\right]$

$\Rightarrow Z_{12} = 1 + \frac{1}{s} + \frac{2s+1}{2s^2-s}$

$$\mathbf{Z} = \begin{bmatrix} (3 + s + \frac{1}{s} + \frac{2s+1}{2s^2-s}) & (1 + \frac{1}{s} + \frac{2s+1}{2s^2-s}) \\ (1 + \frac{1}{s} + \frac{s+1}{2s^2-s}) & (2 + \frac{1}{s} + \frac{s+1}{2s^2-s}) \end{bmatrix}$$

16.3
$$\begin{bmatrix} 3\angle 0^\circ \\ 2\angle -90^\circ \end{bmatrix} = \begin{bmatrix} j2+1 & 2 \\ 2 & 3-j4 \end{bmatrix} \begin{bmatrix} I_1 \\ I_2 \end{bmatrix}$$

$\Rightarrow I_1(\sqrt{5}\angle 63.435^\circ) + 2I_2 = 3\angle 0^\circ$

$2I_1 + I_2(5\angle -53.13^\circ) = 2\angle -90^\circ$

Solving ...

$\Rightarrow I_1 = 1.654\angle -57.59^\circ$ A

$I_2 = 0.3885\angle -150.95^\circ$ A

$\therefore i_1(t) = 1.654\cos(2t - 57.59^\circ)$ A

$i_2(t) = 0.3885\cos(2t - 150.95^\circ)$ A

16.4 (a) Set $I_2=0$: $V_2 = I_1 Z_3 \Rightarrow \boxed{Z_3 = 2}$

$V_1 = I_1 (Z_1+Z_3) \Rightarrow Z_1+Z_3 = s+1$

$\Rightarrow \boxed{Z_1 = s-1}$

Set $I_1=0$: $V_2 = I_2 (Z_2+Z_3) \Rightarrow Z_2+Z_3 = 3-2s$

$\Rightarrow \boxed{Z_2 = 1-2s}$

(b) No, since the $z=-1$ and $z=-2s$ require active components to realize. There are no negative Laplace impedances.

16.5 $y_{11} = \left.\dfrac{I_1}{V_1}\right|_{V_2=0}$ $\quad y_{21} = \left.\dfrac{I_2}{V_1}\right|_{V_2=0}$

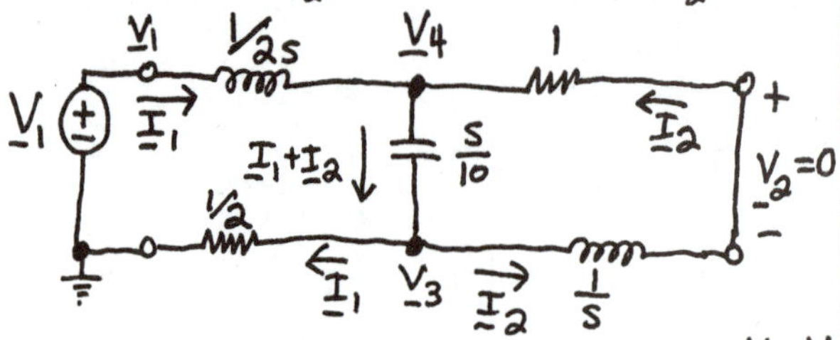

$$I_1 = (V_1 - V_4)\left(\frac{1}{2s}\right) = V_3\left(\frac{1}{2}\right) \Rightarrow V_3 = \frac{V_1 - V_4}{s}$$

$$I_2 = (V_3 - V_4)\frac{\left(\frac{1}{s}\bullet 1\right)}{\left(\frac{1}{s}+1\right)} = (V_3 - V_4)\left(\frac{1}{s+1}\right)$$

$$I_1 + I_2 = (V_4 - V_3)\left(\frac{s}{10}\right)$$

$$\Rightarrow I_1 = (V_4 - V_3)\left(\frac{s}{10} + \frac{1}{s+1}\right)$$

$$\Rightarrow I_2 = \frac{-I_1}{\left(\frac{s}{10}+\frac{1}{s+1}\right)}\left(\frac{1}{s+1}\right) = \frac{-10 I_1}{s(s+1)+10}$$

$$V_3 = \frac{V_1}{s} - \frac{V_4}{s} \Rightarrow V_1 = sV_3 + V_4$$

$$\frac{-10 I_1}{s(s+1)+10} = \frac{V_3}{s+1} - \frac{V_4}{s+1} \Rightarrow V_3 - V_4 = \frac{-10(s+1)I_1}{s(s+1)+10}$$

$$V_1 = sV_3 + V_4 \Rightarrow V_1 = sV_4 - \frac{10s(s+1)}{s(s+1)+10} I_1 + V_4$$

$$2sI_1 = V_1 - V_4 \Rightarrow V_4 = V_1 - 2sI_1$$

$$\Rightarrow V_1 = (s+1)[V_1 - 2sI_1] - \left[\frac{10s(s+1)}{s(s+1)+10}\right] I_1$$

$$\Rightarrow V_1(-s) = I_1\left[-2s(s+1) - \frac{10s(s+1)}{s(s+1)+10}\right]$$

$$y_{11} = \frac{I_1}{V_1} = \frac{-s}{\left[-2s(s+1) - \frac{10s(s+1)}{s(s+1)+10}\right]}$$

$$\Rightarrow \boxed{y_{11} = \frac{s^2+s+10}{2s^3+4s^2+32s+30}}$$

$$\boxed{y_{21} = \frac{I_2}{V_1} = \frac{-10}{s(s+1)+10}\left(\frac{I_1}{V_1}\right) = \frac{-10}{2s^3+4s^2+32s+30}}$$

$y_{12} = \left.\dfrac{I_1}{V_2}\right|_{V_1=0}$ $\quad y_{22} = \left.\dfrac{I_2}{V_2}\right|_{V_1=0}$

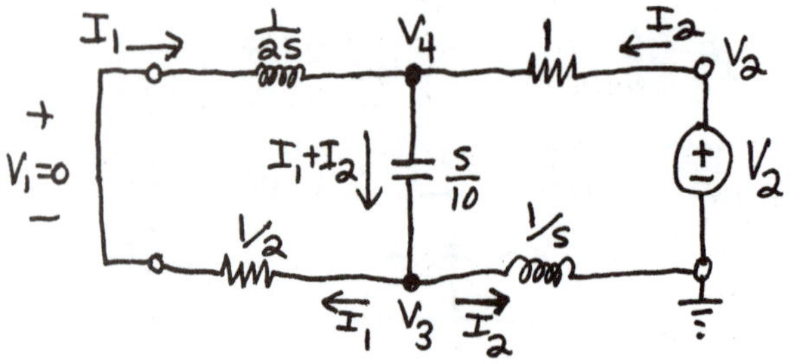

$$I_2 = V_3\left(\frac{1}{s}\right) = (V_2 - V_4)(1) \Rightarrow V_2 = V_4 + \frac{V_3}{s}$$

$$I_1 = (V_3 - V_4)\left(\frac{1}{2+2s}\right)$$

$$I_1 + I_2 = (V_4 - V_3)\left(\frac{s}{10}\right)$$

$$\Rightarrow \frac{V_3 - V_4}{2+2s} + \frac{V_3}{s} = \frac{s}{10}V_4 - \frac{s}{10}V_3$$

$$\Rightarrow 10sV_3 - 10sV_4 + (20+20s)V_3 = (2s^2+2s^3)V_4 - (2s^2+2s^3)V_3$$

$$\Rightarrow V_3(s^3+s^2+15s+10) = V_4(s^3+s^2+5s)$$

$$V_2 = V_4 + \frac{1}{s}\left(\frac{s^3+s^2+5s}{s^3+s^2+15s+10}\right)V_4$$

$$= V_4\left[\frac{s^3+2s^2+16s+15}{s^3+s^2+15s+10}\right]$$

$$(2+2s)I_1 = V_3 - V_4 = V_4\left[\frac{s^3+s^2+5s}{s^3+s^2+15s+10} - 1\right]$$

$$= V_2\left[\frac{s^3+s^2+15s+10}{s^3+2s^2+16s+15}\right]\left[\frac{-10s-10}{s^3+s^2+15s+10}\right]$$

$$\Rightarrow \boxed{y_{12} = \frac{I_1}{V_2} = \frac{-5}{s^3+2s^2+16s+15}}$$

$$= y_{21}$$

16.5 (cont'd)

$$I_1+I_2=(V_4-V_3)\left(\frac{s}{10}\right)$$

$$\Rightarrow I_2=(V_4-V_3)\left(\frac{s}{10}+\frac{1}{2+2s}\right)$$

$$\Rightarrow \frac{I_2}{I_1}=\frac{\frac{s}{10}+\frac{1}{2+2s}}{\left(\frac{-1}{2+2s}\right)}=-\frac{s^2}{5}-\frac{s}{5}-1$$

$$\Rightarrow y_{22}=\frac{I_2}{V_2}=\left(-\frac{s^2}{5}-\frac{s}{5}-1\right)y_{12}$$

$$\boxed{y_{22}=\frac{s^2+s+5}{s^3+2s^2+16s+15}}$$

16.6 Since v_1 and v_2 are at different frequencies, superposition must be used:

$$\begin{bmatrix} I_{1A} \\ I_{2A}\end{bmatrix} \equiv \text{response when } V_2=0$$

$$\begin{bmatrix} I_{1B} \\ I_{2B}\end{bmatrix} \equiv \text{response when } V_1=0$$

$$\Rightarrow \begin{bmatrix} I_1 \\ I_2\end{bmatrix}=\begin{bmatrix} I_{1A} \\ I_{2A}\end{bmatrix}+\begin{bmatrix} I_{1B} \\ I_{2B}\end{bmatrix}$$

$\omega=3$

$$\mathbf{I}_{1A}=\begin{bmatrix}2 & j6\end{bmatrix}\begin{bmatrix}6\angle 20^\circ \\ 0\end{bmatrix}=12\angle 20^\circ$$

$$\mathbf{I}_{2A}=\begin{bmatrix}1+j9 & -j\tfrac{1}{3}\end{bmatrix}\begin{bmatrix}6\angle 20^\circ \\ 0\end{bmatrix}=6\angle 20^\circ+54\angle 110^\circ=54.33\angle 103.66^\circ$$

$\omega=2$

$$\mathbf{I}_{1B}=\begin{bmatrix}2 & j4\end{bmatrix}\begin{bmatrix}0 \\ 6\angle 20^\circ\end{bmatrix}=24\angle 110^\circ$$

$$\mathbf{I}_{2B}=\begin{bmatrix}1+j6 & -j\tfrac{1}{2}\end{bmatrix}\begin{bmatrix}0 \\ 6\angle 20^\circ\end{bmatrix}=3\angle -70^\circ$$

$$\therefore i_1(t)=12\cos(3t+20^\circ)+24\cos(2t+110^\circ)\text{ A}$$

$$i_2(t)=54.33\cos(3t+103.66^\circ)+3\cos(2t-70^\circ)\text{ A}$$

16.7

$$\mathbf{y}=\frac{1}{\left(6s+3-\frac{1}{s^2}\right)}\begin{bmatrix}3 & -\frac{1}{s}\\ -\frac{1}{s} & 2s+1\end{bmatrix}$$

$$\mathbf{h}=\frac{1}{3}\begin{bmatrix}6s+3-\frac{1}{s^2} & 1/s\\ -1/s & 1\end{bmatrix}$$

$$\mathbf{g}=\frac{1}{2s+1}\begin{bmatrix}1 & -\frac{1}{s}\\ \frac{1}{s} & 6s+3-1/s^2\end{bmatrix}$$

$$\mathbf{t}=s\begin{bmatrix}2s+1 & 6s+3-1/s^2\\ 1 & 3\end{bmatrix}$$

$$\mathbf{t}'=s\begin{bmatrix}3 & -(6s+3-1/s^2)\\ -1 & 2s+1\end{bmatrix}$$

16.8

$$\mathbf{z}=s\begin{bmatrix}2s+1 & 6s+3-\frac{1}{s^2}\\ 1 & 3\end{bmatrix}$$

$$\mathbf{y}=s\begin{bmatrix}3 & -(6s+3-\frac{1}{s^2})\\ -1 & 2s+1\end{bmatrix}$$

$$\mathbf{h}=\frac{1}{3}\begin{bmatrix}1/s & 6s+3-1/s^2\\ -1 & 1/s\end{bmatrix}$$

$$\mathbf{g}=\frac{1}{2s+1}\begin{bmatrix}1/s & -(6s+3-1/s^2)\\ 1 & 1/s\end{bmatrix}$$

$$\mathbf{t}'=\frac{1}{6s+3-1/s^2}\begin{bmatrix}3 & -1/s\\ -1/s & 2s+1\end{bmatrix}$$

16.9 (a)

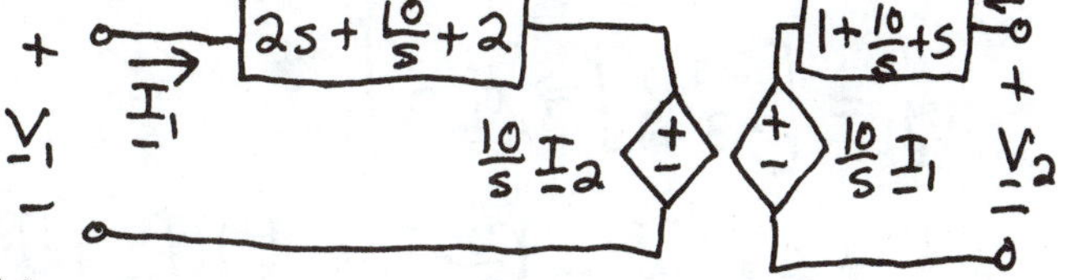

(b)

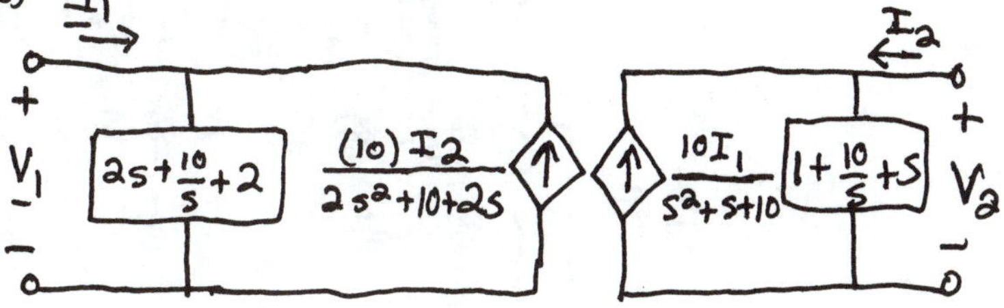

16.10
$$\underline{V}_1 = s\,\underline{I}_1 + 3s\,\underline{I}_2$$
$$\underline{V}_2 = 3s\,\underline{I}_1 + 2s\,\underline{I}_2$$
$$\Rightarrow \underline{z} = \begin{bmatrix} s & 3s \\ 3s & 2s \end{bmatrix}$$

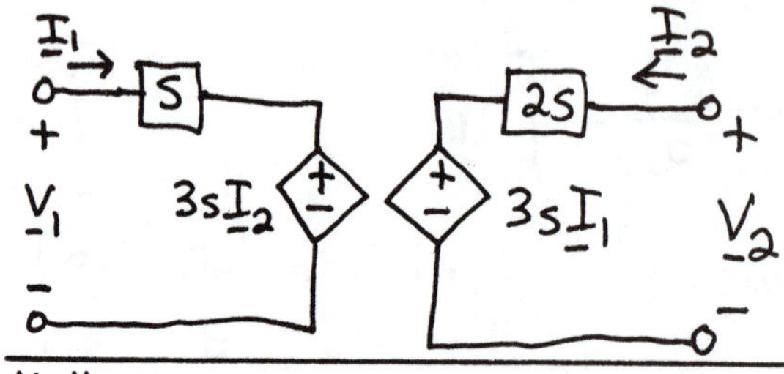

16.11
$$z_{11} - z_{12} = (s - 3s) = -2s$$
$$z_{22} - z_{12} = (2s - 3s) = -s$$
$$z_{22} - z_{21} = (2s - 3s) = -s$$

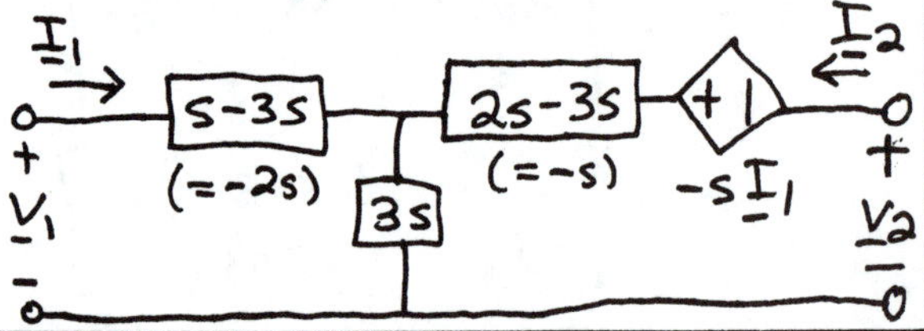

16.12

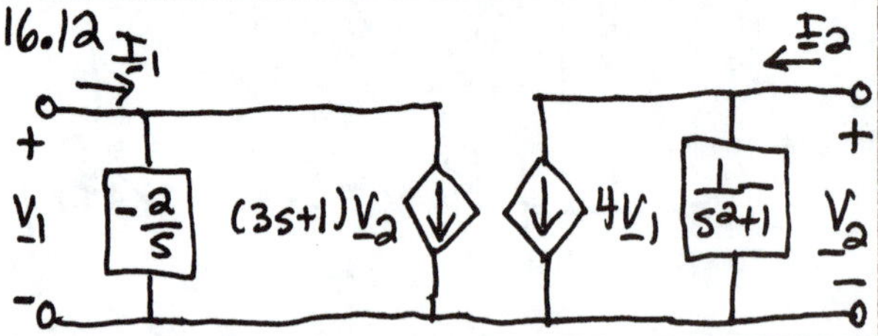

16.13
$$\underline{z}_a = \begin{bmatrix} s & 3 \\ 3 & 1/s \end{bmatrix} \Rightarrow \underline{t}_a = \begin{bmatrix} \frac{s}{3} & -\frac{8}{3} \\ \frac{1}{3} & \frac{1}{3s} \end{bmatrix}$$
$$\underline{y}_b = \begin{bmatrix} 2s & 4 \\ -3 & 2 \end{bmatrix} \Rightarrow \underline{t}_b = \begin{bmatrix} 2/3 & 1/3 \\ \frac{4}{3}s + 4 & \frac{2}{3}s \end{bmatrix}$$
$$\therefore \underline{t} = \underline{t}_a \underline{t}_b = \begin{bmatrix} \left(\frac{2s}{9} - \frac{32s}{9} - \frac{32}{3}\right) & \left(\frac{s}{9} - \frac{16}{9}s\right) \\ \left(\frac{2}{9} + \frac{4}{9} + \frac{4}{3s}\right) & \left(\frac{1}{9} + \frac{2}{9}\right) \end{bmatrix}$$
$$\underline{t} = \begin{bmatrix} \left(-\frac{10}{3}s - \frac{32}{3}\right) & -\frac{15}{9}s \\ \left(\frac{2}{3} + \frac{4}{3s}\right) & \frac{1}{3} \end{bmatrix}$$

16.14
$$\underline{y}_A = \frac{1}{-8}\begin{bmatrix} 1/s & -3 \\ -3 & s \end{bmatrix} = \begin{bmatrix} -\frac{1}{8s} & 3/8 \\ 3/8 & -5/8 \end{bmatrix}$$
$$\underline{y}_B = \begin{bmatrix} 2s & 4 \\ -3 & 2 \end{bmatrix}$$
$$\underline{y} = \underline{y}_A + \underline{y}_B = \begin{bmatrix} 2s - \frac{1}{8s} & \frac{35}{8} \\ -\frac{21}{8} & 2 - \frac{5}{8} \end{bmatrix}$$

16.15
$$\underline{z}_A = \begin{bmatrix} s & 3 \\ 3 & 1/s \end{bmatrix}$$
$$\underline{z}_B = \frac{1}{4s+12}\begin{bmatrix} 2 & -4 \\ 3 & 2 \end{bmatrix} = \begin{bmatrix} \frac{1}{2s+6} & -\frac{1}{s+3} \\ \frac{3}{4s+12} & \frac{1}{2s+6} \end{bmatrix}$$
$$\therefore \underline{z} = \underline{z}_A + \underline{z}_B = \begin{bmatrix} \left(s + \frac{1}{2s+6}\right) & \left(3 - \frac{1}{s+3}\right) \\ \left(3 + \frac{3}{4s+12}\right) & \left(\frac{1}{s} + \frac{1}{2s+6}\right) \end{bmatrix}$$

16.16

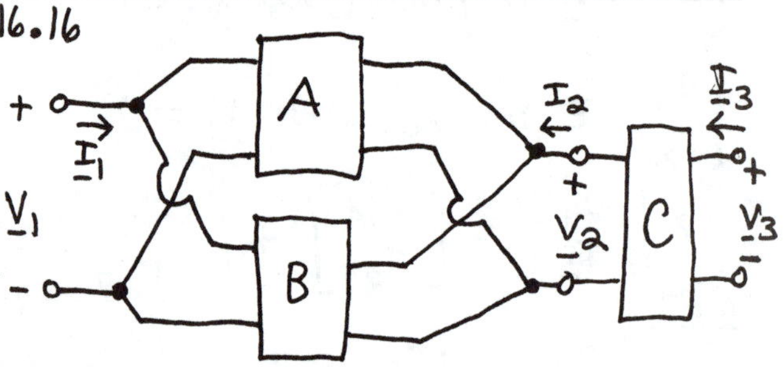

Circuit A:

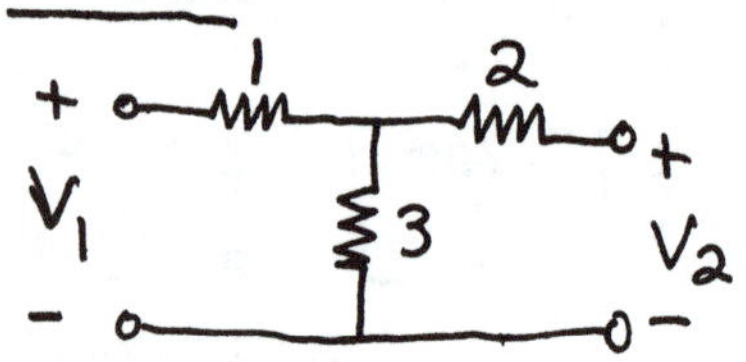

$$\underline{z}_A = \begin{bmatrix} 4 & 3 \\ 3 & 5 \end{bmatrix}$$

Circuit B:

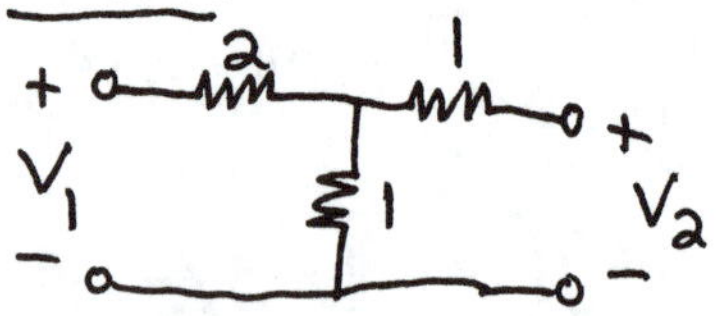

$$\underline{z}_B = \begin{bmatrix} 3 & 1 \\ 1 & 2 \end{bmatrix}$$

Circuit C:

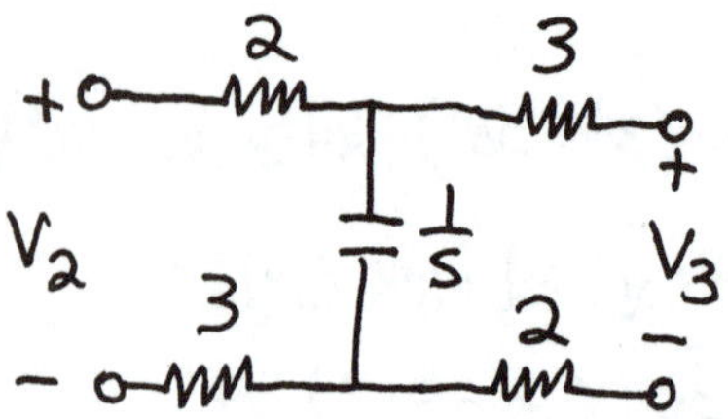

$$\underline{z}_C = \begin{bmatrix} \left(5 + \frac{1}{s}\right) & \frac{1}{s} \\ \frac{1}{s} & \left(5 + \frac{1}{s}\right) \end{bmatrix}$$

16.16 (cont'd)

$$\Rightarrow \underline{y}_A = \frac{1}{11}\begin{bmatrix} 5 & -3 \\ -3 & 4 \end{bmatrix} = \begin{bmatrix} 5/11 & -3/11 \\ -3/11 & 4/11 \end{bmatrix}$$

$$\Rightarrow \underline{y}_B = \frac{1}{5}\begin{bmatrix} 2 & -1 \\ -1 & 3 \end{bmatrix} = \begin{bmatrix} 2/5 & -1/5 \\ -1/5 & 3/5 \end{bmatrix}$$

Parallel combination of A and B:

$$\underline{y}_P = \underline{y}_A + \underline{y}_B = \begin{bmatrix} 47/55 & -26/55 \\ -26/55 & 53/55 \end{bmatrix}$$

Overall circuit is parallel combination cascaded with circuit C:

$$\underline{t} = \underline{t}_P \underline{t}_C$$

$$\underline{t}_P = -\frac{55}{26}\begin{bmatrix} -53/55 & -1 \\ -363/605 & -\frac{47}{55} \end{bmatrix} = \begin{bmatrix} \frac{53}{26} & \frac{55}{26} \\ \frac{33}{26} & \frac{47}{26} \end{bmatrix}$$

$$\underline{t}_C = s\begin{bmatrix} 5+1/s & (5+1/s)^2 - 1/s^2 \\ 1 & 5+1/s \end{bmatrix} = \begin{bmatrix} 5s+1 & 25s+10 \\ s & 5s+1 \end{bmatrix}$$

$$\therefore \underline{t} = \begin{bmatrix} \frac{53}{26} & \frac{55}{26} \\ \frac{33}{26} & \frac{47}{26} \end{bmatrix}\begin{bmatrix} (5s+1) & (25s+10) \\ s & (5s+1) \end{bmatrix}$$

$$\Rightarrow \underline{t} = \begin{bmatrix} \left(\frac{160}{13}s + \frac{53}{26}\right) & \left(\frac{800}{13}s + \frac{585}{26}\right) \\ \left(\frac{106}{13}s + \frac{33}{26}\right) & \left(\frac{530}{13}s + \frac{377}{26}\right) \end{bmatrix}$$

$$\begin{bmatrix} \underline{V}_1 \\ \underline{I}_1 \end{bmatrix} = \underline{t}\begin{bmatrix} \underline{V}_3 \\ -\underline{I}_3 \end{bmatrix}$$

16.17 SPICE input for analysis at $\omega = 1$ rad/s:

```
PROBLEM 16.17: OMEGA = 1
* USE A UNIT CURRENT SOURCE FOR
* I1 AND I2, AND CHECK V1, V2
I1 1 0 AC 1 180
I2 5 4 AC 1 180
L1 1 3 2
R1 0 2 2
R2 3 5 1
C1 3 2 0.1
RDUMMY 3 2 100G
L2 2 4 1
.AC LIN 1 0.159155 0.159155
.PRINT AC VM(I1) VP(I1) VM(I2) VP(I2)
.END
```

SPICE output:

```
*************************************

  FREQ          VM(I1)         VP(I1)

1.592E-01    1.811E+01   -8.366E+01

                VM(I2)         VP(I2)

             1.903E+01   -8.699E+01
```

SPICE input for analysis at $\omega = 10$ rad/s:

```
PROBLEM 16.17: OMEGA = 10
* USE A UNIT CURRENT SOURCE FOR
* I1 AND I2, AND CHECK V1, V2
I1 1 0 AC 1 180
I2 5 4 AC 1 180
L1 1 3 2
R1 0 2 2
R2 3 5 1
C1 3 2 0.1
RDUMMY 3 2 100G
L2 2 4 1
.AC LIN 1 1.59155 1.59155
.PRINT AC VM(I1) VP(I1) VM(I2) VP(I2)
.END
```

SPICE output:

```
*************************************

  FREQ          VM(I1)         VP(I1)

1.592E+00    1.811E+01    8.366E+01

                VM(I2)         VP(I2)

             8.062E+00    8.288E+01
```

16.17 (continued)

Analysis: In the case where $\mathbf{I}_1 = \mathbf{I}_2 = 1\angle 0°$ A,

$$\mathbf{V}_1 = 2j\omega - j\tfrac{10}{\omega} + 2 - j\tfrac{10}{\omega}$$
$$= 2 + j\left(2\omega - \tfrac{20}{\omega}\right)$$

$$\mathbf{V}_2 = -j\tfrac{10}{\omega} + 1 - j\tfrac{10}{\omega} + j\omega$$
$$= 1 + j\left(\omega - \tfrac{20}{\omega}\right)$$

At $\omega = 1$ rad/s:

$$\mathbf{V}_1 = 2 - 18j = 18.11\angle -83.66° \text{ V}$$
$$\mathbf{V}_2 = 1 - 19j = 19.03\angle -86.99° \text{ V}$$

At $\omega = 10$ rad/s:

$$\mathbf{V}_1 = 2 + 18j = 18.11\angle 83.66° \text{ V}$$
$$\mathbf{V}_2 = 1 + 8j = 8.06\angle 82.87° \text{ V}$$

This is in agreement with the SPICE outputs above.

16.18 SPICE input for analysis at $\omega = 1$ rad/s:

```
PROBLEM 16.18: OMEGA = 1
* USE A UNIT CURRENT SOURCE FOR
* I1 AND I2, AND CHECK V1, V2
I1 1 0 AC 1 180
I2 6 0 AC 1 180
R1 1 2 1
L1 2 3 1
G1 0 3 5 0 0.5
R2 3 4 1
R3 4 5 1
C1 5 0 1
RDUMMY 5 0 100G
R6 4 6 1
.AC LIN 1 0.159155 0.159155
.PRINT AC VM(I1) VP(I1) VM(I2) VP(I2)
.END
```

SPICE output:

```
************************************

  FREQ          VM(I1)        VP(I1)

1.592E-01    3.256E+00   -4.251E+01

                VM(I2)        VP(I2)

             3.000E+00   -5.313E+01
```

SPICE input for analysis at $\omega = 10$ rad/s:

```
PROBLEM 16.18: OMEGA = 10
* USE A UNIT CURRENT SOURCE FOR
* I1 AND I2, AND CHECK V1, V2
I1 1 0 AC 1 180
I2 6 0 AC 1 180
R1 1 2 1
L1 2 3 1
G1 0 3 5 0 0.5
R2 3 4 1
R3 4 5 1
C1 5 0 1
RDUMMY 5 0 100G
R6 4 6 1
.AC LIN 1 1.59155 1.59155
.PRINT AC VM(I1) VP(I1) VM(I2) VP(I2)
.END
```

SPICE output:

```
************************************

  FREQ          VM(I1)        VP(I1)

1.592E+00    1.039E+01    6.748E+01

                VM(I2)        VP(I2)

             3.000E+00   -5.725E+00
```

16.18 (continued)

Analysis: In the case where $\mathbf{I}_1 = \mathbf{I}_2 = 1\angle 0°$ A,

$$\mathbf{V}_1 = 3 + j\omega - j\frac{1}{\omega} + \frac{1}{2j\omega - 1} + 1 - j\frac{1}{\omega} + \frac{2j\omega + 1}{-2\omega^2 - j\omega}$$

$$= 4 + j\left(\omega - \frac{2}{\omega}\right) + \frac{-1 - 2j\omega}{1 + 4\omega^2} + \frac{-4\omega^2 + j(\omega - 4\omega^3)}{4\omega^4 + \omega^2}$$

$$= 4 - \frac{5}{1 + 4\omega^2} + j\left(\omega - \frac{2}{\omega} + \frac{1 - 4\omega^2}{\omega + 4\omega^3} - \frac{2\omega}{1 + 4\omega^2}\right)$$

$$\mathbf{V}_2 = 1 - j\frac{1}{\omega} + \frac{j\omega + 1}{-2\omega^2 - j\omega} + 2 - j\frac{1}{\omega} + \frac{j\omega + 1}{-2\omega^2 - j\omega}$$

$$= 3 - j\frac{2}{\omega} + 2\left(\frac{-3\omega^2 + j(\omega - 2\omega^3)}{4\omega^4 + \omega^2}\right)$$

$$= 3 - \frac{6}{4\omega^2 + \omega} + j\left(\frac{\omega - 2\omega^3}{4\omega^4 + \omega^2} - \frac{2}{\omega}\right)$$

At $\omega = 1$ rad/s:

$$\mathbf{V}_1 = 3 - 2j = 3.61\angle -33.69° \text{ V}$$
$$\mathbf{V}_2 = 1.8 - 2.2j = 2.84\angle -50.71° \text{ V}$$

At $\omega = 10$ rad/s:

$$\mathbf{V}_1 = 3.99 + 9.65j = 10.44\angle 67.54° \text{ V}$$
$$\mathbf{V}_2 = 2.99 + 0.25j = 3.00\angle 4.78° \text{ V}$$

This is in agreement with the SPICE outputs above.

16.19 SPICE input for analysis at $\omega = 1$ rad/s:

```
PROBLEM 16.19: OMEGA = 1
* USE A UNIT VOLTAGE SOURCE FOR
* V1 AND V2, AND CHECK I1, I2
V1 0 1 AC 1 180
V2 4 5 AC 1 180
L1 1 3 2
R1 0 2 2
R2 3 5 1
C1 3 2 0.1
RDUMMY 3 2 100G
L2 2 4 1
.AC LIN 1 0.159155 0.159155
.PRINT AC IM(V1) IP(V1) IM(V2) IP(V2)
.END
```

SPICE output:

```
************************************

 FREQ          IM(V1)        IP(V1)

1.592E-01    3.562E-02    8.591E+01
               IM(V2)        IP(V2)

             7.125E-02    8.591E+01
```

SPICE input for analysis at $\omega = 10$ rad/s:

```
PROBLEM 16.19: OMEGA = 10
* USE A UNIT VOLTAGE SOURCE FOR
* V1 AND V2, AND CHECK I1, I2
V1 0 1 AC 1 180
V2 4 5 AC 1 180
L1 1 3 2
R1 0 2 2
R2 3 5 1
C1 3 2 0.1
RDUMMY 3 2 100G
L2 2 4 1
.AC LIN 1 1.59155 1.59155
.PRINT AC IM(V1) IP(V1) IM(V2) IP(V2)
.END
```

SPICE output:

```
************************************

 FREQ          IM(V1)        IP(V1)

1.592E+00    5.842E-02   -8.329E+01
               IM(V2)        IP(V2)

             1.168E-01   -8.329E+01
```

Analysis: In the case where $\mathbf{V}_1 = \mathbf{V}_2 = 1\angle 0°$ V,

$$\mathbf{I}_1 = \frac{j\omega - \omega^2}{-2j\omega^3 - 4\omega^2 + 32j\omega + 30}$$

$$\mathbf{I}_2 = \frac{j\omega - \omega^2}{-j\omega^3 - 2\omega^2 + 16j\omega + 15}$$

16.19 (continued)

At $\omega = 1$ rad/s:

$$\mathbf{I}_1 = \frac{j-1}{-2j-4+32j+30} = \frac{j-1}{26+30j}$$
$$= 0.0356\angle 85.91°\ \text{A}$$

$$\mathbf{I}_2 = \frac{j-1}{-j-2+16j+15} = \frac{j-1}{13+15j}$$
$$= 0.0712\angle 85.91°\ \text{A}$$

At $\omega = 10$ rad/s:

$$\mathbf{I}_1 = \frac{10j-100}{-2000j-400+320j+30} = \frac{10j-100}{-370-1680j}$$
$$= 0.0584\angle -83.29°\ \text{A}$$

$$\mathbf{I}_2 = \frac{10j-100}{-1000j-200+160j+15} = \frac{10j-100}{-185-840j}$$
$$= 0.1168\angle -83.29°\ \text{A}$$

This is in agreement with the SPICE outputs above.

16.20 SPICE input:

```
PROBLEM 16.20
V1 0 1 AC 1 180
R1 1 2 1
R2 2 4 2
R3 2 0 3
R4 1 3 2
R5 3 0 1
R6 3 4 1
R7 4 5 2
R8 0 6 3
C1 5 7 1
RDUMMY 5 7 100G
VDUMMY 7 6 AC 0 0
.AC LIN 1 0.31831 0.31831
.PRINT AC IM(VDUMMY) IP(VDUMMY) IM(V1)
IP(V1)
.END
```

SPICE output:

```
**************************************
  FREQ         IM(VDUMMY)   IP(VDUMMY)
  3.183E-01    8.097E-02    4.734E+00
               IM(V1)       IP(V1)
               6.622E-01    2.836E-01
```

Analysis:

$$\mathbf{V}_1 = (j\tfrac{320}{13} + \tfrac{53}{26})\mathbf{V}_3 = (24.70\angle 85.27°)\mathbf{V}_3$$
$$\Rightarrow \mathbf{V}_3 = 0.0405\angle -85.27°\ \text{V}$$

As a check, plug in the SPICE output value:

$$\mathbf{V}_3 = -\tfrac{1}{2}j(0.08097\angle 4.734°) = 0.0405\angle -85.27°$$

Thus, this value checks.

$$\mathbf{I}_1 = (j\tfrac{212}{13} + \tfrac{33}{26})\mathbf{V}_3 = (16.36\angle 85.55°)\mathbf{V}_3$$
$$\Rightarrow \mathbf{I}_1 = 0.662\angle 0.28°\ \text{A}$$

This is in agreement with the SPICE output above.

Chapter 17

Fourier Series and Transform

17.1 Periodic Functions

17.1 Express the following function in the form:

$$f(t) = \sum_{n=-\infty}^{\infty} f_{\mathrm{T}}(t + n\mathrm{T})$$

using the smallest period T possible. Hint: the function is a combination of scaled and shifted unit steps and sinusoids with frequency $f = 1/2\,\mathrm{Hz}$.

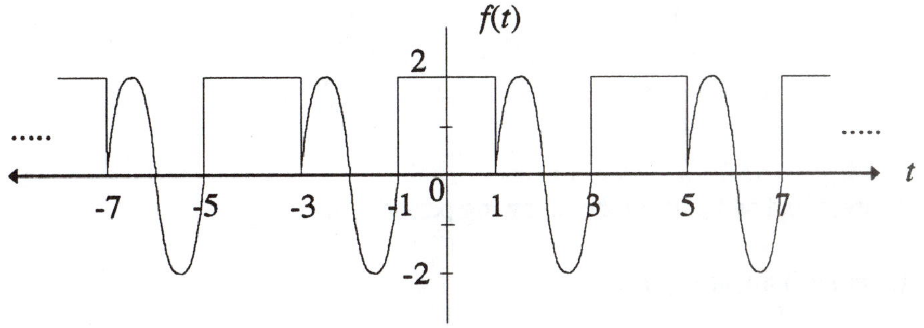

PROBLEM 17.1

17.2 For each of the following functions, determine if the function is periodic and, if it is, find the smallest period T.

(a)

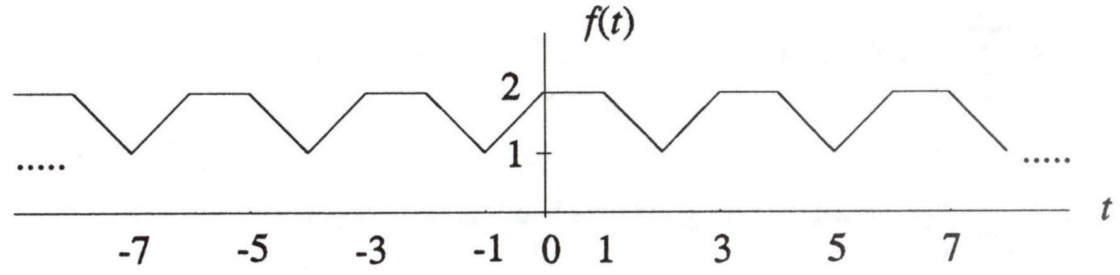

(b)

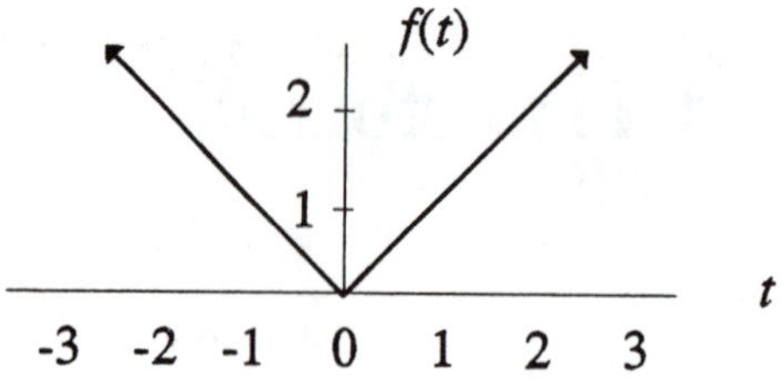

(c)

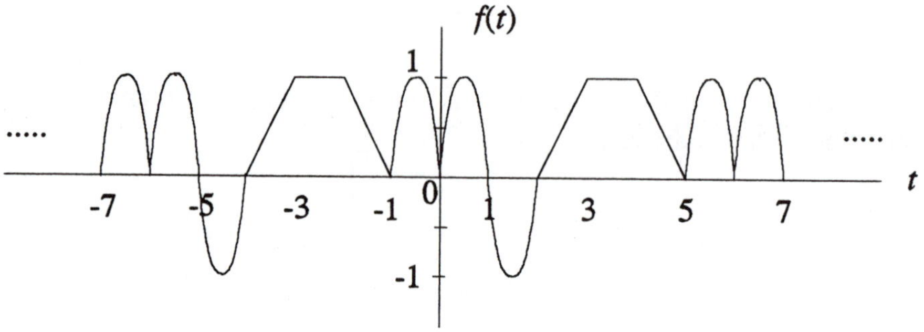

17.3 Find the even and odd parts of the following periodic functions:

(a) $f(t) = 1 + 3\sin(4t + 60°)$

(b) $f(t) = 2\cos(4t - 20°) + 5\cos(t + 20°)$

17.4 Find the average value of each of the following functions:

(a) $f(t) = 1 + 3\sin(4t + 60°)$

(b) $f(t) = 2\cos(4t - 20°) + 5\cos(t + 20°)$

(c) $f(t) = \sum_{n=-\infty}^{\infty} g_T(t + n)$, where $g_T(t) = e^{-t}\,[u(t) - u(t-1)]$

(d) $f(t) = \sum_{n=-\infty}^{\infty} g_T(t + 2n)$, with $g_T(t)$ defined as in part (c).

17.2 Trigonometric Fourier Series

17.5 Find the trigonometric Fourier series for the function $f(x)$ with a period of 4, with one period given by:

$$f(x) = 2, \quad -2 < x \le 0$$
$$= x, \quad 0 < x \le 2$$

17.6 Find the trigonometric Fourier series for the function $f(x)$ with a period of 2π, with one period given by:

$$f(x) = x^2, \quad -\pi < x \le \pi.$$

17.7 Find the trigonometric Fourier series for the function $f(x)$ with a period of 2π, with one period given by:

$$f(x) = 3\pi + 2x, \quad -\pi < x \le 0$$
$$= \pi + 2x, \quad 0 < x \le \pi$$

17.8 Find the trigonometric Fourier series for the function $f(x)$ with a period of 1/4, with one period given by:

$$f(x) = 3\cos(8\pi x), \quad 0 \le x < \tfrac{1}{4}.$$

17.9 Find the trigonometric Fourier series for the function

$$f(x) = 0, \quad 0 < x < \tfrac{1}{2}$$
$$= 1, \quad \tfrac{1}{2} \le x \le 1$$

where $f(x)$ is an odd function with a period T = 2.

17.10 Find the trigonometric Fourier series for the function

$$f(x) = x(2 - x), \quad 0 \le x < 2$$

where $f(x)$ is an even function with a period T = 4.

17.11 Find the trigonometric Fourier series for the function

$$
\begin{aligned}
f(x) &= -1, \quad -1 \le x < -\tfrac{1}{2} \\
&= 1, \quad -\tfrac{1}{2} \le x < \tfrac{1}{2} \\
&= -1, \quad \tfrac{1}{2} \le x < 1
\end{aligned}
$$

where $f(x)$ is an even function with a period T = 2.

17.12 Find the trigonometric Fourier series for the function

$$
\begin{aligned}
f(x) &= -1, \quad -1 \le x < 0 \\
&= 1, \quad 0 \le x < 1
\end{aligned}
$$

where $f(x)$ is an odd function with a period T = 2.

17.3 The Exponential Fourier Series

17.13 Find the exponential Fourier series for the current

$$
\begin{aligned}
i &= I_{\text{m}}, \quad -1 \le t < 1 \\
&= 0, \quad 1 \le t < 3
\end{aligned}
$$

where $i(t)$ is periodic with period T = 4.

17.14 Find the rms value for $i(t)$ in Prob. 17.13.

17.15 Find the exponential Fourier series for the function $f(t)$, where

$$
\begin{aligned}
f(t) &= e^{-t} - 1, \quad -1 \le t < 0 \\
&= e^{t} - 1, \quad 0 \le t < 1
\end{aligned}
$$

$$f(t+2) = f(t)$$

17.16 Find the exponential Fourier series for the function $f(x)$, where

$$f(x) = x, \quad 0 \le x < 2$$

$$f(x+2) = f(x)$$

17.17 Find the exponential Fourier series for the function $f(x)$, where

$$f(x) = 2\cos(2\pi x), \quad 0 \le x < 1$$

$$f(x+1) = f(x)$$

17.18 Find the exponential Fourier series for the function $f(t)$, where

$$\begin{aligned} f(t) &= 2, \quad 0 \le t < 3 \\ &= 0, \quad 3 \le t < 4 \end{aligned}$$

$$f(t+4) = f(t)$$

17.19 Find the exponential Fourier series for the function of Prob. 17.5.

17.20 Find the exponential Fourier series for the function of Prob. 17.9.

17.4 Response to Periodic Inputs

17.21 Find the forced voltage across the inductor if $v(t) = f(t)$ from Prob. 17.6.

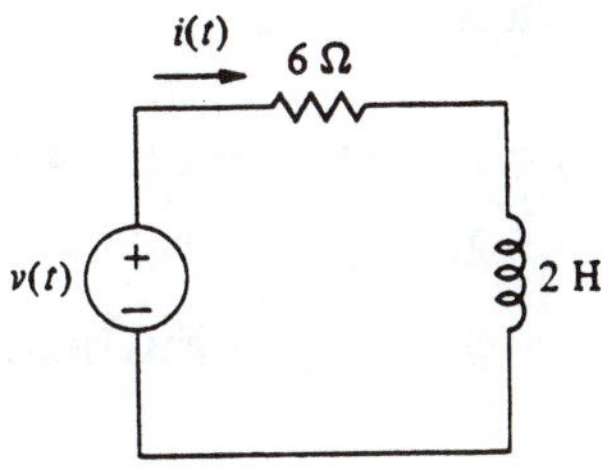

PROBLEM 17.21

17.22 Find the forced voltage across the capacitor if $v(t) = f(x)$ from Prob. 17.10.

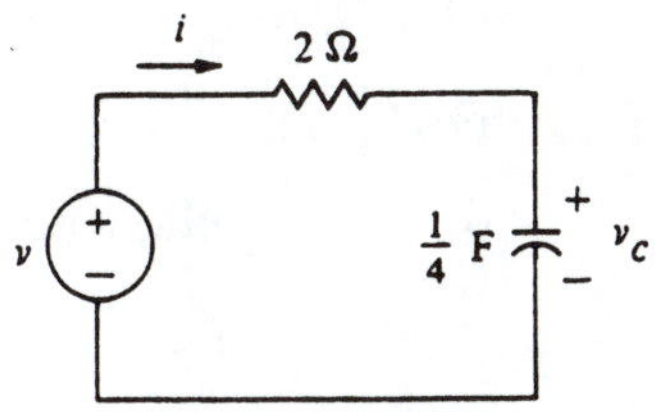

PROBLEM 17.22

17.23 Find the forced current in Prob. 17.21 if

$$v(t) = 1 + t, \quad -1 \le t < 0$$
$$= 1 - t, \quad 0 \le t < 1$$

where $v(t)$ is a periodic function with period $T = 2$.

17.5 Discrete Spectra and Phase Plots

17.24 Plot the discrete amplitude spectra for Prob. 17.5.

17.25 Plot the discrete amplitude spectra for Prob. 17.6.

17.26 Plot the discrete amplitude spectra for Prob. 17.7.

17.27 Plot the discrete amplitude spectra for Prob. 17.8.

17.28 Plot the discrete amplitude spectra for Prob. 17.16.

17.29 Plot the discrete amplitude spectra for Prob. 17.17.

17.6 The Fourier Transform

17.30 Find the Fourier transform of the function

$$f(t) = 2(t+1), \quad -1 \le t < 0$$
$$= 2(1-t), \quad 0 \le t < 1$$
$$= 0, \quad \text{elsewhere}$$

17.31 Find the Fourier transform of the function

$$f(t) = \cos(\pi t), \quad -1 < t < 1$$
$$= 0, \quad \text{elsewhere}$$

17.32 Find the Fourier transform of the function

$$f(t) = \cos^2(\pi t), \quad -1 < t < 1$$
$$= 0, \quad \text{elsewhere}$$

17.33 Find the Fourier transform of the function

$$\begin{aligned} f(t) &= e^t, \quad 0<t<10 \\ &= 0, \quad \text{elsewhere} \end{aligned}$$

17.34 Find the Fourier transform of the function

$$\begin{aligned} f(t) &= t, \quad 0<t<1 \\ &= 2, \quad 1<t<3 \\ &= 0, \quad \text{elsewhere} \end{aligned}$$

17.35 Find the Fourier transform of the function

$$f(t) = \cos(\pi t)u(t) - \cos(\pi t)u(t-10)$$

17.36 Find the Fourier transform of the function

$$f(t) = \cos(\pi t)u(t+5) - \cos(\pi t)u(t-5)$$

17.37 Find the Fourier transform of the function

$$\begin{aligned} f(t) &= -1, \quad -1<t<0 \\ &= 1, \quad 0<t<1 \\ &= 0, \quad \text{elsewhere} \end{aligned}$$

17.38 Find the Fourier transform of the function

$$\begin{aligned} f(t) &= -1, \quad -1<t<-1/2 \\ &= 1, \quad -1/2<t<1/2 \\ &= -1, \quad 1/2<t<1 \\ &= 0, \quad \text{elsewhere} \end{aligned}$$

17.39 Find the Fourier transform of the function

$$\begin{aligned} f(t) &= 2\cosh t, \quad -2<t<2 \\ &= 0, \quad \text{elsewhere} \end{aligned}$$

17.40 Find the Fourier transform of the function

$$\begin{aligned} f(t) &= 2\sinh t, \quad -2<t<2 \\ &= 0, \quad \text{elsewhere} \end{aligned}$$

17.7 Fourier Transform Properties

17.41 Solve Prob. 17.37 using the properties of even and odd functions.

17.42 Solve Prob. 17.38 using the properties of even and odd functions.

17.43 Using the properties of even and odd functions, find the Fourier transform of the function

$$\begin{aligned} f(t) &= |t|, \quad -2 < t < 2 \\ &= 0, \quad \text{elsewhere} \end{aligned}$$

17.44 Using the properties of even and odd functions, find the Fourier transform of the function

$$\begin{aligned} f(t) &= t, \quad -2 < t < 2 \\ &= 0, \quad \text{elsewhere} \end{aligned}$$

17.45 Using the properties of even and odd functions, find the Fourier transform of the function

$$\begin{aligned} f(t) &= t^2, \quad -2 < t < 2 \\ &= 0, \quad \text{elsewhere} \end{aligned}$$

17.46 Using the properties of even and odd functions, find the Fourier transform of the function

$$\begin{aligned} f(t) &= t|t|, \quad -2 < t < 2 \\ &= 0, \quad \text{elsewhere} \end{aligned}$$

17.47 Find the Fourier transform of $f(t) = \left(5e^{-2t} + 4e^{-3t}\right)u(t)$.

17.48 Find the Fourier transform of $f(t) = 6e^{-3(t+6)}u(t+6)$.

17.49 Find the Fourier transform of $f(t) = 2e^{-t/6}u(t)$.

17.50 Find the Fourier transform of $f(t) = \dfrac{d}{dt}2e^{-3t}u(t)$.

17.51 Find the Fourier transform of

$$\begin{aligned} f(t) &= t, \quad -1 < t < 1 \\ &= 0, \quad \text{elsewhere} \end{aligned}$$

17.52 Find the inverse Fourier transform of $F(\omega) = \left(\dfrac{\sin\omega}{\omega}\right)e^{-j\,\omega/2}$.

17.53 Find the inverse Fourier transform of $F(\omega)=\dfrac{-\omega^2}{1+j\omega}$.

17.54 If $x(t)$ is the input to an electrical system and $y(t)$ is the output, find the transfer function of the system using Fourier transforms if $4y''+6y'+y=2x$.

17.55 If $x(t)=e^{-(t-2)}u(t-2)$ is applied to the system in Prob. 17.54, find $Y(j\omega)$.

17.56 If $x(t)$ is the input to an electrical system and $y(t)$ is the output, find the transfer function of the system using Fourier transforms if $y'''+3y'=4x$.

17.57 If $x(t)=1$, for $-1/2<t<1/2$ in Prob. 17.56, find $Y(j\omega)$.

17.8 SPICE and Fourier Analysis

17.58 Use SPICE to find the Fourier coefficients of the waveform shown:

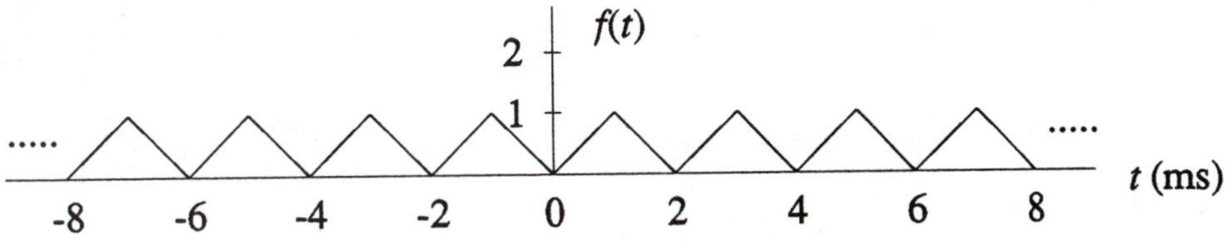

17.59 Use SPICE to find the Fourier coefficients of the waveform shown:

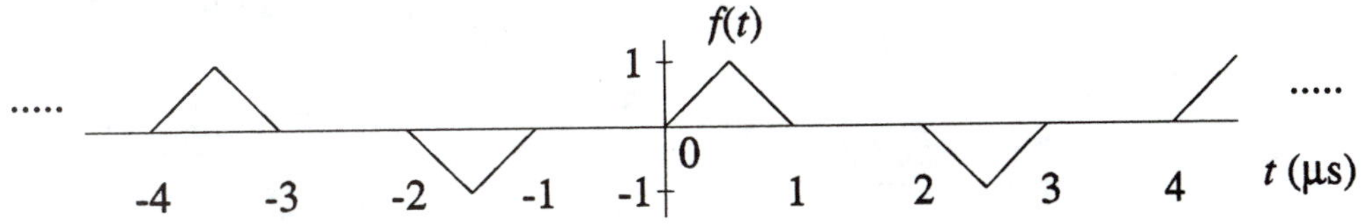

17.60 Use SPICE to find the Fourier coefficients of the waveform shown:

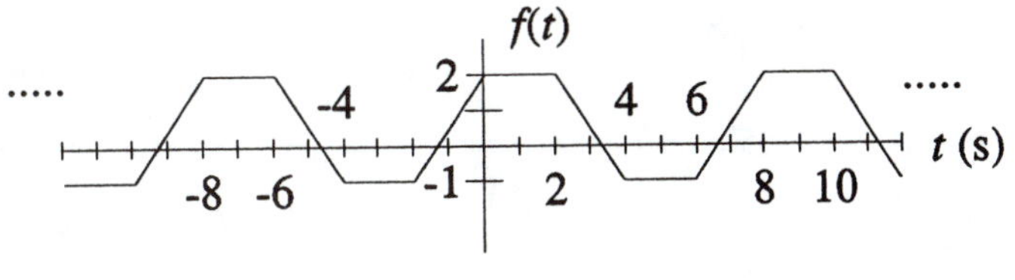

CHAPTER 17 SOLUTIONS

17.1 $T=4 \Rightarrow f(t)=\sum_{n=-\infty}^{\infty} f_T(t+4n)$

where $f_T(t)=2u(t+1)-2u(t-1)$
$+2\sin(\pi(t-1))[u(t-1)-u(t-3)]$

17.2 (a) Yes, T=3 (b) No (c) Yes, T=6

17.3 (a)

$$f_e(t)=\frac{[1+3\sin(4t+60°)]+[1+3\sin(-4t+60°)]}{2}$$

$$=1+\frac{3}{2}[\sin 4t\cos 60°+\cos 4t\sin 60°]$$
$$+\frac{3}{2}[-\sin 4t\cos 60°+\cos 4t\sin 60°]$$
$$=1+\frac{3\sqrt{3}}{2}\cos 4t$$

$$f_o(t)=\frac{[1+3\sin(4t+60°)]-[1+3\sin(-4t+60°)]}{2}$$

$$=\frac{3}{2}[\sin 4t\cos 60°+\cos 4t\sin 60°]$$
$$-\frac{3}{2}[-\sin 4t\cos 60°+\cos 4t\sin 60°]$$
$$=\frac{3}{2}\sin 4t$$

(b)
$$f_e(t)=\frac{2\cos(4t-20°)+5\cos(t+20°)}{2}+\frac{2\cos(-4t-20°)+5\cos(-t+20°)}{2}$$

$$=\cos 4t\cos 20°+\sin 4t\sin 20°$$
$$+\frac{5}{2}[\cos t\cos 20°-\sin t\sin 20°]$$
$$+\cos 4t\cos 20°-\sin 4t\sin 20°$$
$$+\frac{5}{2}[\cos t\cos 20°+\sin t\sin 20°]$$
$$=2(0.9397)\cos 4t+5(0.9397)\cos t$$
$$=1.879\cos 4t+4.698\cos t$$

$$f_o(t)=\frac{2\cos(4t-20°)+5\cos(t+20°)}{2}-\frac{[2\cos(-4t-20°)+5\cos(-t+20°)]}{2}$$

$$=\cos 4t\cos 20°+\sin 4t\sin 20°$$
$$-\cos 4t\cos 20°+\sin 4t\sin 20°$$
$$+\frac{5}{2}[\cos t\cos 20°-\sin t\sin 20°]$$
$$-\frac{5}{2}[\cos t\cos 20°+\sin t\sin 20°]$$
$$=0.684\sin 4t-1.710\sin t$$

17.4

(a)
$$f_{av}=\frac{1}{(\frac{\pi}{2})}\int_0^{\pi/2}[1+3\sin(4t+60°)]dt$$
$$=\frac{2}{\pi}\int_0^{\pi/2}dt+\frac{2}{\pi}\int_0^{\pi/2}3\sin(4t+60°)dt$$
$$=\frac{2}{\pi}\left(\frac{\pi}{2}\right)+0=1$$

(b)
$$f_{av}=\frac{1}{2\pi}\int_0^{2\pi}[2\cos(4t-20°)+5\cos(t+20°)]dt$$
$$=\frac{1}{\pi}\int_0^{2\pi}\cos(4t-20°)dt+\frac{5}{2\pi}\int_0^{2\pi}\cos(t+20°)dt$$
$$=0+0=0$$

(c)
$$f_{av}=\frac{1}{1}\int_0^1 e^{-t}[u(t)-u(t-1)]\,dt$$
$$=\int_0^1 e^{-t}dt=-e^{-t}\Big|_0^1$$
$$=1-\frac{1}{e}\approx 0.63212$$

(d)
$$f_{av}=\frac{1}{2}\int_0^2 e^{-t}[u(t)-u(t-1)]dt$$
$$=\frac{1}{2}\int_0^1 e^{-t}dt$$
$$=\frac{1}{2}\left[1-\frac{1}{e}\right]=\frac{1}{2}-\frac{1}{2e}$$
$$\approx 0.31606$$

17.5 $T = 4,\ \omega_o = \frac{2\pi}{T} = \frac{\pi}{2}$

$$a_o = \frac{2}{4}\left[\int_{-2}^{0} 2\,dx + \int_{0}^{2} x\,dx = \underline{3}\right.$$

$$a_n = \frac{2}{4}\left[\int_{-2}^{0} 2\cos\frac{n\pi x}{2}\,dx + \int_{0}^{2} x\cos\frac{n\pi x}{2}\,dx\right]$$

$$= \frac{1}{2}\left[\frac{2\sin\frac{n\pi x}{2}}{n\pi/2}\Bigg|_{-2}^{0}\right] + \frac{1}{2}\left[\frac{\cos n\pi x/2}{(n\pi/2)^2} + \frac{x\sin n\pi x/2}{n\pi/2}\right]_0^2$$

$$= \underline{\frac{2}{n^2\pi^2}\left[(-1)^n - 1\right]}$$

$$b_n = \frac{1}{2}\left[\int_{-2}^{0} 2\sin\frac{n\pi x}{2}\,dx + \int_{0}^{2} x\sin\frac{n\pi x}{2}\,dx\right]$$

$$= \frac{1}{2}\left\{\frac{-2\cos\frac{n\pi x}{2}}{n\pi/2}\Bigg|_{-2}^{0} + \left[\frac{\sin n\pi x/2}{(n\pi/2)^2} - \frac{x\cos n\pi x/2}{n\pi/2}\right]_0^2\right\}$$

$$= \underline{-\frac{2}{n\pi}}$$

17.6 $T = 2\pi,\ \omega_o = \frac{2\pi}{T} = 1$

$$a_o = \frac{2}{2\pi}\int_{-\pi}^{\pi} x^2\,dx = \frac{1}{\pi}\frac{x^3}{3}\Bigg|_{-\pi}^{\pi} = \underline{\frac{2\pi^2}{3}}$$

$$a_n = \frac{1}{\pi}\int_{-\pi}^{\pi} x^2\cos nx\,dx$$

$$= \frac{1}{\pi}\left[\frac{2x\cos nx}{n^2} + \frac{n^2x^2 - 2}{n^3}\sin nx\right]_{-\pi}^{\pi}$$

$$= \underline{\frac{4(-1)^n}{n^2}}$$

$$b_n = \frac{1}{\pi}\int_{-\pi}^{\pi} x^2\sin nx\,dx$$

$$= \frac{1}{\pi}\left[\frac{2x\sin nx}{n^2} - \frac{n^2x^2 - 2}{n^3}\cos nx\right]_{-\pi}^{\pi}$$

$$= \underline{0}$$

17.7 $T = 2\pi,\ \omega_o = \frac{2\pi}{T} = 1$

$$a_o = \frac{1}{\pi}\left[\int_{-\pi}^{0}(3\pi + 2x)\,dx + \int_{0}^{\pi}(\pi + 2x)\,dx\right]$$

17.7 (cont'd) $a_o = 2\pi + 2\pi = \underline{4\pi}$

$$a_n = \frac{1}{\pi}\left[\int_{-\pi}^{0}(3\pi + 2x)\cos nx\,dx + \int_{0}^{\pi}(\pi + 2x)\cos nx\,dx\right]$$

$$= \frac{1}{\pi}\left[\frac{-3\pi\sin nx}{n}\Bigg|_{-\pi}^{0} + 2\left(\frac{\cos nx}{n^2} + \frac{x}{n}\sin nx\right)\Bigg|_{-\pi}^{0} - \frac{\pi}{n}\sin nx\Bigg|_{0}^{\pi} + 2\left(\frac{\cos nx}{n^2} + \frac{x}{n}\sin nx\right)\Bigg|_{0}^{\pi}\right]$$

$$= \underline{0}$$

$$b_n = \frac{1}{\pi}\left[\int_{-\pi}^{0}(3\pi + 2x)\sin nx\,dx + \int_{0}^{\pi}(\pi + 2x)\sin nx\,dx\right]$$

$$= \frac{1}{\pi}\left\{\frac{-3\pi\cos nx}{n}\Bigg|_{-\pi}^{0} + 2\left(\frac{\sin nx}{n^2} - \frac{x}{n}\cos nx\right)\Bigg|_{-\pi}^{0} - \frac{\pi\cos nx}{n}\Bigg|_{0}^{\pi} + 2\left(\frac{\sin nx}{n^2} - \frac{x}{n}\cos nx\right)\Bigg|_{0}^{\pi}\right\}$$

$$= \underline{-\frac{2}{n}\left[1 + (-1)^n\right]}$$

$b_n = \underline{0,\ n \text{ odd}}$; $b_n = \underline{-\frac{4}{n},\ n \text{ even}}$

17.8 $\omega_o = \frac{2\pi}{T} = \frac{2\pi}{(1/4)} = 8\pi$

$$a_o = \frac{2}{T}\int_{0}^{1/4} 3\cos 8\pi x\,dx$$

$$= 24\,\frac{\sin 8\pi x}{8\pi}\Bigg|_{0}^{1/4} = \underline{0}$$

$$a_n = 8\int_{0}^{1/4} 3\cos 8\pi x\cos 8n\pi x\,dx$$

$a_1 = \underline{3}$; $a_n = \underline{0},\ n \neq 1$

$$b_n = 8\int_{0}^{1/4} 3\cos 8\pi x\sin 8n\pi x\,dx$$

$$= \underline{0}$$

$f(x) = \underline{3\cos 8\pi x}$

17.9 $T=2, \omega_0 = 2\pi/2 = \pi$

$f(x)$ odd $\Rightarrow a_n = \underline{0}$

$$b_n = \frac{4}{2}\int_{1/2}^{1} \sin n\pi t\, dt$$

$$= 2\left[-\frac{1}{n\pi}\cos n\pi t\right]_{1/2}^{1}$$

$$= -\frac{2}{n\pi}\left[\cos n\pi - \cos\frac{n\pi}{2}\right]$$

$$= \underline{-\frac{2}{n\pi}\left[(-1)^n - \cos\frac{n\pi}{2}\right]}$$

17.10 $\omega_0 = \frac{2\pi}{4} = \frac{\pi}{2}$

$$a_0 = \frac{4}{4}\int_0^2 x(2-x)dx$$

$$= x^2 - \frac{x^3}{3}\Big|_0^2 = \frac{4}{3}$$

$$a_n = \frac{4}{4}\int_0^2 (2x - x^2)\cos\frac{n\pi x}{2}dx$$

$$= \left\{2\left[\frac{4\cos\frac{n\pi x}{2}}{n^2\pi^2} + \frac{2x}{n\pi}\sin\frac{n\pi x}{2}\right] - \left[\frac{\frac{n^2\pi^2 x^2}{4} - 2}{n^3\pi^3/8}\sin\frac{n\pi x}{2} + \frac{4x}{n^2\pi^2}\cos\frac{n\pi x}{2}\right]\right\}\Bigg|_0^2$$

$$= \frac{-8}{n^2\pi^2}\left[1 + (-1)^n\right]$$

$a_n = \underline{0}$, $\underline{n \text{ odd}}$; $a_n = \underline{-\frac{16}{n^2\pi^2}}$, $\underline{n \text{ even}}$

$b_n = \underline{0}$ since $f(x)$ is even.

17.11 $T=2, \omega_0 = \frac{2\pi}{2} = \pi$

$$a_0 = \frac{4}{2}\left[\int_0^{1/2} dx + \int_{1/2}^{1}(-1)dx\right] = \underline{0}$$

$$a_n = 2\left[\int_0^{1/2}\cos n\pi x\, dx - \int_{1/2}^{1}\cos n\pi x\, dx\right]$$

$$= 2\left[\frac{\sin n\pi x}{n\pi}\Big|_0^{\frac{1}{2}} - \frac{\sin n\pi x}{n\pi}\Big|_{\frac{1}{2}}^{1}\right]$$

$$= \underline{\frac{4}{n\pi}\sin\frac{n\pi}{2}}$$

$b_n = 0$ since $f(x)$ is even

17.12

$T=2$, $\omega_0 = 2\pi/2 = \pi$

$a_n = 0$ since $f(x)$ is odd.

$$b_n = \frac{4}{2}\int_0^1 (1)\sin n\pi x\, dx$$

$$= 2\,\frac{-\cos n\pi x}{n\pi}\Big|_0^1 = \frac{2}{n\pi}\left[1 - (-1)^n\right]$$

$b_n = \underline{0, n \text{ even}}$; $\underline{b_n = \frac{4}{n\pi}, n \text{ odd.}}$

17.13

$\omega_0 = \frac{2\pi}{4} = \frac{\pi}{2}$, $T=4$

$$c_0 = \frac{1}{4}\int_{-1}^{1} I_m\, dt = \underline{\frac{1}{2}I_m}$$

$$c_n = \frac{1}{4}\int_{-1}^{1} I_m e^{-jn\pi t/2}dt$$

$$= -\frac{I_m}{j2n\pi}\left[e^{-j\frac{n\pi}{2}} - e^{j\frac{n\pi}{2}}\right]$$

$$= \underline{\frac{I_m}{n\pi}\sin\frac{n\pi}{2}, \quad n \neq 0}$$

17.14

$$I_{rms} = \left[|c_0|^2 + \sum_{\substack{n=-\infty\\ n\neq 0}}^{\infty} |c_n|^2\right]^{1/2}$$

$$I_{rms}^2 = \frac{I_m^2}{4} + \sum_{\substack{n=-\infty\\ n\neq 0}}^{\infty} \frac{I_m^2}{n^2\pi^2}\sin^2\frac{n\pi}{2}$$

$\sin\frac{n\pi}{2} = 0$ for n even

$\sin(2n-1)\frac{\pi}{2} = (-1)^{n+1}$

$$\underline{I_{rms} = I_m\left[\frac{1}{4} + \frac{1}{\pi^2}\sum_{n=-\infty}^{\infty}\frac{1}{(2n-1)^2}\right]^{\frac{1}{2}}}$$

$$\underline{I_{rms} = I_m\left[\frac{1}{4} + \frac{2}{\pi^2}\sum_{n=1}^{\infty}\frac{1}{(2n-1)^2}\right]^{\frac{1}{2}}}$$

17.15 $T = 2,\ \omega_o = \frac{2\pi}{2} = \pi$

$$c_o = \frac{1}{2}\left[\int_{-1}^{0}(e^{-t}-1)dt + \int_0^1 (e^{t}-1)dt\right]$$
$$= \int_0^1 (e^{t}-1)dt = e^{t}-t\Big|_0^1$$
$$= e-1-1 = \underline{e-2}$$

$$c_n = \frac{1}{2}\left[\int_{-1}^{0}(e^{-t}-1)e^{-jn\pi t}dt + \int_0^1 (e^{t}-1)e^{-jn\pi t}dt\right]$$
$$= \frac{1}{2}\left[\frac{e^{-(jn\pi+1)t}}{-(jn\pi+1)} - \frac{e^{-jn\pi t}}{jn\pi}\Big|_{-1}^{0} - \frac{e^{-(jn\pi-1)t}}{jn\pi-1} + \frac{e^{-jn\pi t}}{jn\pi}\Big|_0^1\right]$$
$$= \frac{1}{2}\left[\frac{-1+e(-1)^n}{jn\pi+1} - \frac{e(-1)^n-1}{jn\pi-1}\right]$$
$$= \underline{\frac{e(-1)^n-1}{1+n^2\pi^2}}$$

17.16 $T = 2,\ \omega_o = \pi$

$$c_o = \frac{1}{2}\int_0^2 x\,dx = \underline{1}$$
$$c_n = \frac{1}{2}\int_0^2 x e^{-jn\pi x}dx$$
$$= \frac{1}{2(-jn\pi)^2}e^{-jn\pi x}(-jn\pi x-1)\Big|_0^2$$
$$= \frac{-1}{2n^2\pi^2}\left[e^{-j2n\pi}(-j2n\pi-1)-(-1)\right]$$
$$e^{-j2n\pi} = 1$$
$$c_n = \underline{\frac{j}{n\pi}},\quad n \neq 0$$

17.17 $T = 1,\ \omega_o = 2\pi$

$$c_o = \int_0^1 2\cos 2\pi x\,dx$$
$$= \frac{1}{\pi}\sin 2\pi x\Big|_0^1 = \underline{0}$$
$$c_n = \int_0^1 2\cos 2\pi x\, e^{-j2n\pi x}dx$$

17.17 (cont'd)

$$c_n = \frac{2e^{-j2n\pi x}}{(2\pi)^2+(-j2n\pi)^2}\cdot\left[-j2n\pi\cos 2\pi x + 2\pi\sin 2\pi x\right]_0^1$$
$$c_n = \underline{0} \text{ for } n^2 \neq 1$$
$$c_1 = 2\int_0^1 \cos 2\pi x\, e^{-j2\pi x}dx$$
$$= 2\int_0^1 \cos^2 2\pi x\,dx - j2\int_0^1 \sin 2\pi x\cos 2\pi x\,dx$$
$$= \int_0^1 (1+\cos 4\pi x)dx - j\int_0^1 \sin 4\pi x\,dx$$
$$= x + \frac{\sin 4\pi x}{4\pi} + j\frac{\cos 4\pi x}{4\pi}\Big|_0^1 = \underline{1}$$
$$c_{-1} = c_1^* = \underline{1}$$
$$f(x) = e^{j2\pi x} + e^{-j2\pi x}$$
$$= \underline{2\cos 2\pi x}$$

17.18

$T = 4,\ \omega_o = \pi/2$

$$c_o = \frac{1}{4}\int_0^3 2\,dt = \underline{\frac{3}{2}}$$
$$c_n = \frac{1}{4}\int_0^3 2e^{-jn\pi t/2}dt$$
$$= \frac{1}{2}\,\frac{e^{-jn\pi t/2}}{-jn\pi/2}\Big|_0^3$$
$$= \frac{1}{-jn\pi}\left[e^{-j3n\pi/2}-1\right]$$
$$= \underline{\frac{1-e^{-j3n\pi/2}}{jn\pi}}$$

17.19 $T=4, \omega_0=\pi/2$

$$c_0=\frac{1}{4}\left[\int_{-2}^{0} 2dx+\int_0^2 x dx\right]$$
$$=\frac{1}{4}\left\{2x\Big|_{-2}^{0}+\frac{x^2}{2}\Big|_0^2\right]$$
$$=\frac{1}{4}[4+2]=\underline{\frac{3}{2}}$$

$$c_n=\frac{1}{4}\left[\int_{-2}^{0} 2e^{-j\frac{n\pi x}{2}}dx + \int_0^2 x e^{-j\frac{n\pi x}{2}}dx\right]$$
$$=\frac{1}{4}\left\{\frac{2e^{-jn\pi x/2}}{-jn\pi/2}\Big|_{-2}^{0} + \frac{e^{-jn\pi x/2}}{(-jn\pi/2)^2}\left(-j\frac{n\pi x}{2}-1\right)\Big|_0^2\right\}$$
$$=\frac{1}{-jn\pi}\left[1-e^{jn\pi}\right] - \frac{1}{n^2\pi^2}\left[e^{-jn\pi}(-jn\pi-1)-(-1)\right]$$
$$=\frac{1}{n^2\pi^2}\left[e^{-jn\pi}-1\right]-\frac{1}{jn\pi}$$
$$=\underline{\frac{1}{n^2\pi^2}\left[(-1)^n-1)\right]-\frac{1}{jn\pi}}$$

17.20 $T=2, \omega_0=\frac{2\pi}{2}=\pi$

$f(x)$ odd $\Rightarrow a_n=0, n=0,1,2,3,\cdots$

$$c_0=\frac{a_0}{2}=\underline{0};\ c_n=\frac{a_n-jb_n}{2}$$
$$c_n=-\frac{j}{2}b_n=-\frac{j}{2}\frac{4}{2}\int_{1/2}^{1}\sin n\pi x\,dx$$
$$=-j\int_{1/2}^{1}(1)\sin n\pi x\,dx$$
$$=-j\,\frac{-\cos n\pi x}{n\pi}\Big|_{1/2}^{1}$$
$$=\frac{j}{n\pi}\left[\cos n\pi-\cos\frac{n\pi}{2}\right]$$
$$=\underline{\frac{j}{n\pi}\left[(-1)^n-\cos\frac{n\pi}{2}\right]}$$

17.21

$$v=\frac{\pi^2}{3}+\sum_{n=1}^{\infty}\frac{4(-1)^n}{n^2}\cos nt$$
$$\mathbf{V}_n=\frac{4(-1)^n}{n^2}\angle 0^\circ,\ \omega_n=n$$
$$\mathbf{Z}(j\omega_n)=6+j2n$$
$$\mathbf{I}_n=\frac{4(-1)^n}{n^2(6+j2n)}$$
$$\mathbf{V}_{Ln}=j2n\,\mathbf{I}_n=\frac{j4(-1)^n}{n(3+jn)}$$
$$=\frac{4(-1)^n}{n\sqrt{9+n^2}}\angle 90^\circ-\tan^{-1}\frac{n}{3}$$
$$\mathbf{V}_{L0}=0$$
$$v_L(t)=\sum_{n=1}^{\infty}v_{Ln}(t)$$
$$=\underline{4\sum_{n=1}^{\infty}\frac{(-1)^n\cos\left(nt+90^\circ-\tan^{-1}\frac{n}{3}\right)}{n\sqrt{n^2+9}}}$$

17.22

$$v=\frac{2}{3}-\frac{4}{\pi^2}\sum_{n=1}^{\infty}\frac{1}{n^2}\cos n\pi t$$
$$\mathbf{V}_0=\frac{2}{3};\ \mathbf{V}_n=-\frac{4}{n^2\pi^2},\ \omega_n=n\pi$$
$$\mathbf{Z}(j\omega_n)=2-j\frac{4}{n\pi};\ \mathbf{V}_{c0}=\mathbf{V}_0=\frac{2}{3}$$

Voltage division:

$$\mathbf{V}_{cn}=\frac{-j4/\omega_n}{2-j\frac{4}{\omega_n}}\mathbf{V}_n=\frac{-j2}{n\pi-j2}\mathbf{V}_n$$
$$=\frac{8}{n^2\pi^2\sqrt{n^2\pi^2+4}}\angle\theta_n$$
$$\theta_n=90^\circ-\tan^{-1}\frac{-2}{n\pi}$$
$$=90^\circ+\tan^{-1}\frac{2}{n\pi}$$
$$v_c=v_0+\sum_{n=1}^{\infty}v_{cn}$$
$$\underline{v_c=\frac{2}{3}+\frac{8}{\pi^2}\sum_{n=1}^{\infty}\frac{\cos(n\pi t+\theta_n)}{n^2\sqrt{n^2\pi^2+4}}}$$

17.23 $T=2, \omega_0 = 2\pi/2 = \pi$

$b_n = 0$ since v is even

$a_0 = \frac{4}{2}\int_0^1 (1-t)dt = 1$

$a_n = \frac{4}{2}\int_0^1 (1-t)\cos n\pi t\, dt$

$= 2\left[\frac{\sin n\pi t}{n\pi} - \left(\frac{\cos n\pi t}{n^2\pi^2} + \frac{t \sin n\pi t}{n\pi}\right)\right]_0^1$

$= \frac{2}{n^2\pi^2}[1-(-1)^n]$

$a_n = 0$, n even

$a_n = \frac{4}{n^2\pi^2}$, n odd

$$v = \frac{1}{2} + \sum_{n=1}^{\infty} \frac{4\cos(2n-1)\pi t}{(2n-1)^2\pi^2}$$

$\omega_n = (2n-1)\pi$; $\mathbf{Z}(j\omega_n) = 6 + j2\omega_n$

$\mathbf{I}_0 = \frac{V_0}{6} = \frac{1}{12}$

$\mathbf{I}_n = \frac{V_n}{\mathbf{Z}(j\omega_n)} = \frac{4/(2n-1)^2\pi^2}{6+j2(2n-1)\pi}$

$$|\mathbf{I}_n| = \frac{2}{(2n-1)^2\pi^2\sqrt{9+(2n-1)^2\pi^2}}$$

$$i = \frac{1}{12} + \sum_{n=1}^{\infty} |\mathbf{I}_n| \cos[(2n-1)\pi t + \theta_n]$$

where $\theta_n = -\tan^{-1}\frac{(2n-1)\pi}{3}$

17.24

$|c_0| = \frac{3}{2}$

$|c_n|^2 = \left[\frac{(-1)^n - 1}{n^2\pi^2}\right]^2 + \left[\frac{1}{n\pi}\right]^2$

$= \frac{[(-1)^n-1]^2 + n^2\pi^2}{n^4\pi^4}$

$|c_n| = \frac{1}{n^2\pi^2}\sqrt{[(-1)^n-1]^2 + n^2\pi^2}$

17.25 $c_0 = \frac{a_0}{2} = \frac{1}{2}\,\frac{2\pi^2}{3} = \frac{\pi^2}{3}$

$|c_0| = \frac{\pi^2}{3}$, $\phi_0 = 0$

$c_n = \frac{1}{2}(a_n - jb_n) = \frac{1}{2}\,\frac{4(-1)^n}{n^2}$

$= \frac{2(-1)^n}{n^2}$

$|c_n| = \frac{2}{n^2}$

$\phi_n = 0$, n even

$= 180°$, n odd, $n>0$

17.26 $c_0 = \frac{a_0}{2} = \frac{1}{2}(4\pi) = 2\pi$

$|c_0| = 2\pi$, $\phi_0 = 0$

$c_n = \frac{1}{2}(a_n - jb_n) = -\frac{j}{2} b_n$

$= \frac{j}{n}[1 + (-1)^n]$

$c_n = 0$ for n odd

$c_n = j\frac{2}{n} \Rightarrow |c_n| = \frac{2}{|n|}$, n even

$\phi_n = 90°$, n even $+$ $n>0$

$= -90°$, n even $+$ $n<0$

17.27 $c_0 = \frac{a_0}{2} = 0$

$c_n = \frac{1}{2}(a_n - jb_n)$

$= 0$ for $n^2 \neq 1$

$c_1 = c_{-1} = 3$

$|c_1| = |c_{-1}| = 3$

$\phi_n = 0$

17.28 $c_0 = 1 \Rightarrow |c_0| = 1$, $\phi_0 = 0$

$c_n = \frac{j}{n\pi} \Rightarrow |c_n| = \frac{1}{|n|\pi}$

$\phi_n = 90°$, $n>0$

$= -90°$, $n<0$

17.29 $c_0 = 0$

$c_n = 0$ for $n^2 \neq 1$

$c_1 = c_{-1} = 1$

$|c_1| = |c_{-1}| = \underline{1}$

$\phi_1 = \phi_{-1} = \underline{0}$

17.30

$$F(j\omega) = \int_{-1}^{0} 2(t+1)e^{-j\omega t}dt + \int_{0}^{1} 2(-t+1)e^{-j\omega t}dt$$

$$= 2\left[\frac{e^{-j\omega t}}{-\omega^2}(-j\omega t-1) + \frac{e^{-j\omega t}}{-j\omega}\right]_{-1}^{0} + 2\left[\frac{e^{-j\omega t}}{\omega^2}(-j\omega t-1) + \frac{e^{-j\omega t}}{-j\omega}\right]_{0}^{1}$$

$$= 2\left[\frac{1}{\omega^2} + \frac{e^{j\omega}}{\omega^2}(j\omega-1) + \frac{1}{-j\omega} + \frac{e^{j\omega}}{j\omega} + \frac{e^{-j\omega}}{\omega^2}(-j\omega-1) + \frac{1}{\omega^2} + \frac{1}{j\omega}\right]$$

$$= \underline{\frac{4}{\omega^2}(1-\cos\omega)}$$

17.31

$$F(j\omega) = \int_{-1}^{1} \cos\pi t\, e^{-j\omega t}dt$$

$$= \frac{e^{-j\omega t}}{\pi^2-\omega^2}\left[j\omega\cos\pi t + \pi\sin\pi t\right]_{-1}^{1}$$

$$= \frac{1}{\pi^2-\omega^2}\left[e^{-j\omega}(-j\omega) - e^{j\omega}(-j\omega)\right]$$

$$= \frac{j\omega}{\pi^2-\omega^2}\left[e^{j\omega} - e^{-j\omega}\right]$$

$$= \underline{\frac{-2\omega}{\pi^2-\omega^2}\sin\omega}$$

17.32

$$F(j\omega) = \int_{-1}^{1} \cos^2\pi t\, e^{-j\omega t}dt$$

$$= \frac{1}{2}\int_{-1}^{1}(1+\cos 2\pi t)e^{-j\omega t}dt$$

$$= \frac{1}{2}\left\{\frac{e^{-j\omega t}}{-j\omega} + \frac{e^{-j\omega t}}{4\pi^2-\omega^2}[-j\omega\cos 2\pi t + 2\pi\sin 2\pi t]\right\}\Big|_{-1}^{1}$$

$$= \frac{1}{-2j\omega}\left[e^{-j\omega} - e^{j\omega}\right] + \frac{e^{-j\omega}(-j\omega) - e^{j\omega}(-j\omega)}{2[4\pi^2-\omega^2]}$$

$$F(j\omega) = \frac{\sin\omega}{\omega} - \frac{\omega\sin\omega}{4\pi^2-\omega^2}$$

$$= \underline{\frac{2\sin\omega}{4\pi^2-\omega^2}\left[\frac{2\pi^2}{\omega} - \omega\right]}$$

17.33

$$F(j\omega) = \int_{0}^{10} e^{t}e^{-j\omega t}dt$$

$$= \frac{e^{-(j\omega-1)t}}{-j\omega+1}\Big|_{0}^{10}$$

$$= \underline{\frac{e^{-10(j\omega-1)}}{-j\omega+1}}$$

17.34

$$F(j\omega) = \int_{0}^{1} te^{-j\omega t}dt + \int_{1}^{3} 2e^{-j\omega t}dt$$

$$= \frac{e^{-j\omega t}}{-\omega^2}(-j\omega t-1)\Big|_{0}^{1} + \frac{2e^{-j\omega t}}{-j\omega}\Big|_{1}^{3}$$

$$= \frac{e^{-j\omega}}{-\omega^2}(-j\omega-1) - \frac{1}{\omega^2} + \frac{2e^{-j3\omega}}{-j\omega} + \frac{2e^{-j\omega}}{j\omega}$$

$$= \underline{\frac{e^{-j\omega}(j\omega+1)-1}{\omega^2} + \frac{2(e^{-j3\omega}-e^{-j\omega})}{-j\omega}}$$

17.35 $F(j\omega)=\int_0^{10}(\cos\pi t)e^{-j\omega t}\,dt$

$$=\frac{e^{-j\omega t}}{\pi^2-\omega^2}\Big[-j\omega\cos\pi t+\pi\sin\pi t\Big]_0^{10}$$

$$=\frac{j\omega}{\pi^2-\omega^2}\left(1-e^{-j10\omega}\right)$$

$$=\frac{2\omega e^{-j5\omega}}{\omega^2-\pi^2}\left(\frac{e^{j5\omega}-e^{-j5\omega}}{2j}\right)$$

$$=\underline{\frac{2\omega e^{-j5\omega}\sin 5\omega}{\omega^2-\pi^2}}$$

17.36 $F(j\omega)=\int_{-5}^{5}(\cos\pi t)e^{-j\omega t}\,dt$

$$=\frac{e^{-j\omega t}}{\pi^2-\omega^2}\Big[-j\omega\cos\pi t+\pi\sin\pi t\Big]_{-5}^{5}$$

$$=\frac{j\omega}{\pi^2-\omega^2}\left(e^{-j5\omega}-e^{j5\omega}\right)$$

$$=\underline{\frac{2\omega}{\omega^2-\pi^2}\sin 5\omega}$$

17.37

$$F(j\omega)=\int_{-1}^{0}-e^{-j\omega t}dt+\int_0^1 e^{-j\omega t}dt$$

$$=\frac{e^{-j\omega t}}{j\omega}\Bigg|_{-1}^{0}+\frac{e^{-j\omega t}}{-j\omega}\Bigg|_0^1$$

$$=\frac{1}{j\omega}\left(1-e^{j\omega}-e^{-j\omega}+1\right)$$

$$=\underline{\frac{2-2\cos\omega}{j\omega}}$$

17.38 $F(j\omega)=\int_{-1}^{-1/2}-e^{-j\omega t}dt+\int_{-1/2}^{1/2}e^{-j\omega t}dt-\int_{1/2}^{1}e^{-j\omega t}dt$

$$=\frac{1}{j\omega}\left[e^{-j\omega t}\Big|_{-1}^{-1/2}-e^{-j\omega t}\Big|_{-1/2}^{1/2}+e^{-j\omega t}\Big|_{1/2}^{1}\right]$$

$$=\frac{2\left(e^{j\omega/2}-e^{-j\omega/2}\right)-\left(e^{j\omega}-e^{-j\omega}\right)}{j\omega}$$

$$=\underline{\frac{4\sin\frac{\omega}{2}-2\sin\omega}{\omega}}$$

17.39

$$F(j\omega)=\int_{-2}^{2}\left(e^{t}+e^{-t}\right)e^{-j\omega t}dt$$

$$=\frac{e^{(1-j\omega)t}}{1-j\omega}-\frac{e^{-(1+j\omega)t}}{1+j\omega}\Bigg|_{-2}^{2}$$

$$=\frac{e^{2(1-j\omega)}-e^{-2(1-j\omega)}}{1-j\omega}-\frac{e^{-2(1+j\omega)}-e^{2(1+j\omega)}}{1+j\omega}$$

$$=e^2\left(\frac{e^{-j2\omega}}{1-j\omega}+\frac{e^{j2\omega}}{1+j\omega}\right)-e^{-2}\left(\frac{e^{j2\omega}}{1-j\omega}+\frac{e^{-j2\omega}}{1+j\omega}\right)$$

$$=2\,\mathrm{Re}\left[e^2\left(\frac{e^{j2\omega}}{1+j\omega}\right)-e^{-2}\left(\frac{e^{-j2\omega}}{1+j\omega}\right)\right]$$

$$=\frac{2}{1+\omega^2}\Big\{e^2\,\mathrm{Re}\left[(1-j\omega)(\cos 2\omega+j\sin 2\omega)\right]-e^{-2}\,\mathrm{Re}\left[(1-j\omega)(\cos 2\omega-j\sin 2\omega)\right]\Big\}$$

$$=\frac{2}{1+\omega^2}\left[e^2(\cos 2\omega+\omega\sin 2\omega)-e^{-2}(\cos 2\omega-\omega\sin 2\omega)\right]$$

$$F(j\omega)=\frac{4\omega}{1+\omega^2}\left[\frac{e^2-e^{-2}}{2}\,\frac{\cos 2\omega}{\omega}+\frac{e^2+e^{-2}}{2}\sin 2\omega\right]$$

$$=\underline{\frac{4}{1+\omega^2}\left(\sinh 2\cos 2\omega+\omega\cosh 2\sin 2\omega\right)}$$

17.40 $F(j\omega) = \int_{-2}^{2} (e^{t} - e^{-t}) e^{-j\omega t} dt$

$$= \frac{e^{(1-j\omega)t}}{1-j\omega} + \frac{e^{-(1+j\omega)t}}{1+j\omega}\Bigg|_{-2}^{2}$$

$$= \frac{e^{2}e^{-j2\omega}}{1-j\omega} + \frac{e^{-2}e^{-j2\omega}}{1+j\omega} - \frac{e^{-2}e^{j2\omega}}{1-j\omega} - \frac{e^{2}e^{j2\omega}}{1+j\omega}$$

$$= \left[\frac{e^{2}e^{-j2\omega}}{1-j\omega} - \frac{e^{2}e^{j2\omega}}{1+j\omega}\right] + \left[\frac{e^{-2}e^{-j2\omega}}{1+j\omega} - \frac{e^{-2}e^{j2\omega}}{1-j\omega}\right]$$

$$= 2j\, Im\left[\frac{e^{2}e^{-j2\omega}}{1-j\omega} + \frac{e^{-2}e^{-j2\omega}}{1+j\omega}\right]$$

$$= \frac{2j\, Im}{1+\omega^2}\Big[e^{2}(1+j\omega)(\cos 2\omega - j\sin 2\omega) + e^{-2}(1-j\omega)(\cos 2\omega - j\sin 2\omega)\Big]$$

$$= \frac{4j}{1+\omega^2}(\omega\cos 2\omega \sinh 2 - \sin 2\omega \cosh 2)$$

17.41 $f(t)$ is odd.

$$F(j\omega) = -2j\int_0^{\infty} f(t)\sin\omega t\, dt$$

$$= -2j\int_0^{1} \sin\omega t\, dt$$

$$= -2j\left[-\frac{\cos\omega t}{\omega}\right]_0^1$$

$$= \frac{-2j}{\omega}(-\cos\omega + 1)$$

$$= \frac{2(1-\cos\omega)}{j\omega}$$

17.42 $f(t)$ is even.

$$F(j\omega) = 2\int_0^{\infty} f(t)\cos\omega t\, dt$$

$$= 2\left[\int_0^{1/2}\cos\omega t\, dt - \int_{1/2}^{1}\cos\omega t\, dt\right]$$

$$= 2\left[\frac{\sin\omega t}{\omega}\Bigg|_0^{1/2} - \frac{\sin\omega t}{\omega}\Bigg|_{1/2}^{1}\right]$$

$$= \frac{2}{\omega}\left[\sin\frac{\omega}{2} - \sin\omega + \sin\frac{\omega}{2}\right]$$

$$= \frac{4\sin\frac{\omega}{2} - 2\sin\omega}{\omega}$$

17.43 $f(t)$ is even.

$$F(j\omega) = 2\int_0^{3} t\cos\omega t\, dt$$

$$= 2\left[\frac{1}{\omega^2}\cos\omega t + \frac{t}{\omega}\sin\omega t\right]_0^3$$

$$= 2\left[\frac{\cos 3\omega - 1}{\omega^2} + \frac{3\sin 3\omega}{\omega}\right]$$

17.44 $f(t)$ is odd.

$$F(j\omega) = -j2\int_0^{3} t\sin\omega t\, dt$$

$$= -j2\left[\frac{1}{\omega^2}\sin\omega t - \frac{t}{\omega}\cos\omega t\right]_0^3$$

$$= -j2\left[\frac{\sin 3\omega}{\omega^2} - \frac{3\cos 3\omega}{\omega}\right]$$

17.45 $f(t)$ is even.

$$F(j\omega) = 2\int_0^{2} t^2\cos\omega t\, dt$$

$$= 2\left[\frac{2t\cos\omega t}{\omega^2} + \frac{\omega^2 t^2 - 2}{\omega^3}\sin\omega t\right]_0^2$$

$$= 2\left[\frac{4\cos 2\omega}{\omega^2} + \frac{4\omega^2 - 2}{\omega^3}\sin 2\omega\right]$$

17.46 $f(t)$ is odd.

$$F(j\omega) = -j2\int_0^2 t^2 \sin \omega t \, dt$$

$$= \frac{-j2}{\omega^3}\Big[-\omega^2 t^2 \cos \omega t + 2 \cos \omega t + 2\omega t \sin \omega t\Big]_0^2$$

$$F(j\omega) = -\frac{j2}{\omega^3}\left[-4\omega^2 \cos 2\omega + 2 \cos 2\omega + 4\omega \sin 2\omega - 2\right]$$

$$= \frac{j4}{\omega^3}\left[1 + (2\omega^2 - 1)\cos 2\omega - 2\omega \sin 2\omega\right]$$

17.47 Since $e^{-at}u(t) \longleftrightarrow \frac{1}{a + j\omega}$,

$$F(j\omega) = \frac{5}{2 + j\omega} + \frac{4}{3 + j\omega}$$

17.48 From $f(t - \tau) \longleftrightarrow F(j\omega)e^{-j\omega\tau}$,

$$F(j\omega) = \frac{6e^{j6\omega}}{3 + j\omega}$$

17.49 $$F(j\omega) = \frac{2}{\frac{1}{6} + j\omega} = \frac{12}{1 + j6\omega}$$

17.50 $$F(j\omega) = j\omega\, \mathcal{F}\left[2e^{-3t}u(t)\right]$$

$$= j\omega \frac{2}{3 + j\omega} = \frac{j2\omega}{3 + j\omega}$$

17.51 Let $f_1(t) = 1, \; -1 < t < 1$

$= 0$, elsewhere

Then $f(t) = t f_1(t)$; by (18.16)

$$F_1(j\omega) = \mathcal{F}[f_1(t)] = \frac{2 \sin \omega}{\omega}$$

$$t f_1(t) \longleftrightarrow -\frac{d}{d(j\omega)} F_1(j\omega)$$

or

$$f(t) \longleftrightarrow -\frac{d}{d(j\omega)}\left(\frac{2 \sin \omega}{\omega}\right)$$

$$= j\frac{d}{d\omega}\left(\frac{2 \sin \omega}{\omega}\right)$$

$$= \frac{j2}{\omega^2}(\omega \cos \omega - \sin \omega)$$

17.52 Let $g(t) = \mathcal{F}^{-1}\left[\frac{\sin \omega}{\omega}\right] = \frac{1}{2}, \; -1 < t < 1$

$= 0$, elsewhere

Then $f(t) = \mathcal{F}^{-1}\left[\frac{\sin \omega}{\omega} e^{-j\omega/2}\right]$

$= g(t - \frac{1}{2})$

$= \frac{1}{2}, \; -1 < t - \frac{1}{2} < 1$

$= 0$, elsewhere

or

$f(t) = \frac{1}{2}, \; -\frac{1}{2} < t < 3/2$

$= 0$, elsewhere

17.53 $F(j\omega) = \frac{(j\omega)^2}{1 + j\omega}$; since

$\mathcal{F}^{-1}\left[\frac{1}{1 + j\omega}\right] = e^{-t}u(t)$, then

$$f(t) = \frac{d^2}{dt^2}\left[e^{-t}u(t)\right]$$

17.54 $$\left[4(j\omega)^2 + 6j\omega + 1\right]Y(j\omega) = 2X(j\omega)$$

$$H(j\omega) = \frac{Y(j\omega)}{X(j\omega)} = \frac{2}{-4\omega^2 + 6j\omega + 1}$$

17.55 $$X(j\omega) = \frac{e^{-j2\omega}}{1 + j\omega}$$

$$Y(j\omega) = H(j\omega)X(j\omega)$$

$$= \frac{2e^{-j2\omega}}{(-4\omega^2 + 6j\omega + 1)(1 + j\omega)}$$

17.56 $$\left[(j\omega)^3 + 3j\omega\right]Y(j\omega) = 4X(j\omega)$$

$$H(j\omega) = \frac{Y(j\omega)}{X(j\omega)} = \frac{4}{-j\omega^3 + 3j\omega}$$

17.57 $$X(j\omega) = \frac{2}{\omega} \sin \frac{\omega}{2}$$

$$Y(j\omega) = H(j\omega)X(j\omega)$$

$$= \frac{4\left(\frac{2}{\omega} \sin \frac{\omega}{2}\right)}{-j\omega^3 + 3j\omega}$$

$$= \frac{8 \sin \frac{\omega}{2}}{j\omega^2(-\omega^2 + 3)}$$

17.58 SPICE input:

```
PROBLEM 17.58 FOURIER SERIES
I 0 1 PWL(0 0A 0.001 1A 0.002 0A)
R 1 0 1
.TRAN 0.00005 0.002
.FOUR 500 V(1)
.END
```

SPICE output:

```
PROBLEM 17.58 FOURIER SERIES

 ****      FOURIER ANALYSIS                       TEMPERATURE =    27.000 DEG C
******************************************************************************

FOURIER COMPONENTS OF TRANSIENT RESPONSE V(1)

 DC COMPONENT =    5.000000E-01

 HARMONIC    FREQUENCY    FOURIER     NORMALIZED     PHASE        NORMALIZED
    NO          (HZ)      COMPONENT    COMPONENT     (DEG)        PHASE (DEG)

     1      5.000E+02    4.054E-01    1.000E+00    -9.000E+01     0.000E+00
     2      1.000E+03    3.488E-09    8.604E-09     9.742E+01     1.874E+02
     3      1.500E+03    4.517E-02    1.114E-01    -9.000E+01    -3.345E-06
     4      2.000E+03    7.750E-09    1.912E-08    -1.770E+02    -8.700E+01
     5      2.500E+03    1.635E-02    4.032E-02    -9.000E+01    -1.459E-06
     6      3.000E+03    4.382E-09    1.081E-08     1.165E+02     2.065E+02
     7      3.500E+03    8.406E-03    2.073E-02    -9.000E+01     2.759E-05
     8      4.000E+03    2.154E-09    5.313E-09    -9.366E+01    -3.661E+00
     9      4.500E+03    5.139E-03    1.268E-02    -9.000E+01     3.873E-06

     TOTAL HARMONIC DISTORTION =    1.209417E+01 PERCENT

          JOB CONCLUDED
```

17.59 SPICE input:

```
PROBLEM 17.59 FOURIER SERIES
I 0 1 PWL(0 0A 5E-7 1A 1E-6 0A 2E-6 0A 2.5E-6 -1A 3E-6 0A 4E-6 0A)
R 1 0 1
.TRAN 4E-8 4E-6
.FOUR 2.5E5 V(1)
.END
```

17.59 (continued) SPICE output:

```
PROBLEM 17.59 FOURIER SERIES

 ****     FOURIER ANALYSIS                          TEMPERATURE =    27.000 DEG C
******************************************************************************

FOURIER COMPONENTS OF TRANSIENT RESPONSE V(1)

 DC COMPONENT =   -1.960669E-04

 HARMONIC   FREQUENCY    FOURIER     NORMALIZED     PHASE        NORMALIZED
    NO         (HZ)     COMPONENT    COMPONENT      (DEG)       PHASE (DEG)

     1      2.500E+05   4.748E-01    1.000E+00    4.499E+01    0.000E+00
     2      5.000E+05   2.884E-04    6.074E-04   -1.346E+02   -1.796E+02
     3      7.500E+05   3.075E-01    6.477E-01   -4.501E+01   -9.000E+01
     4      1.000E+06   6.126E-06    1.290E-05    1.773E+02    1.323E+02
     5      1.250E+06   1.107E-01    2.332E-01   -1.350E+02   -1.800E+02
     6      1.500E+06   2.477E-04    5.217E-04   -4.339E+01   -8.838E+01
     7      1.750E+06   9.651E-03    2.033E-02    1.354E+02    9.045E+01
     8      2.000E+06   4.042E-04    8.513E-04   -8.121E+01   -1.262E+02
     9      2.250E+06   5.823E-03    1.226E-02    4.438E+01   -6.093E-01

     TOTAL HARMONIC DISTORTION =    6.887753E+01 PERCENT
```

17.60 SPICE input:

```
PROBLEM 17.60 FOURIER SERIES
I 0 1 PWL(0 2A 2 2A 4 -1A 6 -1A 8 2A)
R 1 0 1
.TRAN 0.1 8
.FOUR 0.125 V(1)
.END
```

SPICE output:

```
PROBLEM 17.60 FOURIER SERIES

 ****     FOURIER ANALYSIS                          TEMPERATURE =    27.000 DEG C
******************************************************************************

FOURIER COMPONENTS OF TRANSIENT RESPONSE V(1)

 DC COMPONENT =    5.000000E-01

 HARMONIC   FREQUENCY    FOURIER     NORMALIZED     PHASE        NORMALIZED
    NO         (HZ)     COMPONENT    COMPONENT      (DEG)       PHASE (DEG)

     1      1.250E-01   1.720E+00    1.000E+00    4.500E+01    0.000E+00
     2      2.500E-01   7.891E-09    4.588E-09    9.647E+01    5.147E+01
     3      3.750E-01   1.916E-01    1.114E-01    1.350E+02    9.000E+01
     4      5.000E-01   1.610E-09    9.360E-10   -1.287E+02   -1.737E+02
     5      6.250E-01   6.935E-02    4.032E-02    4.500E+01    8.902E-07
     6      7.500E-01   7.332E-09    4.263E-09    9.549E+01    5.049E+01
     7      8.750E-01   3.566E-02    2.073E-02    1.350E+02    9.000E+01
     8      1.000E+00   1.651E-09    9.600E-10   -9.159E+01   -1.366E+02
     9      1.125E+00   2.180E-02    1.268E-02    4.500E+01    3.179E-06

     TOTAL HARMONIC DISTORTION =    1.209417E+01 PERCENT
```

Chapter 18

Design of Linear Filters

18.1 Passive Filters

18.1 (a) Design a passive circuit using only resistors and capacitors which realizes the lowpass filter whose uncorrected Bode plot is shown below. Assume that there is no loading at the output.
(b) Repeat part (a) using only resistors and inductors.
(c) Repeat part (a), but design for a 10 ohm resistive load at the output of the circuit.
(d) Verify parts (a), (b), and (c) using SPICE.

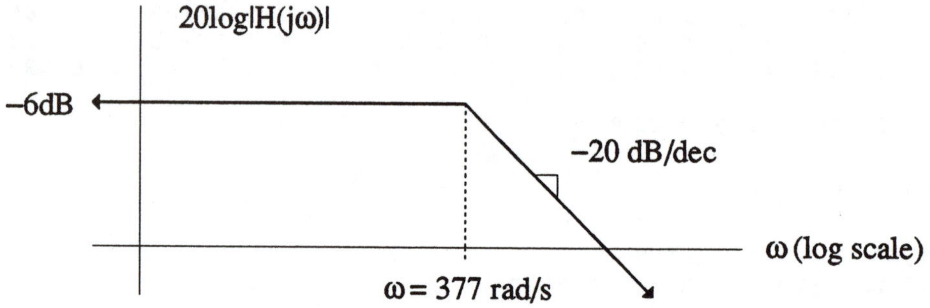

PROBLEM 18.1

18.2 Redo Prob. 18.1 if the uncorrected Bode plot is:

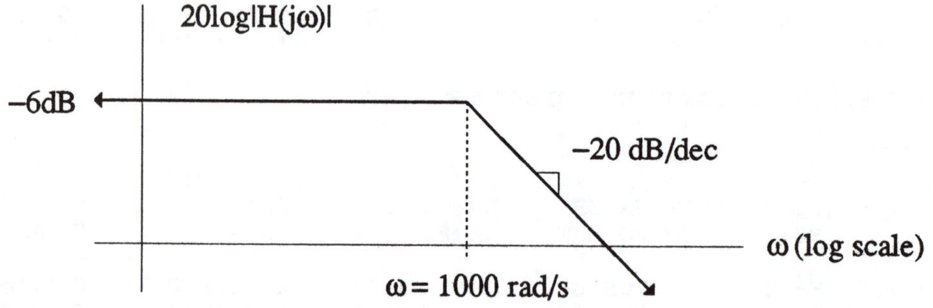

PROBLEM 18.2

18.3 Redo Prob. 18.1 if the circuit is a passive highpass filter whose uncorrected Bode plot is:

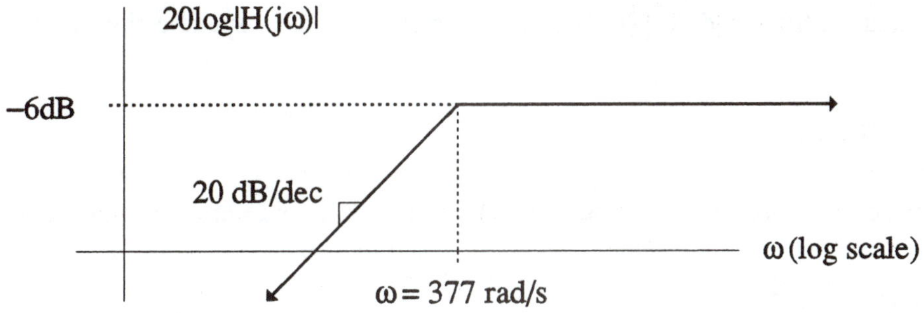

PROBLEM 18.3

18.4 (a) Design a passive bandpass filter whose Bode plot is:

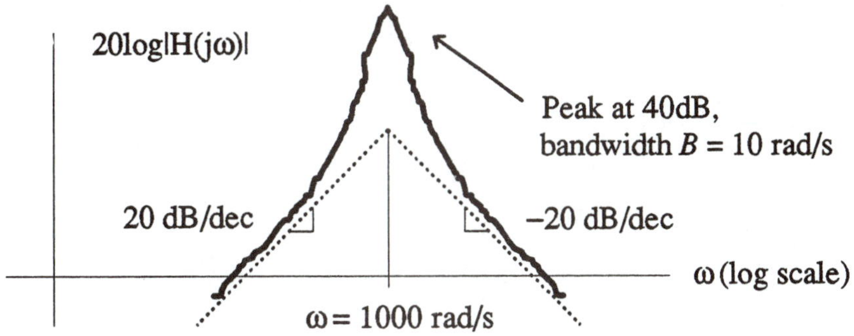

PROBLEM 18.4

(b) Use SPICE to verify that your design meets these specifications.

18.5 (a) Design a passive bandstop (notch) filter with (approximately) unity gain at low and high frequencies, but at most a –20dB gain at the center frequency of the stopband, $\omega_c = 500$ rad/s. The notch half-power bandwidth should be no greater than $B = 200$ rad/s. (b) Use SPICE to verify that your design meets these specifications.

18.6 Design a passive crossover network for the following two-loudspeaker system. Assume that the woofer and tweeter speakers may be both modeled as 4Ω resistors within the frequency range in which they operate most efficiently. The woofer transduces electrical energy to acoustic energy most efficiently for frequencies below 1200 Hz, and the tweeter for those above 1200Hz. The power amplifier which the two speakers will be used with has an output impedance of 4Ω. Design the circuit so that the power delivered to each speaker at the crossover frequency is equal, and there is an exact impedance match at this frequency. The output impedance of the crossover network should vary no more than ±10% of the exact matching impedance from 100Hz to 15kHz.

18.7 (a) Design a (bandpass) power coupling filter using a passive resonant circuit which passes the power waveform with a fundamental frequency $f = 50$ Hz but suppresses the harmonics (at frequencies that are multiples of the fundamental) by at least 20 dB. Assume that the filter will be connected to a 10Ω load. (b) Verify that the design meets these specifications using SPICE.

18.2 Active Filters

18.8 Design active filters (using op amp circuits) which realize the following voltage transfer functions:

(a) $H(s) = \dfrac{2s+3}{s+4}$

(b) $H(s) = \dfrac{-2s+3}{s+4}$

(c) $H(s) = \dfrac{-2s-3}{s+4}$

(d) $H(s) = \dfrac{2s-3}{s+4}$

(e) $H(s) = \dfrac{2s}{s+4}$

(f) $H(s) = \dfrac{-3}{s+4}$

(g) $H(s) = \dfrac{-2s}{s+4}$

(h) $H(s) = \dfrac{3}{s+4}$

18.9 Design active filters (using op amp circuits) which realize the following voltage transfer functions:

(a) $H(s) = \dfrac{1}{s^2 + s + 2}$

(b) $H(s) = \dfrac{s^2}{s^2 + s + 2}$

(c) $H(s) = \dfrac{5s^2 - 1}{s^2 + 3s + 2}$

(d) $H(s) = \dfrac{9s^2 + 8s - 3}{s^2 + 17s + 3}$

(e) $H(s) = \dfrac{13s}{s^2 - 27s + 1}$

(f) $H(s) = \dfrac{3}{s^2 + 2}$

18.10 Design active filters which realize the following uncorrected Bode plots:

(a)

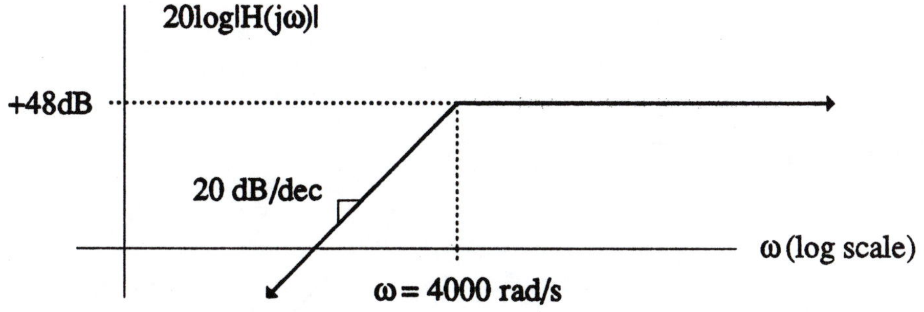

(b)

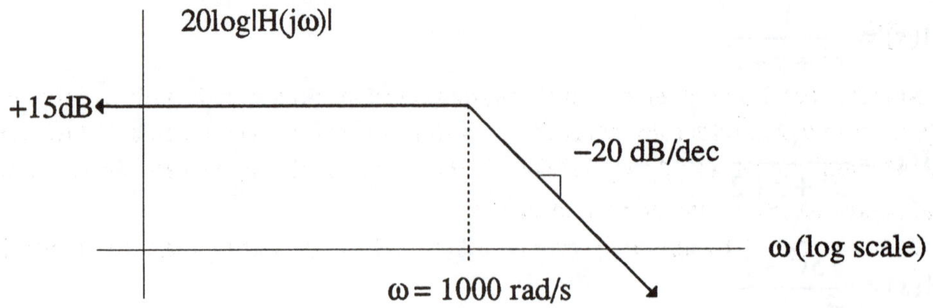

(c)

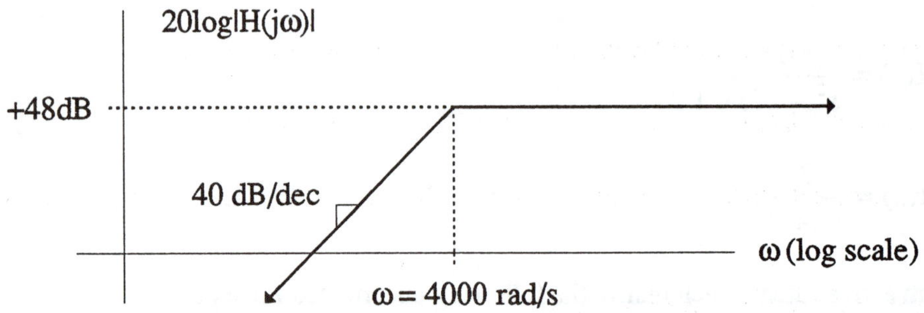

(d)

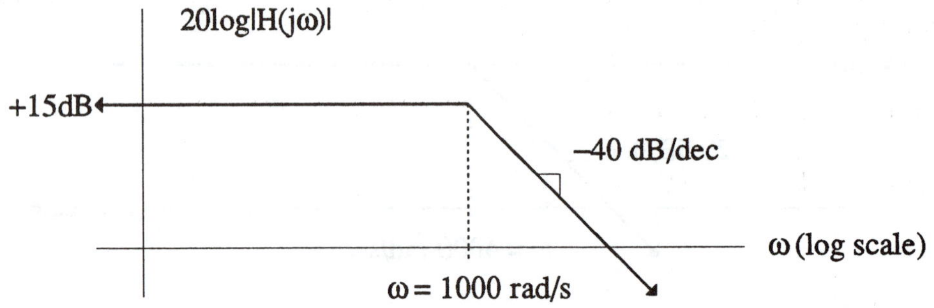

18.11 Design higher-order active filters which realize the following transfer functions by cascading appropriate first-order and second-order filters:

(a) $H(s) = \dfrac{-3}{(s+4)(s^2+s+2)^2}$

(b) $H(s) = \left(\dfrac{2s+3}{s+4}\right)^3$

18.12 A certain digital communications system transmits a sinusoidal voltage with frequency $f_0 = 1\,\text{kHz}$ to represent a zero (0) bit, and a sinusoidal voltage with frequency $f_1 = 10\,\text{kHz}$ to represent a one (1) bit.

(a) Assuming that the receiver side of this system receives both signals with an equal magnitude, design an active bandpass filter using only resistors, capacitors, and op amps which boosts the "zero received" signals (at $f_0 = 1\,\text{kHz}$) by 12dB but attenuates signals an octave (or more) higher and an octave (or more) lower by at least 8dB.
(b) Design a circuit with similar specifications around f_1 for passing the "one received" signals.

18.3 Classical Filters and Scaling

18.13 Design a normalized third-order Butterworth lowpass filter.

18.14 Design a normalized second-order Chebyshev (1/2 dB) lowpass filter.

18.15 Design a normalized third-order Chebyshev (1 dB) lowpass filter.

18.16 Design a second-order Bessel lowpass filter with a +24dB dc gain and unity −3dB bandwidth.

18.4 Filter Transformation and Scaling

18.17 (a) Using Prob. 18.16, design a second-order Bessel highpass filter with a cutoff (−3dB) frequency $\omega_c = 1$ rad/s and a +24dB high-frequency (passband) gain.
(b) Using the results in (a) and frequency scaling, design a similar highpass filter with a cutoff (−3dB) frequency $\omega_c = 100$ rad/s.

18.18 Design a normalized second-order Butterworth bandpass filter.

18.19 Design a normalized second-order Butterworth bandstop filter.

18.20 Given the following circuit, replace all of the *RLC* elements such that the smallest resistor is 10kΩ but the original frequency response is unchanged.

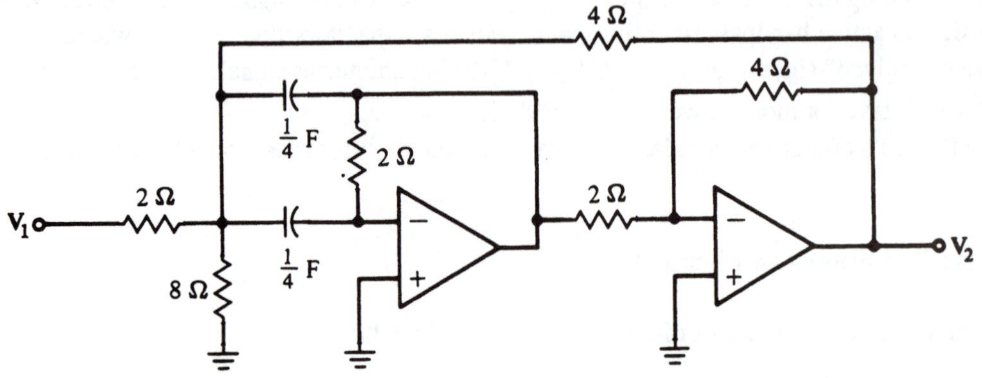

PROBLEM 18.20

18.21 (a) Given the circuit in Prob. 18.20, replace all of the *RLC* elements such that the smallest resistor is 10kΩ and the original frequency response is shifted exactly 2 decades (scaled 100 times) higher.
(b) Given the circuit in Prob. 18.20, replace all of the *RLC* elements such that the smallest resistor is 10kΩ and the original frequency response is shifted exactly 2 decades (scaled 100 times) lower.

18.22 A certain circuit has the following transfer function:

$$H(s) = \frac{9s^2 + 8s - 3}{2s^3 + s^2 + 17s + 3}.$$

(a) Find the new transfer function if each impedance value is scaled by $\alpha = 3$.
(b) Find the new transfer function if each impedance value is scaled by $\alpha = 1/3$.
(c) Find the new transfer function if the frequency response is scaled by $\alpha = 3$.
(d) Find the new transfer function if the frequency response is scaled by $\alpha = 1/3$.

18.1 (a) From Bode plot, circuit is a first-order LPF with DC output gain $-6dB$ (multiplication by $\frac{1}{2}$) and $\omega_c = 377 \frac{rad}{s} = \frac{1}{\tau}$

$\Rightarrow 377 = \frac{1}{R_{eq}C}$

Use a voltage divider to get $\frac{1}{2}$ input voltage at DC as the output:

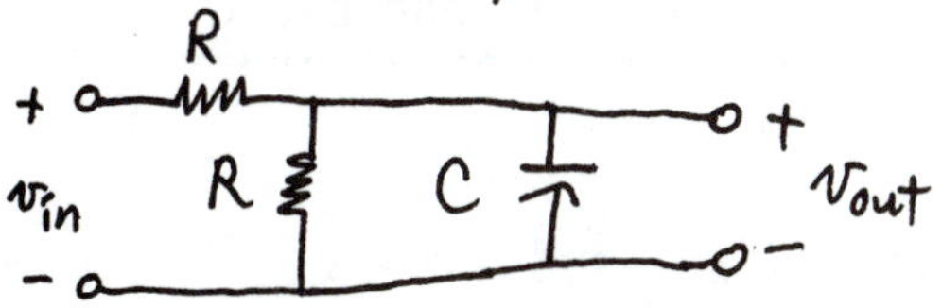

$$H(s) = \frac{V_{out}(s)}{V_{in}(s)} = \frac{\left[\frac{R\left(\frac{1}{Cs}\right)}{R+\frac{1}{Cs}}\right]}{R+\left[\frac{R\left(\frac{1}{Cs}\right)}{R+\frac{1}{Cs}}\right]}$$

Simplify $\Rightarrow H(s) = \frac{1}{2+RCs}$

$\therefore H(j\omega) = \frac{1}{2+j\omega RC}$

$R_{eq}C = \left(\frac{R}{2}\right)(C) = \frac{1}{377} \Rightarrow RC = 5.305\times 10^{-3}$

I'll choose $\underline{R = 100\,k\Omega}$, $\underline{C = 0.05305\,\mu F}$

(b) Reverse the orientation (order) of the storage & dissipative elements:

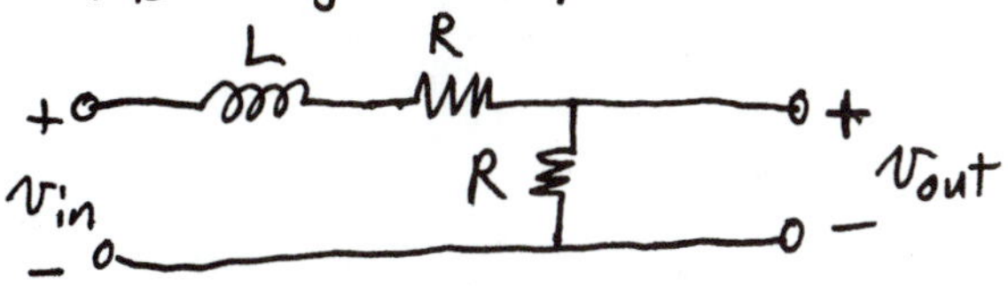

$\omega_c = 377 = \frac{1}{\tau} = \frac{R_{eq}}{L} = \frac{2R}{L}$

I'll choose $\underline{R = 1\,k\Omega}$, $\underline{L = 5.305\,H}$

Check: $H(s) = \frac{V_{out}(s)}{V_{in}(s)} = \frac{R}{2R+Ls}$

$= \frac{1}{\left(2+\frac{L}{R}s\right)} = \frac{1}{2+0.005305\,s}$

which is the same $H(s)$ as in part (a) ✓

(c) Use the results of part (a):

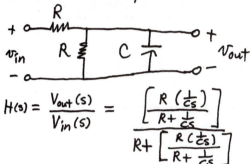

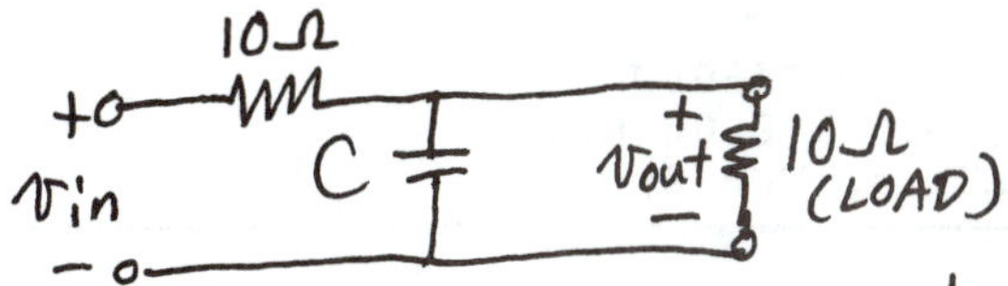

$R_{eq} = \frac{10\Omega}{2} = 5\Omega \Rightarrow R_{eq}C = 5C = \frac{1}{377}$

$\Rightarrow \underline{C = 530.5\,\mu F}$

(d) See SPICE code included below.

18.2 (a) Same circuit as in 18.1(a) except $R_{eq}C = \frac{R}{2}C = \frac{1}{1000}$

$\Rightarrow$ I'll choose $\underline{R = 100\,k\Omega}$, $\underline{C = 0.02\,\mu F}$

(b) Same circuit as in 18.1(b) except $\frac{R_{eq}}{L} = \frac{2R}{L} = 1000$

$\Rightarrow$ I'll choose $\underline{R = 1\,k\Omega}$, $\underline{L = 2\,H}$

(c) Same circuit as in 18.1(c) except $R_{eq}C = 5C = \frac{1}{1000}$

$\Rightarrow \underline{C = 200\,\mu F}$

(d) See SPICE code included below.

18.1(d) SPICE code and outputs:

```
PROBLEM 18.1(A)
VIN 1 0 AC 1 0
R1 1 2 100K
R2 2 0 100K
C1 2 0 0.05305U
.AC DEC 3 0.06 6000
.PLOT AC VDB(2)
.END
```

```
****     AC ANALYSIS                          TEMPERATURE =   27.000 DEG C
******************************************************************************

  FREQ          VDB(2)

(*)----------    -6.0000E+01  -4.0000E+01  -2.0000E+01  -3.5527E-15
2.0000E+01
                         - - - - - - - - - - - - - - - - - - - - - - - - - -
  6.000E-02 -6.021E+00 .           .           .        *  .           .
  1.293E-01 -6.021E+00 .           .           .        *  .           .
  2.785E-01 -6.021E+00 .           .           .        *  .           .
  6.000E-01 -6.021E+00 .           .           .        *  .           .
  1.293E+00 -6.023E+00 .           .           .        *  .           .
  2.785E+00 -6.030E+00 .           .           .        *  .           .
  6.000E+00 -6.064E+00 .           .           .        *  .           .
  1.293E+01 -6.218E+00 .           .           .        *  .           .
  2.785E+01 -6.868E+00 .           .           .        *  .           .
  6.000E+01 -9.031E+00 .           .           .      *    .           .
  1.293E+02 -1.353E+01 .           .           .   *       .           .
  2.785E+02 -1.955E+01 .           .           *           .           .
  6.000E+02 -2.606E+01 .           .       *   .           .           .
  1.293E+03 -3.270E+01 .           .   *       .           .           .
  2.785E+03 -3.936E+01 .           *           .           .           .
  6.000E+03 -4.602E+01 .       *   .           .           .           .
                         - - - - - - - - - - - - - - - - - - - - - - - - - -
```

```
PROBLEM 18.1(B)
VIN 1 0 AC 1 0
L1 1 2 5.305
R1 2 3 1K
R2 3 0 1K
.AC DEC 3 0.06 6000
.PLOT AC VDB(3)
.END
```

```
****     AC ANALYSIS                           TEMPERATURE =   27.000 DEG C
******************************************************************************

  FREQ          VDB(3)

(*)----------     -6.0000E+01  -4.0000E+01  -2.0000E+01  -3.5527E-15
2.0000E+01

                       - - - - - - - - - - - - - - - - - - - - - - - - - - -
  6.000E-02 -6.021E+00 .            .             .        *   .            .
  1.293E-01 -6.021E+00 .            .             .        *   .            .
  2.785E-01 -6.021E+00 .            .             .        *   .            .
  6.000E-01 -6.021E+00 .            .             .        *   .            .
  1.293E+00 -6.023E+00 .            .             .        *   .            .
  2.785E+00 -6.030E+00 .            .             .        *   .            .
  6.000E+00 -6.064E+00 .            .             .        *   .            .
  1.293E+01 -6.218E+00 .            .             .        *   .            .
  2.785E+01 -6.868E+00 .            .             .        *   .            .
  6.000E+01 -9.031E+00 .            .             .      *     .            .
  1.293E+02 -1.353E+01 .            .             .   *        .            .
  2.785E+02 -1.955E+01 .            .             *            .            .
  6.000E+02 -2.606E+01 .            .        *    .            .            .
  1.293E+03 -3.270E+01 .            .     *       .            .            .
  2.785E+03 -3.936E+01 .            *             .            .            .
  6.000E+03 -4.602E+01 .        *   .             .            .            .
                       - - - - - - - - - - - - - - - - - - - - - - - - - - -
```

```
PROBLEM 18.1(C)
VIN 1 0 AC 1 0
R1 1 2 10
R2 2 0 10
C1 2 0 530.5U
.AC DEC 3 0.06 6000
.PLOT AC VDB(2)
.END
```

```
****     AC ANALYSIS                           TEMPERATURE =   27.000 DEG C
******************************************************************************

  FREQ          VDB(2)

(*)----------     -6.0000E+01  -4.0000E+01  -2.0000E+01  -3.5527E-15
2.0000E+01

                       - - - - - - - - - - - - - - - - - - - - - - - - - - -
  6.000E-02 -6.021E+00 .            .             .        *   .            .
  1.293E-01 -6.021E+00 .            .             .        *   .            .
  2.785E-01 -6.021E+00 .            .             .        *   .            .
  6.000E-01 -6.021E+00 .            .             .        *   .            .
  1.293E+00 -6.023E+00 .            .             .        *   .            .
  2.785E+00 -6.030E+00 .            .             .        *   .            .
  6.000E+00 -6.064E+00 .            .             .        *   .            .
  1.293E+01 -6.218E+00 .            .             .        *   .            .
  2.785E+01 -6.868E+00 .            .             .        *   .            .
  6.000E+01 -9.031E+00 .            .             .      *     .            .
  1.293E+02 -1.353E+01 .            .             .   *        .            .
  2.785E+02 -1.955E+01 .            .             *            .            .
  6.000E+02 -2.606E+01 .            .        *    .            .            .
  1.293E+03 -3.270E+01 .            .     *       .            .            .
  2.785E+03 -3.936E+01 .            *             .            .            .
  6.000E+03 -4.602E+01 .        *   .             .            .            .
                       - - - - - - - - - - - - - - - - - - - - - - - - - - -
```

18.2(d) SPICE code and outputs:

```
PROBLEM 18.2(A)
VIN 1 0 AC 1 0
R1 1 2 100K
R2 2 0 100K
C1 2 0 0.02U
.AC DEC 3 0.159155 15915.5
.PLOT AC VDB(2)
.END
```

```
****     AC ANALYSIS                           TEMPERATURE =   27.000 DEG C
******************************************************************************

  FREQ          VDB(2)

(*)----------    -6.0000E+01  -4.0000E+01  -2.0000E+01  -3.5527E-15
2.0000E+01

                          - - - - - - - - - - - - - - - - - - - - - - - - - - -
  1.592E-01 -6.021E+00 .             .             .        *    .             .
  3.429E-01 -6.021E+00 .             .             .        *    .             .
  7.387E-01 -6.021E+00 .             .             .        *    .             .
  1.592E+00 -6.021E+00 .             .             .        *    .             .
  3.429E+00 -6.023E+00 .             .             .        *    .             .
  7.387E+00 -6.030E+00 .             .             .        *    .             .
  1.592E+01 -6.064E+00 .             .             .        *    .             .
  3.429E+01 -6.218E+00 .             .             .        *    .             .
  7.387E+01 -6.868E+00 .             .             .        *    .             .
  1.592E+02 -9.031E+00 .             .             .      *      .             .
  3.429E+02 -1.353E+01 .             .             .   *         .             .
  7.387E+02 -1.955E+01 .             .             *             .             .
  1.592E+03 -2.606E+01 .             .        *    .             .             .
  3.429E+03 -3.270E+01 .             .   *         .             .             .
  7.387E+03 -3.936E+01 .             *             .             .             .
  1.592E+04 -4.602E+01 .        *    .             .             .             .
                          - - - - - - - - - - - - - - - - - - - - - - - - - - -
```

```
PROBLEM 18.2(B)
VIN 1 0 AC 1 0
L1 1 2 2
R1 2 3 1K
R2 3 0 1K
.AC DEC 3 0.159155 15915.5
.PLOT AC VDB(3)
.END
```

```
****     AC ANALYSIS                            TEMPERATURE =   27.000 DEG C
******************************************************************************

  FREQ          VDB(3)

(*)----------     -6.0000E+01  -4.0000E+01  -2.0000E+01  -3.5527E-15
2.0000E+01

                         - - - - - - - - - - - - - - - - - - - - - - - - - - -
  1.592E-01 -6.021E+00 .            .            .        *   .            .
  3.429E-01 -6.021E+00 .            .            .        *   .            .
  7.387E-01 -6.021E+00 .            .            .        *   .            .
  1.592E+00 -6.021E+00 .            .            .        *   .            .
  3.429E+00 -6.023E+00 .            .            .        *   .            .
  7.387E+00 -6.030E+00 .            .            .        *   .            .
  1.592E+01 -6.064E+00 .            .            .        *   .            .
  3.429E+01 -6.218E+00 .            .            .        *   .            .
  7.387E+01 -6.868E+00 .            .            .        *   .            .
  1.592E+02 -9.031E+00 .            .            .      *     .            .
  3.429E+02 -1.353E+01 .            .            .   *        .            .
  7.387E+02 -1.955E+01 .            .            *            .            .
  1.592E+03 -2.606E+01 .            .        *   .            .            .
  3.429E+03 -3.270E+01 .            .    *       .            .            .
  7.387E+03 -3.936E+01 .            *            .            .            .
  1.592E+04 -4.602E+01 .        *   .            .            .            .
                         - - - - - - - - - - - - - - - - - - - - - - - - - - -
```

```
PROBLEM 18.2(C)
VIN 1 0 AC 1 0
R1 1 2 10
R2 2 0 10
C1 2 0 200.0U
.AC DEC 3 0.159155 15915.5
.PLOT AC VDB(2)
.END
```

```
****     AC ANALYSIS                            TEMPERATURE =   27.000 DEG C
******************************************************************************

  FREQ          VDB(2)

(*)----------     -6.0000E+01  -4.0000E+01  -2.0000E+01  -3.5527E-15
2.0000E+01

                         - - - - - - - - - - - - - - - - - - - - - - - - - - -
  1.592E-01 -6.021E+00 .            .            .        *   .            .
  3.429E-01 -6.021E+00 .            .            .        *   .            .
  7.387E-01 -6.021E+00 .            .            .        *   .            .
  1.592E+00 -6.021E+00 .            .            .        *   .            .
  3.429E+00 -6.023E+00 .            .            .        *   .            .
  7.387E+00 -6.030E+00 .            .            .        *   .            .
  1.592E+01 -6.064E+00 .            .            .        *   .            .
  3.429E+01 -6.218E+00 .            .            .        *   .            .
  7.387E+01 -6.868E+00 .            .            .        *   .            .
  1.592E+02 -9.031E+00 .            .            .      *     .            .
  3.429E+02 -1.353E+01 .            .            .   *        .            .
  7.387E+02 -1.955E+01 .            .            *            .            .
  1.592E+03 -2.606E+01 .            .        *   .            .            .
  3.429E+03 -3.270E+01 .            .    *       .            .            .
  7.387E+03 -3.936E+01 .            *            .            .            .
  1.592E+04 -4.602E+01 .        *   .            .            .            .
                         - - - - - - - - - - - - - - - - - - - - - - - - - - -
```

18.3 (a) Reverse order of R's and C's to get highpass version:

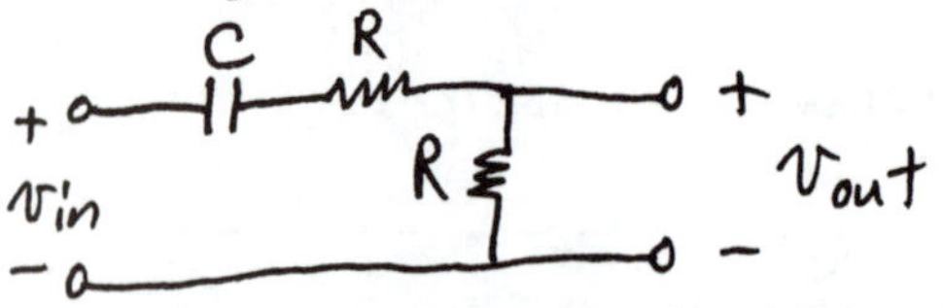

Now $R_{eq} = 2R \Rightarrow \tau = R_{eq}C = 2RC = \frac{1}{377}$

I'll choose $\underline{R = 100k\Omega}$, $\underline{C = 0.0132626\mu F}$

(b) Again, reverse order:

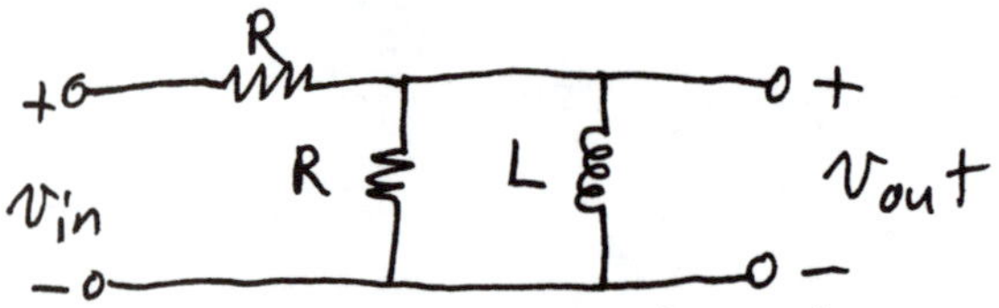

Now $R_{eq} = R||R = \frac{1}{2}R \Rightarrow \frac{R_{eq}}{L} = \frac{R}{2L} = 377$

I'll choose $\underline{R = 1k\Omega}$, $\underline{L = 1.32626H}$

(c) Reverse order:

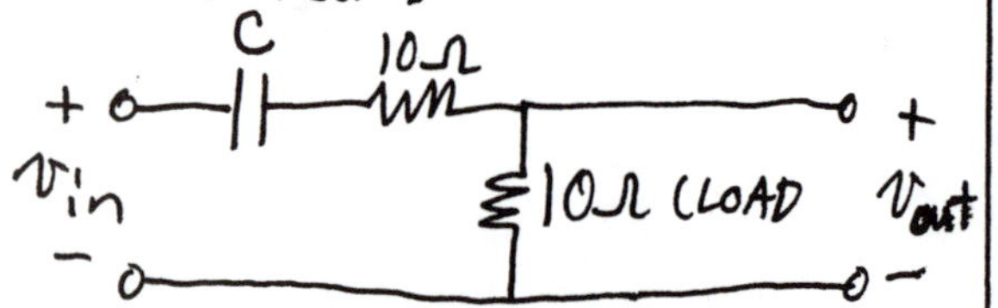

$R_{eq} = 20\Omega \Rightarrow 20C = \frac{1}{377}$

$\Rightarrow \underline{C = 132.626\mu F}$

(d) See SPICE code included below.

18.4 (a) Parallel RLC version:
Input is $i(t)$, Output is $v(t)$

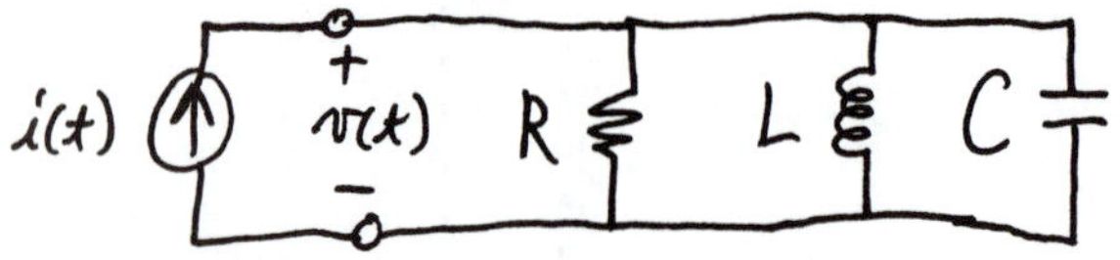

$\omega_n = 1000 = \frac{1}{\sqrt{LC}}$, $B = \frac{1}{RC} = 10$

$20 \log R = 40dB \Rightarrow \log R = 2 \Rightarrow \boxed{R = 100\Omega}$

$\Rightarrow \boxed{C = \frac{1}{10R} = 1mF} \Rightarrow \boxed{L = 1mH}$

(b) See SPICE code included below.

18.5 (a) Using a series RLC circuit with input and output as shown:

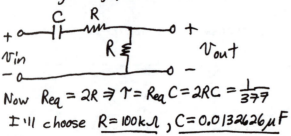

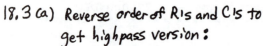

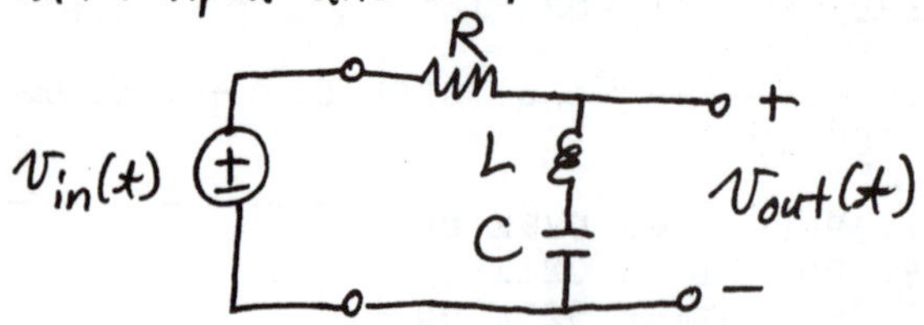

Choose $\frac{1}{\sqrt{LC}} = \omega_c = 500 \Rightarrow LC = 4\times10^{-6}$

[At $\omega = \omega_c$, gain should ideally be zero.]

At DC, capacitor is open-ckt
$\Rightarrow$ unity gain.
At high freq ($\omega \to \infty$), inductor is open-ckt $\Rightarrow$ unity gain.

$B = \frac{R}{L} \leq 200 \Rightarrow$ Choose $\boxed{L = C = 2\times10^{-3}}$

$\Rightarrow \boxed{R \leq 0.4\Omega}$

(b) See SPICE code included below.

18.3(d) SPICE code and outputs:

```
PROBLEM 18.3(A)
VIN 1 0 AC 1 0
C1 1 2 0.01326U
R1 2 3 100K
R2 3 0 100K
.AC DEC 3 0.06 6000
.PLOT AC VDB(3)
.END
```

```
****     AC ANALYSIS                              TEMPERATURE =    27.000 DEG C
******************************************************************************

  FREQ         VDB(3)

(*)----------    -8.0000E+01  -6.0000E+01  -4.0000E+01  -2.0000E+01
0.0000E+00
                        - - - - - - - - - - - - - - - - - - - - - - - - - -
  6.000E-02 -6.602E+01 .       *    .            .            .            .
  1.293E-01 -5.936E+01 .            *            .            .            .
  2.785E-01 -5.269E+01 .            .    *       .            .            .
  6.000E-01 -4.602E+01 .            .        *   .            .            .
  1.293E+00 -3.936E+01 .            .            *            .            .
  2.785E+00 -3.270E+01 .            .            .    *       .            .
  6.000E+00 -2.607E+01 .            .            .        *   .            .
  1.293E+01 -1.955E+01 .            .            .            *            .
  2.785E+01 -1.354E+01 .            .            .            .   *        .
  6.000E+01 -9.032E+00 .            .            .            .      *     .
  1.293E+02 -6.868E+00 .            .            .            .        *   .
  2.785E+02 -6.218E+00 .            .            .            .        *   .
  6.000E+02 -6.064E+00 .            .            .            .        *   .
  1.293E+03 -6.030E+00 .            .            .            .        *   .
  2.785E+03 -6.023E+00 .            .            .            .        *   .
  6.000E+03 -6.021E+00 .            .            .            .        *   .
                        - - - - - - - - - - - - - - - - - - - - - - - - - -
```

```
PROBLEM 18.3(B)
VIN 1 0 AC 1 0
R1 1 2 1K
R2 2 0 1K
L1 2 0 1.3263
.AC DEC 3 0.06 6000
.PLOT AC VDB(2)
.END
```

```
****     AC ANALYSIS                          TEMPERATURE =   27.000 DEG C
******************************************************************************

  FREQ          VDB(2)

(*)----------    -8.0000E+01  -6.0000E+01  -4.0000E+01  -2.0000E+01
0.0000E+00

  6.000E-02 -6.602E+01 .  -------*-------------------------------------------
  1.293E-01 -5.935E+01 .           *           .            .            .
  2.785E-01 -5.269E+01 .           .    *      .            .            .
  6.000E-01 -4.602E+01 .           .       *   .            .            .
  1.293E+00 -3.936E+01 .           .           *            .            .
  2.785E+00 -3.270E+01 .           .           .    *       .            .
  6.000E+00 -2.606E+01 .           .           .        *   .            .
  1.293E+01 -1.955E+01 .           .           .            *            .
  2.785E+01 -1.353E+01 .           .           .            .   *        .
  6.000E+01 -9.031E+00 .           .           .            .      *     .
  1.293E+02 -6.868E+00 .           .           .            .        *   .
  2.785E+02 -6.218E+00 .           .           .            .        *   .
  6.000E+02 -6.064E+00 .           .           .            .        *   .
  1.293E+03 -6.030E+00 .           .           .            .        *   .
  2.785E+03 -6.023E+00 .           .           .            .        *   .
  6.000E+03 -6.021E+00 .           .           .            .        *   .
                        -------------------------------------------------
```

```
PROBLEM 18.3(C)
VIN 1 0 AC 1 0
C1 1 2 132.63U
R1 2 3 10
R2 3 0 10
.AC DEC 3 0.06 6000
.PLOT AC VDB(3)
.END
```

```
****     AC ANALYSIS                          TEMPERATURE =   27.000 DEG C
******************************************************************************

  FREQ          VDB(3)

(*)----------    -8.0000E+01  -6.0000E+01  -4.0000E+01  -2.0000E+01
0.0000E+00

  6.000E-02 -6.602E+01 .  -------*-------------------------------------------
  1.293E-01 -5.935E+01 .           *           .            .            .
  2.785E-01 -5.269E+01 .           .    *      .            .            .
  6.000E-01 -4.602E+01 .           .       *   .            .            .
  1.293E+00 -3.936E+01 .           .           *            .            .
  2.785E+00 -3.270E+01 .           .           .    *       .            .
  6.000E+00 -2.606E+01 .           .           .        *   .            .
  1.293E+01 -1.955E+01 .           .           .            *            .
  2.785E+01 -1.353E+01 .           .           .            .   *        .
  6.000E+01 -9.031E+00 .           .           .            .      *     .
  1.293E+02 -6.868E+00 .           .           .            .        *   .
  2.785E+02 -6.218E+00 .           .           .            .        *   .
  6.000E+02 -6.064E+00 .           .           .            .        *   .
  1.293E+03 -6.030E+00 .           .           .            .        *   .
  2.785E+03 -6.023E+00 .           .           .            .        *   .
  6.000E+03 -6.021E+00 .           .           .            .        *   .
                        -------------------------------------------------
```

18.4 SPICE code and output:

```
PROBLEM 18.4
I 0 1 AC 1 0
R 1 0 100
L 1 0 1M
C 1 0 1M
.AC DEC 6 0.159155 15915.5
.PLOT AC VDB(1)
.END
```

```
****     AC ANALYSIS                        TEMPERATURE =   27.000 DEG C
******************************************************************************

  FREQ        VDB(1)

(*)----------    -1.0000E+02  -5.0000E+01   0.0000E+00   5.0000E+01
1.0000E+02
                       - - - - - - - - - - - - - - - - - - - - - - - - - - -
  1.592E-01 -6.000E+01 .         *  .            .            .            .
  2.336E-01 -5.667E+01 .          * .            .            .            .
  3.429E-01 -5.333E+01 .           *.            .            .            .
  5.033E-01 -5.000E+01 .            *            .            .            .
  7.387E-01 -4.667E+01 .            .*           .            .            .
  1.084E+00 -4.333E+01 .            . *          .            .            .
  1.592E+00 -4.000E+01 .            .  *         .            .            .
  2.336E+00 -3.666E+01 .            .  *         .            .            .
  3.429E+00 -3.333E+01 .            .   *        .            .            .
  5.033E+00 -2.999E+01 .            .    *       .            .            .
  7.387E+00 -2.665E+01 .            .     *      .            .            .
  1.084E+01 -2.329E+01 .            .      *     .            .            .
  1.592E+01 -1.991E+01 .            .       *    .            .            .
  2.336E+01 -1.648E+01 .            .        *   .            .            .
  3.429E+01 -1.292E+01 .            .         *  .            .            .
  5.033E+01 -9.085E+00 .            .          * .            .            .
  7.387E+01 -4.559E+00 .            .           *.            .            .
  1.084E+02  2.085E+00 .            .            .*           .            .
  1.592E+02  4.000E+01 .            .            .         *  .            .
  2.336E+02  2.085E+00 .            .            .*           .            .
  3.429E+02 -4.559E+00 .            .           *.            .            .
  5.033E+02 -9.085E+00 .            .          * .            .            .
  7.387E+02 -1.292E+01 .            .         *  .            .            .
  1.084E+03 -1.648E+01 .            .        *   .            .            .
  1.592E+03 -1.991E+01 .            .       *    .            .            .
  2.336E+03 -2.329E+01 .            .      *     .            .            .
  3.429E+03 -2.665E+01 .            .     *      .            .            .
  5.033E+03 -2.999E+01 .            .    *       .            .            .
  7.387E+03 -3.333E+01 .            .   *        .            .            .
  1.084E+04 -3.666E+01 .            .  *         .            .            .
  1.592E+04 -4.000E+01 .            .  *         .            .            .
                       - - - - - - - - - - - - - - - - - - - - - - - - - - -
```

18.5 SPICE code and output:

```
PROBLEM 18.5
VIN 1 0 AC 1 0
R 1 2 0.2
L 2 3 2M
C 3 0 2M
RDUMMY 3 0 1000G
.AC DEC 3 0.079577 7957.75
.PLOT AC VDB(2)
.END
```

```
****      AC ANALYSIS                         TEMPERATURE =   27.000 DEG C
******************************************************************************

  FREQ        VDB(2)

(*)----------    -1.5000E+02  -1.0000E+02  -5.0000E+01  -7.1054E-15
5.0000E+01

  7.958E-02 -3.440E-07 -  -  -  -  -  -  -  -  -  -  -  -  -  -  *  -  -  -  -  -
  1.714E-01 -7.468E-07 .          .            .            *            .
  3.694E-01 -3.505E-06 .          .            .            *            .
  7.958E-01 -1.731E-05 .          .            .            *            .
  1.714E+00 -8.082E-05 .          .            .            *            .
  3.694E+00 -3.759E-04 .          .            .            *            .
  7.958E+00 -1.772E-03 .          .            .            *            .
  1.714E+01 -8.858E-03 .          .            .            *            .
  3.694E+01 -6.038E-02 .          .            .            *            .
  7.958E+01 -8.455E+01 .          .   *        .            .            .
  1.714E+02 -6.038E-02 .          .            .            *            .
  3.694E+02 -8.858E-03 .          .            .            *            .
  7.958E+02 -1.772E-03 .          .            .            *            .
  1.714E+03 -3.759E-04 .          .            .            *            .
  3.694E+03 -8.082E-05 .          .            .            *            .
  7.958E+03 -1.731E-05 .          .            .            *            .
                       -  -  -  -  -  -  -  -  -  -  -  -  -  -  -  -  -  -  -  -
```

18.6 (a) Using the approach outlined in the text, put the woofer in series with an inductor and the tweeter in series with a capacitor. Place these two circuits in parallel across the input (from the power amplifier):

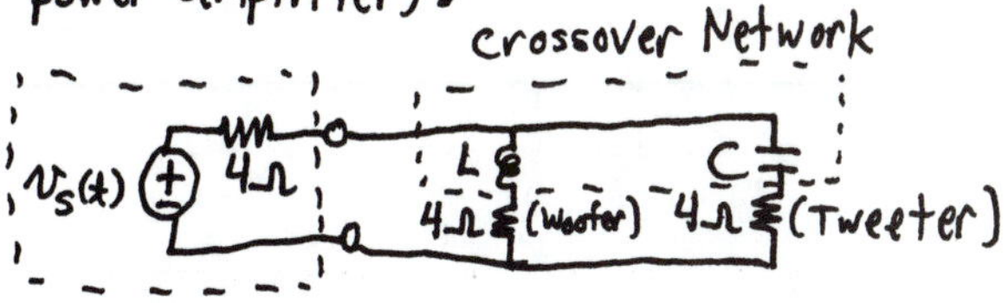

$$\omega_r = \frac{1}{\sqrt{LC}} = 2\pi(1200) = 7539.82 \ \frac{rad}{s}$$

$$\Rightarrow \boxed{\frac{1}{LC} = 5.684892\times10^7}$$

$$Z_L = j\omega L \Rightarrow \begin{cases} \text{At } f = 100\text{Hz}, \ Z_L = j\,628.3L \\ \text{At } f = 15\text{kHz}, \ Z_L = j(9.425\times10^4)L \end{cases}$$

$$Z_C = \frac{-j}{\omega C} \Rightarrow \begin{cases} \text{At } f = 100\text{Hz}, \ Z_C = -\frac{j}{628.3C} \\ \text{At } f = 15\text{kHz}, \ Z_C = -\frac{j}{(9.425\times10^4)C} \end{cases}$$

±10% criterion:

$$4 < \sqrt{4^2+(\omega L)^2} < 4.4 \Rightarrow 16+\omega^2L^2 < 19.36$$

$$\Rightarrow \boxed{\omega^2L^2 < 3.36} \text{ for any } \omega$$

$$3.6 < \sqrt{4^2 - \frac{1}{\omega^2C^2}} < 4 \Rightarrow -3.04 < -\frac{1}{\omega^2C^2}$$

$$\Rightarrow \boxed{\omega^2C^2 > 0.328947} \text{ for any } \omega$$

$$\Rightarrow L^2 < \frac{3.36}{\omega^2} \text{ and } C^2 > \frac{.328947}{\omega^2}$$

At $\omega = 628.3 \Rightarrow C > 0.000912843$

At $\omega = 9.425\times10^4 \Rightarrow L < 1.94486\times10^{-5}$

Pick $C \Rightarrow L = \dfrac{1}{C(5.684892\times10^7)}$

Choosing

$$\boxed{C = 1\text{F}, \ L = 1.759\times10^{-8}\text{ H}}$$

is one possible solution

18.7 (a)

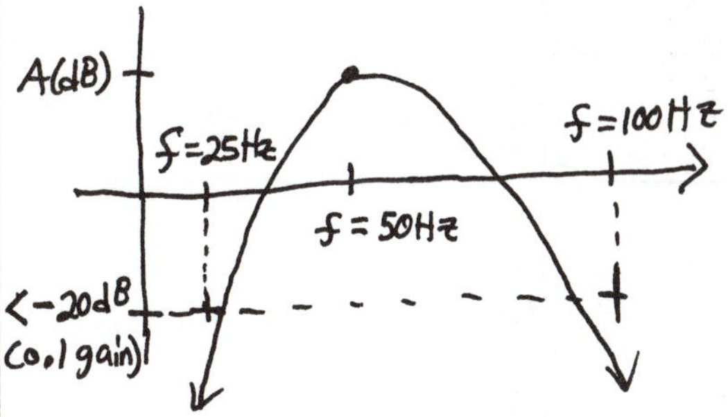

A single RLC circuit yeilds only a ± 20 dB/dec rolloff at the low and high freqeuncy extremes. We want ± 20dB/octave = 100dB/dec at least. ⇒ 5 or more resonant circuit filters cascaded toget this.

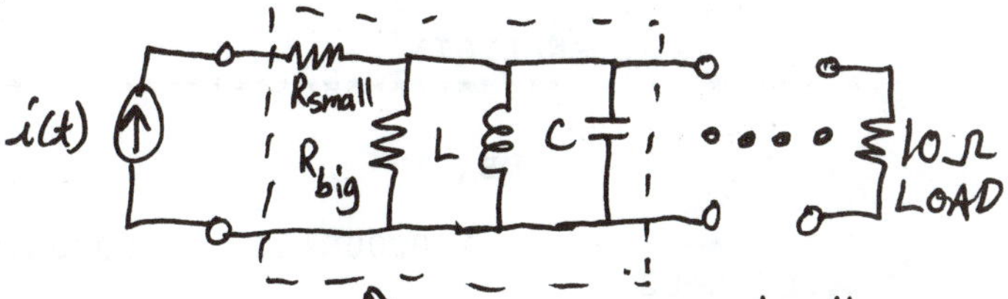

$$\frac{1}{\sqrt{LC}} = \omega_r = 314.16 \Rightarrow LC = 1.0132\times10^{-5}$$

I'll choose $\boxed{C = 1\mu\text{F}, \ L = 10.132\text{ H}}$

Make R_{big} huge to reduce loading effects and adjust using SPICE.

⇒ $R_{big} = 1\text{M}\Omega$ and $R_{small} = 10\text{k}$ works (for example).

(b) See SPICE code included below.

18.7 SPICE code and output:

```
PROBLEM 18.7
IIN 0 1 AC 1 0
R1SMALL 1 2 10K
R1LARGE 2 0 1MEG
L1 2 0 10.132
C1 2 0 1U
R2SMALL 2 3 10K
R2LARGE 3 0 1MEG
L2 3 0 10.132
C2 3 0 1U
R3SMALL 3 4 10K
R3LARGE 4 0 1MEG
L3 4 0 10.132
C3 4 0 1U
R4SMALL 4 5 10K
R4LARGE 5 0 1MEG
L4 5 0 10.132
C4 5 0 1U
R5SMALL 5 6 10K
R5LARGE 6 0 1MEG
L5 6 0 10.132
C5 6 0 1U
R6SMALL 6 7 10K
R6LARGE 7 0 1MEG
L6 7 0 10.132
C6 7 0 1U
RLOAD 7 0 10
.AC LIN 20 12.5 200
.PLOT AC VDB(7)
.END
```

```
****     AC ANALYSIS                          TEMPERATURE =   27.000 DEG C
******************************************************************************

  FREQ        VDB(7)

(*)----------    -1.0000E+02  -5.0000E+01    0.0000E+00    5.0000E+01
1.0000E+02

                       - - - - - - - - - - - - - - - - - - - - - - - - - - -
  1.250E+01 -8.789E+01 .  *         .            .            .            .
  2.237E+01 -5.797E+01 .          * .            .            .            .
  3.224E+01 -3.435E+01 .            .   *        .            .            .
  4.211E+01 -1.052E+01 .            .         *  .            .            .
  5.197E+01  9.017E+00 .            .            . *          .            .
  6.184E+01 -1.486E+01 .            .        *   .            .            .
  7.171E+01 -2.826E+01 .            .     *      .            .            .
  8.158E+01 -3.804E+01 .            .  *         .            .            .
  9.145E+01 -4.578E+01 .            .*           .            .            .
  1.013E+02 -5.219E+01 .           *.            .            .            .
  1.112E+02 -5.767E+01 .          * .            .            .            .
  1.211E+02 -6.245E+01 .         *  .            .            .            .
  1.309E+02 -6.670E+01 .        *   .            .            .            .
  1.408E+02 -7.052E+01 .       *    .            .            .            .
  1.507E+02 -7.400E+01 .      *     .            .            .            .
  1.605E+02 -7.719E+01 .     *      .            .            .            .
  1.704E+02 -8.015E+01 .    *       .            .            .            .
  1.803E+02 -8.290E+01 .   *        .            .            .            .
  1.901E+02 -8.547E+01 .   *        .            .            .            .
  2.000E+02 -8.789E+01 .  *         .            .            .            .
                       - - - - - - - - - - - - - - - - - - - - - - - - - - -
```

18.8(a) $H(s) = \dfrac{2s+3}{s+4} \Rightarrow$ Bilinear with $b_1 = 2$, $b_0 = 3$, $a_0 = 4$

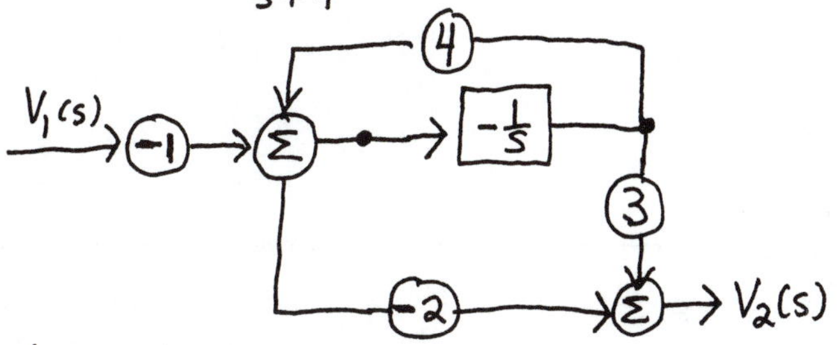

Realization:

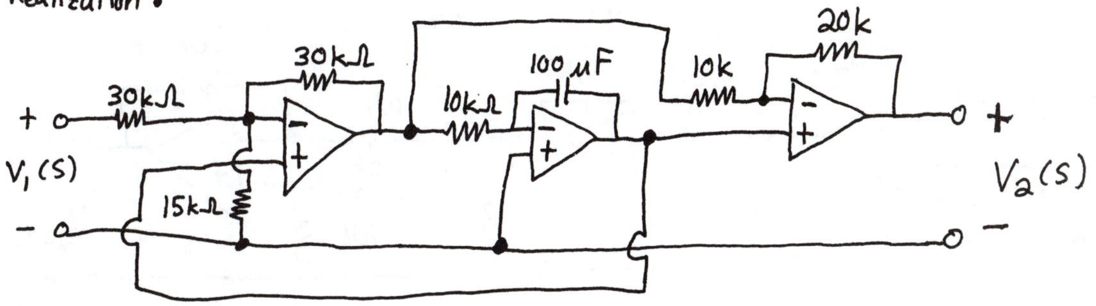

(b) $H(s) = \dfrac{-2s+3}{s+4} \Rightarrow$ Bilinear with $b_1 = -2$, $b_0 = 3$, $a_0 = 4$

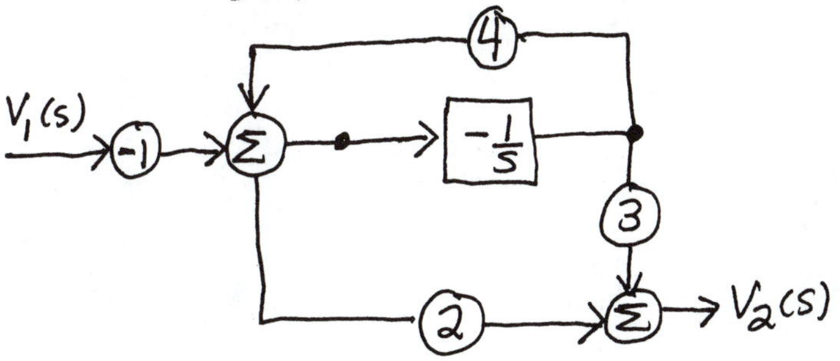

Realization:

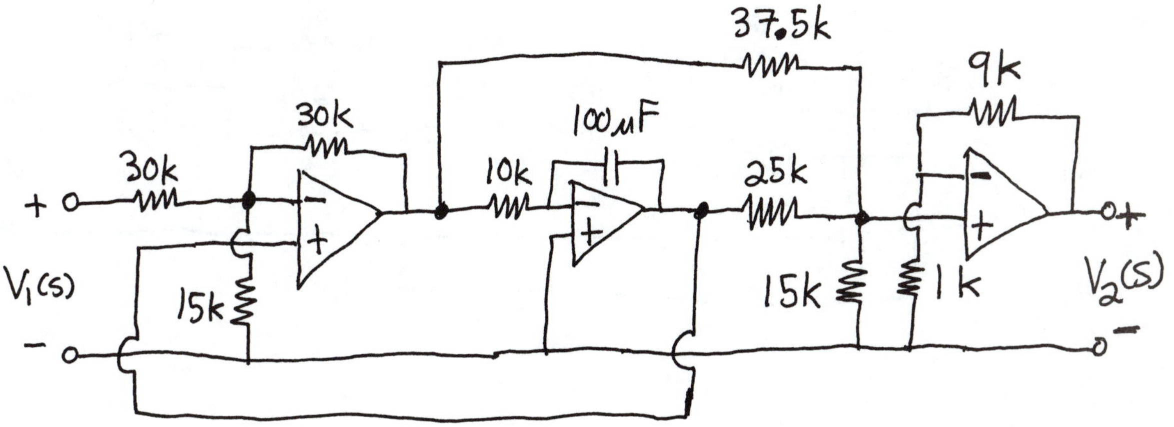

18.8(c) $H(s) = \dfrac{-2s-3}{s+4} \Rightarrow$ Bilinear with $b_1=-2$, $b_0=-3$, $a_0=4$

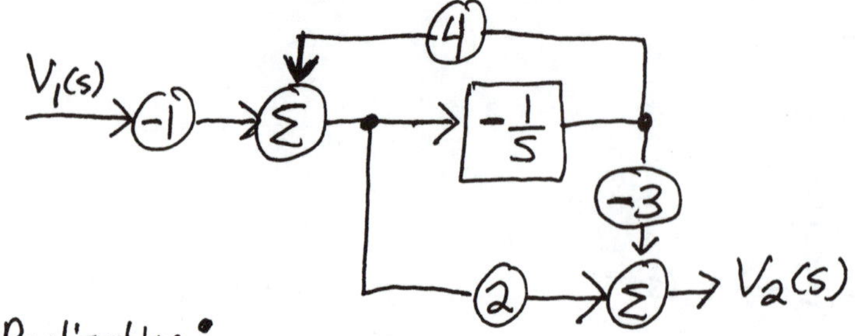

Realization:

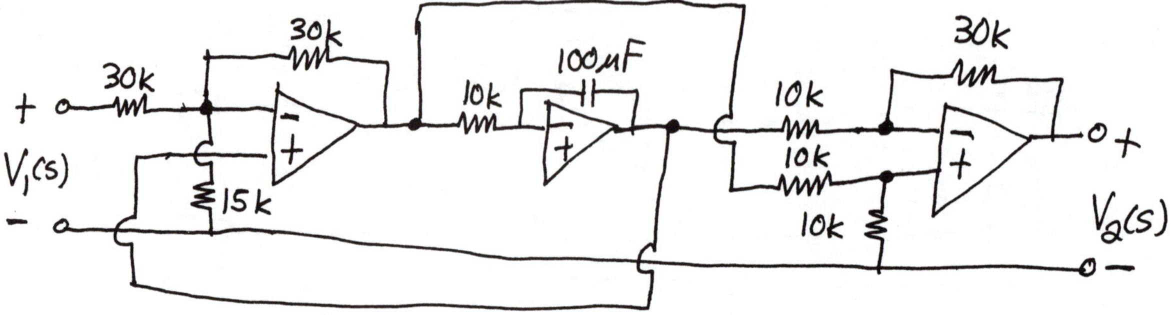

(d) $H(s) = \dfrac{2s-3}{s+4} \Rightarrow$ Bilinear with $b_1=2$, $b_0=-3$, $a_0=4$

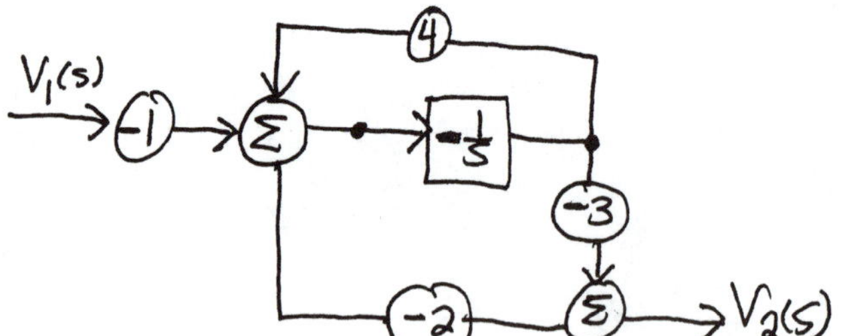

Realization:

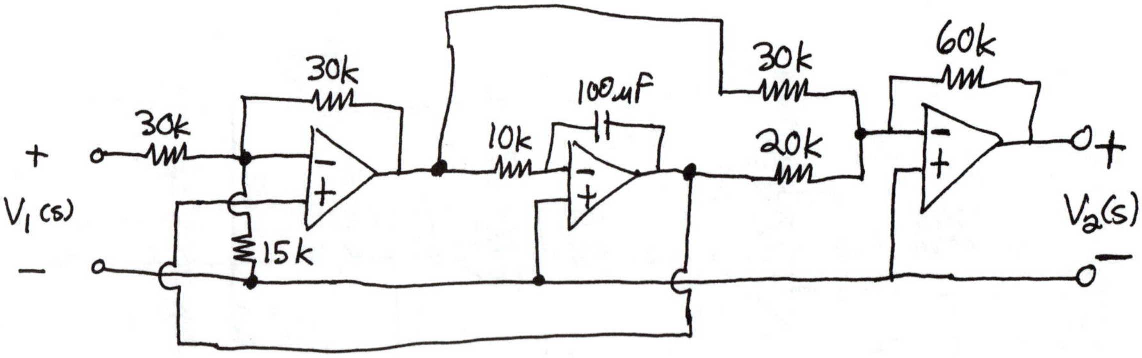

18.8 (e) $H(s) = \frac{2s}{s+4}$ ⇒ Bilinear with $b_1 = 2$, $b_0 = 0$, $a_0 = 4$

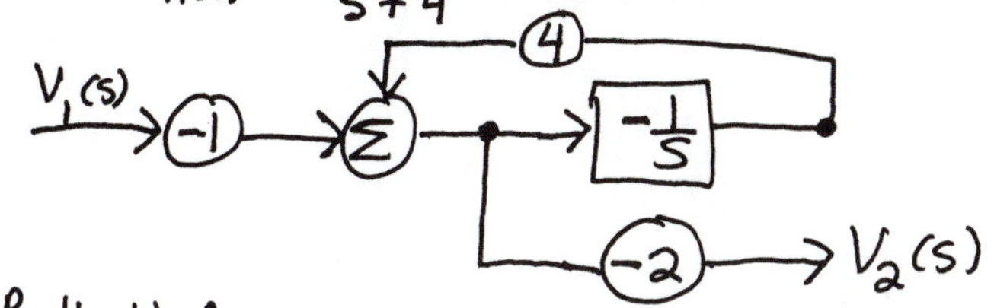

Realization:

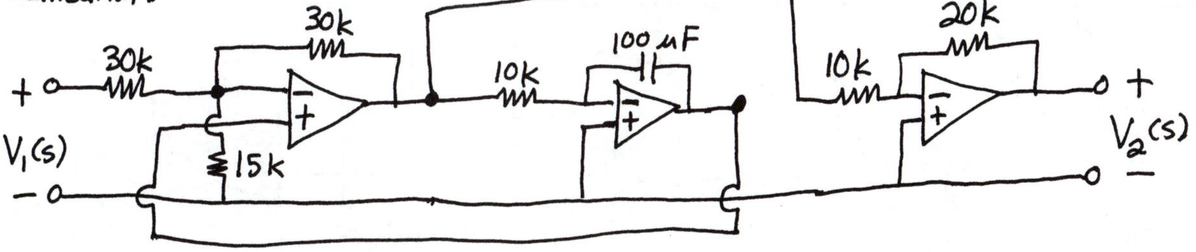

(f) $H(s) = \frac{-3}{s+4}$ ⇒ Bilinear with $b_1 = 0$, $b_0 = -3$, $a_0 = 4$

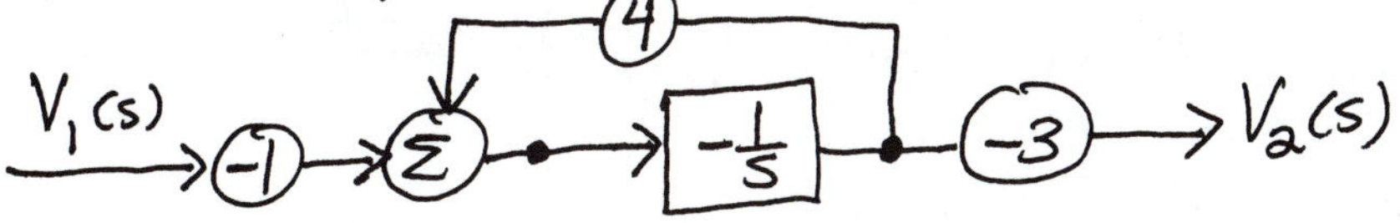

Realization:

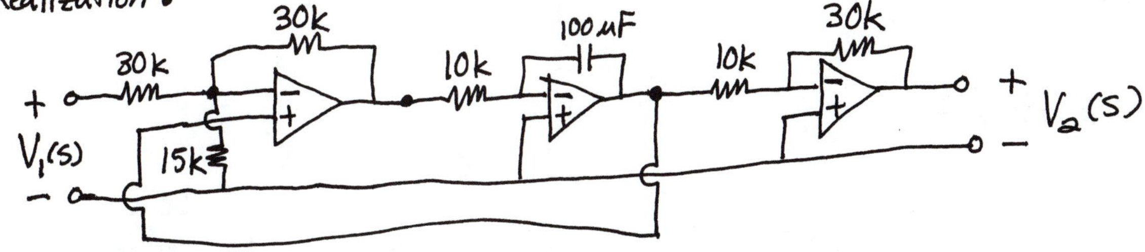

Alternative: $H(s) = -3\left(\frac{1}{s+4}\right)$ ⇒ LPF w/ $\omega_c = 4$, DC gain $= -3$

⇒ $RC = \tau = \frac{1}{4}$

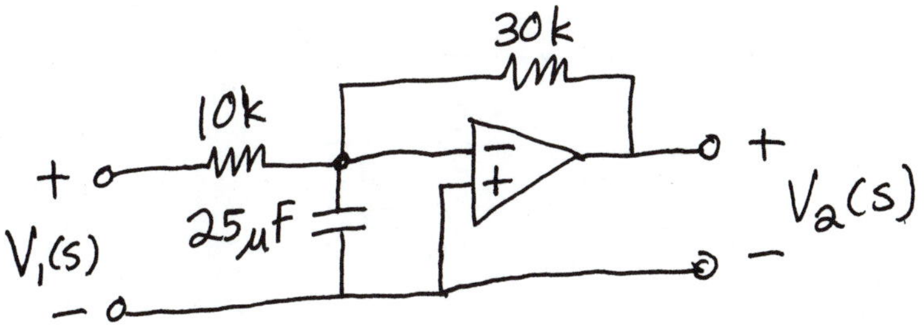

18.8 (g) $H(s) = \frac{-2s}{s+4} \Rightarrow$ Bilinear with $b_1 = -2$, $b_0 = 0$, $a_0 = 4$

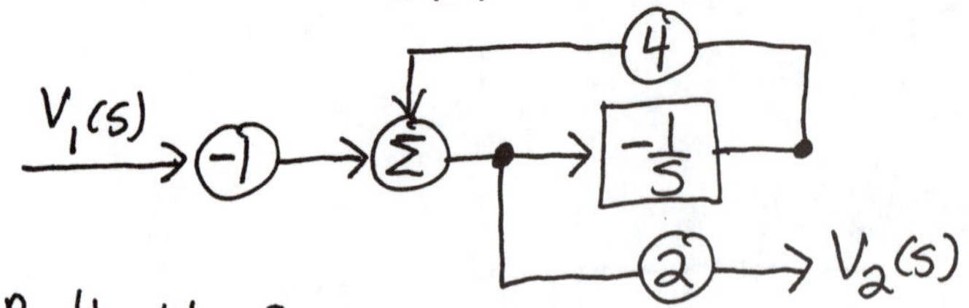

Realization:

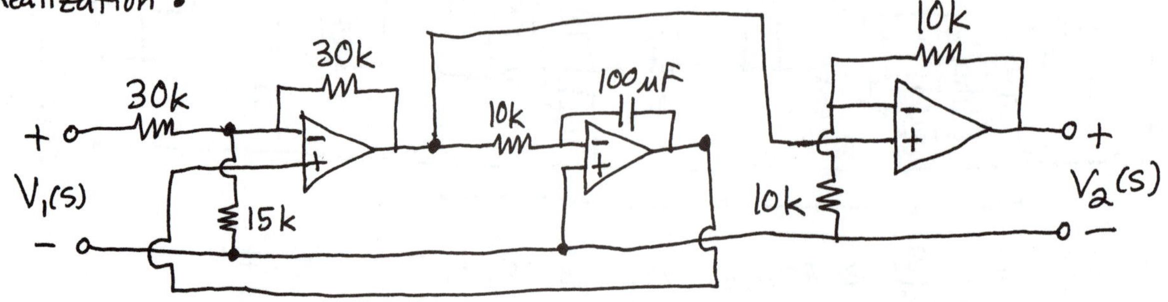

(h) $H(s) = \frac{3}{s+4} \Rightarrow$ Bilinear with $b_1 = 0$, $b_0 = 3$, $a_0 = 4$

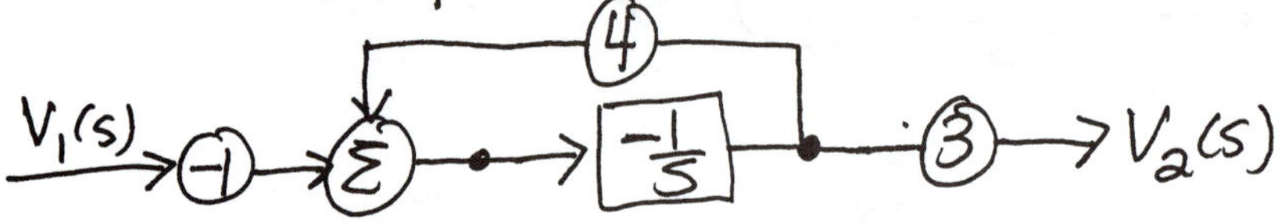

Realization:

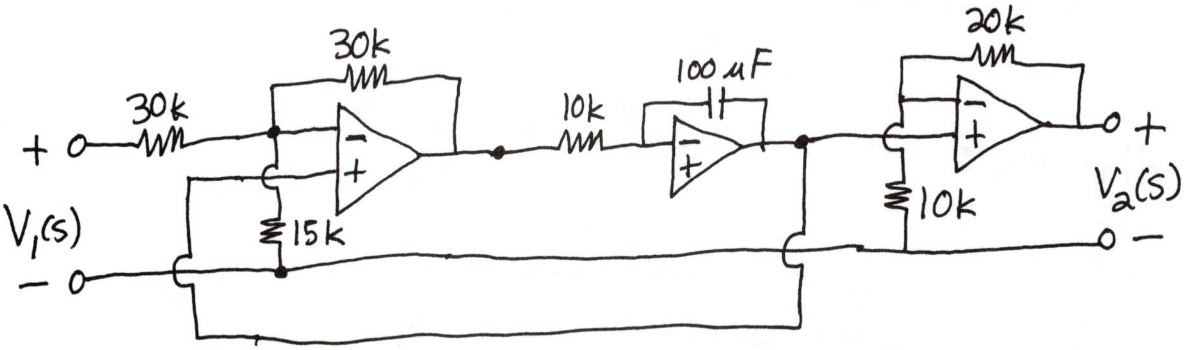

18.9(a) $H(s) = \frac{1}{s^2+s+2} \Rightarrow$ Biquad with $b_2 = b_1 = 0$, $b_0 = 1$, $a_1 = 1$, $a_0 = 2$

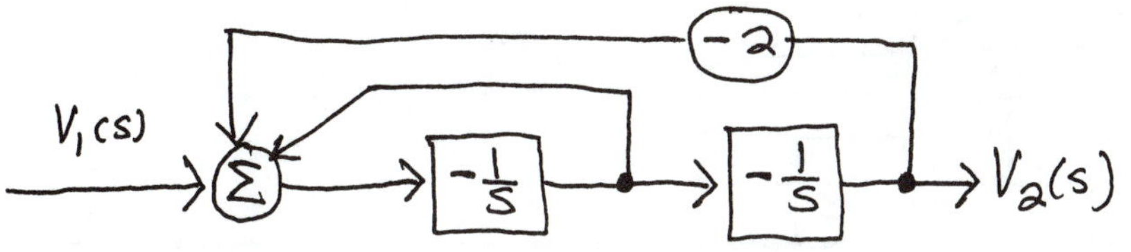

Realization:

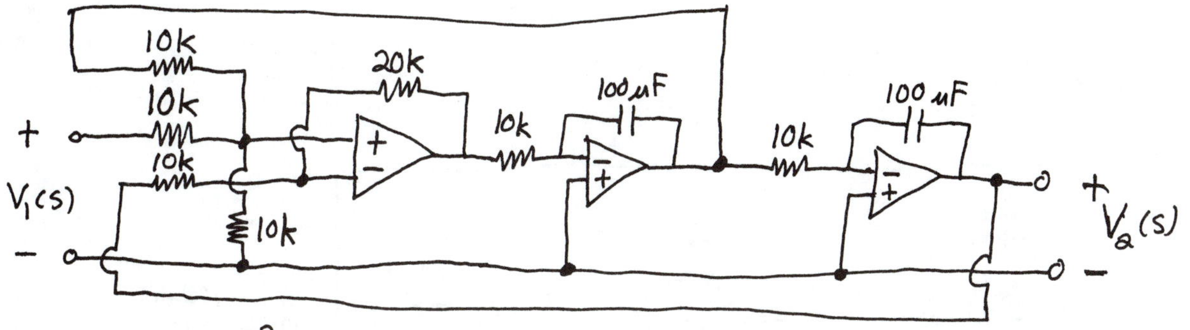

(b) $H(s) = \frac{s^2}{s^2+s+2} \Rightarrow$ Biquad with $b_2 = 1$, $b_1 = b_0 = 0$, $a_1 = 1$, $a_0 = 2$

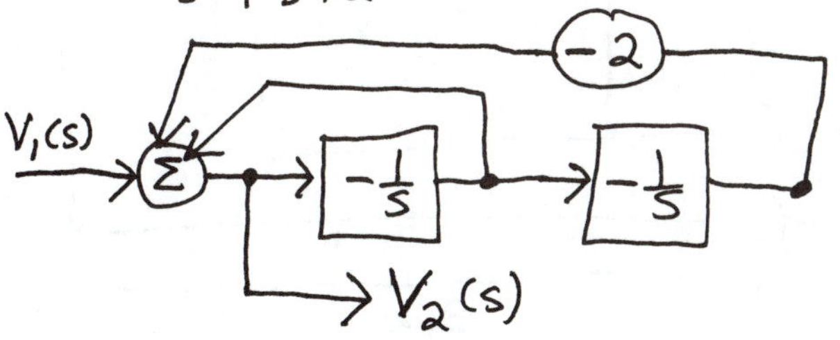

Realization:

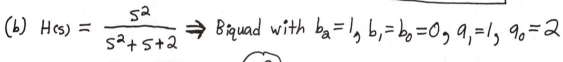

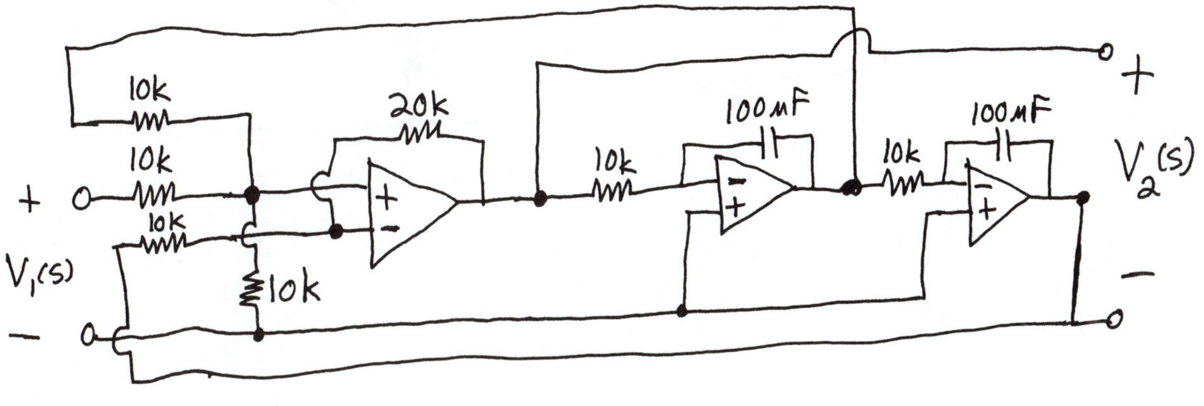

18.9 (c) $H(s) = \dfrac{5s^2-1}{s^2+3s+2} \Rightarrow$ Biquad with $b_2=5$, $b_1=0$, $b_0=-1$, $a_1=3$, $a_0=2$

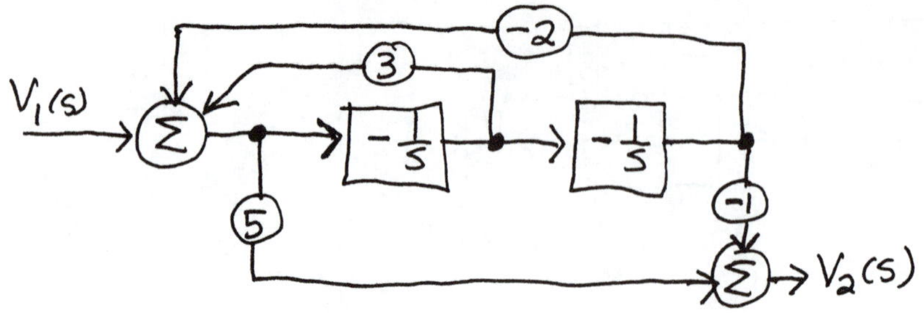

Realization:

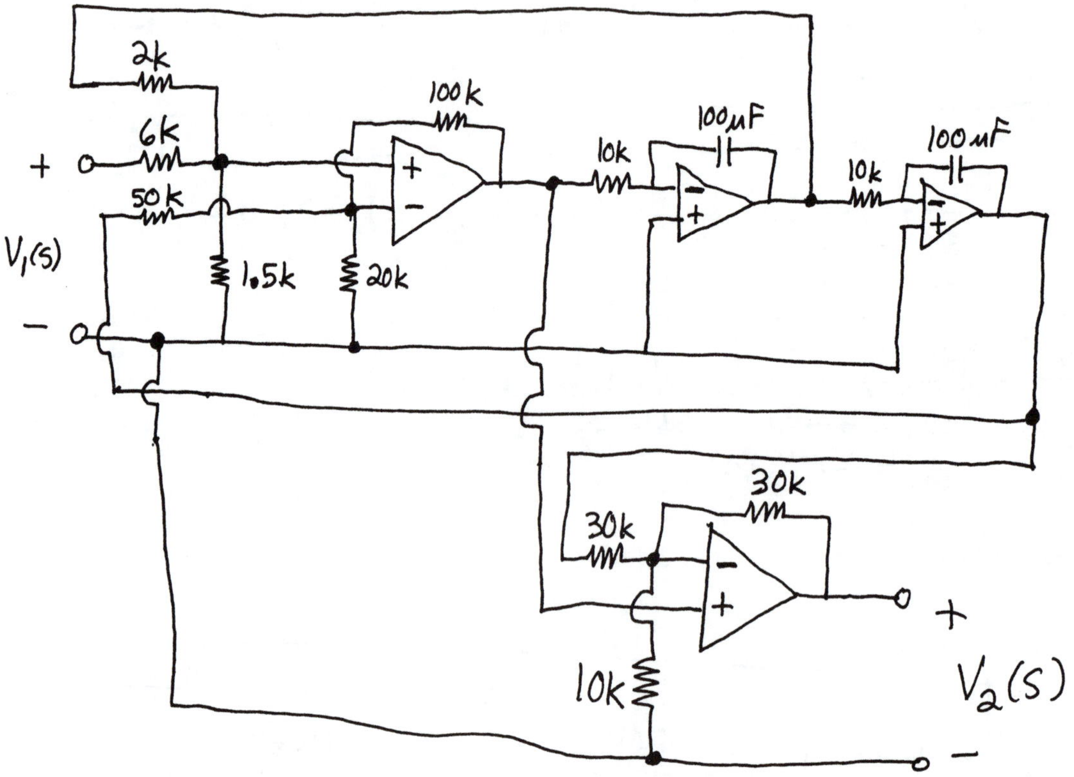

18.9(d) $H(s) = \frac{9s^2+8s-3}{s^2+17s+3} \Rightarrow b_2=9, b_1=8, b_0=-3, a_1=17, a_0=3$

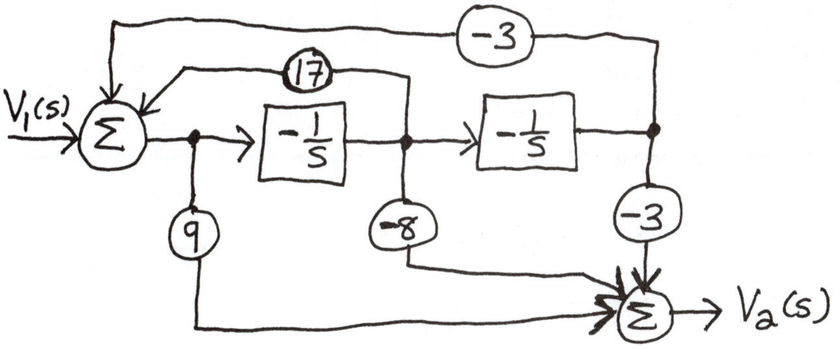

Realization:

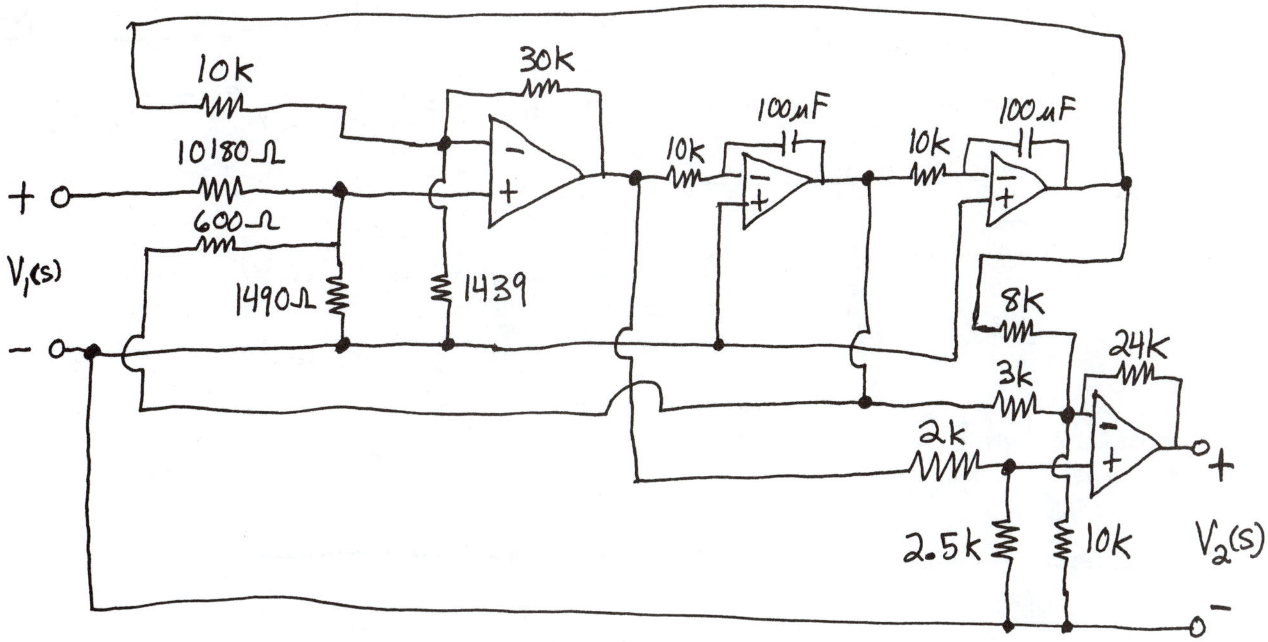

18.9(e) $H(s) = \dfrac{13s}{s^2 - 27s + 1} \Rightarrow$ Biquad with $b_2 = b_0 = 0$, $b_1 = 13$, $a_1 = -27$, $a_0 = 1$

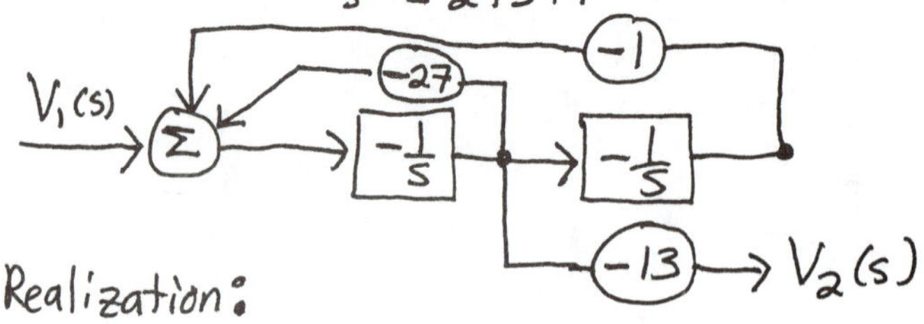

Realization:

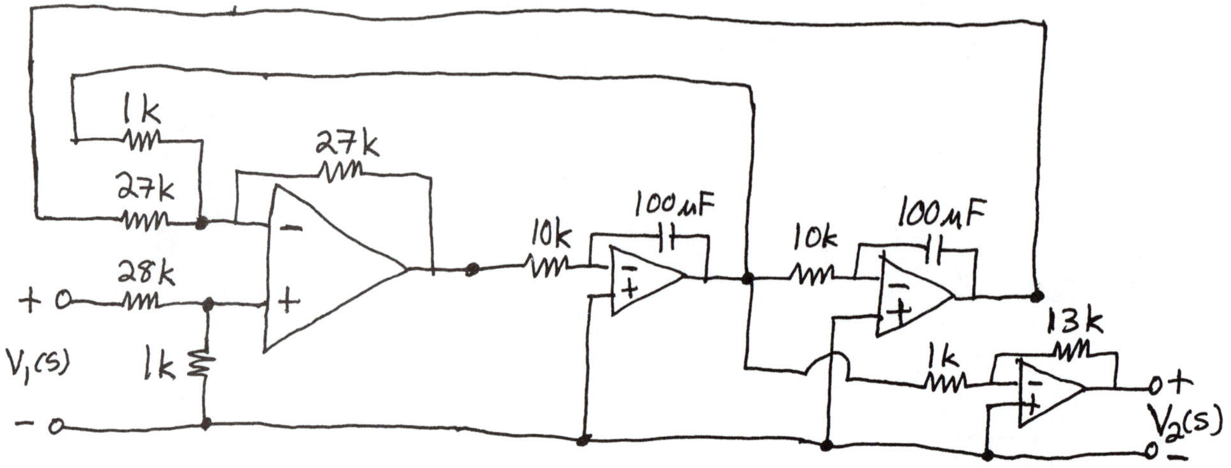

(f) $H(s) = \dfrac{3}{s^2 + 2} \Rightarrow$ Biquad with $b_2 = b_1 = 0$, $b_0 = 3$, $a_0 = 2$, $a_1 = 0$

Realization:

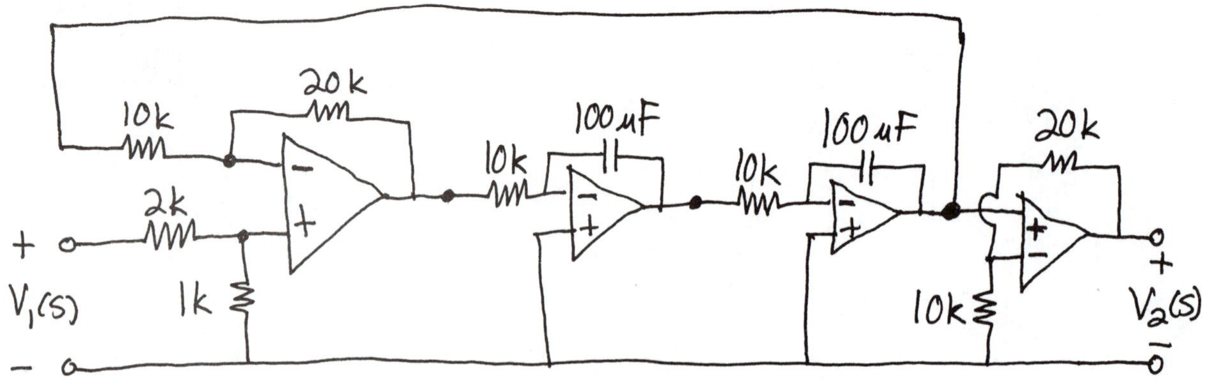

18.10(a) $RC = \frac{1}{\omega_c} = \frac{1}{4000}$, High freq. gain $= +48\,dB = 256$ times

$\Rightarrow$ I'll choose $R = 10\,k\Omega$, $C = 25\,nF$. First-order HPF.

Realization:

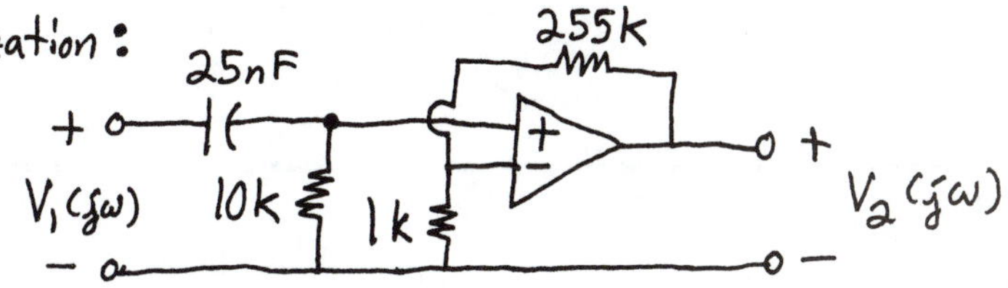

(b) First-order LPF. $RC = \frac{1}{\omega_c} = \frac{1}{1000} \Rightarrow$ Choose $R = 1\,k\Omega$, $C = 1\,\mu F$

DC gain of $+15\,dB = 5.623$ times larger.

Realization:

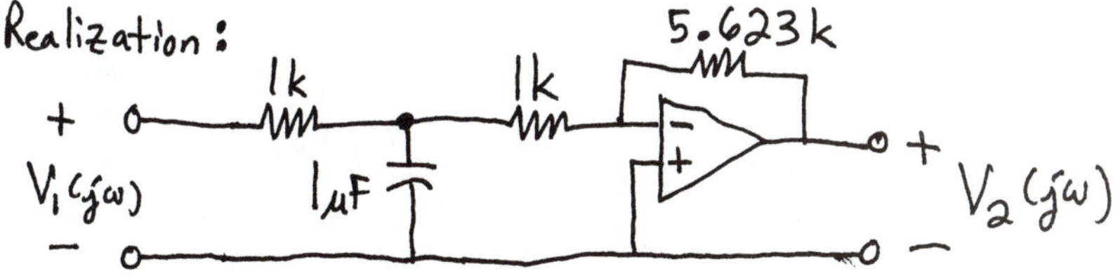

(c) Second-order HPF. $RC = \frac{1}{4000} \Rightarrow$ Choose $C = 25\,nF$, $R = 10\,k\Omega$

High freq. gain $= +48\,dB = 256$ times.

Realization:

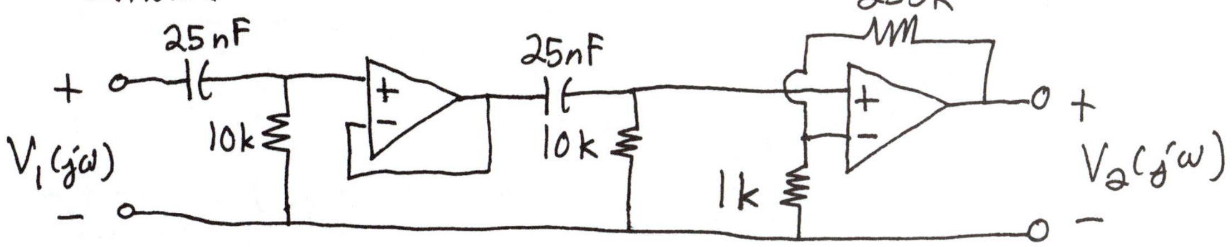

(d) Second-order LPF. $RC = \frac{1}{1000} \Rightarrow$ Choose $C = 1\,\mu F$, $R = 1\,k$

DC gain $= 5.623$ times

Realization:

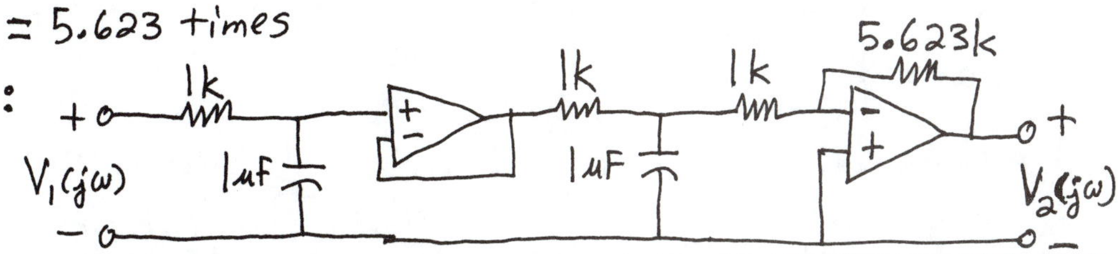

18.11(a) $H(s) = \left(-\frac{3}{s+4}\right)\left(\frac{1}{s^2+s+2}\right)\left(\frac{1}{s^2+s+2}\right)$

$=$ [Circuit from Prob. 18.8(f)] cascaded with [Circuit from Prob. 18.9(a)] cascaded with [circuit from Prob. 18.9(a)].

(b) $H(s) = \left(\frac{2s+3}{s+4}\right)\left(\frac{2s+3}{s+4}\right)\left(\frac{2s+3}{s+4}\right)$

$=$ Three circuits from Prob. 18.8(a) cascaded together.

18.12 (a)

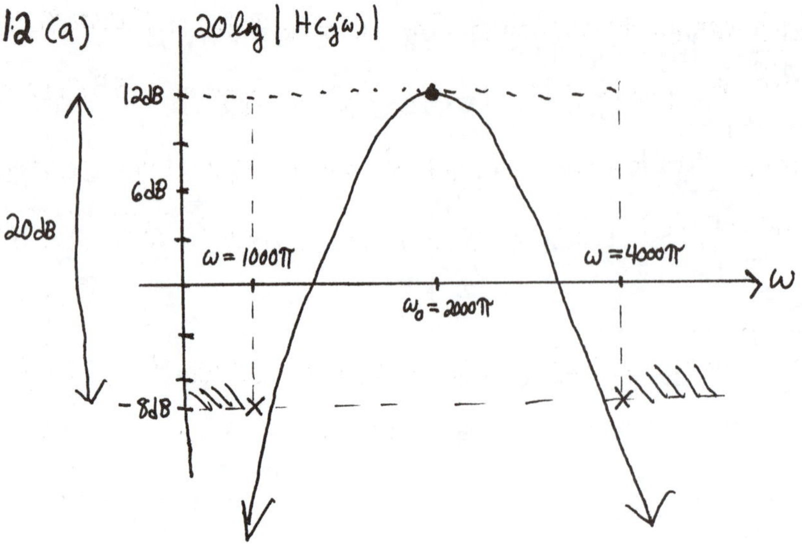

One approach is to try cascading a lowpass and a highpass filter:

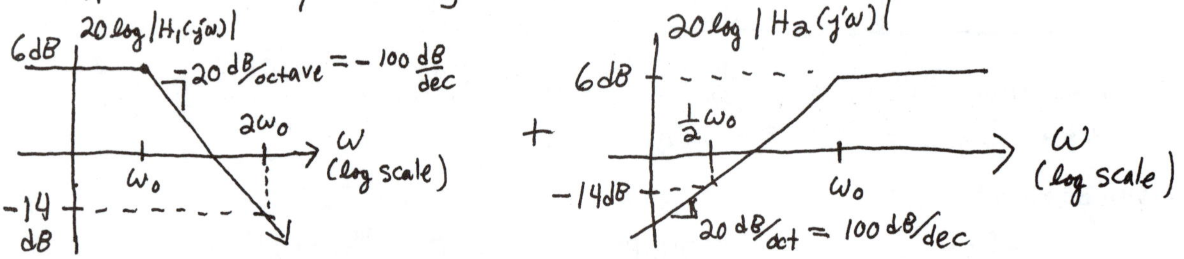

$\pm 100 \frac{dB}{dec} \Rightarrow 5^{th}$ order LPF and HPF

$$H_1(s) = \frac{2\omega_0^5}{(s+\omega_0)^5}$$

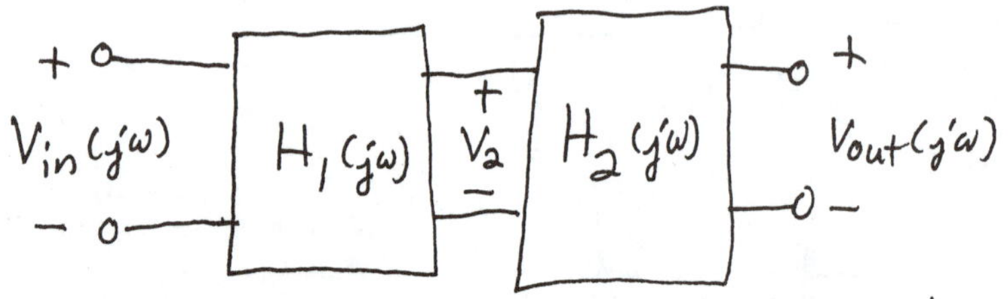

$$H_2(s) = \frac{2\omega_0^5}{(\frac{1}{s}+\omega_0)^5}$$

A possible realization for $H_1(s)$ is: $RC = \frac{1}{\omega_0} = 1.59155 \times 10^{-4}$

$\Rightarrow$ Use $C = 1\mu F$, $R = 159.155\,\Omega$

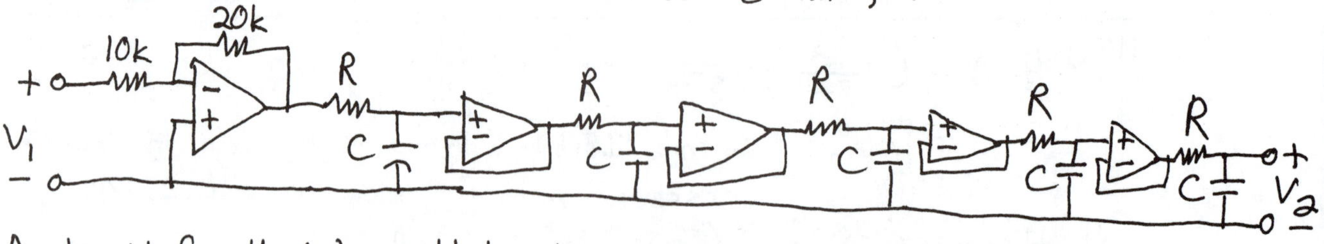

A circuit for $H_2(s)$ would be the same, except swap the locations of the R's and C's to get a HPF.

18.12 (b) Same idea, but design the circuit for $\omega_1 = 2\pi f_1 = 20000\pi$ as the center frequency $\Rightarrow$ Make $RC = \frac{1}{\omega_1} = 1.59155 \times 10^{-5}$

$\Rightarrow$ We can use the same circuit designs for the cascaded $H_1(s)$ and $H_2(s)$, but use (for example) $R = 159.155\,\Omega$, $C = 0.1\,\mu F$

[In other words, frequency scale the previous circuits by a factor of 10 $\Rightarrow R_s = R,\ C_s = \frac{C}{10}$.]

18.13 $$H(s) = \frac{1}{s^3 + 2s^2 + 2s + 1}$$

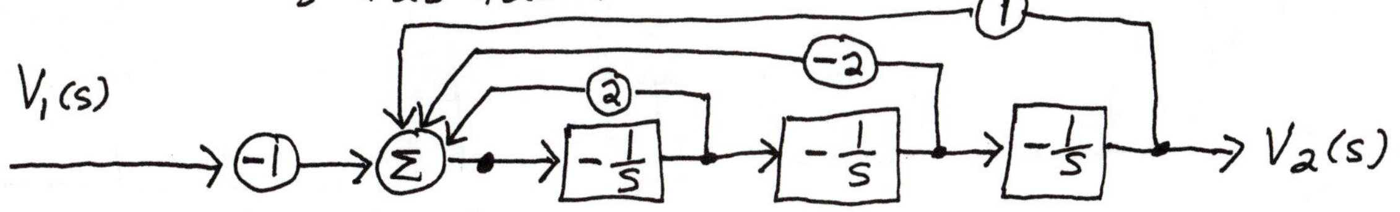

Realization:

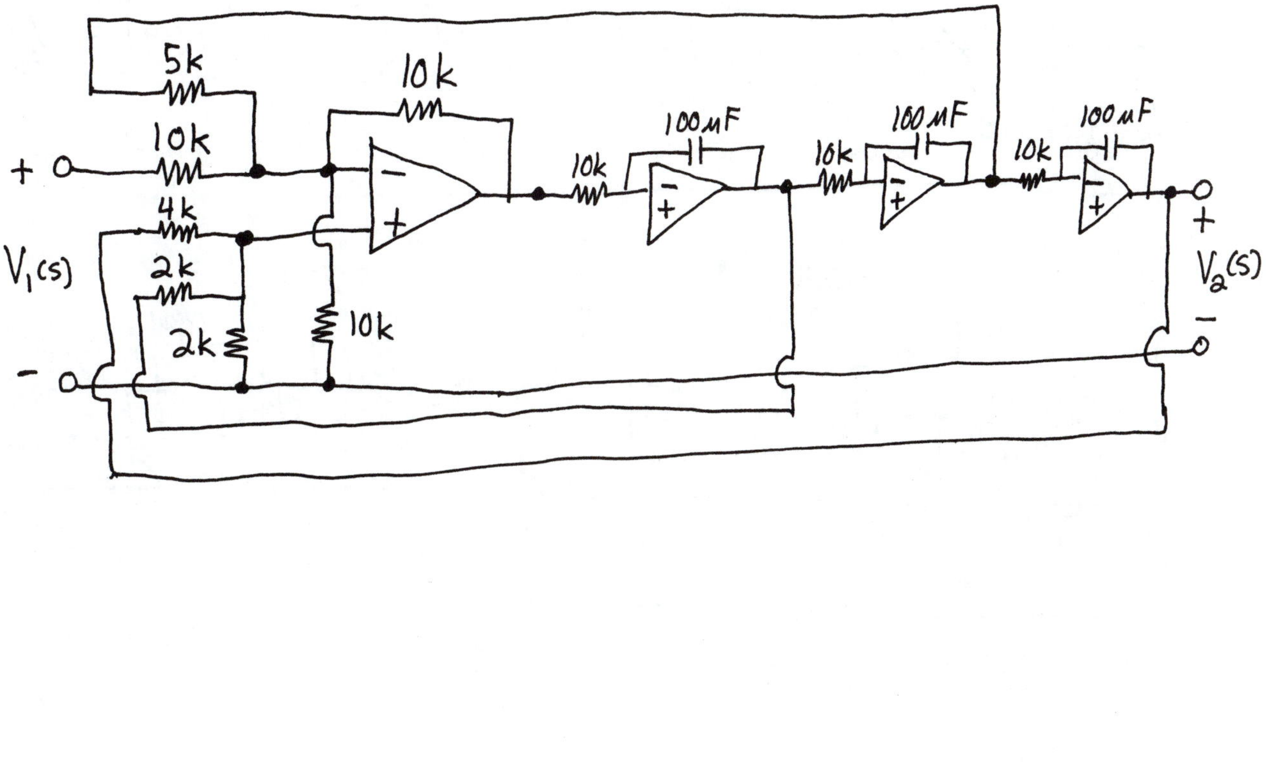

18.14 $H(s) = \dfrac{1.43}{s^2 + 1.43s + 1.52}$ ⇒ Biquad with $b_2 = b_1 = 0$, $b_0 = 1.43$, $a_1 = 1.43$, $a_0 = 1.52$

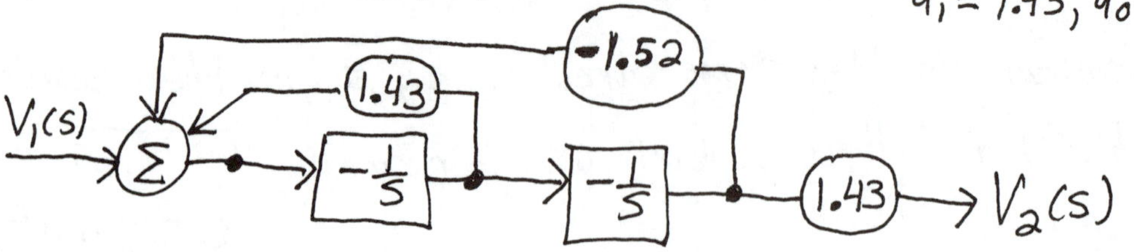

Realization:

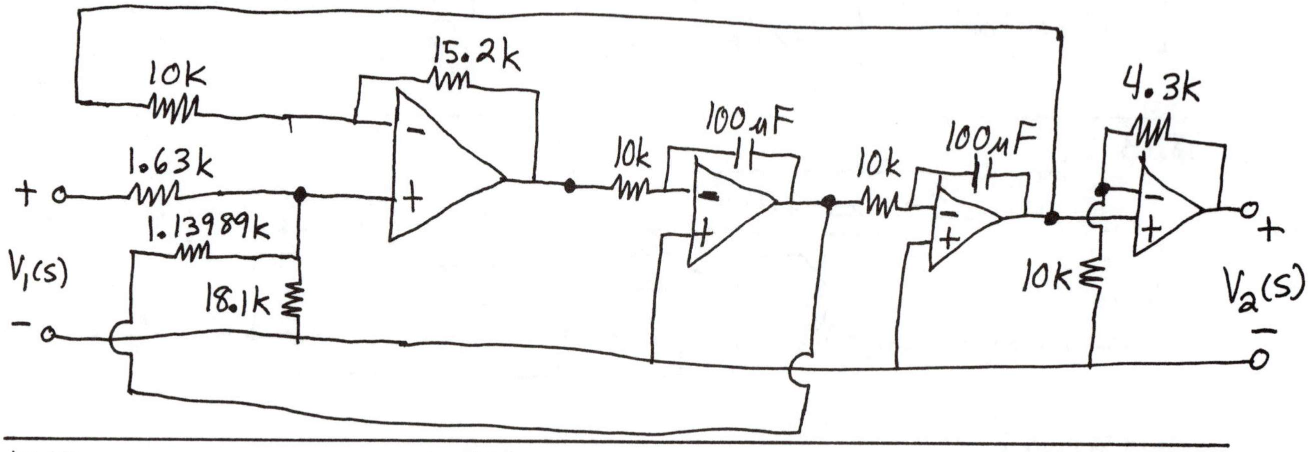

18.15 $H(s) = \dfrac{0.491}{s^3 + 0.988s^2 + 1.24s + 0.491}$

Realization:

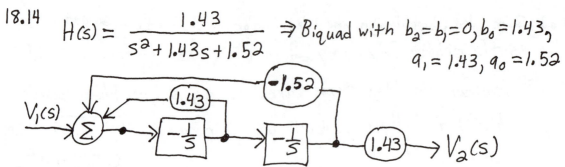

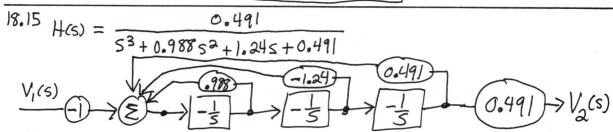

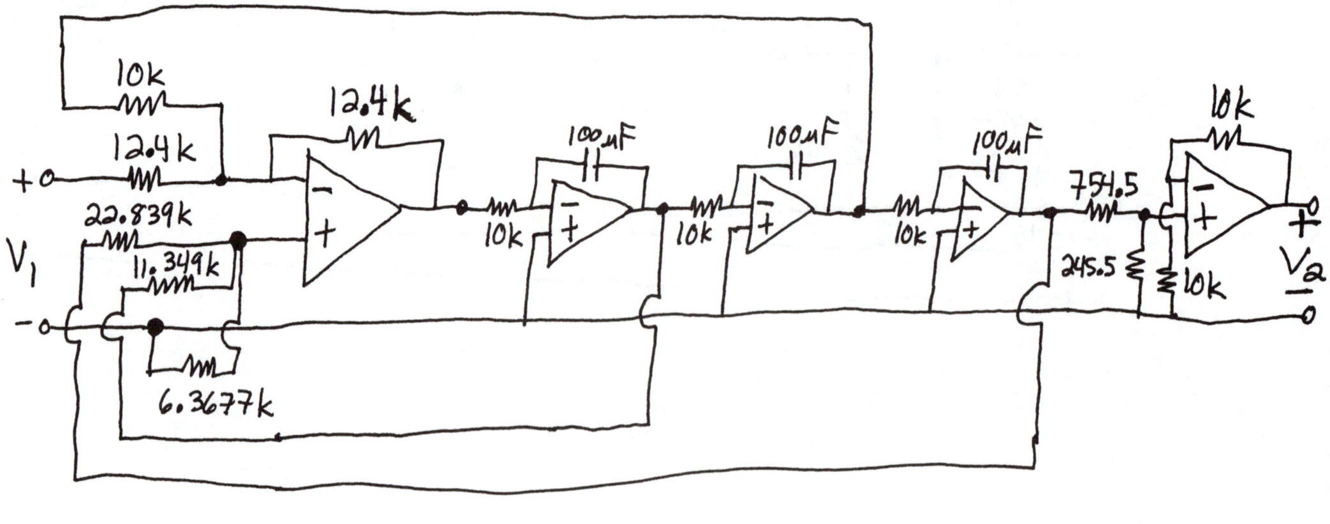

18.16 $H(s) = \left[\frac{3}{s^2+3s+3}\right](2^4) = \frac{48}{s^2+3s+3}$

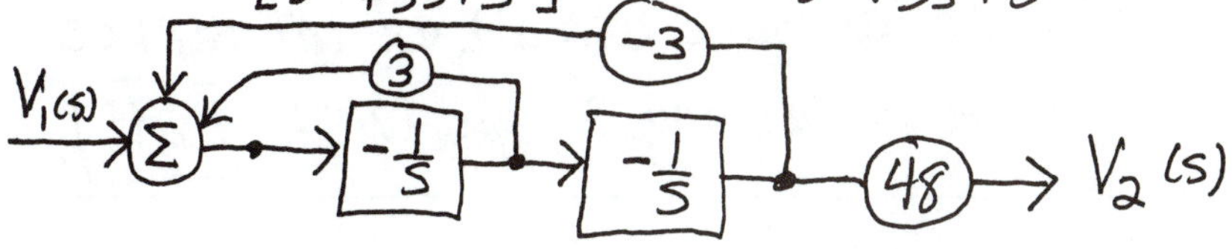

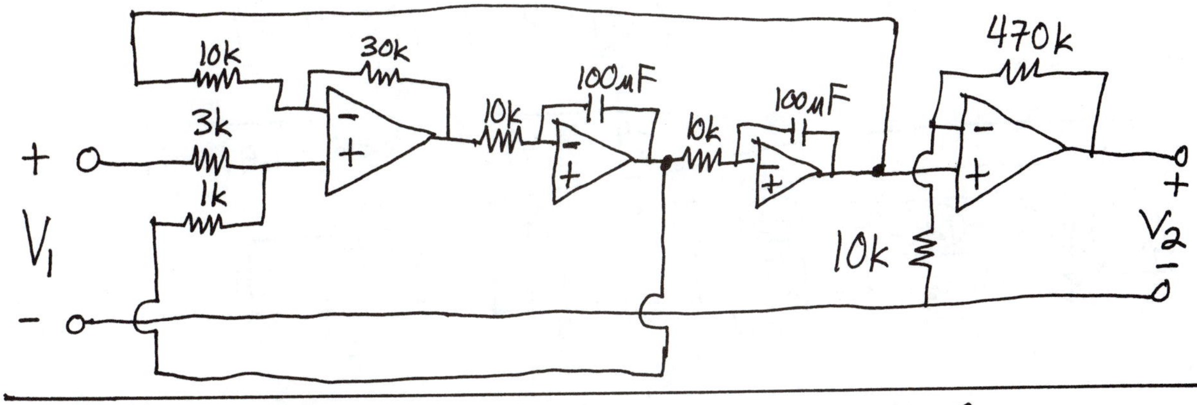

18.17 (a) $H_{HP}(s) = H_{LP}\left(\frac{1}{s}\right) = \frac{48}{\left(\frac{1}{s}\right)^2 + 3\left(\frac{1}{s}\right) + 3} = \frac{16s^2}{s^2 + s + \frac{1}{3}}$

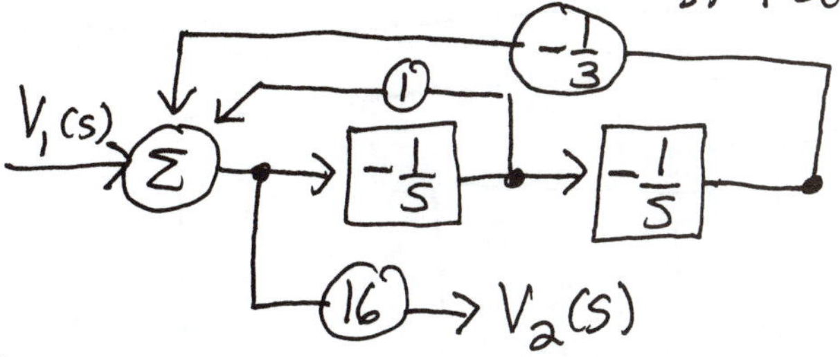

Realization:

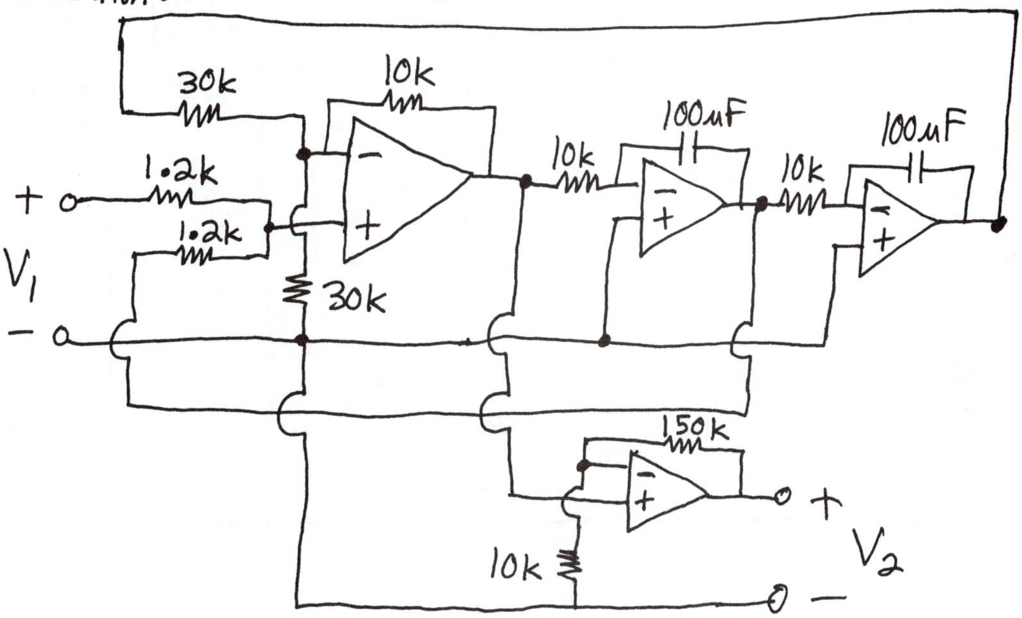

18.17 (b) Frequency scale the circuit in part (a) by $\alpha = 100$:

$$R_s = R, \quad C_s = \frac{C}{100} \Rightarrow \text{Same circuit, except replace all } 100\mu F \text{ caps with } \boxed{1\mu F}.$$

18.18

$$H_{BP}(s) = H_{LP}\left(\frac{s^2+1}{s}\right) = \frac{1}{\left(\frac{s^2+1}{s}\right)^2 + 1.41\left(\frac{s^2+1}{s}\right)+1}$$

$$= \frac{s^2}{s^4 + 1.41s^3 + 3s^2 + 1.41s + 1}$$

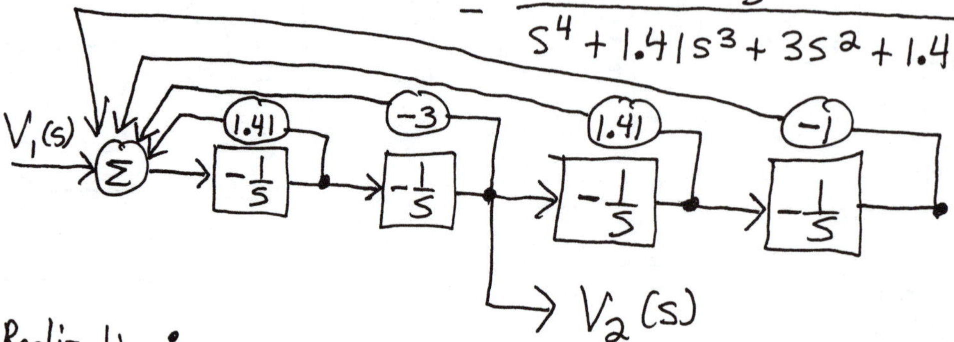

Realization:

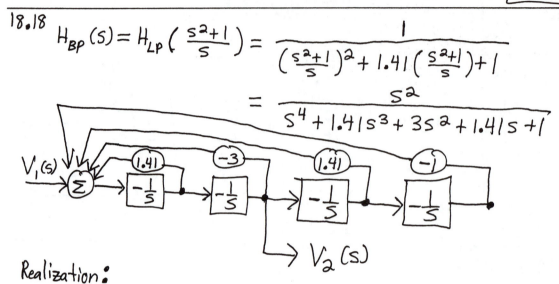

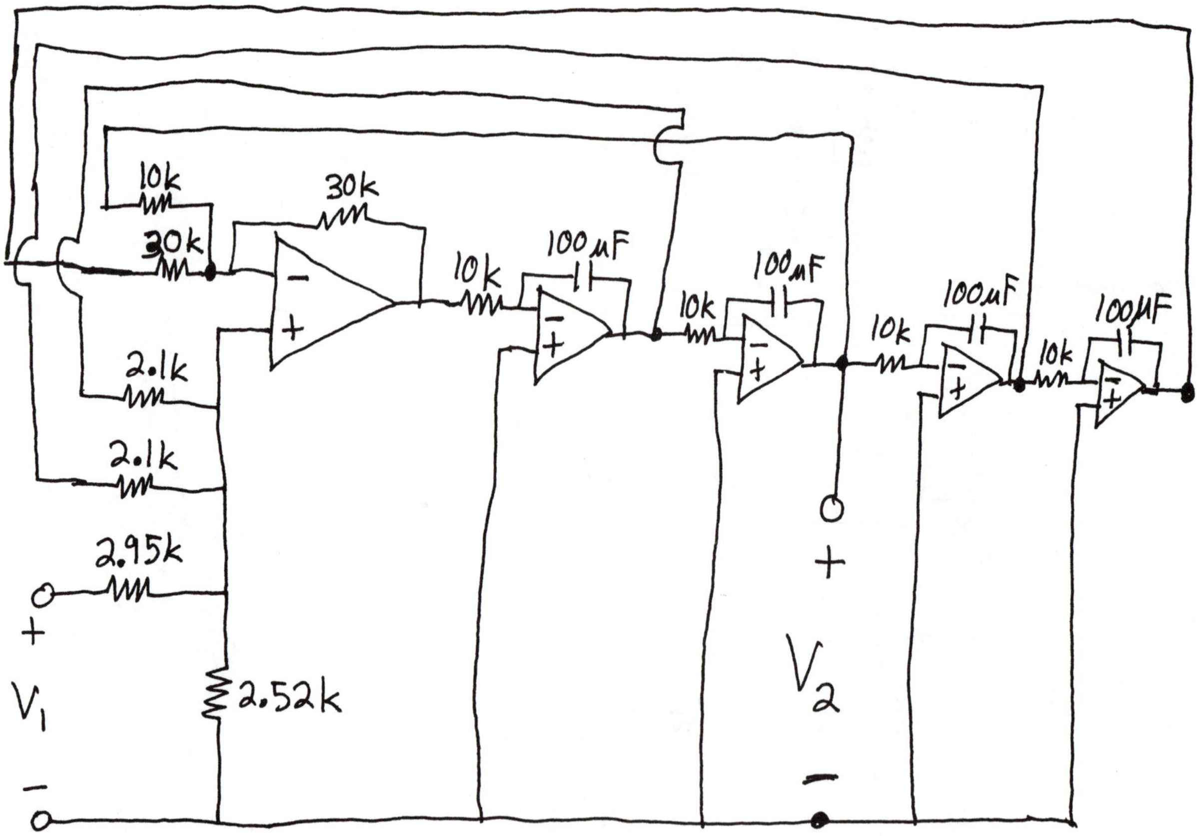

18.19 $H_{BS}(s) = H_{LP}\left(\frac{s}{s^2+1}\right) = \frac{1}{\left(\frac{s}{s^2+1}\right)^2 + 1.41\left(\frac{s}{s^2+1}\right) + 1} = \frac{s^4+2s^2+1}{s^4+1.41s^3+3s^2+1.41s+1}$

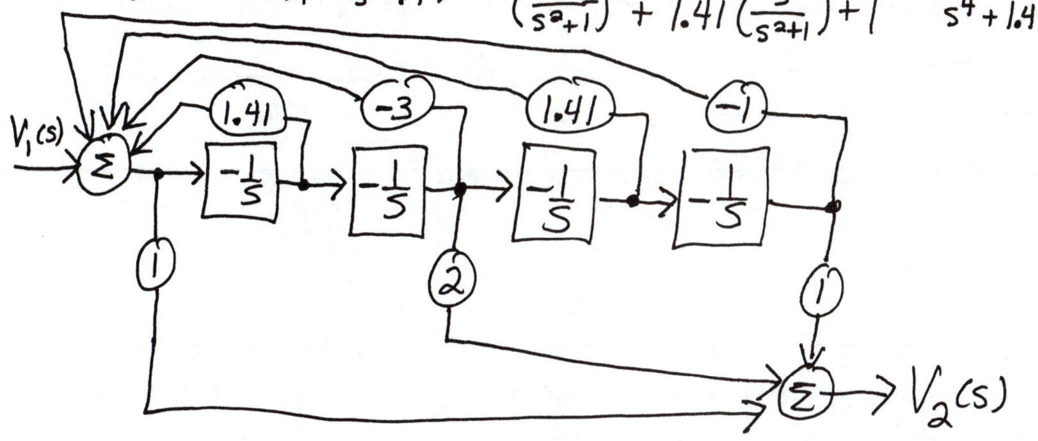

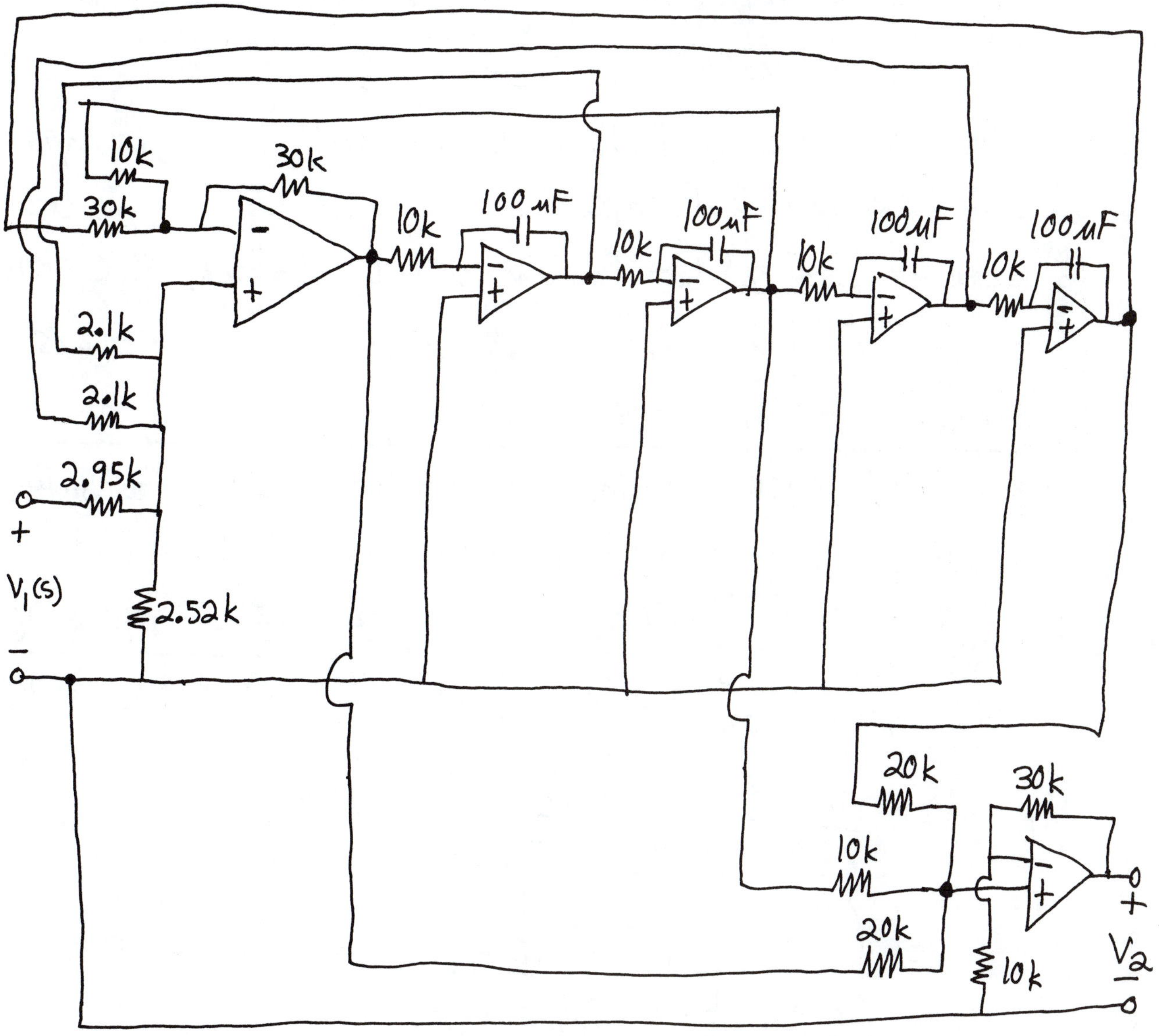

18.20 Impedance scale the circuit by $\alpha = 5000$

$\Rightarrow R_S = 5000R,\ L_S = 5000L,\ C_S = \frac{C}{5000}$

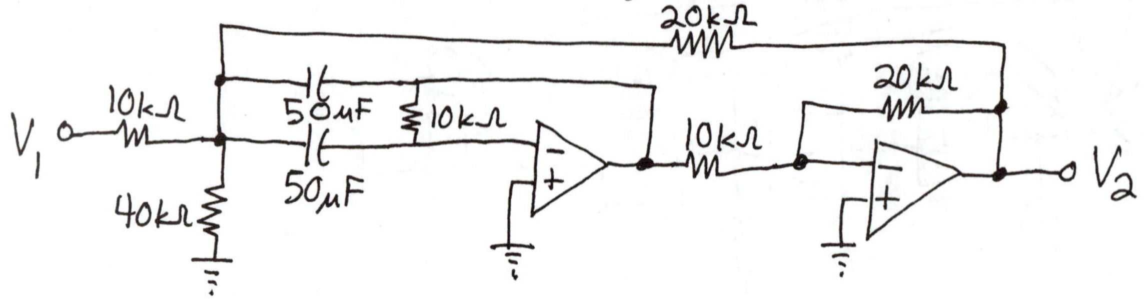

18.21 (a) First impedance scale by $\alpha = 5000$ to get the required resistor values (as in Prob. 18.20). Then frequency scale by 100

$\Rightarrow R_S = R,\ L_S = \frac{L}{100},\ C_S = \frac{C}{100}$

$\Rightarrow$ Same circuit as in Prob. 18.20, but with the 50μF caps replaced by 0.5μF caps.

(b) Same as part (a) but frequency scale by $\frac{1}{100} \Rightarrow R_S = R,\ C_S = 100C$

$\Rightarrow$ Same circuit as in Prob. 18.20, but with the 50 μF caps replaced by 5 mF caps.

18.22 (a) $H(s)$ is unchanged since impedance scaling has no effect on $H(s)$.

(b) $H(s)$ is unchanged (see part (a)).

(c) $$H\left(\frac{s}{3}\right) = \frac{9\left(\frac{s^2}{9}\right) + 8\left(\frac{s}{3}\right) - 3}{2\left(\frac{s^3}{27}\right) + \frac{s^2}{9} + 17\frac{s}{3} + 3} \cdot \frac{54}{54} = \frac{54s^2 + 144s - 162}{s^3 + 6s^2 + 306s + 162}$$

(d) $$H(3s) = \frac{9(9s^2) + 8(3s) - 3}{2(27s^3) + (9s^2) + 17(3s) + 3} \cdot \frac{\left(\frac{1}{54}\right)}{\left(\frac{1}{54}\right)}$$

$$= \frac{\left(\frac{3}{2}s^2 + \frac{4}{9}s - \frac{1}{18}\right)}{\left(s^3 + \frac{1}{6}s^2 + \frac{51}{54}s + \frac{1}{18}\right)}$$